Dramatic narrative, arresting analysis and original research are combined in this history of one of the world's biggest oil businesses between 1950 and 1975. Assessing BP's comparative performance, the book focuses on how BP responded politically, economically and culturally to the rise of new competitors, the decline of Britain's imperial power, and the determination of nation states to assert national sovereignty over the vital commodity, oil.

Climaxing with the OPEC crisis which shook the world in the 1970s, the book – authorised by BP with uniquely unrestricted access to its records – has wide appeal and relevance, especially for those interested in big business, globalisation and nationalism, international affairs, OPEC, the Middle East and oil.

JAMES BAMBERG is an authority on the history of the world oil industry. Author of *The History of The British Petroleum Company: Volume II, The Anglo-Iranian Years, 1928–1954*, he has for some years been the official historian of BP. He is also a visiting fellow at the Centre for International Business History in the Economics Department, University of Reading, and a research associate at the Faculty of History, University of Cambridge.

THE HISTORY OF THE
BRITISH PETROLEUM COMPANY

BRITISH PETROLEUM AND GLOBAL OIL 1950–1975

The Challenge of Nationalism

JAMES BAMBERG

CAMBRIDGE
UNIVERSITY PRESS

PUBLISHED BY THE PRESS SYNDICATE OF THE UNIVERSITY OF CAMBRIDGE
The Pitt Building, Trumpington Street, Cambridge, United Kingdom

CAMBRIDGE UNIVERSITY PRESS
The Edinburgh Building, Cambridge CB2 2RU, UK   www.cup.cam.ac.uk
40 West 20th Street, New York, NY 10011-4211, USA   www.cup.org
10 Stamford Road, Oakleigh, Melbourne 3166, Australia
Ruiz de Alarcón 13, 28014, Madrid, Spain

First published 2000

Printed in the United Kingdom at the University Press, Cambridge

Typeface Adobe Sabon 10/12pt   *System* QuarkXPress™   [SE]

A catalogue record for this book is available from the British Library

Library of Congress Cataloguing in Publication data

Bamberg, J. H.
British Petroleum and global oil, 1950–1975 : the
challenge of nationalism / James Bamberg.
p.   cm.
Includes bibliographical references and index.
ISBN 0 521 25951 7 (hardback)   0 521 78515 4 (paperback)
1. British Petroleum Company–History.   2. Petroleum industry and trade–Great
Britain–History.   I. Title.
HD9571.9.B73 B36   2000
338.7′622338′0941–dc21

ISBN 0 521 25951 7 hardback
ISBN 0 521 78515 4 paperback

Contents

Contents

Colour plates

xi

Illustrations

All the colour plates and most of the illustrations are from the BP Archive
(University of Warwick), the BP Photographic Library (London) and the BP
Chemicals Photographic Section (Saltend, Hull). The other text illustrations
were reproduced by kind permission of Associated Press (no. 29); Hulton
Getty (nos. 10, 11, 26, 28 and 65); the Trustees of the Imperial War Museum,
London, (nos. 12 and 13); Francisco R. Parra (nos. 22, 23, 24 and 25); Shell
UK Ltd (no. 9, from the Shell-Mex & BP archive, c/o the BP Archive); and
United Distillers & Vintners (no. 55).

Maps, graphs and diagrams

Maps

Graphs and diagrams

Tables

Preface

This book is the sequel to two earlier volumes on the history of BP. The first, written by R. W. Ferrier, was *The History of The British Petroleum Company: Volume I, The Developing Years, 1901–1932* (Cambridge University Press, 1982); the second, which I wrote, was *The History of The British Petroleum Company: Volume II, The Anglo-Iranian Years, 1928–1954* (Cambridge University Press, 1994).

Breaking with precedent, I avoided calling this book Volume 3 because I did not want to imply that it was best approached by reading the earlier volumes first. People with different interests should, I felt, be able to approach this book from whatever angle suited them, and via whatever literature they chose, undeterred by the thought that they had first to undergo a specific initiation process. The book, therefore, has a free-standing title and can be read either on its own, or as a sequel to the earlier volumes.

A large cast of people contributed to the book in diverse ways and are owed more recognition and thanks than these acknowledgements can convey. They include several researchers, who dug deeply into rich veins of state records, corporate archives and personal papers, sifting out and helping to analyse the most valuable material. Depending on individual circumstances, they worked part time or full time, short term or long term, on a variety of assignments. I would particularly like to thank Frances Bostock for her work at the Public Record Office; Valerie Johnson for her research on management cultures, marketing, photographs and maps (drawn by Malcolm Barnes, cartographer); Christine Shaw for her contribution to the chapter on nutrition and Jenny Ward for delving into the relations between the oil companies and OPEC. They and I were helped by archivists, librarians and others at institutions too numerous to be acknowledged individually. Special thanks are, however, due to the staff of the BP Archive.

While the primary sources are rich, this book also draws on the published work of many authors in many fields and countries. The extent of their contributions is apparent in the notes and references that follow the text, and they are listed in the bibliography.

Still others have contributed, not with the written word, but by allowing me to call upon their memories, or by making suggestions on those parts of the text covering matters in which they were involved, or of which they have special knowledge. A list of those who have helped in these ways would be too long to include here, and they are therefore shown under interviews in the select bibliography.

I would like to thank, in addition, the members, past and present, of the BP History Committee, who read successive drafts of the book and offered welcome comment and advice. They were Rodney Chase, Professor Donald Coleman, Dr Chris Gibson-Smith, Lord Greenhill of Harrow, Professor Geoffrey Jones, Professor Peter Mathias, Professor Paul Stevens and Lord Wright of Richmond.

These people, and others unmentioned, have helped to make the book better than it otherwise might have been. BP has funded this book, but I alone am responsible for any errors, and for all interpretations and judgements.

Abbreviations

ADMA	Abu Dhabi Marine Areas Ltd
AGIP	Azienda Generale Italiana Petroli
AIOC	Anglo-Iranian Oil Company
Aramco	Arabian American Oil Company
ARCO	Atlantic Richfield Company
BHC	British Hydrocarbon Chemicals
bpd	barrels per day
BPX	BP Exploration
BRP	Bureau de Recherches de Pétrole
CDPD	Central Developmental Planning Department
CENTO	Central Treaty Organisation
CESP	Central European Supply Programme
CFP	Compagnie Française des Pétroles
CIVO	Centraal Instituut voor Voedingsonderzoek
Conoco	Continental Oil Company
CSS	Consiglio Superiore della Sanita
DCL	Distillers Company Limited
DEA	Deutsche Erdöl AG
DEUCE	Digital Electronic Universal Calculating Engine
DUMA	Dubai Marine Areas Ltd
ENI	Ente Nazionale Idrocarburi
ERAP	Entreprise de Recherches et d'Activités Pétrolières
ERSP	European Refineries Supply Programme
FNCB	First National City Bank
FPSC	Foreign Petroleum Supply Committee
GRAM	Group Resource Allocation Model
ICI	Imperial Chemical Industries

ICT	International Computers and Tabulators
IMR	Integrated Marketing and Refining
INOC	Iraq National Oil Company
IOP	Iranian Oil Participants
IPC	Iraq Petroleum Company
ISS	Istituto Superiore della Sanita
JPDC	Japan Petroleum Development Corporation
KOC	Kuwait Oil Company
LAM	Local Area Model
LP	Linear Programming
MEEC	Middle East Emergency Committee
MSG	Manpower Study Group
NIOC	National Iranian Oil Company
NPRI	Net Profits Royalty Interest
OEEC	Organisation for European Economic Co-operation
OELAC	Oil Emergency London Advisory Committee
OPEC	Organisation of Petroleum Exporting Countries
OPEG	OEEC Petroleum Emergency Group
OR	Operational Research
ORDG	Operational Research Directing Group
ORPDG	Operational Research Policy Directing Group
OSAC	Oil Supply Advisory Committee
p.	pence
PCD	Petroleum Chemical Developments
ppm	parts per million
S&D	Supply and Development
SFPBP	Société Française des Pétroles BP
SGHP	Société Générale des Huiles de Pétrole
SLIM	Simplified Linear Integrated Model
SMBP	Shell-Mex and BP
Socal	Standard Oil Company of California
Socony	Standard Oil Company of New York
Sohio	Standard Oil Company of Ohio
Standard Oil (NJ)	Standard Oil Company (New Jersey)
Tapline	Trans-Arabian Pipeline
TAPS	Trans Alaska Pipeline System
TNO	Technische Nederland Organisatje
TRC	Texas Railroad Commission
UAE	United Arab Emirates
UAR	United Arab Republic

UGP	Union Générale des Pétroles
UK	United Kingdom
UN	United Nations
UOP	Universal Oil Products
US	United States
VLCC	Very Large Crude Carrier

A note on the text

Some of the country names used in the period covered by this book have gone out of use, and others will no doubt follow. To preserve historical context, the country names that appear in this book are generally those which were current at the time of the events described, with later names following in parentheses. For example, Rhodesia, which adopted the name Zimbabwe in 1980, is shown as Rhodesia (Zimbabwe).

Although the retrospective use of modern names is generally avoided, an exception is made in the case of Persia, which adopted the name Iran in 1935. This country is mentioned frequently in the text, sometimes in historical generalisations which cut across the change of name in 1935. The name Iran has therefore generally been used throughout the text, except in quotations, in which the original wording is unchanged.

COMPANY NAMES: THE OIL MAJORS

Three of the majors – Royal Dutch-Shell, Socal and Gulf Oil – held to the same names throughout the period covered by this book. The other four adopted new names: Anglo-Iranian became British Petroleum in 1954; Socony-Vacuum became Socony Mobil in 1955, and changed again to Mobil in 1966; the Texas Company became Texaco in 1959; and Standard Oil (NJ) became Exxon in 1972.

For the most part, these changes are reflected in the text, which (as with the country names) uses names that were current in their historical context. An exception is made in the case of British Petroleum, which until 1954 was called first Anglo-Persian, then Anglo-Iranian. To avoid the confusion that might be caused by frequent switching between these names, 'the Company' is generally used for the period up to 1954, and thereafter BP.

INTRODUCTION

In the last third of the nineteenth century, before The British Petroleum Company was formed, a new type of capitalism began to take hold in the industrialised world. In a range of industries, giant corporations arose to establish powerful positions from which they could not easily be shaken. These corporations, prime movers in their respective industries, harnessed new technologies to the mass production and distribution of their products. They developed organisational capabilities to plan and co-ordinate their operations, to mobilise capital, technology and information and to expand into new markets and products. Managed by ranks of salaried executives rather than the owner–managers of earlier generations of firms, the largest of them deployed resources on a scale comparable to many a medium-sized country. Operating like planned economies, they performed many of the functions of resource allocation which would have been left to the market in a purely free market economic model. Of course, market forces did not disappear, and many small, owner-managed firms continued to spring up, survive and grow, but the new giant corporations were a class apart, representing the new form of capitalism, corporate capitalism, as distinct from the more personal capitalism of an earlier era. Standing out like mountains in the economic landscape, these giants were regarded with mixed feelings. Some saw them, then and later, as engines of economic growth, while to others they represented sinisterly powerful concentrations of monopolistic power. Few, though, could deny that they exercised great influence, for good or evil.[1]

No firms symbolised these trends more than the oil majors. Between 1870, when John D. Rockefeller formed the Standard Oil Company of Ohio, and the mid-twentieth century, seven majors – the 'Seven Sisters' – became the dominant players in the international oil industry.[2] By the mid-twentieth century they accounted for nearly 90 per cent of crude oil production, and

1

Table 0.1 *The oil majors' rankings among the world's largest industrial firms, 1956*

	Gross sales (including excise taxes) ($billions)	Ranking (by sales) among world's largest industrial firms
Standard Oil (NJ)	7·13	2
Royal Dutch-Shell	6·50	3
Socony Mobil	2·75	8
Gulf	2·34	14
Texas Company	2·05	16
British Petroleum	2·02	17
Socal	1·45	23

over 70 per cent of refining and marketing capacity in the international oil industry.[3] One of them, called Anglo-Persian when it was formed in 1909, was renamed Anglo-Iranian in 1935, and in 1954 took the name of its original UK marketing subsidiary as the name of the parent company: British Petroleum.[4]

BP and the six other majors all ranked at that time among the world's twenty-five largest industrial corporations, measured by gross sales (see table 0.1). They had climbed by different, but frequently intersecting, paths to the commanding heights of the international oil industry. Three of them originated in the Standard Oil group built up by Rockefeller, the original prime mover in large-scale organisation in the global oil industry. Broken up for violation of the US antitrust laws by order of the US Supreme Court in 1911, the Standard group had by then become so vast that the larger fragments were giant corporations in their own right.[5] The biggest, Standard Oil (New Jersey) (later Exxon), was one of the two largest oil companies in the world, with extensive international operations, to which it would later add.

The other two international majors that emerged from Rockefeller's Standard were the Standard Oil Company of New York (Socony) and the Standard Oil Company of California (Socal). Socony had large market outlets overseas, especially in the Far East, and would merge in 1931 with the Vacuum Oil Company, a leader in lubricants, to form Socony-Vacuum. Its name was changed to Socony Mobil in 1955 to identify it more closely with the brand name, Mobil, under which its products were marketed; and in 1966 this was taken further when the name Socony Mobil was changed

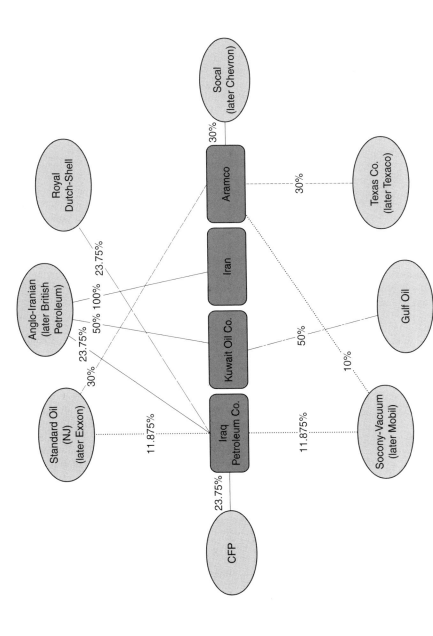

Figure 0.1   The oil majors' shares in the big four Middle East oil concessions, 1950

This revolution in the energy market had far-reaching international ramifications. Most of the world's coal was produced for local consumption, and only about 7 per cent of world coal production entered into international trade in 1950–65. Oil had a far greater international reach. By 1965, 60 per cent of the world's oil production was internationally traded, accounting for 89 per cent of total world energy exports.[19] Oil, in short, ruled international energy exchanges, with the oil majors at their epicentre.

Inevitably, as oil became vital to the economic and military security of oil-exporting and oil-importing countries alike, the majors came under growing pressure from governments pursuing nationalistic oil policies on all sides. In the USA, mounting oil imports in the 1950s, with consequent growing dependence on foreign supplies, raised fears that the nation's security was being undermined. This, plus pressure from domestic producers of oil and coal, led to the introduction of import controls on oil, at first voluntary, but mandatory from 1959.[20]

In other oil-importing countries there were similar concerns about security. Britain had long sought to guarantee its supplies through the ownership of crude oil by British companies, principally Anglo-Persian, in territories within Britain's sphere of influence. Indeed, that had been the British government's chief motive in financing, and taking a majority shareholding, in Anglo-Persian.[21] Later, the French, Italians, Germans and Japanese also set out to acquire crude oil supplies under their own national control. By World War II France had already established a highly *dirigiste* system of controls on oil companies operating in France, and had formed the state-owned CFP to hold France's 23¾ per cent share in the IPC. The French government later went further, sponsoring French companies to search for oil in the French colony of Algeria (before it attained independence in 1962). After large Algerian oil reserves were discovered in the 1950s, Algeria became a major exporter, especially to France, where under a growing panoply of state controls, oil companies were required to purchase specified quotas of the new 'franc' oil.[22]

In Italy, the state oil company, Ente Nationale Idrocarburi (ENI), enjoyed a privileged status and set out, under the vigorous leadership of Enrico Mattei, to challenge the dominance of the majors. One of ENI's chief aims was to find its own sources of crude oil, to which end it offered oil-producing countries more generous terms for new concessions than the majors were willing to do. In Germany and Japan, the story was different, but the theme was the same. Both countries encouraged and supported national, but privately owned, companies to search for their own supplies of crude oil.[23]

This growth in economic nationalism among the industrialised countries not only fettered the freedom of the majors, it also introduced unwelcome (for the majors) new competition into the international oil industry. This came not only from the state oil companies of France and Italy, and the private national companies of Japan and Germany, but also from US private independents. A number of these – firms such as Marathon, Amerada, Occidental and Continental – had ventured abroad to seek crude oil supplies for their downstream operations in the USA before the introduction of US mandatory import controls. Some of them were successful in finding large oil reserves, especially in Libya. Seeking markets for their new production, which was largely shut out from the USA, they became significant competitors to the majors, especially in Western European markets. Owing to a wealth of new oil discoveries by the majors and their competitors in the established oil-producing countries and in new ones – principally Abu Dhabi, Libya, Algeria and Nigeria – there was no shortage of supply. On the contrary, oil supply grew even faster than demand, with the result, under conditions of growing competition, that prices were cut. Thus the power of the majors was seriously undermined as their previous control over production and prices slipped.[24]

But a still greater threat to the dominance of the majors came from the rise of nationalism in the developing world. For centuries, the European imperial powers had extended their dominion over Asia and Africa. Convinced of their political, economic and cultural superiority, the European powers brought the Afro-Asian world under political tutelage and economic domination, while also proclaiming the civilising influence of European colonial expansion. The main threat to their hegemony lay less in the resistance of their subjects than in the risk that the imperialists might fall out between themselves. Indeed, they did, and after igniting two world wars in the first half of the twentieth century, they had all but destroyed one another.

World War II, in particular, brought seismic changes to the international order. Heralding the arrival of the USA and the USSR as the two new 'superpowers', the war precipitated the collapse of the old European spheres of influence, the largest of which was Britain's. Quite rapidly, after 1945, the British and other European empires were broken up, to be replaced, nominally at least, by a world of independent nations. Many of these were determined to assert their political, economic and cultural independence from the old imperial powers, and from the new superpowers which, locked in mutual antagonism, could be played off against each other.[25]

The arrival of political independence in much of the less developed world became the signal not only for economic nationalism, but also for the reduction of expatriate colonial communities. Even where substantial numbers of expatriates remained, their social and economic positions were sharply modified as their status declined from that of a privileged class to that of tolerated aliens. By the mid-1960s it was only in the last bastions of white rule in southern Africa that white settlers could hope for the permanency and elevated positions that had seemed a more universal birthright in the old colonial world.[26]

In this new international political economy, foreign enterprises associated with imperialism became much less secure than they had been in the bygone imperial era.[27] Anglo-Iranian, widely seen as an instrument of Britain's imperial power, was one of the most prominent, particularly in the Middle East where, as elsewhere, Britain's power would recede rapidly in the 1950s and 1960s.[28] The Middle East was, indeed, the crucial theatre for Anglo-Iranian. Its position in the international oil industry had been founded, as noted, on its being a first mover in that region. Whereas Standard Oil (NJ) and Royal Dutch-Shell had reached out internationally from their early starts in the USA and the Far East respectively, in 1950 Anglo-Iranian's oil production and refining remained concentrated almost entirely in the Middle East, especially in Iran. It was absolutely clear that if Anglo-Iranian were to lose its Middle East concessions, the bottom would fall out of the Company.

The fear of that happening would underlie the Company's actions in almost every sphere of its operations from 1950 to the 1970s. Of course, the Company had long been conscious of its susceptibility to upsets in the Middle East, especially Iran. But it had never before been shaken by tremors as violent as those which it would experience between 1950 and the mid-1970s. The first was not long coming. In 1951 the Iranian government, under its fiercely anti-British Prime Minister, Muhammad Musaddiq, nationalised the Company's operations in Iran and threw the Company into a crisis which would turn out to be a harbinger of things to come.

PART I

IN A RISING TIDE OF NATIONALISM

I

'The structure and sinews of the Company'

'Safety and certainty in oil lie in variety and in variety alone.'[1] This prescient maxim, spoken by Winston Churchill in 1913, resonated through the Company in 1951. Since its foundation in 1909 to develop the giant oil field at Masjid-i-Suleiman in Iran, the Company, un-Churchillian in this aspect, had kept most of its eggs in one basket. Its first concern had consistently been to preserve and develop its position as sole oil concessionaire in Iran. Retarded in the development of its non-oil economy, labyrinthine in its politics, gripped by conflicting forces of tradition and modernisation, this unfathomable country of seemingly inexhaustible oil reserves commanded and absorbed the Company's attention to a degree that did not permit equivalent developments elsewhere.

By comparison with the leading oil majors, Royal Dutch-Shell and Standard Oil (NJ), the Company had achieved little geographical diversification. Iran remained the main centre of its operations, in which much more of the Company's capital, management and labour were employed than in any other country. Although the Company held shares in oil-producing concessions in Iraq, Kuwait and Qatar, its oil fields in Iran accounted for three quarters of its 1950 crude oil production. In refining, the Company's refinery at Abadan was Britain's biggest single overseas investment, and the largest refinery in the world. It dwarfed the Company's other refineries, such as that at Llandarcy in the UK, in the scale and variety of its plant and in its vast infrastructure of transport, utilities, housing, schools, hospitals and social amenities. In 1950 Abadan alone accounted for more than three quarters of the Company's refinery throughputs.[2]

This position was the outcome of a wholly deliberate strategy of concentration rather than diversification, based on the justifiable belief that the retention of the Company's most prized asset, its concession in Iran, should have primacy over anything else. The Company knew very well that the security of its concession in Iran depended upon its ability to expand its operations there and to maintain Iran's position as the leading oil producer in the Middle East; and this could not possibly be achieved if the Company allowed other interests to compete with those of Iran. For the Company, therefore, the security and safety of its most valued possession seemed to lie not in diversification, but in concentration.

The primacy of Iran was, however, embedded much more deeply in the Company than can be conveyed by top-level strategy or by statistics on the Company's hardware. Harder to measure, but no less important, was the imprint of Iran on the software of the Company, its managerial culture.

By 1950 the Company had made little progress towards the culture of internationalism, which had been an early hallmark of Royal Dutch-Shell,[3] or towards sexual equality. On the contrary, the most distinctive feature of the Company's management was its retention of an essentially male British national identity conditioned, and if anything accentuated, by the experience of expatriate 'eastern service' in Iran.

The male British bias was made explicit in the Company's staff manual. Emphasising that the Company was 'essentially a British company', the manual stipulated that recruitment should be restricted to British subjects by birth, and of European origin or descent, except in foreign countries where the Company had legal or concessional obligations to employ local nationals.[4] It was also laid down in the Company's Articles of Association that every director must be a British subject, a restriction that would not be removed until 1978.[5]

As far as women were concerned, it was not the Company's policy 'to employ women on work which is normally regarded as men's work', though exceptions could be made for females with exceptional qualifications, experience or abilities. Female staff who married were required to resign (with generous grants) and 'no exceptions to this practice can be contemplated'.[6]

While nationality and gender were matters of discrimination, political and religious beliefs were not. Indeed, the staff manual noted that employees should not be asked about their political or religious affiliations, and if known, these were not to be recorded. The same applied to membership of trade unions, which employees were free to join.

Once in the fold, the British staff enjoyed far-ranging benefits. According to an outside survey of British university students graduating in 1950, the oil industry offered the highest average starting salaries of all categories of first employment.[7] The Company also ran a pension scheme, provided *ex gratia* sickness payments, and offered loans for house purchases to male

would not attempt to finish the year with a surplus, (any more than would the people in Llandarcy, Abadan or this office). The official would get no kudos for economy, while our staff receive no reprimand for over-expenditure. There seems to me to be an utter lack of commercial outlook on the 5th and 6th floors in Britannic House, and generally in Iran, from all I hear.

Jackson emphasised the importance of distinguishing between expenditure on revenue-earning equipment and other facilities 'desirable in themselves but offering no profit'. He went on:

> Here I come to a point which gives me personally a good deal of concern and which I feel is largely responsible for the lack of commercial outlook. With few exceptions, the 5th and 6th floors are almost exclusively staffed with ex-Persian employees. Indeed, there are three Managing Directors and one Deputy-Director whose experience is predominantly Persian. I suggest that . . . Persian influence is overweighted. If this influence were entirely beneficial I should have nothing to say, but clearly and obviously it is nothing of the kind. There are admirable exceptions, but it is hard to avoid the conclusion that in general Persian experience does not produce an economical and commercial minded business man. This is a position which has gradually developed and which is not easy to alter. Old friends are offered jobs at home at the <u>end</u> of their Persian service (and <u>not</u> after say five years in that area), and a faction has been created whose influence is strong and difficult in many cases to resist.[19]

Jackson observed that there was much overlapping and duplication in the Production and Administration Departments and 'enormous preparation and discussion before any action is taken'. Moreover, data on costs were not widely circulated and in any case did not seem to matter. In Jackson's view, 'Men cannot become commercially minded unless they carry operating costs in their heads. Ask any American refiner what the [US] Gulf price of his products is and he will answer promptly. Try asking our Refinery General Manager! It appears to be a matter of no importance.'

The Production and Administration Departments were not the only ones with which Jackson found fault. The much smaller Concessions Department, which was concerned mainly with the Iranian concession, was, Jackson thought, 'very much a one-tune band. A great knowledge of Persia and experience, but insufficiently in tune with world affairs.' However, other departments, in which the Iranian influence was less pervasive, came in for less criticism. Thus Jackson thought that the Company's shipping subsidiary, the British Tanker Company, was run by an 'efficient, cohesive and economical management, overconservative, but not a bad fault, and decidedly inbred'. The Distribution Department, which was responsible for the supply and distribution of crude oil and products, was, on the other hand, 'pretty darned good – easily in my view the best department. There are individuals who could do more and have more responsibility placed upon them, and there should be a very strong, able and aggressive number 2 to the Deputy Director [Harold Snow]. Nothing is wrong here that is not being put gradually right. The process might be speeded up, that is all.'[20]

With old Iranian hands so deeply entrenched in key areas of the Company's London management the chances of reform of the Company culture from within were, of course, greatly diminished. Moreover, the Company was at the time achieving record levels of production and profits, making it difficult to justify the case for radical measures.[21] Conditions were not propitious for a new broom like Jackson to make sweeping changes.

But within the space of a few months in 1951 events in Iran changed all that, and negated the whole basis of the Company's longstanding strategy of concentration. In late April, Dr Muhammad Musaddiq, fervent nationalist and virulent opponent of the Company, became the Prime Minister of Iran. At the beginning of May an Iranian law nationalising the Company's operations in Iran received the royal assent of the Shah. The Company, contesting the validity of the nationalisation, kept its Iranian operations going as the dispute escalated into a major international crisis. A series of diplomatic initiatives failed to resolve matters. When the Iranians attempted physically to take over the Company's assets the Company withdrew its tankers from Abadan, British personnel were withdrawn from the oil fields, and women and children were evacuated from Abadan. The withdrawal of tankers meant that there could be no oil exports and on 31 July the Abadan refinery was shut down, its storage capacity having been filled. Just over two months later, on 3–4 October, the Company's remaining British staff were evacuated from Abadan.[22]

All of a sudden, change was imperative, and it had to be fast. The Company was out of Iran and did not know whether it would be able to return. Its supplies of crude and refined products from that country were cut off, and the retention of managers with long service in that country could no longer be justified on the grounds that their experience and knowledge were indispensable. For those who looked back, it was a loss; but for the forward-looking it was a marvellous opportunity for change.

THE EXPANSION OF NON-IRANIAN SUPPLIES

While the drama unfolded in Iran, Sir William Fraser, the Company's chairman, initiated a review of the Company's supply position. In July 1951, stipulating that the Company should aim to keep its marketing outlets fully supplied despite the stoppage of exports from Iran, he set up a senior inter-departmental Future Programme Committee to propose appropriate short- and long-term action.[23]

Within three weeks of being formed, the Committee completed an interim report. It estimated that without Iran the Company would be able to produce 490,000 barrels per day (bpd) of crude in 1952, mainly from Kuwait. This would be more than enough to cover the Company's projected refining capacity of about 350,000 bpd of crude (yielding 315,000 bpd of marketable products) by mid-1952. However, another 165,000 bpd of crude were needed to fulfil the Company's high-volume long-term crude supply contracts with Standard Oil (NJ) and Socony-Vacuum. Deliveries under these contracts, which ironically had been devised specifically to increase offtake from Iran, were due to commence at the beginning of 1952. On top of that, some 100,000 bpd of crude would be required in 1952 to meet obligations to supply other oil companies. Overall, therefore, the Committee estimated that in 1952 the Company would have about 490,000 bpd of crude oil available, against requirements amounting to nearly 615,000 bpd, i.e. a deficit close to 125,000 bpd. The crude shortage was, however, expected to lessen quickly after 1952, as the Company's production outside Iran was stepped up.[24]

The projected deficit of refined products presented more serious problems. The estimated requirements of the Company's own marketing outlets were 480,000 bpd in 1952, against the forecast output of 315,000 bpd from the Company's refineries, excluding Abadan. The resultant shortage of products would be felt most acutely in markets east of Suez, where the Company had previously relied almost entirely on supplies of products from Abadan. West of Suez, the position was not so bad. The Company's European refining capacity had increased rapidly since the end of World War II, rising to 150,000 bpd in 1950. Since the end of 1950, substantial new capacity had come into operation at Grangemouth and the Company had also acquired a majority interest in the Oelwerke Julius Schindler refinery at Hamburg in March 1951.[25] Moreover, new refineries were under construction at Antwerp (in partnership with Petrofina) and on the Isle of Grain in Kent.[26] Projected product deficits in western markets were therefore much less than in the east, and in the case of fuel oil there was actually a projected surplus of supply west of Suez (see table 1.1).[27]

The Future Programme Committee's report covered only main products, but there were also shortages of other, more specialised products, especially aviation spirit (often described as avgas). The Company's business in aviation

2 Sir William Fraser (later Lord Strathalmond), the Company's chairman and chief executive, with the Queen Mother, visiting Grangemouth refinery in 1952

spirit depended for its supplies on the Abadan refinery, which had become by far the largest producer of aviation spirit in the Eastern Hemisphere after the wartime installation of alkylation plant for the manufacture of this product.[28] Indeed, in 1950 Abadan's production represented more than 90 per cent of all the aviation spirit produced in the Eastern Hemisphere.[29] When the Abadan refinery was shut down in mid-1951, the Company had no internal supplies of aviation spirit with which to maintain, let alone expand, its aviation business.

Acting on the recommendations of the Future Programme Committee, the Company took immediate steps to make up for the cessation of oil supplies from Iran. In crude production, the Company's non-Iranian output was stepped up very rapidly, especially in Kuwait, where the Company's share of production increased from 205,000 bpd in June 1951 to nearly 350,000 bpd

3 Electric welding during the erection of Kent refinery in 1952

Table 1.1 *The Company's products
shortfall after nationalisation in Iran*

	Estimated availability as percentage of requirements in 1952	
	East of Suez	West of Suez
Gasoline	3%	81%
Kerosenes	0%	54%
Gas/diesel oils	7%	79%
Fuel oils	10%	134%

in July.[30] Increased output in Iraq and Qatar also contributed to the expansion of non-Iranian crude production (see figure 1.1).

At the same time, the Company opened discussions with Standard Oil (NJ) and Socony-Vacuum for a reduction in the quantities of crude deliverable under the long-term supply contracts. After negotiations, it was agreed that the Company would supply only 83,000 bpd in 1952, little more than half the quantities which the Company had earlier contracted to supply on the assumption that Iranian crude would be available. The two US companies agreed not only to accept reduced quantities in 1952 as a whole, but also to the phasing of deliveries so that the Company could in effect defer making supplies which would otherwise have been due in the first half of the year.[31] This helped to relieve the immediate pressure on the Company's crude supplies pending the further expansion of non-Iranian production.

Notwithstanding these measures, the Company was unable to meet its crude requirements from its own sources of production in the second half of 1951 and the first few months of 1952. Resort was therefore made to purchasing crude from other companies. The emphasis on finding a sufficient volume of supplies sometimes gave rise to quality problems, especially with some cargoes of extremely heavy crudes from the US Gulf and Venezuela.[32] However, most of the Company's purchases consisted of Saudi and Kuwaiti oil, supplied by various companies of which the most important were the Texas Company, Gulf and Standard Oil (NJ). Crude purchases reached a peak in November 1951, when they amounted to about 25 per cent of the Company's total crude supplies for that month. Purchases continued on a significant scale until early spring 1952, making an important contribution to the Company's crude supplies.[33]

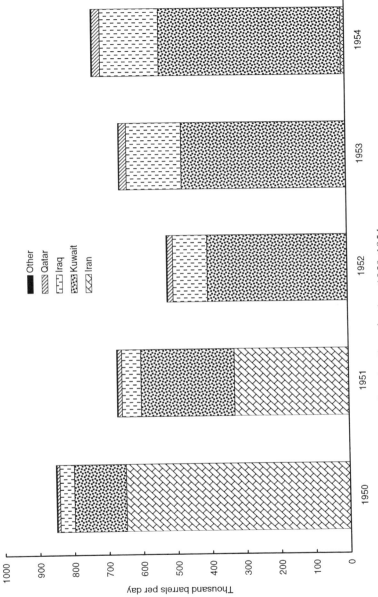

Figure 1.1 The Company's sources of crude oil production, 1950–1954

4   A 99-mile rig skid in Kuwait in 1954

Meanwhile, discussions were held, economic assessments prepared and physical investigations made of possible locations for the erection of new refining capacity east of Suez. At the end of July 1951 it was agreed that a new refinery should be constructed in Australia and, after further investigation, it was decided that the refinery, which was to have a capacity of just over 60,000 bpd, should be located at Kwinana, near Fremantle.

The choice of location for another new refinery to provide supplies for the Company's large marine bunker operations at Aden and for other outlets around the Red Sea and in eastern, central and southern Africa was also carefully deliberated. Comparisons were made between locations at the sources of the Company's crude production, such as Kuwait and Qatar, and the alternative of Aden, which was well situated for imported crude supplies and for export shipments of products to markets east of Suez and to Europe via the Suez Canal. As a major marine bunkering port Aden also offered a ready outlet for a bulk product adjacent to the refinery, and, as Britain's only colony in the Middle East, it was considered to be more secure than alternative locations in the Persian Gulf area.[34]

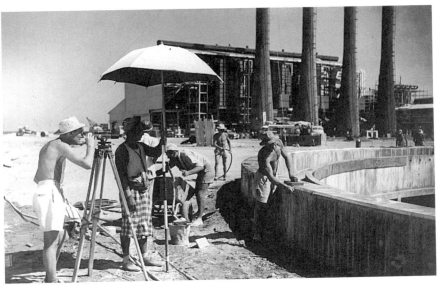

5   Surveyor stripped to the waist and native boy shielding him from the sun with an umbrella, at work on the power station at Aden refinery in the 1950s

Persuaded by this combination of economic and strategic factors, the Company decided to construct a new refinery with a capacity of 100,000 bpd at Aden. Official clearances were duly obtained and construction completed in time for the refinery to come into operation in July 1954. A few months later, at the beginning of February 1955, the new Kwinana refinery in Australia also came on stream. The Company's smaller Australian refinery at Laverton, near Melbourne, was then closed down.[35]

The erection of new refining capacity at Aden and Kwinana offered a medium-term solution to the shortage of refined products, but it did nothing to increase the supply of products until the new capacity began to come on stream. During the intervening period the Company took short-term measures to increase supplies. In Europe, Company refineries were run at peak throughputs and two new refineries, whose construction had commenced before the Iranian crisis, were brought on stream.[36] The first, in October 1951, was the Antwerp refinery, owned jointly with Petrofina. The second was the large Kent refinery, which was commissioned early in 1953.[37] By this combination of maximising throughputs and commissioning new capacity, the Company's refinery throughputs in the UK and the rest of Europe were approximately doubled between 1951 and 1954 (see figure 1.2). These

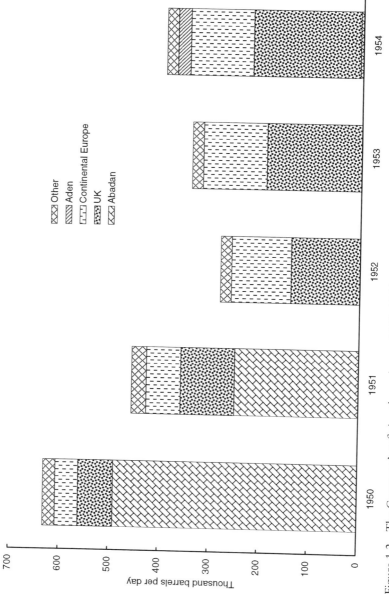

Figure 1.2   The Company's refining throughputs, 1950–1954

increases in throughput were not, however, nearly enough to make up for the loss of output from Abadan, without which the Company's overall refinery throughputs fell dramatically, from 632,000 bpd in 1950 to 280,000 bpd in 1952. Other measures were clearly necessary to boost supplies of refined products.

To that end, the Company negotiated numerous processing agreements under which other oil companies processed crude oils supplied by the Company and provided the Company with pre-arranged yields of products. The quantities of crude oil processed under these arrangements were substantial, rising from 8,000 bpd in 1950 to over 60,000 bpd in 1952. The largest processors of Company crude were Royal Dutch-Shell and the French oil company, Compagnie Française des Pétroles. Otherwise, apart from very small quantities processed by Standard Oil (NJ)'s French subsidiary, the crude refined for the Company under processing agreements was by smaller refiners, mainly in Europe.[38]

Despite the processing agreements, it remained necessary for the Company to supplement its products supplies by making purchases from other oil companies. In the case of aviation spirit the Company purchased supplies in the USA, initially on a 'spot' basis, but later through supply contracts with the US companies, Cities Service and Pan-Am Southern.[39] A large proportion of the Company's products purchases was obtained through the Royal Dutch-Shell subsidiary, Asiatic Petroleum Corporation, which acted as the Company's purchasing agent in the USA, where the Company's own New York office had no purchasing organisation, although it did perform a valuable intermediary role.[40] The purchasing arrangement with Asiatic Petroleum remained in being until the beginning of October 1953, when the Company's New York office set up its own purchasing organisation.[41]

By these various measures the Company quickly developed a pattern of operations which was quite outside its previous experience, as the old, well-practised routines of the Iran-based business were replaced with great speed by a much more complicated pattern of supply from diverse sources.

The dislocation of established patterns of trade inevitably affected shipping operations. After the shutdown of the Abadan refinery the Company had to move more crude than ever before from the Persian Gulf to refineries in Europe, from where some cargoes of finished products had to be back-hauled to markets east of Suez. Up to a point, the Company was able to meet its shipping requirements by redeploying its own fleet, but it was also necessary to charter additional tonnage. This was not easy in the winter of 1951–2, when there was a world shortage of tanker tonnage available for charter. The shortage forced up charter rates and imposed a bottleneck on the Company's supply operations. In February 1952 Snow reported that the supply position 'at present was primarily controlled by the shipping factor. We had insufficient ships to lift all the Crude Oil available to us at Kuwait and it had accordingly been necessary to ease back on the throughput rate at Llandarcy.' Although the charter market subsequently eased, the Company's shipping routes would never return to the pre-crisis pattern.[42]

HOLDING ON TO MARKETS

When Fraser laid down that the Company should aim to meet the requirements of its marketing outlets, he was setting a target which would have been laughably easy before the nationalisation crisis. Traditionally, one of the Company's most persistent problems had been its excess of productive capacity over markets. That imbalance between supply and demand was turned on its head in the crisis, when the Company suddenly found itself without the supplies that were needed to meet the demands of its customers. No longer could it be taken for granted that the Company had more than enough supplies to match its market share.

Because crude production fell less precipitously, and could be increased more quickly, than refining capacity, the Company soon had crude oil once again available for sale. With the exception of a minor setback in 1951, crude oil sales continued the upward trend which had started before the crisis, rising to 280,000 bpd in 1954, compared with 175,000 bpd in 1950. A large proportion of these sales was taken up by deliveries under the long-term contracts with Standard Oil (NJ) and Socony-Vacuum.

In products supplies, on the other hand, the Company's outside purchases failed to compensate for the fall in its refinery throughputs. Starved of supplies, the Company suffered a sharp drop in products sales, which fell from 615,000 bpd in 1950 to less than 450,000 bpd in 1952 before recovering to 527,000 bpd in 1954. The decline did not, however, damage the Company's own marketing business. By reducing sales to other major oil companies, which were considerable before the crisis, and to the British armed services (the Admiralty contract), the Company succeeded in maintaining an increasing flow of products to its marketing subsidiaries and associates. The only exceptions were in the Middle East, where the Company ceased its marketing operations in Iran in 1951 and in Iraq in 1952.[43]

Apart from the Middle East, the only notable exception to the continued expansion of sales through the Company's own outlets was in marine bunkering. Before the crisis the Company was very strongly established in the bunker market in the Eastern Hemisphere, where it held a market share of about 37 per cent.[44] But in the autumn of 1951 its position became untenable, and Freddie Morris, director of marketing, decided not to tender for the renewal of some bunker contracts.[45] The curtailment of one of the Company's most successful marketing activities was apparently achieved without loss of customer goodwill. As a member of the Distribution Department's Marine Fuel Branch reported in 1952: 'The process of relinquishing contracts has been extremely difficult: it is fair to say that generally the [ship] Owners concerned have shown themselves very disinclined to leave us, not mainly because they could encounter difficulty in securing contracts with other suppliers, but largely because they have been well satisfied with our service and relationships. They leave us more in sorrow than in anger, appreciating the abnormal conditions.'[46]

Inland, however, the Company's marketing subsidiaries and associates continued to expand their sales in all classes of main products in most areas. Even in aviation spirit, the volume sold through the Company's outlets increased.[47]

STAFF AND ORGANISATION

While the Company strove to maintain its market position, another priority was the impact of the crisis on its staff and organisation. In the middle of 1951 the Company employed about 2,500 British staff in Iran. A similar number worked in the UK head office, about a third of them being engaged on work concerned solely with operations in Iran.[48] With the cessation of operations in Iran and the evacuation of British staff from that country the Company therefore had many staff who were surplus to immediate requirements.

Pattinson, by this time deputy director of production, thought that 'the sooner most of them are out of it the better, and I cannot think that any of the better men would want to sit around doing nothing for long when there is so much to be done and so many good opportunities to be seized in other businesses'.[49] There were others, however, who seemed more aware of the difficulties facing surplus staff, especially those who had returned to Britain from Iran. For example, Lewis Baxter of the Staff Department sympathised with evacuees from Iran who had been 'suddenly transported with their families to face an uncertain future on another continent'. Their biggest problem was usually accommodation, for staff were naturally reluctant to buy houses or take long-term rented accommodation while they remained unsure about their future employment. Many moved in with relatives, but as Baxter put it: 'Even those who get on well with their relatives find sharing a house trying after the accommodation and amenities of Persia.' The alternative was to stay in hotels or short-term rented accommodation, but that was 'ruinously expensive' and often 'far beyond their means'. Apart from accommodation, there was also the problem that 'boredom, uncertainty as to the future, and, in some cases, financial embarrassment, make them restless . . . When idle in the UK they tend to spend a great deal more than they would if they were working and are therefore dissipating resources which they may need later to tide them over a period of unemployment. Many of them, too, are living at a higher standard than they would normally maintain here if they were permanently domiciled, e.g. they send their children to boarding schools.'[50]

To deal with the staff problem, Fraser asked Edward Elkington, director of administration, to set up a Staff Redeployment Committee under Sir Humfrey Gale, the Company's adviser on organisation.[51] Coinciding with the evacuation of the last British staff from Abadan, the Committee held its first meeting on 3 October 1951. Just nine days later it set out the principles to be followed in staff redeployment. The overriding principle was that the best men, whether ex-Iran or London, should be retained within an overall personnel structure that would be balanced with regard to age, length of service and experience. With the female staff the guideline was, other things being equal, to retain single, in preference to married, women. A balanced team of about 1,100 of the best men ex-Iran was to be retained as a reserve which could be sent back to Iran to restart operations in the event of a settlement of the crisis being reached and the Company again having a part to play in the Iranian oil industry. Otherwise, the review of the Company's staff was to be made on the basis that the Company was out of Iran, and had no plans for expansion. The Company would do what it could to help redundant staff find other jobs, and to that end a Resettlement Panel was set up under Gale.[52]

These principles were quickly applied and by the end of 1952 the position of nearly all of the 2,479 British staff who had been withdrawn from Iran had been decided. Of that number, 686 were redeployed in the Company or its associates, mainly the Iraq Petroleum Company and the Kuwait Oil Company, and nearly 1,700 were made redundant. At the same time, 647 staff were removed from the UK head office payroll. Of that number, 91 were redeployed in the Company or its associates and 556 were made redundant. However, this reduction in head office staff was partly offset by new engagements (which included some of the staff withdrawn from Iran) so that the net reduction in head office numbers came to only 376 staff.[53] In the meantime, as the nationalisation dispute continued without a settlement in sight, the number of staff in the 'Persian Reserve' was steadily whittled down to the point where the reserve was dispersed.[54]

Staff who were made redundant were offered help in finding other employment by the Resettlement Panel, which held its first meeting on 29 October 1951. Before then, the Staff Department had already written to all the principal oil companies and other firms such as ICI and Unilever asking them for information regarding any vacancies in their organisations. After the Panel was set up, the search for vacancies was intensified by press advertisements. The Panel, which itself had a staff strength of eighteen, tried to match redundant staff to suitable vacancies, introduced staff to potential employers, helped in drawing up job applications and monitored the progress of each individual applicant. The two main factors governing the speed and success of resettlement were age and technical qualifications. For those in their twenties a lack of qualifications did not prove a serious handicap, but there were fewer opportunities for higher age groups. Moreover, it was found that men who had been overseas for some years 'lacked the knowledge of the labour market and the Trade Unions necessary to men of foreman or supervisory level'. Overall, however, the Panel claimed a high degree of success. By the time it was disbanded in July 1953, 94 per cent of the staff who had requested its assistance had found jobs. A majority of 52 per cent had found jobs for themselves, compared with 42 per cent who were resettled by the Panel.[55]

Although there was no relish in laying off staff, the redundancies provided the Company with an opportunity to cut out dead wood while retaining the most capable and promising employees. They also helped to weaken the influence of the ex-Iran element of which Jackson had earlier been so critical. Before the nationalisation crisis he had already taken steps to reduce the power of the Production Department, in which the influence of old Iranian hands was most strongly concentrated. In February–March 1950 he had made an extensive tour of Iran and the Middle East, and in July, shortly after his return, he was appointed deputy chairman.[56] A few months later, in March 1951, he announced the formation of a new Refineries Department in head office. The new Department consisted of three divisions, the heads of which were directly responsible to Jackson.[57] By this move, responsibility for refining was taken out of the hands of the Production Department, which was reduced to two divisions – the Geological and Geophysical Division and the Development and Engineering Division – and three smaller branches: the General Branch, the Petroleum Engineering Branch and the Kirklington Hall research station.[58]

The work of the Production Department was further affected by the nationalisation crisis. All the major oil fields operated by the Company were in Iran, and when operations there came to a halt the core of the Production

Department's work was removed. There was still some work to do operating the Company's small onshore fields in the UK, and providing technical support for the D'Arcy Exploration Company, the wholly owned subsidiary through which the Company managed its other exploration and production interests, which were in shared concessions. But these activities could nothing like make up for the loss of operations for which the Production Department had previously been responsible. The fact was, as Pattinson commented, 'most of the work of the department has disappeared'.[59] The Development and Engineering Division of the Department was therefore wound up, leaving only the Geological and Geophysical Division and the three branches already mentioned. It was a measure of the reduced scale of the Production Department that by May 1952 its head office staff numbered only 64, compared with 340 in the Refineries Department and 262 in the Distribution Department.[60] The power of the old Iranian hands in the Production Department had clearly been greatly weakened.

The Distribution Department, on the other hand, was in a period of ascendancy. At the onset of the nationalisation crisis this department, under Snow as deputy director reporting to Morris on the board, was responsible for the twin functions of supply and marketing. To perform those functions, the department consisted of four divisions: the Trade Division, the Supply Division and two regional marketing divisions covering western and eastern markets respectively.[61] In November 1951 these last two were amalgamated into a single Marketing Division. More significant, though, were the changes that were shortly to be made to the Supply Division.

The primary role of the Supply Division was to match the Company's supply and demand for oil so that crude and products were available in the quantities and qualities required to meet the marketers' estimates of demand. Before the crisis, when Iran was the principal source of oil supplies, this was a relatively simple task. As sole owner and operator of its producing and refining interests in Iran, the Company was free to conduct its operations without the need to consult partners, and the Abadan refinery offered unparalleled flexibility in matching products supplies to demand on account of its great variety of plant, ample tankage and the facility for pumping surplus products back into the oil fields for later recovery.

The cessation of supplies from Iran brought an immediate end to this simple and flexible system of supply programming, and placed a great additional workload on the Supply Division.[62] To help in coping with the burden, two new branches were created in the Division at about the end of 1952. One, Procurement Branch, was to handle all purchasing and processing arrangements and to provide the rest of the Division with medium- and long-term forecasts of oil supply. The other, Co-ordination and General Branch, had various functions, including general co-ordination of supply and liaison with the Refineries Department.[63]

Further measures were, however, required to clarify the increasingly blurred lines of demarcation between supply, marketing and planning. Within the Distribution Department the dual functions of supply and marketing had traditionally been easily separable, with supply focused on Iran, and marketing focused on markets. But with the growing trend towards market-located refining, the marketers were becoming much more involved in supply than formerly. Snow, who succeeded Morris on the board at the end of July 1952, recognised that 'the two must work together'. He thought that a clearer allocation of responsibilities was needed and that 'we should aim to restore the position that Marketing is primarily concerned with the sale and distribution of products in the markets'. Subsidiaries and associates with refineries in their territories should, Snow suggested, refer to the Supply Division and not the Marketing Division on refinery matters.[64]

Snow also recommended a change in the relationship between the supply and planning functions. Since its formation under George Coxon in 1943, the Central Planning Department had acted as a small, independent think-tank within the Company.[65] It advised Fraser and the board on long-term issues, and enjoyed a high standing, largely because of Coxon's high intellectual calibre. However, in the summer of 1953 Coxon, who had been working extra hard as chairman of the Future Programme Committee, suffered a stroke and there were doubts about his fitness to return to work.[66] This provided a natural moment at which to reassess the relationship between the Central Planning and Distribution Departments with a view to integrating forward planning more closely with the supply function. Snow thought that this was logical, as planning was 'in fact primarily concerned with future supply availability'. He therefore suggested that the Central Planning Department should be amalgamated with the Supply Division of the Distribution Department to form a single supply organisation.[67]

His ideas were put into effect in January 1954, when a new Supply and Development Division was formed by merging the Supply Division, the Central Planning Department and some of the Trade Division.[68] The new division dealt with all aspects of supply, from current programming and procurement to long-term forward planning.[69] It was much more powerful than the old Supply Division had been and it was realised that the general manager of Supply and Development would have a very important role in the Company. The person selected for the job was Eric Drake who, since leaving Abadan in the heat of the nationalisation crisis in June 1951, had been the Company's negotiator with the Australian authorities over the agreement to build the Kwinana refinery, and then been posted to New York as the Company's representative in North America. He returned from there to take up his new post in charge of Supply and Development in 1954.[70] Meanwhile, Coxon, the pioneer of corporate planning in the Company, retired on medical advice.[71]

Drake was perfect for the role of the 'very strong, able and aggressive number 2' to Snow, the need for which had earlier been identified by Jackson. In Drake's early years in Iran it was thought by some that he was 'inclined to be somewhat aggressive' and his impatience had prompted Elkington to comment that Drake wished 'to obtain the plums before the tree has had time to grow'.[72] But there were no doubts about his ability or about the importance of his job at the head of the new Supply and Development Division, which had a central role in the Company's operations. It was the Company's equivalent to Standard Oil (NJ)'s Coordination Department, which had been set up in the 1920s to co-ordinate the flow of oil through the major functions of production, transportation, refining and marketing.[73] Just as the Coordination Department was a 'vitally important unit' in Standard Oil (NJ), so too was the Supply and Development Division in the Company. It soon acquired elite status and became one of the focal points for bright young managers on their way up through the Company hierarchy.

FINANCE

Although the Company's marketing subsidiaries were largely successful in maintaining their shares of inland markets for oil, the cuts in supplies to other oil companies and to the Admiralty, coupled with the reduction in marine bunker sales, inevitably resulted in a fall in the overall volume of the Company's trade. At the same time, costs were increased by the extra expenses incurred in purchasing crude and products from external suppliers, in processing arrangements with other refiners, and in uneconomic shipping operations. The inescapable result was a fall in profits.

These are shown in table 1.2 and require a little explanation. Until 1950, the main concession agreements in the Middle East provided for the main part of the payments to the producing countries to be made as royalties, which were generally set at a fixed amount per ton of oil. Under the Company's accounting procedures royalties were treated as part of the cost of oil, which was deducted from trading income in order to arrive at pre-tax profits. Accordingly, the data given for pre-tax profits in 1950 were arrived at after taking account of royalties.

Table 1.2 *The Company's profits, 1950–1954 (£millions)*

	Pre-tax profits at current prices	Deduction for overseas taxation	Profits after overseas taxation	Deduction for UK taxation before overseas tax relief	Overseas tax relief	Post-tax profits	Post-tax profits adjusted for price inflation (1950 prices)
1950	86	–	86	51	–	35	35
1951	59	2	57	28	–	29	27
1952	76	27	49	23	–	26	22
1953	78	31	47	20	–	27	22
1954	96	43	53	28	1	26	21

However, after 1950 there was an important shift in the basis on which oil companies made their payments to producing countries. The major part of the oil companies' payments ceased to be in the form of royalties based on tonnage and instead took the form of taxes representing an agreed share of profits. This new system was adopted in the Middle East in 1950, when the US-owned Aramco, holder of the prime oil concession in Saudi Arabia, agreed to pay 50 per cent of its profits as tax (inclusive of royalties) to the Saudi government. The 50:50 system increased Saudi Arabia's revenue without any real cost to Aramco because the US authorities allowed US companies to offset their tax payments to foreign governments against their tax liabilities in the USA.[74]

The example of Aramco in Saudi Arabia was quickly followed elsewhere. The 50:50 profit-sharing principle was adopted in Kuwait in 1951, Iraq and Qatar in 1952, and Iran, under the terms of the settlement of the nationalisation dispute, in 1954.[75] As payments to producing countries shifted to the new system, the Company pleaded with the UK tax authorities that it should be afforded the same rights as the US oil companies to offset overseas tax payments against domestic tax liabilities. The Inland Revenue agreed to this principle in relation to Iranian taxation in 1954, but it was not until the following year that relief for overseas tax was extended to cover tax payments to Kuwait, Iraq and Qatar.[76]

The effects of overseas taxation on the Company's profits in 1950–4 are shown in table 1.2, from which it can be seen that profits after deductions of taxes due to producing governments fell very sharply indeed from £86 million in 1950 to £47 million in 1953 before staging a modest recovery in 1954. Profits after UK taxation fell somewhat less, but in 1954 were still only some two thirds of the 1950 level.

The drop in profits coincided with great increases in capital expenditure, mainly on the construction and extension of refineries and the building of tankers. With rising expenditure and falling profits, the Company was unable to cover its spending with funds generated from its operations. The financing of the deficit was not, however, a problem as the Company had entered the crisis in an extremely strong financial position, with liquid assets of £100 million or more and virtually no debt. It therefore had ample financing capacity to sustain its heavy capital expenditure programme, which it achieved partly by drawing on its liquid resources and partly by raising new long-term debt by issuing £20 million 5 per cent debentures in January 1953.[77] This was the first time since 1932 that the Company had turned to the capital market for funds, ending a remarkable unbroken run of more than twenty years in which the finance for expansion was entirely self-generated. The draw-down of liquid resources and the debenture issue did not, however, undermine the Company's financial strength. At the end of 1954 it held £90 million in liquid resources and its debt:debt-plus-equity ratio was less than 12 per cent, leaving it with plenty of spare financial capacity.

Indeed, so strong was the Company's financial position that Fraser was keen to raise the dividend to the ordinary shareholders even in the depth of the nationalisation crisis. He was not, however, the only party interested in the Company's dividend policy. So too was the British government whose relationship with the Company was a sensitive one, rife with contradictions. Some of them stemmed from the special relationship formed in 1914, when the government had acquired a majority shareholding in the Company, with the right to appoint two of the directors, and the power to veto board decisions, which the government had pledged not to use except in special circumstances, mainly of a national strategic nature.[78]

This characteristically British compromise had ever since been the cause of much misapprehension on all sides. Although the veto had never been used, its very existence tended to heighten the sensitivity of the Company's management to the slightest hint of unwelcome government interference; while in the world at large the Company was widely perceived as the instrument of the government on the understandable assumption that what the state owned, it surely controlled.

Even without the majority shareholding and the veto, it was inconceivable that the government would have adopted an entirely hands-off attitude towards the Company. There could be no ducking the fact that the Company was a vital strategic concern, whose activities could affect Britain's foreign relations and economic position. When that happened, the government could not possibly stand aside. The necessity for the government to step in and take over the British side of the negotiations during the Iranian nationalisation crisis was the most obvious example.

But it was not the only one. After World War II, when the state played a far greater role in the economy than it had done in 1914, there were fundamental contradictions in the arrangement whereby the government as majority shareholder denied itself the right to interfere in the normal affairs of the Company when, with or without the shareholding, the Company's decisions could be of close concern to the government. The Company's dividend policy was just such an issue.

Since 1948 it had been Fraser's practice to consult the Treasury before putting dividend proposals to the Company's board. The initial consultation of 1948 arose out of one of the provisions of the 1933 concession agreement whereby Iran received annual payments equal to 20 per cent of the dividends paid to the Company's ordinary shareholders in excess of 5 per cent of the Company's 1933 ordinary share capital. In 1948, when the final dividend for the previous year of 1947 was decided, the Company felt in a strong enough financial position to pay an increased dividend, which would, of course, have benefited Iran. However, an increase would have been contrary to the policy of Clement Attlee's Labour government which, anxious to win trade union backing for wage restraint, was urging British companies not to increase dividends. The Chancellor of the Exchequer, Sir Stafford Cripps, accordingly informed Fraser in May 1948 that an increase in dividends would be undesirable.[79] The Company therefore left its dividend unchanged and for the five years from 1947 to 1951 it remained constant at 30 per cent of the ordinary share capital. After deducting income tax withheld by the Company the annual net payment to the shareholders was, in decimal currency, 16p. per share in these years.[80]

In July 1951 Hugh Gaitskell, who had succeeded Cripps as Chancellor, announced that the government intended to introduce a parliamentary bill in the autumn for the statutory limitation of dividends. This measure would, Gaitskell told the House of Commons, 'help to ensure the acceptance of a policy of reasonable restraint in the field of wages'.[81] However, the bill had not been introduced by the time of the October general election, in which a new Conservative government was returned to power with Winston Churchill as Prime Minister and R. A. Butler as Chancellor of the Exchequer. The change from Labour to Conservative government did not, however, bring radical changes in economic policy. In fact, Butler's policies appeared so indistinguishable from Gaitskell's that the term 'Butskellism' was invented to cover both.

Although one of Butler's first statements on coming to power was to reject Labour's idea of statutory limitation of dividends, he at the same time emphasised 'the absolute necessity, in the interests of the economy as a whole and of the companies themselves, of a very cautious policy on dividend distribution'.[82] The absence of statutory powers did not, in other words, mean that the government was prepared to adopt a hands-off policy on dividends. On the contrary, as Fraser was to find out, the Treasury was prepared to exert great informal pressure behind the scenes to restrain dividend increases.

The government's interest in the Company's dividend policy did not, however, arise solely from the Treasury's concern with the management of the domestic British economy. There was, in addition, the Foreign Office's interest in restraining Fraser's hand on dividend increases for fear of damaging relations with Iran in the changed circumstances of the nationalisation dispute. In the Foreign Office's view, dividends were a most sensitive point, as increases would add to the income of shareholders, mostly British, at the very same time as the oil revenues of Iran had dried up because of the stoppage of Iranian oil exports and the cessation of payments under the 1933 agreement. By highlighting the contrast between the prosperity of the Company's shareholders and the penury of Iran, dividend increases would, it was thought, anger the Iranians and so make it more difficult to arrive at a settlement of the nationalisation crisis.

Such concerns were expressed in 1953 when Fraser proposed that the dividend for the previous year of 1952 should be increased from the 30 per cent at which it had stood for the previous five years. His proposal was unwelcome to the government, who thought that it was 'likely to affect adversely' the negotiations with Iran. In the event, the dividend was kept at 30 per cent, but shareholders also received a bonus of 5p. per share, which effectively raised the rate to 35 per cent. After deducting income tax withheld by the Company the net payment to the shareholders was 19p. per share, an increase of 3p. over the previous year.[83]

In April 1954 dividend policy again came to the fore in discussions between the Company and the government. This time, it was the final dividend for 1953 that had to be agreed before the Company publicly declared the dividend in early May. Fraser wanted to raise the 1953 dividend to 50 per cent of the nominal value of the Company's issued ordinary capital. The Treasury was assured by Frederic Harmer, one of the two government directors on the Company's board, that such an increase was fully justified on commercial grounds. He felt that 'there was no escape from increasing the dividend'. The market expected it and if Fraser kept the dividend at 35 per cent 'he would be in grave difficulties: the market would be flabbergasted and the shareholders indignant'.[84]

Politically, however, a large increase in the dividend was difficult, and was made still more so by Butler's public appeal for companies to exercise voluntary dividend restraint in his budget speech to the House of Commons on 19 April.[85] Harmer insisted that any firm in the Company's position would increase its dividend to 50 per cent 'even after taking into account what the Chancellor has said'. The Company was, he went on, 'right out of line' with other oil companies in dividend policy. Shell, for example, had recently increased its dividend which was already much higher than the Company's. In Harmer's view, the only way that Fraser could justify holding the dividend down was by openly stating that he was doing so at the request of the government as the Company's principal shareholder, which, he added, was 'unthinkable'. If, on the other hand, the Treasury vetoed Fraser's proposed dividend increase without giving him a public directive, then, as a Treasury official admitted, 'we are rubbing in all the disadvantages of public ownership without taking public responsibility'.

After interdepartmental consultations in Whitehall, and with the Bank of England, a letter was sent to Fraser by Sir Edward Bridges, the Permanent Secretary of the Treasury, arguing that a large rise in the dividend would conflict with domestic economic policy. Fraser was reportedly 'greatly distressed' by the letter and argued that, for the first time in his experience, the Treasury was interfering in the commercial affairs of the Company, contrary to the government's 1914 pledge not to use the veto. Within the Treasury it was admitted that there was 'a good deal in Sir William's [Fraser's] point' and that Bridges' letter had 'the shadow of the veto on it'.

As an alternative to the Treasury view, the government fell back on the Foreign Office line that an increase in the dividend was ill-timed in relation to the Iranian negotiations. This, it was thought, was 'a point which we have every right to make, now that we have taken on responsibility for fighting this battle with the Persians'. Bridges accordingly told Fraser that an increase in the dividend 'at this juncture would be disastrous' because of its effect on the Iranian negotiations. This, however, 'did not seem to placate Sir William Fraser at all'. He took the view that the Company's profits in 1953 'did not take into account a single gallon of Persian oil. He thought that it would be a help in the negotiations and would strengthen our position, if it could be shown that AIOC could do without Persian Oil and would even pay an increased dividend without any help from Persia'. According to Bridges 'the argument continued and at moments got quite heated'. However, he and Fraser eventually agreed on a dividend of 30 per cent plus a bonus of 12½p. per share, equivalent to 42½ per cent in total.[86] In May, that was duly declared as the 1953 dividend, which, after deducting income tax withheld by the Company, was 23p. per share.[87] As will be seen in chapter five, the settlement of the 1953 dividend did not by any means bring to an end friction between Fraser and the government over dividend policy.

In the meantime, efforts to resolve the nationalisation dispute continued in negotiations with the Iranian government under General Fazlullah Zahidi, who had become Prime Minister after Musaddiq was overthrown in an American- and British-backed coup in August 1953.[88] By late August 1954 the negotiations were entering their final stages and the large delegation of oil company negotiators would soon return home from Tehran. Before they left, Sir Roger Stevens, the British Ambassador to Tehran, held a party for them in the gardens of his summertime residence in the cool of the mountains to the north of Tehran. 'It was', wrote Stevens, 'a perfect night & the scene was a very pretty one, with flood lights showing up the white bark of the giant plane trees, little floating trays of flowers with candles in the middle on the ornamental pool, and the chiaroscuro on the lawns of white dresses, white dinner jackets & black coats & trousers. I think most people enjoyed themselves – certainly no one showed any sign of wishing to leave till well after 1 am.'[89]

On 19 September the final agreement was signed in Tehran, and was flown via The Hague (for signature by Shell and others) to London Heathrow Airport near where, in the early hours of 20 September, it was signed by Fraser at the Berkeley Arms hotel. This amused Dr Ali Amini, the leader of the Iranian negotiating team, who, 'quick & scintillating as ever', gave Stevens 'a delightful account of how Sir Wm Fraser of AIOC had to spend the night at Heath Row Airport to sign the oil agreement'. Stevens replied: 'I hope this is the last indignity you will demand of the AIOC.' At which Amini 'roared with laughter'.[90]

The agreement was ratified by the Iranian Majlis and Senate (the lower and upper chambers respectively of the Iranian parliament) in October and received the royal assent of the Shah on the 29th of that month. It upheld the principle of national ownership of the Iranian oil industry, but put the Company's nationalised assets (apart from small-scale domestic operations in Kirmanshah which would be run by the state-owned National Iranian Oil Company) under the management of a new consortium of international oil companies. The Consortium would operate the assets and buy all the output, with each company in the Consortium marketing its share of the oil through its own marketing system. Profits from the Consortium's operations would be divided equally between the Consortium and the Iranian government, in line with the 50:50 profit-sharing principle which had become the norm in concession agreements in the Middle East.

Though the Company's former position in Iran was much reduced by these arrangements, it was still easily the largest foreign stakeholder in the Iranian oil industry, holding a 40 per cent share in the Consortium. Shell had 14 per cent, each of the US majors had 8 per cent initially, and CFP had 6 per cent. After a few months each of the US majors surrendered 1 per cent to allow a 'mini-consortium' of US independents to take a 5 per cent share in the Consortium through a new entity called the Iricon Agency.

The settlement with Iran was the signal for Fraser to call an extraordinary general meeting of shareholders, at which he confirmed the enormous financial strength of the Company. Throughout the Iranian crisis the Company had retained its Iranian assets and liabilities in its books, but now that the dispute was settled the Company intended to write off its assets in Iran from its balance sheet. This could be accomplished by the simple expedient of drawing on the Special Contingencies Account in which the Company had made provision for increased royalties to Iran, which would now not be payable. The amount placed to this account was nearly £50 million, enough to cover the write-off of the Iranian assets without so much as touching the Company's general financial reserves.[91]

At the end of 1953 these reserves stood at the enormous sum of £111 million, more than five times the Company's issued ordinary capital of about £20 million.[92] Fraser therefore proposed to capitalise £80,550,000 of the Company's reserves by making a bonus issue of four new ordinary shares for every ordinary share held by shareholders on 23 November 1954.[93] Fraser had earlier discussed this capitalisation with the Treasury and had agreed to defer it until after the Iranian crisis had been settled because it would have been inopportune to display such 'evidence of tremendous wealth' while seeking to negotiate compensation from Iran for the nationalisation.[94] Fraser also proposed to change the Company's name from Anglo-Iranian Oil Company to The British Petroleum Company. This too had been discussed with the Treasury, where the new name was thought to be 'dignified and sensible'.[95]

In proposing the bonus issue and the change of name, Fraser also mentioned that with the crisis over and the Company's main capital projects nearing completion 'It should . . . be possible in future to distribute as dividends a larger proportion of the profits available.'[96] In the less guarded language of the City correspondent of the *Daily Express*, Fraser told the shareholders 'what they already knew – that the company is in the money and that he will be more generous with the bawbees in future'.[97] The resolutions for the bonus issue and the name change were passed unanimously and duly enacted.[98] When dealings commenced in the 'Little Persians', as the shares were described after their number was increased fivefold by the four-for-one bonus issue, the price per share was £3·75.[99] For each share with a mid-price of £6·34 in 1950, BP shareholders now held five shares worth, together, £18·75. The value of their shareholdings had tripled. No other oil major came close to this share performance, and nor did the index of British industrial shares.[100] Fraser, much criticised by government ministers and officials on both sides of the Atlantic for his tough attitude in the Iranian negotiations, could certainly not be faulted for the increase in shareholder value which reflected the turn-around of the Company. He may have lacked finesse in foreign relations, but he was certainly not short of business acumen at an absolutely critical juncture.[101]

STRATEGY

The terms of the Consortium agreement have generally been seen, with justification, as a defeat for Iran in its attempt to challenge the existing order in which the international oil industry was controlled by the major oil companies operating under the concessionary system. By conceding only the principle of nationalisation without the substance of effective national control the Consortium agreement preserved the status quo and did not fundamentally alter the relationship between the international oil companies and the oil-exporting nations.

However, although the concessionary system remained broadly intact, the settlement of the dispute did not signal a return to the *status quo ante* for the Company. In the comparatively short period of the crisis, the Company, spurred by the imperatives of the emergency, had taken steps which amounted to a major watershed in its development. The central element in this brief, but critical period of transformation was a shift away from the previous concentration on Iran towards a more multinational orientation.

The change of the Company's name was only a surface indication of the great changes that had taken place. Inside the Company, the reduced influence of the old hands, steeped in expatriate service in Iran, signified a break with the Company's past management culture. In the Company's balance sheet, the Iranian assets had been written off. In the pattern of its supply operations, the Company had become much more diversified. In refining, the shift from dependence on a single refinery to a much wider spread of more evenly sized refineries is illustrated graphically in figure 1.3. In crude oil production as well, the Company had turned to non-Iranian sources. The Company would never again be as dependent on Iran, or any other single country, for its crude oil as it had been before the crisis.

The only cloud in the sky was that nearly all the Company's crude oil still came from the Middle East. The risks associated with that source of supply had been acknowledged by the Future Programme Committee, who had recommended that the Company should seek other sources of crude oil.[102] Subsequently, in May 1953, Coxon, who was then still head of the Central Planning Department, wrote a characteristically policy-provoking document entitled 'Widening of our sources of crude oil'. Although he could see no prospect of the Company running short of crude from its great share in the huge oil reserves of the Middle East, Coxon argued in favour of searching for new sources of crude elsewhere. His reasons were several. For one thing, the Company's main crude supplies were relatively high in sulphur content, which degraded the quality of the Company's products. Secondly, now that

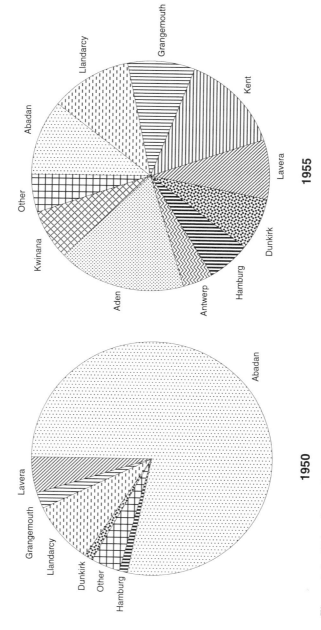

Figure 1.3   BP's refining throughputs by location, 1950–1955

the Company had constructed refineries in a variety of locations outside the Middle East, crude oil from sources outside the Middle East could be shipped to one or other of the Company's refineries without incurring excessive transport costs on long shipping hauls. Thirdly (though not necessarily in order of importance), the Company's high degree of dependence on the Middle East gave it a risky exposure to an area of political instability.[103]

Coxon's memorandum was discussed at a meeting in Fraser's room in June 1953, when Fraser accepted 'as our aim and policy <u>the energetic search for new sources of crude oil</u> both to give us low-sulphur products and to provide supplies outside the Middle East'. A worldwide review of oil prospects, which had already been started, was completed by the end of the year and its contents summarised in a Central Planning Department memorandum which adopted as its theme the quotation from Winston Churchill cited at the beginning of this chapter: 'Safety and certainty in oil lie in variety and in variety alone.' Thus the discovery and development of new sources of crude oil outside the Middle East became a strategic objective for the Company, albeit one which would not come to fruition for a number of years.[104]

In summary, the Company emerged from the nationalisation crisis in a transformed condition. It had a new name, strong finances and had cut out much of the dead wood in its staff and organisation. Its producing and refining interests were much more diversified than before and it had a new strategic direction. The Company's reduced concentration on Iran, and the presence of American, French and Dutch interests in the Consortium, reduced, but did not eliminate, the Company's exposure to political upheaval. It was not seen as a major drawback that the Company no longer had any oil fields under its sole ownership and control, apart from its small, internationally insignificant, onshore fields in the UK. As Jackson summed up the Company's decision at the end of the Iranian nationalisation crisis in October 1954:

> For myself, I think it represents good fortune for this Company compared with 18 months ago when we were out of Persia and Mossadeq was Prime Minister. There are now four countries in the Consortium, which surely must be more difficult to kick out than one. However, I do not think the leopard has changed his spots, and I think the AIOC policy must be to make hay while the sun shines, even though it be rather watery!
>
> For the first time in the history of this Company we have now no single field of our own except in this country. However, the structure and sinews of the Company appear to me to be stronger than they have been in the last thirty years.[105]

2

Management and culture

THE SUCCESSION PROBLEM:
'A MATTER OF PUBLIC IMPORTANCE'

The board of directors who presided over the transformation of the Company during the Iranian crisis did not itself undergo great changes in that time. When the Iranian oil industry was restarted at the end of October 1954, Fraser, autocratic as ever, was showing no obvious sign of wishing to retire, although he had already passed the normal retirement age of sixty-five and was a few days short of his sixty-sixth birthday. He had served on the board for more than thirty-one years, of which more than thirteen had been spent as chairman and chief executive, exercising great personal dominance over the Company's management.

The five other managing directors, though younger than Fraser, were of the same management generation, having all joined the Company in 1919–22, shortly after the end of World War I. Three of them had spent much of their subsequent careers in Iran. One was Neville Gass, the sixty-one-year-old director of concessions, who had spent his first fifteen years with the Company (1919–34) working in Iran, winning Company approval for his unusual ability to promote harmony.[1] In the late 1940s he had conducted the ultimately fruitless negotiations to revise the concession agreement with Iran before the nationalisation of the Company's operations there.[2] Modest and unassuming, shy of publicity, he was, in essence, more a diplomat than businessman.

The other two old 'Iranian hands' on the board were Edward Elkington, the sixty-three-year-old director of administration, who had spent his first sixteen years with the Company (1921–37) in Iran; and John Pattinson, the fifty-five-year-old director of production, an engineering graduate from

Cambridge University who had spent his first twenty-four years with the Company (1922–46) in Iran.

The two other executive directors, Basil Jackson and Harold Snow, had risen to the top of the Company without 'eastern service' in Iran. Jackson, the sixty-two-year-old deputy chairman, had spent much of his career in the Company's New York office. Snow, the fifty-seven-year-old director of distribution, had worked his way up in UK marketing. He gained a higher international profile in 1954 when, with Orville Harden of Standard Oil (NJ) and John Loudon of Shell, he formed the three-man Consortium team which went to Tehran to negotiate the resumption of oil operations in Iran in 1954. Snow, however, was too shy and quiet to enjoy the limelight and he lacked the charisma to make a strong personal impact in the Tehran negotiations. As Sir Roger Stevens, the British Ambassador in Tehran, commented, 'Orville Harden, the leader of the oil negotiators, is a weather beaten hard bitten Texan who seems straightforward & everyone seems to trust. John Loudon, of Shell, is a highly sophisticated & excessively smooth Dutchman who looks as though he had spent all his life trying to prove he was really at Eton. But he has imagination & drive & is obviously a power in his own land. Bill [Harold] Snow of AIOC is a back room boy turned director, a small bald man who scarcely ever speaks & when he does so, speaks in a voice so deep that as a wit recently remarked the ordinary human ear is not attuned to hear it.'[3]

Fraser and the five other managing directors were exactly balanced in number by the six non-executive directors. Two of them were government directors, appointed on account of the British government's majority shareholding in the Company. They were Frederic Harmer and Sir Gordon Munro, from shipping and banking backgrounds respectively.[4] Another two were representatives of the Company's other main shareholder, the Burmah Oil Company. They were Sir Kenneth Harper, Burmah's chairman, and William Abraham, its managing director. Finally, there were two ordinary non-executive directors: William Keswick, chairman of Jardine Matheson, and Desmond Abel Smith, chairman of Dalgety and a director of the National Provincial Bank.

Between 1954 and 1960 there were three changes in the non-executive directors. Of the government directors, Munro was succeeded by Lord Weeks, an influential businessman who, after a long career with Pilkingtons, had spent several years as chairman of Vickers before retiring from that company in May 1956.[5] In the same period, William Eadie and Robert Smith succeeded Harper and Abraham as the Burmah directors.

Much more important than these non-executive changes, indeed one of the most vital issues for the Company, was the search for a suitable successor to

6 Harold Snow, a BP director (centre), with negotiators in Tehran, 1954. The others, from right to left: General Fazlullah Zahidi, Iranian Prime Minister; Orville Harden of Standard Oil (NJ); John Loudon of Royal Dutch-Shell; Abdullah Intizam, Iranian Foreign Minister; Dr Ali Amini, Iranian Finance Minister

Fraser as chairman and chief executive. The main cause of the problem was inadequate succession planning, marked by the absence of a younger generation of new talent on to the board, with the result that there was a dearth of suitable candidates to succeed Fraser.

The exclusion of younger men from the board did not reflect an absence of talented managers, of whom one in particular stood out: the Hon. Maurice Bridgeman, who was both well connected and extremely able. He was the third son of a Conservative politician, the first Viscount Bridgeman, and followed his father to Eton and Trinity College, Cambridge, where he read natural sciences, but left before taking a degree. His father was First Lord of the Admiralty from 1924 to 1929 and knew Sir John Cadman, who was the Company's chairman designate when Bridgeman was recruited. On joining the Company, Bridgeman's first assignment was as private secretary to Cadman on a visit to Tehran in April 1926 for the coronation of Riza

Shah. He subsequently worked in Iran for two years before returning home after contracting hepatitis. A few years later, in 1934, he was posted to New York, where he acquired a first-hand knowledge of the US oil industry and met many of the leading executives of the principal US oil companies before returning to head office in London in 1937.[6]

During the war, Bridgeman strengthened his connections in Whitehall and gained valuable knowledge of government through his work as a temporary civil servant in various capacities, ending with a spell in the Petroleum Division of the Ministry of Fuel and Power from 1944 to 1946. During part of this time he was in Washington and worked closely with Lord Keynes in the lend-lease and loan negotiations with the Americans. According to a later note by Sir Frank Lee, when he was Joint Permanent Secretary of the Treasury in 1960, 'Keynes, as I know from personal experience, thought very highly of him.'[7]

After the war, Bridgeman returned to the Company, but retained wider contacts. He and his wife, Diana, were good friends with Sir Roger Stevens, posted to Tehran as British Ambassador in February 1954, and his wife, Constance.[8] Yet, despite Bridgeman's undoubted ability and his strong connections outside the Company, when he passed his fiftieth birthday in January 1954 he was still excluded from the board.

Bridgeman's position and the larger question of the succession to the chairmanship were matters of concern not only within the Company, but also to the Company's majority shareholder, the British government. This was another matter, like the Company's dividend policy, in which the government's rights to interfere in the Company's affairs were plagued by lack of clarity. The government had undertaken not to use its veto over board decisions any more than was necessary to secure the government's objects in specified areas including 'questions of foreign naval or military policy'.[9] But it could be argued (and was by the Treasury solicitors) that the activities of the Company were of such strategic importance that they were bound to have constant repercussions on such questions. It was, according to that argument, a matter of particular concern to the government 'that the political outlook and conduct of the Company's head should be acceptable'.[10]

Whatever the legal position, the government had in the past taken a keen interest in the chairmanship succession. In 1922 the government had proposed that Cadman should join the board. Some of the Company's directors objected on the grounds that this amounted to government interference in the commercial administration of the Company. The government, however, was politely insistent and Cadman duly became a director in 1923. In 1925, the government was again instrumental in securing Cadman's appointment as chairman designate to succeed Lord Greenway.[11] That succession took place in 1927, when Cadman became chairman and chief executive.

The question of the government's rights to be consulted in the appointment of a new chairman came up again in May 1941, when the Treasury received from the Company a copy of a board minute recording the board's intention to elect Sir William Fraser as chairman when Cadman retired. Sir Horace Wilson, the Permanent Secretary of the Treasury and the Head of the Civil Service, replied that the chairmanship was 'a matter in which H.M. Government is closely concerned and . . . in the event of a vacancy in fact occurring in the Chairmanship, H.M. Government would expect to be consulted before any decision is reached'.[12] In the event, the government did not intervene in the appointment of Fraser as chairman and he was elected to take up that position in June 1941.[13] The government had, however, made it plain that it expected to be consulted over the succession in future.

Between 1954 and 1960 the succession to the Company's chairmanship was a matter in which the Conservative governments of the time took a much keener interest than was conveyed by their repeated public utterances in Parliament that the government did not interfere in the commercial management of the Company.[14] In 1954 the two government directors, Harmer and Munro, discussed the problems of finding a successor to Fraser with top officials in the Treasury and the Ministry of Fuel and Power, and with Lord Cobbold, Governor of the Bank of England. The essential problem was to choose someone to take over from Fraser for long enough to allow a younger man, the favourite being Bridgeman, to gain enough experience at board level to take over the chairmanship.

Various candidates for the immediate succession were considered, but none seemed wholly satisfactory. Jackson, the deputy chairman, was suffering from a heart condition and there were doubts about whether he was fit enough to be chairman. Harmer and Munro were convinced that 'he did not want the job and could not take it on . . . He was now too much of a husk of a man to be given the Chairmanship. He could never get back the necessary drive.'

Of the other existing directors, Elkington, the oldest, was close to retirement and was not in the running. He retired in 1956 and was succeeded on the board by Robert Gillespie, who for some years had been the managing director of the Company's shipping subsidiary, the British Tanker Company (renamed the BP Tanker Company from 1 June 1956). Pattinson, at one time favoured by Harmer because he seemed to be less dominated by Fraser than the other executive directors, was thought on reflection to be 'not up to grade'. The same was said of Gass. More by process of elimination than by positive conviction, Harmer and Munro arrived at the recommendation that Snow should become chairman with Bridgeman as his deputy, with Jackson continuing as 'a wise old man working not at full pressure on the side'.

Casting a great shadow over these deliberations stood the dominant figure of Fraser who, it was feared, might try to continue stamping his overbearing influence on the Company from behind the scenes after he retired. Harmer and Munro were adamant that when Fraser retired 'he must go absolutely: any idea that there was a niche for him to watch over his successor would be fatal'.

The pairing of Snow and Bridgeman as chairman and deputy did not, however, find favour with the government. At the Treasury, Sir Edward Bridges questioned Snow's suitability for the chairmanship, asking 'Was he not too silent? Too secretive? Had he the stature to impose himself and to restore the morale and spirit of the Company?' (presumed to have been crushed by Fraser). To which the government directors replied with singular lack of conviction that although Snow was not ideal, he was 'much the best man'. As if they meant to damn him with faint praise, they said that 'they believed he would do'.

Lord Cobbold, also doubtful about Snow, suggested that perhaps he should be appointed as chief executive under a non-executive chairman, possibly Harmer. But Harmer was convinced that the chairman had to come from inside the Company. 'The oil business', he thought, 'is a most intricate one and it is the practice for the important deals to be negotiated by the Chairman. An outside Chairman who went into negotiations with other oil companies would be very lucky to come out with any portion of his pants still adhering to his anatomy.'[15]

The final stages of discussions on the succession to Fraser remain shrouded, perhaps in Treasury files which at the time of writing remain closed to historians.[16] But it is clear from the outcome that the idea of Snow and Bridgeman as chairman and deputy had been dropped by 12 January 1956, when it was announced that Fraser, ennobled as Lord Strathalmond of Pumpherston in 1955, would retire on 31 March 1956. His successor as chairman and chief executive was to be Jackson, with Gass as his deputy. The vacancy on the board created by Strathalmond's retirement was to be filled by Bridgeman.[17]

It was a measure of the absence of strong candidates to succeed Fraser that Jackson was appointed although it was known that he was not fit for the job on account of his heart condition.[18] When Jackson became chairman the idea was that he and Gass, who were then sixty-two and sixty-three years old respectively, would retire in about two years' time. In the meantime, a

7   Basil Jackson, who succeeded Lord Strathalmond as BP's chairman and chief executive in 1956

second deputy chairman would be appointed with the aim that he would in due course succeed to the chairmanship.[19] As it turned out there was not time for this succession plan to be implemented.

On 9 January 1957 Jackson wrote to the secretary of the board, resigning on grounds of ill health and nominating Gass as his successor.[20] It was planned that the board would meet the next day, when the directors would be asked to ratify the appointment of Gass as chairman. At the same time, Gass was going to propose that Snow should be the deputy chairman.[21]

News of this plan greatly alarmed the government directors, the Treasury and Lord Cobbold, who were convinced that Gass and Snow would make too weak a combination, and be susceptible to the influence of Strathalmond, whose spectre continued to haunt discussions about the succession. Sir Roger Makins (later Lord Sherfield), who was Joint Permanent

Secretary of the Treasury after Bridges' retirement, was so concerned that he intervened to put a stop to discussion of the succession at the Company's board meeting on 10 January.[22] When the board met that day the succession was not discussed and there was no mention of Jackson's resignation in the minutes, which merely recorded that he was 'unable to attend the Meeting'.[23] This was an extraordinary façade. Jackson, despite having sent his letter of resignation the day before, was still nominally, but certainly not in any other sense, the chairman while a decision on the succession arrangements was held up by the government.

After meeting with the government directors, Makins saw Gass and told him that the appointment of the chairman was a matter on which the government expected to be consulted before a decision was taken. The government's views could not, however, be given until the Chancellor of the Exchequer had considered the matter. This would take time because of the Cabinet upheaval caused by the resignation of Sir Anthony Eden who, broken by the Suez crisis (see chapter three), had resigned as Prime Minister on 9 January. Harold Macmillan succeeded Eden as Prime Minister and Peter Thorneycroft took over from Macmillan as Chancellor. As Thorneycroft would need time to take over his new office, the chairmanship of the Company could not be settled immediately. However, Makins assured Gass that in the meantime the Prime Minister 'was being informed of the situation which had arisen'.[24]

After consulting widely with business and government contacts, Makins reported to Thorneycroft that in the unanimous view of those he had consulted, a combination of Gass and Snow in the two top jobs 'would leave the direction weak and under the direction of Lord Strathalmond'. However, it would be 'politically difficult' if the government tried to impose a chairman against the board's wishes, and in any case there was no obvious outside candidate. Harmer might have fitted the bill, but he had said that he was not available. Makins reported that Harmer and Weeks (who had succeeded Munro as the Company's second government director) thought that there was no option but to appoint Gass as chairman and chief executive, while keeping the future succession open by not appointing a deputy for the time being.[25]

Thorneycroft passed that on to Macmillan, who agreed to the appointment of Gass as chairman, without a deputy.[26] Approved by the Prime Minister, this plan was duly enacted, and Gass became chairman on 1 February 1957, having been told by Makins what must have been obvious: that 'the leadership of this great Company is regarded by the Government as a matter of public importance'.[27]

A few months later, the vacant post of deputy chairman was filled twice over when, after consultations between Gass and the Treasury, Snow and Bridgeman were simultaneously appointed joint deputy chairmen in July 1957.[28] It was understood between Gass and the Treasury that Snow was unlikely to be considered for the chairmanship when Gass retired. Although there would be no overt commitment to Bridgeman, 'it would be widely assumed that he was likely to succeed Gass'.[29] Meanwhile, the vacancy on the board created by Jackson's resignation was filled by the forty-four-year-old Bryan Dummett, a modern languages graduate from Cambridge University who had joined the Company in 1936 and had worked his way up on the international marketing side of the Company's business.[30] A further board appointment was made at the beginning of July 1958, when Eric Drake became a managing director at the age of forty-seven. He succeeded Gillespie, who had been on the board for only two years.

Gass, who was knighted at the beginning of 1958, turned out to be a far more able chairman than had been expected. As Harmer reported in March 1959, 'the Company was running better than at any time in his experience' and Gass 'was doing extremely well'.[31] Nearly a year later, in February 1960, Harmer went to see Sir Frank Lee, who had succeeded Makins as Joint Permanent Secretary of the Treasury. They were personal friends of long standing, but this was the first time that Harmer had seen Lee officially as a government director of the Company. Harmer thought that the selection of a successor to Gass was probably the most important issue facing the Company.[32]

Another intensive round of consultations started, but this time in a far more harmonious atmosphere than that which had characterised the hiatus caused by Jackson's resignation. Lee noted approvingly in March that Gass had 'been at pains to bring us into full consultation about the selection of his successor'. Gass had also consulted the small management committee of the Company's board and the government directors, Harmer and Weeks. Lee himself had discussed the succession not only with Harmer, but also with Sir Norman Brook, who was the other Joint Permanent Secretary of the Treasury, and Sir Denis Proctor, the Permanent Secretary of the Ministry of Power (earlier the Ministry of Fuel and Power). Lord Cobbold had also talked with Lee about it. Absolutely everyone was in complete agreement: Bridgeman should succeed Gass as chairman. 'There is no doubt at all', wrote Lee, 'that Mr Bridgeman is a man with outstanding abilities, fine intellectual gifts, and long experience in the oil industry. It is common knowledge that during the long regime of Lord Strathalmond . . . he was "kept down" rigorously and unfairly and was never given the position in the company

8   The Hon. Maurice Bridgeman, who succeeded Sir Neville Gass as BP's chairman and chief executive in 1960

which his abilities warranted. This did not sour him, but he did adopt, as a sort of defence mechanism, a superficial attitude of blasé cynicism which tended (as I and others have thought) to obscure his real qualities.'[33] Derick Heathcoat Amory, who had succeeded Thorneycroft as Chancellor of the Exchequer, approved of Bridgeman, and so did Macmillan.[34]

It only remained to go through the formalities, which took place when the Company's board met on 7 April 1960 and approved the appointment of Bridgeman as chairman with effect from 1 July, when Gass would retire.[35] At the same time, Pattinson was elected as a deputy chairman, joining Snow. It had already been agreed in the consultations with the government that neither of these two was a likely successor to Bridgeman. However, Gass had assured the Treasury that there were one or two younger men in the Company who looked 'like having all the right gifts as potential chairmen'. They could, he added, be moved up to deputy chairmen when Snow and Pattinson retired.[36]

The retirement of Gass created a vacancy on the board which was filled by Maurice Banks, a science graduate from the College of Technology, Manchester University. He had joined the Company in 1924 and had worked in refining, with spells at Llandarcy, Abadan and Lavera (France), rising to become the general manager of the Refineries Department.

The appointment of Bridgeman as chairman finally laid to rest the succession problems which had emerged in Strathalmond's last years as chairman and which had reached crisis proportions in January 1957 when Jackson resigned without consulting the government about his successor. Gass, about whom so many doubts were expressed before his appointment, was initially seen as no more than a stop-gap chairman. But in his unostentatious and unaggressive way he had a soothing, healing influence on the Company's unsettled board. By nature consensus seeking rather than abrasive, he consulted widely and closely with his colleagues and with the government over his succession and smoothed the way for Bridgeman to come to power without a ripple.

It had taken five years since Strathalmond's retirement to rebuild the board in a new image which reflected the Company's evolution. In the Company's early years its executive directors had been drawn predominantly from British managing agencies in India and Burma during Britain's imperial heyday. Later, new accessions to the board had consisted largely of former expatriates to Iran, men such as Gass, Elkington and Pattinson who returned to Britain and became directors after lengthy service in Iran. The ex-Iranian element did not disappear overnight, but it was less marked in Bridgeman's board than it had been in the ageing, Strathalmond-dominated board of the early and mid-1950s.

While the original imperial element had died out, and experience of working in Iran was less prevalent, the board remained close-knit. All six executive directors had spent many years working in the Company, all were British, and – a new factor – nearly all of them shared the same university background. Remarkably, all but one of the executive directors were alumni of the University of Cambridge. They were Bridgeman (who read natural sciences), Pattinson (engineering), Snow (mathematics), Dummett (modern languages) and Drake (modern languages and law).

THE MANAGERIAL HIERARCHY

At the end of the Iranian crisis the board of directors presided over a highly centralised international business, controlled from the London head office. Here, BP's managerial hierarchy was organised in functional departments, of which the most powerful were those controlling the production, refining and distribution of oil. The main functional departments were split into smaller divisions, in which some regional elements were embedded. The most notable example was the Distribution Department, within which there was, even before the Iranian crisis, an area management at divisional level which supervised the Company's international marketing operations.[37] During the Iranian crisis, when the Company was diversifying geographically, the Refineries Department also introduced a regional component into its divisional structure. This took place in 1952, when the Manufacturing Division of the Department was in effect split into four new regional divisions covering refining in the UK, continental Europe, the Middle East and Australia.[38] A similar move was made in the Production Department early in 1953, when regional geologists were appointed to be responsible to the chief geologist for work in their respective regions.[39] However, in all these cases regional responsibilities were carried out only at, or below, divisional level within departments whose primary purposes were functional. The regional element was therefore clearly subordinate to the functions in the management hierarchy and there was no regional co-ordination or authority which was autonomous from the departments.

The shape of BP's pyramid structure was more vertical than horizontal as the functional departments rose steeply, without formal, lateral connections, to the apex of the board of directors. At the top was the chairman who, up to the end of March 1956, was the dominant personality of Lord Strathalmond. He exercised close personal control even over matters of relative detail and kept, as Harmer later recalled, a 'very, very tight concentration of power' in his own hands, so much so that board meetings were a 'totally empty formality'.[40]

His retirement opened the way for Bridgeman to join the board and to become the head of a much more unified and powerful head office department dealing with exploration and production. Previously, the Company's upstream activities had been split three ways. The D'Arcy Exploration Company was the wholly owned subsidiary which dealt with the concessional, legal and financial aspects of the Company's upstream interests outside Iran. It was also responsible for liaison with partners in joint producing companies such as the Kuwait Oil Company and the Iraq Petroleum Company.[41] Separate from the D'Arcy Exploration Company was the small Concessions Department which was concerned mainly with the Iranian concession, though it also dealt with concessionary affairs in relation to the Company's other wholly owned and solely operated oil fields in the UK and at Naftkhana in Iraq.[42] And separate from both the D'Arcy Exploration

Company and the Concessions Department was the Production Department, whose authorised head office staff numbers had been reduced to only fifty-four in March 1952 after much of its work disappeared with the suspension of exploration and production in Iran.[43] However, the Production Department continued to provide technical support for the D'Arcy Exploration Company, mainly in the form of geological, geophysical, drilling and engineering services.[44]

After the cutbacks in early spring 1952 it soon became apparent that the Production Department lacked the staff to provide the necessary technical support to the D'Arcy Exploration Company, whose exploration programme was expanding in new directions. In January 1953 the authorised head office staff strength of the Department was therefore increased to sixty. By the end of 1953 it was clear that this was still not enough and approval was given to increase the numbers to ninety-two.[45] Even that soon looked inadequate.[46]

At the same time as the numbers were being increased Peter Cox, who succeeded Dr George Lees as the Company's chief geologist in 1953, was pressing for the D'Arcy Exploration Company and the Production Department to be amalgamated into a single organisation.[47] His ideas were initially turned down.[48] But eventually, in April 1956, within three weeks of Strathalmond's retirement and Bridgeman's appointment to the board, the D'Arcy Exploration Company was amalgamated with the Production Department. The new combined organisation, which also covered the work earlier done by the Concessions Department, was called the Exploration Department and put under Bridgeman's direction.[49] Its responsibilities covered concessions, exploration, the development of new discoveries and representation of BP at meetings of joint companies such as the KOC, IPC and the Consortium holding company, Iranian Oil Participants (IOP).

A talent which Bridgeman possessed to an uncommon degree was the ability to pick out and develop the most promising staff in a collegiate atmosphere of teamwork, trust and encouragement for the exchange of ideas. Under his aegis, the new Exploration Department became a seedbed for a group of managers who were to have much influence over the later development of BP. These included, on the technical side, Cox, who was the general manager responsible for exploration in the department, and Norman Falcon, who became chief geologist. On the more commercial side, the Hon. William (Billy) Fraser, the son of Lord Strathalmond, was the general manager in charge of concessions and liaison with partners in joint producing companies.[50] A barrister by training and earlier employment, he had joined the Company as a senior assistant in the Concessions Department in 1950 and was soon promoted to become a manager in the D'Arcy

Exploration Company under Bridgeman. Personable and easy-going, Fraser had a different disposition from his father and quickly impressed Bridgeman, who thought that Fraser possessed 'exceptional ability' and that his 'tact in dealing with partners is extremely good'.[51] Under Fraser was David Steel, a lawyer by training who had moved from the law firm, Linklaters and Paines, to join the newly formed legal branch of Concessions Department in 1950. In the new Exploration Department, he was responsible for liaison with the KOC and was presciently regarded as likely to become 'one of the Company's most senior executives'. Alongside him, responsible for liaison with the IPC, was Geoffrey Stockwell, who had joined the Company in 1946 and moved to the Concessions Department in 1951. Another member of their network was Alastair Down, a qualified accountant who in 1954 was posted to Canada as the Company's representative in that country, which was then thought to have great exploration prospects. Also connected with Bridgeman's explorers was Robin Adam, an accountant who, as a manager in the Finance and Accounts Department, took on responsibility for financial matters relating to exploration and production.

The creation of the new Exploration Department strengthened BP's capabilities to carry out the decision, made during the Iranian crisis, to search energetically for new sources of oil. Organisation was thus aligned with strategy in the key area of exploration, with results that will be examined in chapter four.

After the formation of the new Exploration Department there were no significant organisational changes during Jackson's brief term as chairman. However, after Gass became chairman early in 1957 there was a movement away from the largely personal control which Strathalmond had exercised over a small number of functional baronies towards a more mature and impersonal hierarchy which was both deeper and broader than the earlier, more narrow pyramid. These changes partly reflected the personal contrasts between Strathalmond, who had a hands-on style, and Gass, who was much more of a delegator. But they also reflected the need to split up the largest departments before they swelled to outsize proportions as BP's business grew.

The first departmental split during Gass's chairmanship took place at the beginning of September 1957, when BP's growing petrochemicals interests were hived off from the Refineries and Technical Development Department (as the Refineries Department had been renamed in 1956) and placed in a new Petroleum Chemicals Department.[52] About six months later, in May 1958, there was another important split when the three divisions of the Distribution Department became departments in their own right. The old Distribution Department thus disappeared, its functions being performed by the new Markets, Trade Relations, and Supply and Development Departments.[53] The third major departmental split that took place in Gass's term as chairman took effect at the beginning of February 1960, when the work of the Refineries and Technical Department was divided among three new departments, namely Refineries, Research and Technical Development and Engineering.[54] By this series of departmental splits the head office functional departments were kept within manageable proportions and a growing number of managers became heads of departments. This helped to reduce the concentration of power and to broaden the head office hierarchy beneath the managing directors, whose responsibilities for the main departments after Gass's retirement are shown in figure 2.1.

Another important organisational change was the introduction of a system of regional co-ordination to adapt BP's management structure to the widening geographical spread of BP's business. BP was by no means the first large enterprise to face problems of organisation arising from increased product and/or geographic diversity. Such problems had arisen and been solved much earlier in the USA where, in the late nineteenth and early twentieth centuries, a number of large enterprises grew up in oil and other industries.[55] They were often vertically integrated and, if they were, usually adopted a form of organisation characterised by centralised departments, each dealing with a main function. This type of organisation was well suited to firms, such as the major oil companies, in which it was essential to co-ordinate the product flow through a sequence of operations. For that reason, it was adopted by BP and other oil majors.

However, as firms expanded by diversification into new products and/or geographic areas, their operations became increasingly complex, calling for new forms of organisation. The solution pioneered by a few of the largest US corporations in the 1920s was to move towards a more decentralised, multi-divisional form. Its salient characteristic was the devolution of operational responsibility to product or regional divisions which included their own internal functional departments and which were granted a large measure of autonomy, subject to overall strategic planning, appraisal and resource allocation for the enterprise as a whole.

The multi-divisional structure was most commonly adopted by firms that had become diversified in their products. Usually, the divisions served different markets and were largely independent of one another. There was little product flow between them, and the top management, far from aiming to integrate the divisions operationally, would look upon them as quasi-autonomous units amongst which resources could be allocated according to their comparative performance and prospects.

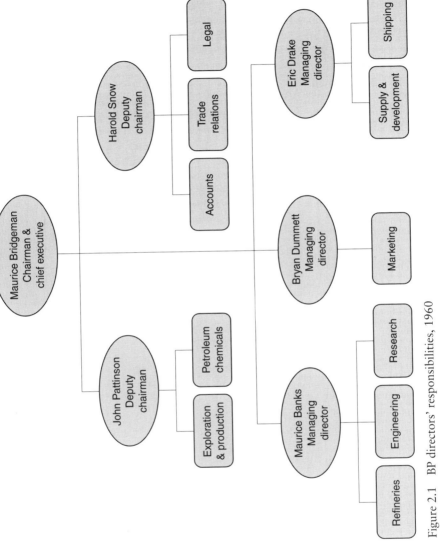

Figure 2.1 BP directors' responsibilities, 1960

STATO

OR THE SHELLMAN'S SNAKES AND LADDERS

RULES

1. Any number of players can take part. To obtain the best results the game should be played with loaded dice.

2. Any player landing on a square already occupied by another aspiring Shellman is allowed to retain his Status.

3. Any player landing at the bottom of a ladder immediately moves to the top rung.

4. Any player landing on a snake's head immediately loses Status to the position of the tail.

(Issued by the Central College of Shellmanship, 1953)

Although the flexible properties of oil had resulted in the development of a variety of products, the international oil companies did not, on the whole, fit the model of product diversification. They had generally moved into petrochemicals, but most of their diversification was by geographic expansion. Standard Oil (NJ) therefore opted in the 1920s for a multi-divisional structure in which operations were organised into largely autonomous, multifunctional regional divisions. Its example was later followed by other major US oil companies such as Standard Oil of California (Socal) and, in 1958, Socony Mobil.[56]

Royal Dutch-Shell was also moving in that direction. In 1957 that company called in McKinsey and Company, consultants, to help in reorganising the dual Anglo-Dutch management structure that was then in need of overhaul. Of most concern, the two main offices in London and The Hague worked as largely separate entities, between which there was considerable

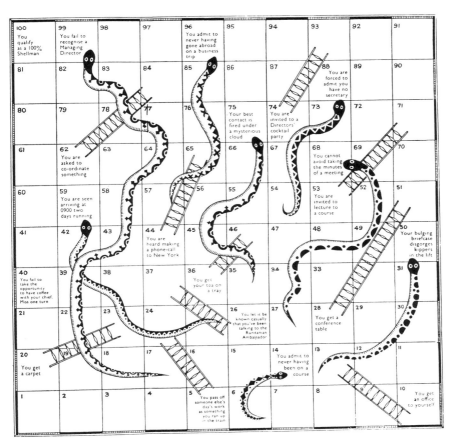

9   Snakes and ladders game from Shell magazine in 1953, illustrating the hierarchical nature of corporate life

confusion, overlapping and duplication in their geographic and functional responsibilities. Under the reorganisation that was introduced in April 1959, duplication was ended, regional boundaries redrawn, and control decentralised to geographic regions within broad lines of central policy and budgets, and subject to the advice of a newly appointed set of regional co-ordinators. The seven managing directors of Royal Dutch-Shell were each given both functional and regional responsibilities.[57]

The organisational dilemma posed by this pattern of development was that while the differing needs of dispersed geographic operations called for decentralisation into regional divisions, the continued integration of supplies and markets called for strong central functional co-ordination. The balance of advantage between regional decentralisation and centralised integration depended upon individual company circumstances. BP, with its traditional concentration on Iran, was a comparative latecomer to geographic diversification, and was therefore slower to move towards regional decentralisation than other oil majors. Its more cautious approach to decentralisation was reflected in the system of regional co-ordination introduced by BP in 1960. This organisational reform stopped short of introducing fully decentralised regional management and continued to lay greater emphasis on the centralised functional roles of the main departments.

Under the new system BP's international interests were divided into seven regions, for each of which a regional co-ordinator was appointed.[58] The central feature of the new system of regional co-ordination was that it introduced a dual responsibility in which regional and functional relationships were interwoven. It was hoped, when the system was introduced, that it would operate through the 'harmonious interplay of regional and functional considerations'.[59] As far as possible, functional managers and regional co-ordinators would, it was envisaged, work together to settle problems and so reduce the workload on the managing directors.[60] Where 'opposing and unconciliable views' were held, reference would be made to the board for a final decision.[61]

In practice, in the early years of regional co-ordination the functional departments retained the upper hand. In June 1961 Robert Belgrave, the acting regional co-ordinator for the Western Hemisphere, had 'no doubt that the idea is right and the job needs doing', but felt that there were weaknesses in the co-ordinators' position. There was, he thought, a lack of definition and clear lines of responsibility in their role and they did not, Belgrave complained, have such good access to the board as the 'overlords' of the functional departments. Moreover, there was a 'tendency of directors still to think mainly as functional managers and only secondarily as regional directors'. It was difficult for the regional co-ordinators 'to discover what is going on', a problem not helped by the 'secretiveness of some departments and individuals'. Belgrave thought that steps needed to be taken to remedy these problems.[62]

The matter was taken up by David Steel, who took over as regional co-ordinator for the Western Hemisphere in the summer of 1961 after returning to London from the USA, where he had spent the past two and a half years, first as senior assistant and later as head of BP's New York office. In September, he suggested that the regional co-ordinators might meet to 'explore common ground and problems' with a view to reaching agreed recommendations for submission to the board.[63] The co-ordinators duly met and in November submitted their recommendations to Bridgeman. Their main proposal was that it would help to solve problems between themselves and functional managers if consultation were made mandatory on defined matters of policy significance. This recommendation was, of course, intended to help the existing system function more effectively. However, the co-ordinators also raised a much larger issue, suggesting that 'there might be merit from a longer term point of view' in examining the advantages of replacing the dual responsibility of regional co-ordinators and functional managers with a system of unified responsibility for all the affairs of each region immediately below board level.[64]

This idea, if adopted, would have meant a substantial change to the way BP managed its international business. The functions would in effect have become subordinate to the regions and BP might have moved away from its centralised, functional organisation towards a more decentralised, multi-divisional form of the type already adopted by other large international oil companies.

Bridgeman, however, rejected the idea. At a meeting with the regional co-ordinators in November, the discussion was mainly on ways in which the existing system could be made to work better. On the larger question of the advantages of unified regional responsibility, Bridgeman 'recalled that the Group's activities had always been run in the past on a functional basis and there was no present intention to make any basic alteration in this respect'.[65] The regional co-ordinators were therefore instructed not to pursue the idea of unified regional responsibility, but to work out improvements in the existing practice of shared responsibility with functional managers. It was characteristic of Bridgeman to place more emphasis on fostering people and talent than on formal organisation structures. When he met the regional co-ordinators again in mid-1962 he was again dismissive of some of the problems they reported in the functioning of dual responsibility. Bridgeman told them that 'such difficulties arose primarily from personal failure rather than from any fault in organisation'.[66]

Bridgeman's preference for retaining a centralised, functional emphasis was a sign of his belief in managing BP as an integrated, indivisible whole rather than as a set of quasi-autonomous divisions. The principle of integration thus remained a fundamental feature of BP's organisation at a time when the more decentralised multi-divisional form was widely adopted not only by other oil majors, but also by the largest companies in other sectors of the British economy.[67]

MANAGEMENT CULTURE

Although BP retained its functional emphasis, the settlement of the Iranian crisis brought a fundamental change in the way that the Company managed its overseas interests. Previously, the Company had performed a dual role. On the one hand, it was an operating company, directly engaged in managing its operations in Iran, where the Company appointed its own management, engaged its own labour and was directly responsible for labour relations. On the other hand, the Company was a holding company which owned, in whole or in part, numerous overseas subsidiaries and associates outside Iran. It was, for example, the parent of the D'Arcy Exploration Company, responsible for exploration and production outside Iran; of the British Tanker Company, engaged in shipping; and of a chain of refining and marketing subsidiaries and associates, too numerous to catalogue, but mainly in Western Europe.

After the Iranian settlement, the new Consortium took over operations in Iran, and the Company, renamed British Petroleum, ceased to be an operating company. Instead, it became a purely holding company, whose operations were carried out by subsidiaries and associates engaged, variously, in exploration and production, transportation, refining, and marketing. The transition to holding company was completed in 1955, when a new wholly owned subsidiary, BP Trading, took over the remaining trading activities of the parent company with effect from the beginning of the year.[68]

This change was more than a mere paper exercise. Previously, as a large-scale direct employer of Iranian labour, the Company had been deeply involved in labour relations, especially during the Iranian strikes and disturbances of 1929, 1946 and 1951.[69] But after the nationalisation of its operations in Iran, the Company ceased to be directly involved in managing a large foreign labour force. As a holding company, its relationships with overseas subsidiaries and associates, which managed their own labour relations, were at a managerial rather than a shopfloor level. Although subsidiaries and associates would experience some local labour problems, there would be no outbreaks of unrest on the seismic scale of previous disturbances in Iran.

More pressing than labour matters was the problem that after shedding many of the staff repatriated from Iran in 1951 BP soon found itself short of the managerial resources that it required to run its international business. In the past, Iran had been the Company's main training ground, where staff had gained experience before being repatriated to London to take up senior positions. At the same time, the expatriate community in Iran had provided a reserve of British managers and technical staff on which the Company had been able to draw to meet requirements elsewhere. Nationalisation had forced on the Company the loss of this system of training and promotion, for which there was no immediate substitute.

BP's directors were acutely aware of this problem, which was made prominent by BP's inability to provide its quota of staff for the newly formed Consortium in Iran, and by shortages of technical staff for refining and production, and for research and development at Sunbury. In production, the situation was so serious that it was thought to pose a danger to BP's competitive position.[70] After downsizing during the Iranian crisis, BP clearly needed to recruit new staff, especially technical staff, as it sought to develop a more international, geographically diversified business, particularly in production and refining.

Recruitment was not, however, easy. There was, at that time, virtually full employment in the British economy, and BP found, in common with other British companies, that there was a particularly severe shortage of science-qualified applicants for jobs in business. This was not a new problem. In 1949, before the Iranian nationalisation crisis, the Company had noted with despair that 147 companies were competing for the five chemistry graduates from Cambridge University who would be available for industry that year. One company alone was looking for 100 chemistry graduates.[71] The essential problem, as Sir Humfrey Gale, the Company's organisation adviser, told a Federation of British Industries conference, was that there was 'a famine of young technologists, and a feast of young arts graduates'.[72]

In 1953 the Company lifted the freeze on new recruitment imposed in 1951 owing to the nationalisation crisis, and hoped to recruit twenty-five science/technical graduates and six non-technical graduates at the end of the academic year in June. But competition for technologists was so strong that only seven of the sought-after science-based graduates were recruited, though the six non-technical graduates were found without apparent difficulty.[73] The situation did not improve in 1954 when, in the peak recruiting season of July–September, the Refineries Department obtained only sixteen graduates against forty-five vacancies.[74] As the Cambridge University Appointments Board reported at the end of the year: 'The demand for scientists and engineers maintained its insatiable level.'[75] On the same theme, in 1956 BP's chairman, Basil Jackson, told shareholders that 'industry as a whole faces a shortage of technologists, and our own growing and widespread operations are taxing severely the BP Group's resources of technical and scientific personnel'.[76]

The Company adopted various measures to ease its recruitment problems. One was the Student Apprenticeships Scheme, which was introduced in

1953 to enable grammar school boys to obtain improved qualifications by attending courses at technical colleges. The 'industrial' side of the scheme was based at the Llandarcy refinery, and was used to train apprentices in chemical, mechanical and instrument engineering for jobs in refining. The 'commercial' side of the scheme was based at the Company's head office in London and concentrated on training student apprentices in senior clerical and administrative skills.[77]

The Student Apprenticeships Scheme was complemented by a higher-level University Apprenticeships Scheme, which was introduced in the mid-1950s to provide financial assistance, vacation employment and future jobs with the Company for university undergraduates.[78] Following an approach from Courtaulds, BP also decided in 1955 to help finance the Industrial Fund for the Advancement of Scientific Education in Schools.[79]

Recruitment difficulties also helped to break down BP's previous resistance to the employment of married women (chapter one), who joined BP in large numbers after BP's ban on their employment was lifted in 1955. At the same time, women were starting to be recruited for more challenging jobs than the usual typing and clerical duties, with the 1956 intake including five female chemists, one cartographical draughtswoman, one female medical officer and a female palaeontologist.[80] For the first time, too, a party of eight female undergraduates were invited to spend two days visiting the Sunbury research centre and the Kent refinery.[81]

The outcome of BP's recruitment drive was a substantial rebuilding of BP's managerial resources in the mid- to late-1950s. The number of staff employed at BP's London headquarters, having been reduced from about 2,250 to less than 1,950 during the Iranian nationalisation crisis, had risen to about 3,000 by 1961.[82] The number of BP's British staff posted overseas had also risen dramatically. After the mass repatriation of staff from Iran in 1951, the number of British staff abroad had fallen to a mere 380 in 1953, but by May 1961 the number had risen to 1,210 staff, working in forty-nine countries.[83]

The large growth in numbers both at home and abroad did not, however, bring major changes in BP's managerial culture. The employment of women to do jobs that might previously have been regarded as 'men's work' (chapter one) made a slight, but only a slight, difference to the predominance of men in activities other than typing and clerical work. More noticeable than this minor change was the continuity of the social composition and culture of BP's management.

One sign of this was that BP's disproportionate employment of men who had been to British public schools (private sector fee-paying schools for the elite) was undiminished; indeed it was increasing. At the end of 1961, 21 per cent of BP's male staff at home and overseas had a public school education. This was three times the national ratio of boys attending public schools. Moreover, the proportion of BP's male staff who had been to public schools rose from 19 per cent for those engaged in 1946–56, to 25 per cent for those engaged in 1957–61.

These unweighted figures tend to understate the influence of public school education in BP's management because they take no account of the fact that the public school element was much more strongly represented in BP's senior management than in its total staff. Working down the managerial hierarchy, in 1962 the percentages of managers who had been to public school were 71 per cent for departmental general managers (just below board level), 48 per cent for assistant general managers and divisional managers, and 36 per cent for branch managers or equivalent. In short, the more senior the manager, the greater the likelihood that he had been to public school.[84]

Many of the BP staff who had been to public school had also been to university. A 1962 BP survey reveals that at that time about one third of BP's graduate staff had arts degrees, while two thirds had technical degrees. In the arts contingent, the preponderance of graduates from Britain's great traditional universities of Oxford and Cambridge was quite remarkable: these two universities accounted for 27 and 35 per cent respectively (62 per cent combined) of BP's arts graduates. BP's technical graduates, on the other hand, came from a much broader base of universities in London, the provinces, and Scotland. Only 13 per cent of them were graduates from Oxford and Cambridge.[85] Altogether, taking arts and technical graduates together, Oxford and Cambridge provided 29 per cent of BP's total graduate staff, while five out of BP's six executive directors were, as mentioned earlier, Cambridge graduates. The odd one out was Maurice Banks, whose degree, significantly in science, was from Manchester University. The unavoidable inference is that education at public school followed by Oxford or Cambridge (preferably Cambridge) was a valuable passport for promotion to the top in BP.

The most valuable passport, however, was a British one. Although BP was internationalising its operations, this was only weakly reflected in its managerial culture, which retained a strongly British national identity. As Pattinson told new recruits in 1955:

> I should perhaps make it clear that the British Petroleum Company is an entirely British Company – we have interests all over the world and many partners who are not British, and associated companies which are not British, but the management control is 100% British; so that while in joining our organisation you may have opportunities for working abroad, that does not involve becoming members of a foreign organisation.[86]

These comments touched on one of the key features of BP's managerial style. It was quite usual for BP's British staff to spend a large part of their careers abroad, but the national staff of overseas subsidiaries and associates rarely worked for any length of time outside their own countries. They might be posted on short-term secondments for training and development in Britain or elsewhere, to enable them to advance to more senior positions within their own companies, but there was no possibility of them rising to the top of BP, the parent company. BP's management was not, therefore, truly internationalised, and there remained, within the BP group of companies, powerful national barriers to management mobility and exchange.

The reasons for BP's Anglocentrism, as put forward in 1961, resonated with imperial themes:

1 Although we are an international Group the parent company is British. It is therefore natural for British staff to go out from the directing centre to the periphery, but for local staff of subsidiary companies to be regarded as having been engaged for service on the periphery only.
2 The Company's origin as Anglo Persian made it necessary until 1951, for it to engage a large part of its staff for service abroad. Its terms of service, and particularly its pension scheme, were (and still are) intended to facilitate the movement of staff between the UK and abroad. By contrast, our subsidiaries abroad normally confine their activities to their own countries. Their terms of service (especially their pension schemes) are, in general, not designed to take account of overseas service.
3 There is an historical tradition in this country of expatriate service (government, armed forces, and commerce), which is uncommon among other nationalities, the Dutch and, (to a lesser extent), the French, excepted.
4 English is the language of the Group's business and is also the commercial language of a large part of the world today.[87]

To some degree, though less than in Iran before nationalisation, BP's expatriate staff continued to resemble colonial societies, carrying British social systems with them. For example, in Aden, still a colony of Britain, the recreational facilities for the refinery staff were divided into A, B and C categories, in order of seniority. There was a Family Pavilion for uses including private functions organised by 'Company residents of A class housing areas'.[88] There were also B and C class bathing facilities, and a sports club 'for use particularly by Company employees of Officer status'. And there were separate wards for different nationalities in the hospital. A BP estimate of staff expenditure in 1957 reflects quintessentially British tastes. British married staff (assumed to have two children) were estimated, on average, to consume per month two bottles of tomato sauce, seven packets of Kelloggs breakfast cereals, three pounds of jam, one bottle of gravy browning, five packets of Birds blancmange powder, and five packets of Chivers jelly crystals. On drinks, the figures were two bottles of whisky, one bottle of gin, one bottle of Harveys dry sherry and thirty large bottles of Tennents beer. For smokers, estimated consumption was forty-five packets of twenty Players cigarettes per month.[89]

A similar pattern could be seen in Papua New Guinea, where BP had for many years been exploring for oil. Here, BP adopted the same style of operations as had been used in Iran, establishing a camp at Badili, a suburb of Port Moresby, to provide administrative and technical services for its exploration activities. The camp, with staff housing, a club and other facilities, was the scene of a small colonial expatriate society, largely isolated from the local people, and unable to communicate easily with them.[90]

The problems of establishing frictionless cross-cultural relations between expatriate staff and local communities were not, of course, unique to BP. There was, in general, a growing awareness among the international oil companies that the best method of reducing frictions was not by keeping a distance between expatriates and locals, but by seeking to develop closer relations. For example, in Nigeria, where Shell provided the staff and management for the jointly owned Shell-BP Petroleum Development Company of Nigeria, there were by 1963 about 300 expatriates at Port Harcourt, Shell-BP's main administrative centre. This expatriate community was, reported a visiting BP executive, 'extremely conspicuous both by its numbers, its relative affluence, and the quality of its housing'. The Shell-BP general office in Port Harcourt was 'by far the largest and most conspicuous building in that place'. It was flanked by workshops, a hospital 'which to the lay observer compares favourably with the most expensive private clinics in the United States', and by an attractive Shell-BP housing estate. 'The contrast between all this and the town of Port Harcourt remains vivid', wrote the visiting BP executive. But he was impressed that, although it was difficult to find Nigerians who were qualified for senior posts, Shell-BP managers were making a conscious effort to make themselves more acceptable to the local community. He wrote that the expatriate staff were 'extremely well schooled in cultivating good relations with the local community whether Nigerian or foreign', and reported that moves were afoot to sell some of the Shell-BP housing to Nigerians, to open the Shell-BP hospital to non-employees and to divest the workshops.[91]

In Iran, the Consortium, aware of the powerful anti-British feeling stirred up during the nationalisation crisis, was also conscious of the need to leave old-style colonialism behind. The questions posed in its standard form for the appraisal of expatriate staff included: 'Is the employee training and developing Iranian personnel to take over as many job functions as possible? Are the employee and his family making an effective social adjustment to community life in Iran? Are the Iranian members of the community accepting this employee and his family into their social life?'[92] A similar lowering of the barriers between foreign business enclaves and host countries was taking place in Latin America, where US companies were making efforts to reduce the isolation of company towns and to strengthen their linkages with local communities.[93]

But despite efforts such as these it remained difficult for the international oil companies to cast off an imperial image. This was a serious problem for BP, which was the only major oil company without substantial oil production in the industrialised West, and therefore had proportionately more at risk in the developing world than any other major. All over the Third World, newly independent nations were springing up with the aim of reducing, to varying degrees, their political and economic ties with the old mother country. BP, associated with the declining imperial power of Britain, ran an increasing risk of losing hold of its concessions as Britain lost hold of its empire.

3

The Suez crisis

No event in the postwar years exposed more starkly the decline in Britain's power, and Western Europe's growing dependence on Middle East oil, than the Suez Canal crisis which broke out in 1956. In April that year the Soviet leaders Nikolai Bulganin and Nikita Kruschev visited London where they had talks with Sir Anthony Eden, the British Prime Minister. Afterwards Eden informed President Eisenhower that 'in our Middle East talk I made plain to them that we had to have our oil and that we were prepared to fight for it'.[1] Those strong words were soon put to the test. On 26 July the Egyptian leader, Colonel Gamal Abdel Nasser, nationalised the Suez Canal, raising fears that Nasser's pan-Arab, nationalist regime would control the supply of Middle East oil to the Western world.

The Suez Canal was one of the world's most strategic waterways, linking the Red Sea with the Mediterranean through the isthmus that joined the continents of Asia and Africa. It had been completed in 1869 and was originally financed by French and Egyptian capital. However, in 1875 the British government acquired a 44 per cent shareholding from the financially troubled Khedive (Ruler) of Egypt. Britain thus gained a large stake in a waterway which provided a lifeline between the mother country and the imperial jewel of India.

The Canal occupied such a strategic position that it was sometimes called the Clapham Junction of the Empire or, as Eden put it in 1929, it was 'the swing-door of the British Empire, which has got to keep continually revolving if our communications are to be what they should'.[2] After India was granted independence in 1947 the Canal lost much of its imperial significance. But by that time it had acquired a new importance as the major route for the westward movement of Middle East oil, particularly to markets in Western Europe.[3]

10   Egyptian leader, Gamal Abdel Nasser, being acclaimed in the streets of
Cairo after he announced the nationalisation of the Suez Canal Company in July
1956

Although patterns of energy use varied greatly between individual
European countries, oil had become an important fuel for Western Europe
by the mid-1950s, accounting for about 20 per cent of the region's energy
consumption. Since the war there had been a great expansion in the
European oil-refining industry so that by 1955 there was, in aggregate,
enough capacity to meet the regional demand for products. However,
Europe's refineries were almost entirely dependent on imports for their supplies of crude oil. The main source of supply was the Middle East, which in
1955 provided about two thirds of Western Europe's oil imports.

Of those oil imports from the Middle East about two thirds were shipped
from the Persian Gulf via the Suez Canal. The rest was transported via the
trunk pipelines linking oil fields in Iraq and Saudi Arabia to loading terminals in the Eastern Mediterranean. In Iraq's case, the Iraq Petroleum
Company (IPC) pipelines connected the oil fields at Kirkuk and Mosul with
the Mediterranean loading terminals at Banias in Syria and Tripoli in the
Lebanon. In Saudi Arabia's case, the US-owned Trans-Arabian Pipeline
(Tapline) brought Saudi oil to the port of Sidon in the Lebanon. The Suez

Canal and the trunk pipelines were thus the arteries for the flow of Middle East oil to Western Europe (see map 3.1).[4]

BP, which produced most of its oil in the Middle East and sold most of it in Western Europe, was a heavy user of the Canal and the IPC pipelines. In 1955, 58 per cent of BP's total crude oil liftings were moved westwards from the Middle East through the Canal in tankers operated by BP or its customers. These movements represented 39 per cent of all the oil shipments passing through the Canal from east to west.[5] BP had no producing interests in Saudi Arabia and was not, therefore, a user of the Tapline. However, with a 23¾ per cent stake in the IPC, BP lifted large quantities of Iraq crude, most of which was transported through the IPC pipelines.[6]

Like the rest of the international oil industry, BP aimed to own or charter enough tankers to cover its shipping requirements on the basis that the Suez Canal and the trunk pipelines would be open. The voyage distance to north-west Europe from the Persian Gulf via the Canal was about 6,500 miles, about twice as far as the voyage from the pipeline terminals in the Eastern Mediterranean. However, if neither the Canal nor the pipelines were available, Europe's supplies of Middle East oil would have to be shipped around Africa via the Cape of Good Hope, a voyage of some 11,000 miles. To transport Western Europe's normal supplies by that route would have required a much larger tanker fleet than was available in mid-1956, when the world tanker fleet was virtually fully employed even with the Suez Canal and the trunk pipelines in operation.[7] If the Canal and/or the pipelines were closed, and tankers were diverted to the Cape route, there would not be enough tanker capacity to maintain normal deliveries of Middle East oil to Western Europe.

There was, however, scope for mitigating a Western European oil shortage by reorganising international oil supply and distribution so that tanker voyages would be minimised by the application of a 'short-haul' policy. This would require the diversion to Western Europe of Middle East oil which would normally have been shipped to the more distant USA; and an increase in oil shipments to Western Europe from Venezuela and the USA, where there was considerable spare crude oil production capacity. These measures would reduce the need for tankers to make the long haul from the Persian Gulf to Western Europe round Africa. The full tanker-saving potential of these measures could not, however, be realised without the international co-operation of oil companies and governments.

PRECAUTIONARY PLANNING

Before Nasser nationalised the Canal, the British government had already examined the problems which would arise in the event of a disruption in the

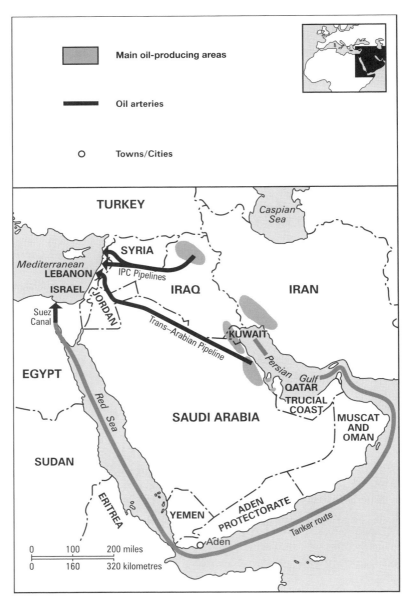

Map 3.1   Arteries for the flow of Middle East oil to Western Europe on the eve of the Suez crisis, 1956

westward movement of Middle East oil. A study by the Ministry of Fuel and Power in April 1956 concluded that if the Canal and the pipelines were closed for more than a very short period, this would 'undoubtedly cause grave dislocation in the economy of Europe through shortages of oil'.[8] The relief of a European oil shortage would, it was realised, depend largely on the availability of increased oil supplies from the USA. Some Anglo-American contingency planning at both industry and government levels was, it was suggested, essential. Consultations with the US government were duly put in hand and in June 1956 US government representatives met with officials of the British embassy in Washington to discuss the two governments' respective calculations on the measures that would need to be taken to maintain Western Europe's oil supplies if the Suez Canal and/or the trunk pipelines were closed.[9]

Meanwhile, the British government also sought the views of the Oil Supply Advisory Committee (OSAC), which had been set up in 1951 as a channel through which the main British or part-British oil companies (BP, Shell and Trinidad Leaseholds) could provide the government with advice on oil matters which might become emergencies. OSAC emphasised that in the event of the closure of the Suez Canal and/or the trunk pipelines 'the setting up of an effective international organisation, able to function promptly and effectively, will be vital'.[10] To meet that requirement OSAC suggested the reactivation of the US Foreign Petroleum Supply Committee (FPSC), which had been formed in 1951 to bring together the leading US oil companies so that they could take co-ordinated action to mitigate supply problems arising from the Iranian nationalisation crisis. For that purpose, the members of the FPSC had been granted special immunity from the US antitrust laws, which were intended to uphold competition in the US economy by outlawing collaborative, anti-competitive action between companies.[11]

Nasser's nationalisation announcement on 26 July immediately injected a new sense of urgency into international consultations on oil. On 27 July, Eden warned Eisenhower that 'the immediate threat is to the oil supplies to Western Europe' and added that 'if the Canal were closed we should have to ask you to help'.[12] In the USA, the Secretary of the Interior called upon the FPSC to prepare a plan of action to solve oil supply problems in an emergency. The FPSC duly met and proposed a 'Plan of Action' providing for the appointment of a Middle East Emergency Committee (MEEC) composed of representatives of US international oil companies who were to formulate plans for dealing with an interruption in the westward flow of Middle East oil. On 10 August the FPSC's recommendations were approved by the US Attorney General and the MEEC was brought into being.[13]

Thus was set up under US government auspices a committee of leading US oil companies, which proceeded to gather data on the oil supply problems that would arise from a closure of the Suez Canal and/or the trans-desert pipelines from Iraq and Saudi Arabia. The MEEC's powers did not, however, extend beyond the gathering of information and the making of recommendations. Further US government approval was required before the MEEC could implement its plans.[14]

The MEEC was an all-American body whose analysis of international oil movements could not possibly be comprehensive without input from the main European oil companies. However, when the MEEC was created there was no parallel European oil industry committee with which it could liaise. The nearest to such a thing was OSAC which, though it performed a valuable intermediary role,[15] was too Anglocentric to deal with matters which were of concern to the whole of Western Europe. BP and Shell therefore felt that a new European industry body should be set up, consisting of themselves and the French oil company, CFP, which held a 23¾ per cent share in the IPC. The smaller Trinidad Leaseholds was in the process of being acquired by the Texas Company and should, it was thought, be excluded from the new European group.[16] This idea of creating a 'European OSAC' was carried through in September, when BP, Shell and CFP formed the Oil Emergency London Advisory Committee (OELAC).[17] Its terms of reference were to advise the British, Dutch and French governments on oil supply problems arising from the Suez Canal crisis, and to collaborate with the MEEC in drawing up and implementing (with the agreement of the governments concerned) emergency supply plans.[18] Arrangements were made for representatives of US oil companies to attend OELAC meetings as observers and, reciprocally, for OELAC members to attend MEEC meetings in the same capacity.[19] During October 1956 these two committees exchanged data in order to gain a unified view on the rearrangement of world oil movements that would minimise shortages in the event of the Suez Canal and/or the main Middle East pipelines being closed.[20]

While the oil companies concerned themselves with the planning of supplies, governments examined the question of how, in the event of an oil shortage, supplies should be allocated between the consuming nations of Western Europe. In the early Anglo-American consultations after the nationalisation of the Canal, it was felt that oil allocations should be based on the principle of 'fair shares' and that they should be settled multilaterally in the Oil Committee of the Organisation for European Economic Co-operation (OEEC), which covered most of Western Europe.[21] At a meeting of the Oil Committee in September its British chairman, Angus Beckett, reported on the likely effects of a disruption in the westward movement of Middle East oil. He stressed that if the oil industry were to make optimum use of the available oil and tankers it would have to be allowed maximum flexibility, free from the interference of governments. However, governments should, Beckett suggested, be ready to introduce oil rationing schemes and agree to share oil shortages equitably, with the Oil Committee taking on the task of recommending fair allocations.[22] Early in October the Oil Committee adopted Beckett's recommendations in making proposals to the Council of the OEEC for measures to be taken in the event of an oil emergency. The Council approved the proposals and recommended that member governments apply them.[23] Thus by the end of October the basic machinery for international government and oil industry co-operation had been set up as a precaution against the closure of the Suez Canal and/or the main Middle East pipelines.

While this elaborate structure was put in place, international negotiations on the Canal dispute continued, as did secret preparations for military action against Egypt by Britain, France and Israel.[24] Haunted by Britain's appeasement of Hitler before World War II, Eden was convinced that if Nasser were allowed to get away with the nationalisation of the Suez Canal, Egypt would gain control of Middle East oil. 'When that moment comes', wrote Eden, 'Nasser can deny oil to Western Europe and we here shall all be at his mercy'.[25] Nasser could not, Eden was convinced, be allowed 'to have his thumb on our windpipe'.[26]

The Treasury, under Harold Macmillan as Chancellor of the Exchequer, made more detailed assessments of the possible effects of the crisis on the British economy. In August 1956 it produced an influential, or at least much-discussed, memorandum entitled 'The Egypt crisis and the British economy', which argued that an interruption of oil supplies would be 'by far the most serious consequence of the present crisis'. If Middle East supplies were completely cut off 'the substitution of American supplies would not be possible on a scale and in time to avoid an insupportable economic situation'.[27] On 26 August Macmillan forwarded the memorandum to Eden with the dramatic comment that 'the conclusion is clear – without the oil both the United Kingdom and Western Europe are lost'.[28]

Realising that the British economy could not withstand a protracted loss of sterling oil supplies from the Middle East, Macmillan might have counselled caution in foreign policy. But in fact he was one of the most hawkish members of the Cabinet, arguing that if Britain took decisive action against Egypt, the crisis could be ended quickly, before it could debilitate the British economy.[29] Britain's economic vulnerability was thus used as an argument

11   British Prime Minister, Sir Anthony Eden (right), and French Premier, Guy Mollet (left), at Downing Street for talks on the Israeli–Egyptian situation on 30 October 1956. Behind them are the British Foreign Secretary, John Selwyn Lloyd, and his French counterpart, François Pineau

in favour of quick military action rather than negotiation. This would prove to be a gross miscalculation.

THE ANGLO-AMERICAN SCHISM

Between 29 October and 6 November, Israel, Britain and France, acting in secret concert, invaded Egypt, where their forces took possession of a twenty-three-mile stretch of the Canal. The USA, which had not been consulted and had been seeking a peaceful solution to the nationalisation dispute, was furious with the three invading powers and introduced resolutions at the United Nations calling for a truce. The resolutions were vetoed by Britain and France.[30]

While the invasion caused a rupture in Anglo-American relations, it also precipitated an international oil crisis as Arab nations reacted against the

12   Blockships in the Suez Canal with burning oil tanks in the background in November 1956

Anglo-French–Israeli aggression. On 1 November Egypt blocked the Suez Canal by sinking blockships in the waterway.[31] On the 2nd, Syria broke off diplomatic relations with Britain and France, and that night units of the Syrian army put the IPC pipeline from Kirkuk to the Mediterranean out of action by blowing up the three pumping stations which were located in Syria.[32] On 6 November Saudi Arabia broke off diplomatic relations with Britain and France, and banned their tankers from loading at Saudi ports or at Sidon. The Saudis also prohibited vessels of other nationalities from carrying Saudi oil to Britain and France, and cut off Saudi oil supplies to the British protectorate of Bahrain.[33]

The resultant disruption of the westward movement of Middle East oil brought about exactly the emergency for which the Western powers had earlier prepared by setting up the MEEC and OELAC. However, Eisenhower was so incensed by the Anglo-French–Israeli military action that he immediately suspended the activities of the MEEC.[34] In his own words, he 'was inclined to think that those who began this operation should be left to work out their own oil problems – to boil in their own oil, so to speak'.[35] In suspending the MEEC he removed a vital component from the machinery for international co-operation on oil, which could not possibly function with full effectiveness without the participation of the US oil companies.

The British government attempted to reopen the Anglo-American dialogue on oil supplies, but found that the US government was uncommunicative.[36] Meanwhile, the pressure on Britain was mounting. When the British government's Egypt Committee met on 4 November, it was reported, to Macmillan's alarm, that there was now talk in New York of oil sanctions against Britain.[37] Nevertheless, the Cabinet agreed later that day to proceed with military operations against Egypt.[38] But by the 6th, when the Cabinet met again, it was clear that something would have to be done quickly to stop the great drain on Britain's foreign currency reserves precipitated by international loss of confidence in Britain, and disapproval of its military action.[39] If Britain's military operations in Egypt were not called off it was, according to the Cabinet records, 'probable that the other Arab States in the Middle East would come actively to the aid of Egypt, and that the United Nations would be alienated to the point of imposing collective measures, including oil sanctions, against the French and ourselves'.[40] Under mounting financial pressure and fearing oil sanctions, the British government ordered its forces in Egypt to cease fire at midnight on 6 November.[41]

The ceasefire may have ended the military hostilities, but it did not resolve the crisis. Despite overtures from Britain, the US government maintained its freeze on Anglo-American relations and made it clear that political and economic co-operation would be forthcoming only when Britain agreed to withdraw its forces from Egypt. Eden's government was not, however, willing to agree to a withdrawal, except on terms which Egypt refused to accept and the USA to support.[42]

Meanwhile the international oil companies, prevented from co-ordinating their supply programmes through the MEEC and OELAC, took individual actions in reaction to the crisis. BP's tanker transits through the Suez Canal continued normally until Israeli troops moved into Egypt on 29 October.[43] The next day, the British Admiralty advised merchant shipping to keep clear of the Canal and Egyptian and Israeli waters.[44] BP's tankers immediately ceased entering the Canal while orders were issued for the redeployment of BP's fleet of about 330 ships.[45] Some were instructed to take the Cape route to or from the Persian Gulf; others were re-routed to the Western Hemisphere in the hope that cargoes could be found to fill them on arrival; and some ballasted vessels which were in the Mediterranean and bound for the Persian Gulf were ordered to load at the Eastern Mediterranean terminals of the IPC pipeline system.[46] However, after the closure of the IPC pipelines by Syrian action no cargoes were available for BP in the Eastern Mediterranean and vessels which had arrived in the area were ordered away.[47] With both the Canal and the IPC pipelines closed, all BP shipments of Middle East oil to Western Europe had to be routed on the long haul via the Cape. So also, of course, did vessels proceeding in ballast in the opposite direction.

While BP and other oil companies rearranged international oil movements (see map 3.2), the last loaded tankers which had cleared the Canal before it was closed discharged their cargoes in Western Europe.[48] In BP's case, the last BP tanker to reach Britain with crude oil carried through the Canal discharged at the Kent refinery on 11 November.[49] There followed a gap of about a fortnight before Cape-routed tankers, delayed by the long haul, arrived in Western Europe with new oil supplies from the Persian Gulf.[50]

In the meantime, BP's liftings of crude oil from the Middle East were very sharply reduced, on a pattern that reflected the political priorities set out in an interdepartmental British government paper. The paper singled out Iran as a country 'where it is in our interests to maintain maximum production' because Iran was allied with Britain in the Baghdad Pact (a regional defence alliance), its economic development depended almost entirely on oil revenues, and the prestige of the Shah and the stability of the government depended on the 'smooth functioning' of the 1954 agreement between Iran and the Consortium.[51] The maintenance of BP's oil liftings from Iran at comparatively high levels during the Suez crisis (see table 3.1) reflected these concerns.

The pro-Western Hashemite regime in Iraq was also an ally of Britain in the Baghdad Pact, but heavy cuts in oil offtake were unavoidable following the closure of the IPC pipelines which provided the only export route for oil from the giant Kirkuk and Mosul fields in the north of the country. Production at the Basra field in southern Iraq, from which oil could be piped to tankers loading in the Persian Gulf, was maintained at a relatively high level, but could not compensate for the loss of production from the northern fields.[52]

Easily the largest source of crude for BP on the eve of the crisis was, however, Kuwait. With a much smaller territory and population than Iran and Iraq, Kuwait was already unable to spend all its oil revenues. It could, noted the government's interdepartmental paper, 'support a certain reduction without serious effects on its economy or its relations with the United Kingdom'.[53] In the Suez crisis, it was, therefore, Kuwait that bore the main

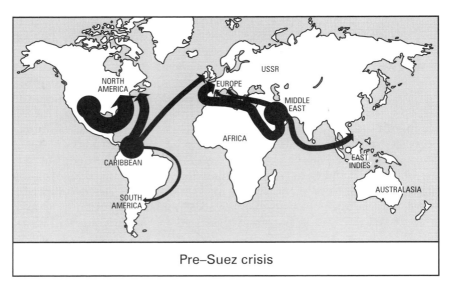

Pre–Suez crisis

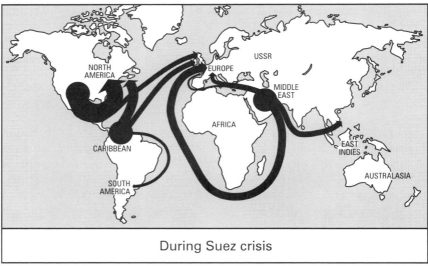

During Suez crisis

Map 3.2   The rearrangement of international oil movements in the Suez crisis, 1956

Table 3.1 *BP's crude oil liftings from the Middle East in the Suez crisis, 1956–1957 (thousand bpd)*

	Iran	Kuwait	Iraq	Qatar	Total
Oct 1956	231	737	170	32	1170
Nov 1956	173	329	54	9	565
Dec 1956	150	337	64	18	569
Jan 1957	225	371	43	15	654
Feb 1957	249	503	43	35	830
Mar 1957	235	315	69	11	630
Apr 1957	264	474	100	32	870
May 1957	202	551	125	28	906
Jun 1957	290	731	87	39	1147

brunt of BP's reduction in crude oil offtake from the Middle East. The much smaller offtake from Qatar was also heavily reduced.

As almost all BP's producing interests were in the Middle East, there was little potential for raising output elsewhere. Small-scale production continued in Britain, Canada and Trinidad, but BP's producing interests outside the Middle East contributed only a minuscule fraction to its overall crude oil supply.

While reducing its oil liftings in the Middle East, BP urgently sought increased oil supplies from short-haul sources across the Atlantic. After deciding to re-route tankers to the Western Hemisphere, the Supply and Development Division in London cabled BP's New York office stressing the 'paramount importance' of obtaining some firm oil cargoes in the USA.[54] The message that came back from New York was disconcerting. The US crude market was reported to be 'very tight'[55] and 'the situation here regarding purchases and diversions could not . . . be more gloomy . . . Our American business friends individually are sympathetic and would I am sure assist if they could. They are however, safeguarding their own positions first.'[56] Despite the grave drain on Britain's dollar reserves that was taking place because of the crisis, the limiting factor on British oil purchases was not so much a shortage of dollars as a lack of physical supplies.[57]

In the face of these difficulties, senior BP managers bemoaned the lack of co-ordination in the international response to the oil crisis. As Billy Fraser, BP's representative in New York, cabled Eric Drake: 'I am getting increasingly worried over the complete inability here to get any kind of co-ordination into this situation. The lack of this co-ordination must in my view be severely hurting our already virtually impossible task'.[58]

Meanwhile, the shortage of crude oil was affecting BP's refining and marketing operations in Europe. In November, the throughputs of BP's European refineries were reduced to an average of about 70 per cent of normal levels.[59] At the same time, the refineries had to deal with the problems of processing American crudes of varying qualities, which differed from their normal supplies. To help the refiners, arrangements were made for a sample of each crude cargo to be analysed in the USA, and for the analysis to be forwarded to the relevant refineries ahead of the main crude shipment. Refineries awaiting deliveries of US crudes thus received preliminary data on, for example, the specific gravities and wax and sulphur contents of the crudes which they would have to process.[60]

Further refining adjustments were made to match product yields as closely as possible to the abnormal pattern of demand during the crisis. The products which were most urgently required were diesel and fuel oils, essential for industry, heating and marine bunkering.[61] Motor spirit, on the other hand, was widely used for private motoring, a luxury which could be cut without inflicting serious harm on the economy.[62] In early November, BP's British refineries were therefore reprogrammed to maximise yields of diesel and fuel oils, and the catalytic cracking units (which increased the yield of motor spirit relative to diesel and fuel oils) at the Kent, Grangemouth and Llandarcy refineries were shut down.[63]

The effects of the crisis were felt not only on BP's producing and refining operations, but also on its marketing activities. The re-routing of world shipping and the need to conserve oil supplies for essential inland uses in Western Europe called for a radical shift in the geographical pattern of BP's marine bunker business. In November BP asked its bunker customers operating vessels between the Western and Eastern Hemispheres to consider restricting their bunker offtake in Europe by bunkering to maximum capacity at Western Hemisphere ports. With the same purpose of conserving fuel oil stocks in Western Europe, BP vessels loading in the Persian Gulf for western destinations took maximum bunkers at their Middle East loading ports, replenishing as necessary at ports in the Canary Islands. BP vessels which had been routed from Western Europe to the Western Hemisphere took on bunkers sufficient only to reach the US Gulf. On loading in the Western Hemisphere for the return voyage, they took on enough bunkers to reach Western Europe and return to the Western Hemisphere without re-bunkering.[64] As a result of these measures and the re-routing of shipping, there were enormous changes in the volume of bunker offtake at individual ports. For example, bunker offtake at Aden, which was easily the largest of BP's bunker outlets in normal times, fell from about 41,500 bpd in July 1956 to 27,000 bpd in October, before collapsing to 12,000 bpd in November and little more than 3,000 bpd in January 1957. On the other hand, there were very large increases in bunker offtake from ports such as Dakar, Mombasa, Durban and Capetown in Africa, and from Las Palmas and Tenerife in the Canary Islands.[65] A graphic picture of the changes in BP's marine bunkering pattern is shown in figure 3.1.

While BP rearranged its international marine bunker business, it also had to deal with a shortage of supplies for its European marketing associates. The closure of the Suez Canal and the IPC pipelines affected BP's supply position more adversely than that of the other major oil companies because, unlike the other oil majors, BP lacked significant production in the Western Hemisphere. On 2 November, the Distribution Department asked the European marketing associates 'to avoid taking on new business and to check any attempt by the public to stock-up unduly'.[66] This was followed on the 6th by the more strongly worded message 'that closure of IPC pipelines to Eastern Mediterranean results in deterioration of supply position. This emphasises absolute necessity to refrain from taking on new commitments of any kind and need for discretion in renewing lapsing contracts.'[67] As BP received its last shipments of Middle East oil via the Suez Canal, still stronger measures of restriction had to be taken. On 13 November, Harold Snow, BP's director of distribution, cabled the general managers of the European marketing associates with the message that they should assume that supplies of products (apart from aviation fuel and bunkers) would be 30 per cent less than normal for the rest of the year.[68]

While the volume of BP's supplies was affected, so was the cost. The diversion of tankers round the Cape, the great increase in freight rates for chartered vessels, the rise in unit costs consequent on the reduction in refinery throughputs, and the high cost of purchased oil all contributed to an increase in the cost of BP's European supplies.[69] From 20 November this increase in costs was passed on as a surcharge to European marketing associates with the instruction that 'we expect you to seek means of recovering this surcharge from consumers'.[70]

Meanwhile the supply position continued to deteriorate, especially in the case of fuel oils, and early in December Snow informed BP's European marketing companies that their supplies, already reduced to 70 per cent of normal, would have to be further reduced to only 60 per cent of normal for products other than fuel oils, which would have to be cut back even more severely, to only 55 per cent of normal.[71]

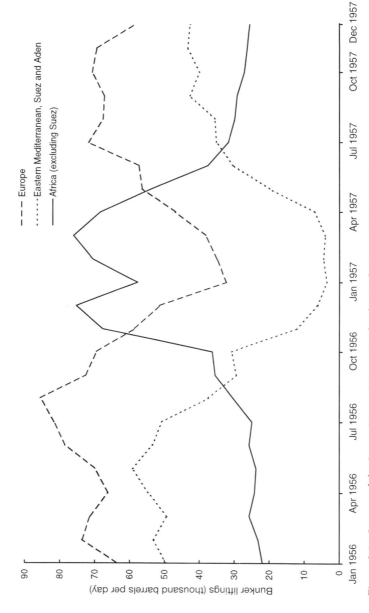

Figure 3.1 Impact of the Suez crisis on BP's marine bunkering business, 1956–1957

By that time, measures to ration consumption had been introduced by governments across Europe.[72] For example, in Britain, the government announced on 7 November an immediate cut of 10 per cent in deliveries of motor spirits, gas oils, diesel and fuel oils.[73] On 20 November further cuts were announced, including, with effect from 17 December, a limit on general purpose motoring (excluding priority motoring, for example by doctors) of 200 miles per month. These cuts were aimed at meeting an estimated overall shortfall of 25 per cent in oil supplies.[74]

As the oil crisis worsened, the US government kept in force its suspension of the MEEC and so continued to prevent the oil companies from co-ordinating their operations through MEEC/OELAC liaison. OELAC remained in being, but was tainted in US eyes by its Anglo-French constitution.[75] To remove the stigma, it was decided to give OELAC a more European and less Anglo-French look. To that end, on 30 November, the Council of the OEEC agreed to establish a new industry body, the OEEC Petroleum Emergency Group (OPEG), which was 'to advise the Oil Committee on the availability of oil supplies for Europe and to assist in the implementation of its recommendations'.[76] In effect, OPEG took over the functions and membership of OELAC's Executive Committee. Its membership consisted of Shell, BP and CFP.[77] The formation of OPEG was, therefore, little more than a facelift operation to give OELAC a Western European rather than an Anglo-French appearance.

By the time that OPEG was formed the international political and economic pressure on Britain had reduced the Cabinet to breaking point. On 23 November, Eden, his health and political career broken, had flown to Jamaica to recuperate, leaving R. A. Butler, the Lord Privy Seal, in charge of the government. On 24 November, the UN General Assembly passed, by sixty-three votes to five, a resolution censuring Britain and France and demanding the immediate withdrawal of their forces from Egypt. The USA voted for the resolution and sought to enforce it by economic pressure. In the words of Robert Rhodes James, Eden's biographer, 'the demands were brutal, and Butler and Macmillan – especially the latter – saw no alternative other than to comply. The stick was further assaults on sterling and the British economy and the denial of oil; the carrot was full assistance and oil supplies if the British and French withdrew unconditionally.' By Friday, 30 November, the Cabinet had crumbled and accepted that a withdrawal of troops from Egypt was inevitable.[78]

That was the signal for the US government to turn from the stick to the carrot by authorising the immediate reactivation of the MEEC.[79] The public announcement of the Anglo-French decision to withdraw their forces followed on Monday, 3 December, and on the same day the MEEC held its first meeting since its suspension.[80] Shortly afterwards it was announced that arrangements had been made, with strong support from the USA, for Britain to draw up to its full quota of $1,300 million from the International Monetary Fund.[81] Anglo-American economic co-operation was at last revived.

THE OIL LIFT

With the creation of OPEG and the reactivation of the MEEC the basic machinery was finally in place for fully co-ordinated action to deal with the oil crisis. Prior to its suspension at the end of October 1956, the MEEC was, as mentioned, purely a fact-finding body, empowered only to investigate and make recommendations on the measures to be taken in the event of an oil emergency. The members of the MEEC were not empowered to implement their recommendations until the relevant 'schedules' were issued by the 'Administrator', namely the US Secretary of the Interior.[82] On 7 December 1956 two such schedules were issued, permitting the members of the MEEC to co-operate in joint arrangements for the most efficient use of tanker tonnage, free from antitrust concerns.[83] This meant, to take a simple example, that if one oil company was planning to ship a cargo of oil east across the Atlantic, and another company was planning to ship a cargo in the opposite direction, they could agree to exchange cargoes at the ports of loading, eliminating the need for marine transportation. In practice, exchanges could be far more complicated than this, involving multilateral exchanges of crude oil and products between companies.

Some inter-company exchange arrangements had already been in operation in November, but the reactivation of the MEEC and the issue of the schedules opened up much greater possibilities for inter-company co-operation in minimising tanker hauls.[84] In the words of John Loudon, the head of Royal Dutch-Shell, 'this work has been going on all the time, but it will clearly be possible to speed things up very much if appropriate representatives of all the companies who are concerned with worldwide movements of oil can sit down together'.[85]

Although the reactivation of the MEEC increased the scope for joint action by the oil companies, BP's supply position remained precarious during December 1956 and January 1957. Its crude oil production in December was only about half the October level and its European refining throughputs, which had been reduced to about 70 per cent of normal in November, were reduced still further to about 65 per cent in December.[86]

The main cause of BP's difficulties was that it was unable to purchase the supplies that it had hoped to obtain in the USA.[87] Disappointment on that score was not restricted to BP. It was a general problem, caused largely by the situation in Texas, which was easily the largest oil-producing state in the USA. Even on a world scale Texas was a major oil producer, with a 21 per cent share of non-communist world oil production in 1955, compared with the Middle East's share of 24 per cent, and Venezuela's of 16 per cent.[88] There was plenty of scope for raising production in Texas, where oil output was pegged well below full capacity by the Texas Railroad Commission (TRC), a state regulatory body responsible for setting the monthly allowable rate of production.[89] However, when the TRC met in mid-December 1956 to settle the level of allowable production in January 1957, it decided that there should be no increase.

The Texas oil industry included numerous independent oil companies, whose interests were local rather than international. They held the international oil majors in deep distrust, mainly because they saw the majors' growing imports of low-cost foreign oil as a threat to the higher-cost domestic oil industry.[90] The independents, who accounted for about 58 per cent of Texan oil production, were a vociferous lobby and they were not inclined to increase production in order to help out the majors during the Suez crisis, only to be left with large stocks and surplus production when the crisis ended.[91] They therefore opposed the increase in allowable production sought by the international oil companies.

While Western European governments expressed concern about the TRC's decision, the international oil companies continued to struggle with supply problems.[92] In January 1957 BP had to cut its European refinery throughputs to only 55 per cent of normal as tankers, which might have picked up crude oil from the USA, were routed on the longer-haul Cape route to the Persian Gulf.[93] Shell had to do the same. Its representative in New York reported to John Loudon that it was 'a dismal picture requiring immediate and definite action which will produce more crude oil from USA sources'.[94]

On 3 January, in an attempt to flush out increased crude oil supplies in Texas, the Humble Oil and Refining Company, a subsidiary of Standard Oil (NJ), increased the prices at which it offered to purchase Texas crudes. This was the first general increase in prices for Texas crude oils since mid-1953.[95] Within a few days, Humble's action was followed by other companies and in other states of the USA.[96] Towards the end of the month the lighter Venezuelan crudes followed the US increases, the prices of heavy Venezuelan crudes having already been increased in late 1956.[97]

However, when the TRC met on 18 January to decide on allowable production in February, it again disappointed the international oil companies and Western European governments by announcing only a small increase.[98] The Petroleum Attaché at the British Embassy in Washington despondently wrote that the TRC's decision 'has cast us all here into the most abject gloom'.[99] However, early in February there was cause for greater optimism after Eisenhower announced that 'we must not allow Europe to go flat on its back for want of oil' and made a veiled threat that if there were no increase in supply, the federal government might have to take some action.[100] After Eisenhower's announcement the TRC met on 19 February and decided on a considerable increase in allowable production in March.[101] At last, a substantial rise in Texas oil production had been authorised.

While the TRC regulated production in Texas, the allocation of supplies between the consuming nations of Western Europe was handled by the OEEC Oil Committee in conjunction with the international oil companies. The first joint meeting of the Oil Committee and OPEG was held in Paris on 6 December 1956. It was a huge gathering, attended by representatives of the seventeen OEEC member states, the US and Canadian governments, OPEG and observers from most of the US oil majors. Altogether, nearly sixty representatives of governments and companies were there.[102]

Agreement was reached on arrangements whereby governments and companies were brought into close international co-operation. Starting in December, representatives of OPEG and the MEEC attended each other's meetings as observers; OPEG members and US oil company observers regularly visited Paris for meetings with the Oil Committee; and operational meetings were held between a programming group of the Oil Committee and OPEG's programming sub-committee.[103] Thus the main international oil companies, represented by OPEG and the MEEC, came together with the oil-consuming nations of Western Europe, represented by the Oil Committee, to allocate oil supplies between Western European countries. OPEG's main functions were to advise the OEEC Oil Committee on the availability of oil supplies and to implement the Oil Committee's decisions on the allocation of supplies between individual countries. The information on which such decisions were taken took the form of oil supply and demand forecasts, known as 'slates', for each country for a period of forty days ahead.[104]

There was still, however, no agreed system by which the Oil Committee was to decide how to allocate scarce oil supplies between countries. On 5 December the Committee set up a small working party to consider this matter.[105] While it deliberated, allocations were dealt with on an ad hoc basis in response to applications from individual countries. For example, Sweden requested increased supplies of fuel oil to provide against the possible cessation of imports due to the icing up of Swedish ports later in the winter; Denmark also asked for additional supplies to build up stocks as a precaution against Danish ports becoming icebound; Greece and Turkey, which were normally supplied from the nearby Eastern Mediterranean pipeline terminals, had low stocks of some products and requested additional supplies; so too did Austria, which was particularly short of fuel oil.[106]

The competition for scarce supplies at times put the principle of co-operation under strain. As Loudon wrote on 18 December, 'yesterday's meeting of OPEG with OEEC was a bit of a shambles . . . we had a number of pleas for special treatment from various countries and it is clear that at the moment the spirit of collaboration between the various European countries is not very highly developed'.[107]

This problem was largely resolved on 4 January 1957, when the Oil Committee adopted principles of allocation, which OPEG agreed to implement.[108] Under the new system, which started at the beginning of February, OPEG estimated for each product the extent to which supplies to the OEEC area as a whole were likely to fall short of normal consumption in each 'slate' period, after setting aside about 145,000 bpd of products as a special reserve. OPEG participants then endeavoured to arrange their operations so that the shortfall of supplies to each country approximated to the overall shortfall for the whole OEEC area, a procedure known as 'equalisation'.

It was recognised, however, that in some countries receipt of only the basic allocation from equalisation might constitute a particular hardship. For example, some countries had especially cold climates and/or lacked alternative sources of energy. It was to provide for such cases that the reserve of 145,000 bpd was set aside for supplementary allocation by the OEEC Oil Committee to countries most in need of extra supplies. The basic and supplementary allocations were supplied by the oil companies, whose operational functions were not subsumed by OPEG. The role of OPEG was rather to provide a forum where the international oil companies could co-ordinate their individual programmes to achieve a common objective.[109]

The introduction of the system of basic and supplementary allocations undoubtedly eased some of the frictions which were inherent in the earlier ad hoc methods. But by the time that the new system came into operation the supply position was in any case beginning to improve. In mid-January 1957 OPEG estimated that no more than 75 per cent of Western Europe's normal oil demand could be met in the first quarter of the year. By 7 February OPEG had raised its estimate to 80 per cent, and by 5 March it had raised it again to 85 per cent.[110]

The easing of the feared oil shortage was due to a combination of factors. On the demand side, the unusually mild winter meant that use of heating fuels was lower than normal. Oil consumption was also curtailed by rationing and by the rise in oil prices, which helped to reduce demand. In addition, the demand for marine bunkers in Western Europe was lessened by the shifting of marine bunker business away from European ports. On the supply side, the increase in Western Hemisphere production helped to ease the pressure, but the main factor, according to the OEEC, was that the co-ordinated activities of the companies working through the MEEC/OPEG association resulted in a greater saving of tanker tonnage than had appeared possible before the crisis.[111] The international oligopoly of the oil majors was perfectly suited to this role. Co-operation between the majors had long been close and during the Suez crisis Western governments were glad to encourage it as an effective way of dealing with the oil shortage. Inter-company collaboration, which was normally outlawed in the USA and regarded with distrust elsewhere, was for once blessed.

Amidst growing confidence that Western Europe was unlikely to go seriously short of oil, there was a relaxation of some of the emergency measures which had earlier been taken. For example, from 15 February BP reduced the emergency surcharge on supplies of oil products to its European marketing associates, and five days later the British government announced a relaxation of rationing.[112] Early in March the outlook brightened further. Following the Anglo-French military withdrawal from Egypt in December 1956, the UN commenced clearing the Canal. Egypt would not grant permission for the final clearance and opening of the waterway until Israel also withdrew its forces. The Israeli withdrawal commenced on 6 March and was completed in two days, whereupon Egypt gave permission for the raising of the two remaining blockships in the Canal.[113] Almost simultaneously, the Syrian government agreed to repairs to the IPC pipelines being put in hand, and Saudi Arabia resumed its oil supplies to Bahrain and lifted its restrictions on oil exports from Saudi Arabian ports and Sidon.[114]

From the third week in March loadings of Iraq crude in the Eastern Mediterranean were resumed, with the pipelines initially operating at about 40 per cent of pre-crisis throughput.[115] This, coupled with the increase in allowable production in Texas, contributed to a further improvement in Western Europe's oil supply position.[116] On 20 March OPEG's Supplies Sub-Committee estimated that even without the Suez Canal in operation, oil supplies in the second quarter of the year should be sufficient to meet Western European demand in full.[117] With regard to national supply positions, the sub-committee 'saw no danger point in any particular country'.[118] When the

13 Wreck of the tug, *Hercule*, in the Suez Canal, with Navy House, gutted by fire, in left background, in 1956

Oil Committee met that afternoon, it decided that all the reserve set aside for supplementary allocations in the first half of April should be made available for OPEG to 'equalise', as no supplementary allocations were necessary.[119] When the Oil Committee and OPEG met again on 2 and 3 April, it was minuted that 'virtually no country wanted to put in a demand for making up its deficit position'.[120] With the reopening of the Suez Canal imminent, expectations of a fall in product prices may have encouraged OEEC countries to hold back requests for additional supplies. Nevertheless, the language of the minutes conveys no sense of crisis.

The threat of a serious oil shortage had, therefore, already disappeared when, on 9 April, it was announced that the Suez Canal was clear. It was indicative of Western Europe's strong supply position that, far from hastening to route vessels through the Canal, the British and French governments kept in force the emergency restrictions which prohibited British and French ships from using the Canal. This action, ostensibly made for technical reasons such as the feared presence of mines in the approaches to Suez, was intended to put economic pressure on Nasser by the denial of Canal dues in

14   The BP Tanker Company's 12,000-ton oil tanker, *British Restraint*, in transit
through the reopened Suez Canal in September 1957

the hope of wringing from him a favourable settlement for the Canal's future
administration. In fact, no advantage was gained by this action and on 13
May Macmillan announced that British ships were free to use the Canal, the
French government following suit on 13 June.[121] BP immediately took steps
to re-route some of its tankers via the Canal, the *British Crown* being the
first BP tanker to transit the reopened waterway, in ballast for the Persian
Gulf, on 20 May.[122]

Meanwhile, the emergency measures for dealing with the crisis were
wound down further. On 18 April the US government suspended the sched-

ules authorising the participants in the MEEC to co-ordinate their world-wide tanker movements.[123] On 2 May, Beckett, chairman of the OEEC Oil Committee, commented that 'in view of the 100% coverage of demand, equalisation had quite naturally become superfluous'. It was therefore agreed that the equalisation procedure and the system of supplementary allocations should be abandoned and that OPEG's day-to-day activities should be suspended with immediate effect.[124] As business returned to normal, BP recommissioned its catalytic cracking units at Grangemouth on 20 March, Kent on 2 April and Llandarcy on 8 May.[125] From 15 May the emergency surcharge on supplies of products to BP's European marketing associates was abolished.[126] From 1 June the last vestiges of British government restrictions on oil consumption were removed.[127] And on 31 July 1957 the MEEC and OPEG were formally disbanded.[128]

The crisis was over, but it left lasting consequences. On the face of it, the international oil companies emerged unscathed. Indeed, they had demonstrated that they still had enough control over the international flow of oil to keep Western Europe supplied in an emergency. But for BP in particular, the political underpinning of its position in the Middle East had been gravely weakened. Before the Suez crisis, Britain had still aspired to the status of a world power, but the crisis made it clear that Britain could no longer muster the resources for an independent world role, unsupported by the USA. In the wake of Britain's humiliation at Suez, the Eisenhower doctrine of 1957 asserted the USA's interest in the security of the Middle East. Henceforth, in Middle East affairs the leading voice among the Western powers would be American, not British. In the Middle East, meanwhile, Nasser, having succeeded where Musaddiq had failed in challenging the old imperial powers, was the hero of radical pan-Arab nationalists. Pouring invective on Britain, he stoked the fires of the nationalists, who stood in outright opposition to imperialism and all that was associated with it, including BP.

= 4 =

'The energetic search for new sources of crude oil'

While BP faced growing radical nationalism in the Middle East, it was at the same time trying to reduce its exposure to the political instability of that region by pursuing the policy, agreed in 1953, of stepping up the search for oil elsewhere. Before the nationalisation of its Iranian oil fields in 1951, the Company had been awash with crude oil. Through its pre-eminent position in the Middle East it owned, according to estimates in the early 1950s, nearly a quarter of the world's proved oil reserves (figure 4.1). It could produce much more oil than it could sell, and had little incentive to seek new reserves which would only add to its surplus. Its upstream portfolio was extremely highly concentrated with just seven giant oil fields in Iran, Iraq, Kuwait and Qatar accounting for 99 per cent of the Company's crude reserves and 94 per cent of production.[1] The unit costs of producing oil from these highly productive reservoirs were among the lowest in the world. No other oil major could match the Company's concentration on low-cost, giant oil fields, which gave it a competitive advantage in the upstream sector of the international oil business.

THE GOLDEN AGE IN THE MIDDLE EAST

The Company's giant fields were the fruits of a golden age of frontier exploration in the Middle East going back half a century to 1901, when William Knox D'Arcy obtained his oil concession in Iran.[2] Like all golden ages, this one had flaws. Exploration proceeded unevenly by fits and starts, punctuated by two world wars, the Great Depression, and the need to mollify the jealous Riza Shah, who did not wish the Company to divert its attention from Iran.[3] Moreover, some serious mistakes were made. The redoubtable Hungarian geologist, Hugo de Böckh, engaged as geological adviser in the

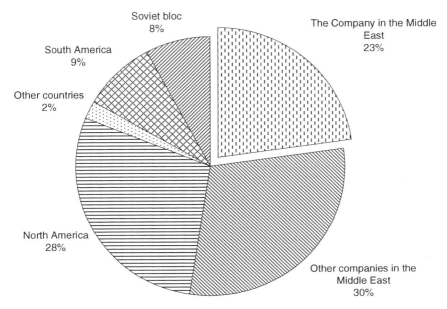

Figure 4.1   The Company's share of the world's published proved oil reserves, 1950

1920s, made some inspired recommendations for drilling locations in Iran. But he also dismissed the Arabian side of the Persian Gulf as an area where 'no great concentration of oil can be expected'.[4]

After de Böckh finished with the Company, his protégé George Lees became chief geologist in 1930 at the age of thirty-two, only nine years after he had taken up geology. In that short time he had participated in geological surveys in Iran and Oman, taken a PhD in geology at the University of Vienna, examined oil prospects in the USA, Canada, Egypt and Germany, and negotiated a personal option on the oil rights of the state of Qatar which he purchased with a jewelled sword.[5] Physically and intellectually adventurous, Lees was a natural frontier explorer, who remained chief geologist for more than two decades, until ill health forced him to retire in 1953. But he made the same mistake as de Böckh about Arabia, counting Saudi Arabia in 1930 as one of the 'countries with little or no [oil] prospects'. Kuwait, he wrote, had 'very poor prospects'.[6]

Notwithstanding these errors, the Company's overall record in finding oil was outstanding. Peter Cox, who joined as a geologist in 1925 and later became the Company's most senior exploration executive, wrote that 'from its formation in 1909 until 1951 the Company was significantly more successful in its exploration than any other in the industry, if quantity of marketable crude oil and cost of finding it are the criteria of success.'[7]

The Company did not achieve its exploration success by technological leadership. Indeed, in geophysics it generally lagged behind other companies, except in the use of seismic refraction methods.[8] Politically and geologically, however, it enjoyed a most favoured position. Britain's authority in the Persian Gulf was long established and when the Ottoman Empire was broken up at the end of World War I Britain became indisputably the paramount imperial power in the Middle East.[9] The British government became the Company's majority shareholder, and helped it to secure oil rights of long duration over vast areas of the most oil-rich region in the world.[10] In Iran, the Company originally held exclusive oil rights over 500,000 square miles for sixty years, reduced to 100,000 square miles with a renewal of the sixty-year term in the 1930s.[11] In Iraq, the Company gained a 23¾ per cent share in the Iraq Petroleum Company (IPC), which obtained three seventy-five-year concessions covering virtually the whole country.[12] In Qatar, the option obtained by Lees was taken up by the Company and assigned to the IPC.[13] In Kuwait, the Company, influenced by the views of de Böckh and Lees, held back at first, but in 1934 gained a 50 per cent share in a concession covering the whole country for seventy-five years.[14] The geological conditions in all these countries were exceptionally favourable for the accumulation of oil. In all of them, giant oil fields had been discovered and brought into production by 1951 (see map 4.1).

One country, however, stood out from the rest as a seminal influence on the Company's exploration. This was Iran, which was both the first country in which the Company explored and the only one where the Company had complete control of operations, without partners. Here, in the foothills of the Zagros mountains in the south-west of the country, the heaving and thrusting of the earth over millions of years had caused the crust to wrinkle and fold, forming hills and valleys on the surface and a similar undulating pattern in the sub-surface layers of rock. Great elongated domes of rock ('anticlines') were thrust up in spectacular folds, protruding boldly from the earth's surface. These provided superb examples of classic oil traps, with the impervious Iranian Lower Fars cap-rock sealing the permeable, porous Asmari limestone in whose interstices oil and gas had accumulated under pressure over millions of years. When the cap-rock of such an anticline was pierced by a drill, oil and gas would be forced upwards by the reservoir's own pressure, erupting at the surface as a spectacular 'gusher'.

Map 4.1   The seven giant oil fields in the Middle East that produced 94 per cent of the Company's crude oil in 1950

15   An anticline in Iran

The Zagros foothills were an almost perfect example of a geological fold-system, marred only by disharmonic movements in the Lower Fars, which made it difficult to locate the crests of the limestone reservoirs beneath. But Lees was so inspired by the bold scale of the anticlines that he thought 'even the vexatious turbulence of the Lower Fars is, like a fault in a truly great man, also greatly conceived'.[15] This geologists' paradise was where the Company's first generation of geologists, who mostly stayed with the Company for the rest of their working lives, gained their formative exploration experience. The association of giant oil fields with anticlinal folds as found in Iran remained for years a powerful influence on the Company's exploration. 'The anticline to the oil-geologist', wrote Lees in 1952, 'is as honey to the bee'.[16]

Its bias in favour of exposed anticlines caused the Company to make some serious errors of judgement not only in underrating the Arabian side of the

Persian Gulf where the gentle anticlinal folding did not show on the surface of the flat, featureless desert, but also later in overrating the anticlinal foothills of the Rocky Mountains in Canada and of the Brooks Range in Alaska. In other respects, though, the influence of the formative years in Iran was more positive, for it encouraged the Company's explorers to concentrate on searching for giant oil fields in previously unexplored areas – in short, to be frontier explorers. Frontier exploration was not in those days spelled out as a formal strategy. It was more in the nature of a culture, instilled by experience in the Company's explorers. These men, mostly geologists, spent long periods on surveys in remote areas, where the terrain was difficult, the living conditions primitive and the climates extreme. Far away from the bureaucratic hierarchies of office life, they adopted an informal style, with 'flat' hierarchies, high motivation and few formal terms of reference.[17]

As knowledge of the world's oil reserves increased, their instinct for frontier exploration was validated by calculable evidence. It was confirmed that in a typical oil province most of the oil was found in a few large fields, with the remainder dispersed in a larger number of smaller fields. Because of their large size, the biggest fields tended to be the easiest to find, with the result that they were often discovered early in the exploration of new territory. The pioneer therefore had a better-than-average chance of winning the largest prizes.[18] If oil was found, the economic advantages of the pioneer were likely to be considerable. The cost of acquiring oil concessions or leases was sure to be lower before oil was found in an area than after. Also, in oil production as in other industries, there were scale economies, which meant that other things being equal, larger fields had lower production costs per barrel than smaller fields. Against these attractions of frontier exploration had to be set, however, the comparatively high risks of failure for the pioneer, who might well undertake a long-term exploration campaign in unproven territory without finding any oil at all. Frontier exploration was thus a high-risk, high-reward strategy. It meant getting access to virgin territory ahead of competitors, committing resources to uncertain long-term prospects, pushing out the margins of exploration into increasingly remote and difficult areas, entering into new territory in every sense of the word – physically, technologically and entrepreneurially.

THE DECISION TO DIVERSIFY OIL SOURCES

By the 1950s the Company's golden age of frontier exploration in the Middle East was coming to an end. The major oil companies, having established virtual dominion over the oil resources of the Middle East in the first half of the century, were coming under increasing challenge in the region. The governments of oil-producing states were becoming more assertive in seeking to reduce the acreage and durations of oil concessions, or even, in the case of Iran, in seeking to end the concessions altogether. At the same time, new competition for concessions was appearing from Western state-owned oil companies and from private US independents venturing abroad for the first time.[19] The Company, which had concentrated on the Middle East more than any other major, was highly vulnerable to these developments.

Although the Company successfully weathered the nationalisation of its assets in Iran, the nationalisation was nevertheless an earthquake which rocked the Company's concessionary foundations. Shaken by the experience, the Company decided, as noted in chapter one, to adopt 'as our aim and policy <u>the energetic search for new sources of crude oil</u> . . . to provide supplies outside the Middle East'.[20]

This decision to try to diversify the Company's sources of crude was a defining moment in the Company's exploration policy. New reserves outside the Middle East acquired a premium rating, and the Company's exploration performance for at least the next quarter-century would be measured not just by straightforward quantity of barrels discovered, but by how many barrels were discovered outside the political hot-spot of the Middle East.

After the policy decision of June 1953, a worldwide review of oil prospects was completed by the end of the year.[21] Meanwhile, exploration and production staff who had been withdrawn from Iran were redeployed. The most senior was Peter Cox, who was appointed exploration manager in London and went on to become managing director of BP's main exploration subsidiary, the BP Exploration Company (formerly D'Arcy Exploration Company), in 1957.[22] He also held the post of chief geologist for a short spell after Lees retired and before the post was taken in 1955 by Norman Falcon, who had also previously worked in Iran for a long spell.[23]

Peter Kent was another Company geologist who was withdrawn from Iran in 1951.[24] Some idea of the life of the frontier explorer is given in his memoirs. After being withdrawn from Iran, he was posted to East Africa, where he made a far-ranging reconnaissance of Somalia, Madagascar, Tanganyika and Zanzibar (later combined into Tanzania) and Kenya, still a British colony. Here he found that British colonial settlers formed a 'free-drinking community', shaken by the anti-colonial Mau Mau movement.[25] From Africa he moved to Papua, where the Company had been exploring for many years in the belief that Papua's large limestone anticlines might hold oil on a Middle East scale.[26] Here Frank Rickwood, an Australian geologist who

16   Seismic survey party in Tanganyika in the 1950s

had joined the Company in 1955, had long experience of exploring in the dense tropical rain forest, where the overhead foliage was so thick that the ground underneath was in constant darkness. Out of the forest, in the town of Port Moresby, Kent shared a house and much jollity with Rickwood, 'an entertaining, erudite but somewhat eccentric bachelor'. The tone of their parties 'was perhaps indicated by a large Gauguin on the wall, depicting two topless Tahitian ladies with their bosoms symbolically meeting on a tray of fruit, which caused raised eyebrows among the more stuffy visitors'.[27]

When Kent left Papua in 1957 his post there was taken by Harry Warman, who had earlier worked under Kent in Iran.[28] Kent moved on to Canada,

17   Laying geophone cables in Papua in the 1950s

where he ventured north from Calgary into little-known country. On one expedition he reached the end of the road and stayed in a small backwoods hotel where he found that 'the beer parlour was occupied by a collection of characters who had been pushed out to the edge of the world – deformed, misshapen or plug-uglies who would have graced a Hollywood horror-film sans make-up'. In the Northwest Territories, he found the opposite to the constant darkness of the Papuan rain forest in an Eskimo village where the barefoot children played all hours of the twenty-four, for there was no night.[29]

From his base in Calgary, Kent took charge of exploration in Alaska before handing that job over, again to Warman. Returning to London, Kent went on to succeed Falcon as chief geologist. After five years in that post, he handed over, yet again to Warman.[30] The line of chief geologists, responsible for the Company's worldwide geological activities, and all with formative experience in Iran, thus ran from Lees (1930–53) to Cox (1953–5) to Falcon (1955–65)[31] to Kent (1966–71) to Warman (1971–3). This was an important post, for the Company's chief geologists were held in much respect and there was much scope for individual insight and interpretation in their work. Under them, the field geologists also enjoyed great prestige, more so than in other oil companies.[32]

The people and countries already mentioned were but a part of the Company's great burst of exploration outside the Middle East after the Iranian nationalisation in 1951. In 1950, the only countries outside the Middle East where the Company was engaged in exploration and/or production were Nigeria (in partnership with Shell, which managed the operations), the UK, Papua and Trinidad.[33] By the end of the 1950s BP had in addition ventured into East Africa (Kenya, Madagascar, Somalia, Tanganyika, Zanzibar); West Africa (Gambia, Senegal, the Cameroons); North Africa (Algeria, Libya, Tunisia); Australasia (Australia, New Zealand); Europe (France, Germany, Malta, Italy, Switzerland); and America (the USA, Canada, Colombia).[34] In 1951 exploration expenditure outside the Middle East was less than £1 million.[35] Ten years later it had increased to more than £13 million and BP had diversified its oil reserves outside the Middle East by adding five new countries in three continents to its reserves portfolio. These were Nigeria and Libya in Africa; Colombia and Trinidad in South America/the Caribbean; and Canada in North America. It had also acquired a foothold in the USA by acquiring the small Kern oil fields in California.

NIGERIA

The discovery and development of oil in Nigeria was reminiscent of the Company's golden age in the Middle East. It started in the first half of the twentieth century with frontier exploration by a pioneer who acquired, with imperial backing, exploration rights covering the whole country. In the third quarter of the century, it progressed, after oil had been discovered, into second-phase exploration by the original pioneer plus new entrants competing in comparatively small blocks of acreage.

Nigeria became a British colony in the late nineteenth century, when the European imperial powers – principally Britain, France and Germany – carved up the African continent into a multitude of colonies in the great 'scramble for Africa'.[36] The colonial Mineral Oil Ordinance of 1914 made the search for oil in Nigeria a British preserve by stipulating that oil licences and leases could be granted only to British companies.[37] In 1937 the Company and Shell formed a partnership to explore for oil in Nigeria and obtained an exploration licence covering the whole country, an area of some 370,000 square miles. The partnership, named Shell/D'Arcy Exploration Parties, carried out some reconnaissance before suspending operations during World War II. Work was resumed at a higher tempo after the war and in 1949 Shell/D'Arcy reduced its licence area to about 60,000 square miles in the southern coastal basin, including the Niger delta. This was a sprawling tropical mangrove swamp, criss-crossed by a maze of shallow creeks and estuaries, where rain fell incessantly for months at a time, disease-spreading insects thrived and the humidity was stifling. Good maps did not exist and the terrain was so difficult to traverse that mapping could not be done on foot. Even mapping by aerial surveys took five years (1948–53) to complete because the mappers were hampered by low clouds in the wet season and, in the dry, by the dust haze caused by the arid Harmattan winds blowing from the deserts to the north.[38]

Meanwhile, in September 1951 a new joint company, the Shell-D'Arcy Petroleum Development Company of Nigeria, was formed to carry on the joint venture. Although Shell-D'Arcy was owned equally by Shell and the Company, it was Shell who provided the management and the technical advice. The Company's interest in Nigerian exploration was, therefore, financial rather than operational. Its involvement was so low-key that there was, according to a Company representative on the spot, 'a general impression that it was an entirely Shell project . . . The name of D'Arcy meant nothing to the Nigerian'.[39]

Less than a fortnight after the formation of the new Shell-D'Arcy company, the first exploration well was spudded in at Ihuo, ten miles northeast of Owerri (see map 4.2). This proved to be a dry hole. Several more prospects were drilled from 1951 to 1954, including one at Akata on the eastern edge of the Niger delta, where oil was discovered but in too small quantities to be commercial. The search moved westwards to the middle of the delta, where in August 1955 a well was spudded in at Oloibiri, in the heart of the mangrove swamps. The well struck oil in January 1956 and the Oloibiri field, though disappointingly small, was developed as Nigeria's first commercial oil field. A second field was soon found at Afam, and in February 1958 Nigeria's first export cargo of crude, produced at Oloibiri and Afam, was loaded at Port Harcourt.[40]

In the meantime, Shell-D'Arcy, which was renamed Shell-BP in 1956, started to relinquish acreage as it converted its exploration licences, allowing only geological and geophysical surveys, for five-year prospecting licences, giving rights to drill for and produce oil over a smaller area. Starting in 1960, the prospecting licences were in turn converted into thirty-year mining leases covering only half the area of the prospecting licences. As Shell-BP relinquished acreage, other companies took it up, with the result that the concessionary map of Nigeria came to resemble a mosaic of competing interests. During this process, new offshore licences were also granted to four companies, of which Shell-BP was one.[41] The other companies which acquired acreage soon began to drill for oil. The first was Socony Mobil, which sank its first exploration well in 1959. Then came Tennessee Nigeria

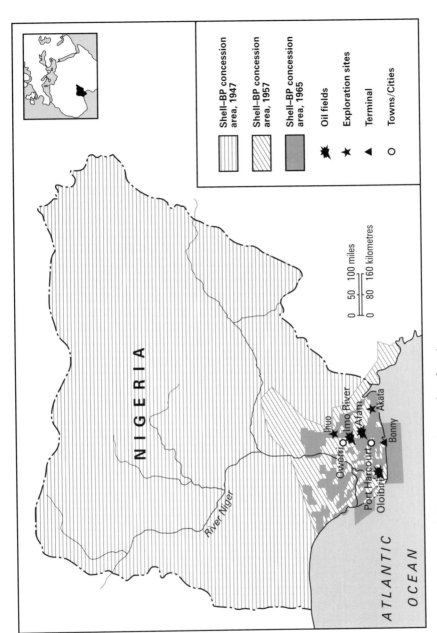

Map 4.2 Nigeria: Shell-BP's concessions and exploration

18   Drilling site at Oloibiri, Nigeria's first commercial oil field, discovered by
Shell-BP in 1956

in 1962; Amoseas and Gulf in 1963; and Safrap and AGIP in 1964. Gulf was
the first to strike oil and began production in 1965.[42]

Shell-BP, however, retained the dominance resulting from its pioneering
investigation of a geologically unexplored territory. By the end of 1965 it had
discovered some fifty oil fields in Nigeria. Five years later the number had
grown to seventy.[43] Despite the entry of competitors, Shell-BP still accounted
for three quarters of Nigeria's crude production in 1970. In that year, Nigeria
ranked eighth in the world league of oil exporters. It provided about 10 per
cent of BP's crude production and held about 4 per cent of BP's worldwide
estimated recoverable reserves.[44]

Most of Shell-BP's Nigerian fields were small- or medium-sized by Middle
East standards and even the largest Nigerian fields, such as Imo River, were
far smaller than the biggest fields in the Middle East.[45] Nigeria thus stood in
contrast to countries like Iran, Iraq, Kuwait and Saudi Arabia, which had
many fewer, but much larger fields. But while Nigerian crude was both less
plentiful and more expensive to produce, it had important advantages over
Middle East oil. It was mostly light low-sulphur oil for which there was strong
demand in increasingly pollution-conscious markets, it was much nearer to

consumers in the USA and Western Europe than the high-sulphur crudes of the Middle East and, most importantly, Nigeria seemed in the early 1960s to be politically stable and moderate. When it was granted independence in 1960 it initially adopted a multi-party democratic system and encouraged foreign private investment. In international relations, the Nigerian leadership was moderate, cautious and pragmatic, without the pan-African fire epitomised by Kwame Nkrumah, leader of the nearby West African state of Ghana.[46] These conditions in Nigeria would not last. But until the mid-1960s, the discovery and development of oil in Nigeria seemed to fit exactly with BP's aim of obtaining secure supplies of low-sulphur crude outside the politically turbulent Middle East, where the pan-Arab President Nasser of Egypt emerged triumphant from the Suez crisis of 1956–7, the pro-British monarchy in Iraq was violently overthrown in 1958, and the Organisation of Petroleum Exporting Countries (OPEC) was formed in 1960.

LIBYA

Libya was another country emerging from colonial rule in the postwar years. It had passed from Turkish to Italian rule in 1912 and remained an Italian colony until the expulsion of Italian and German troops by the wartime Allies in 1943. The two coastal provinces of Tripolitania in the north-west and Cyrenaica in the east were then placed under British military administration, and the Fezzan in the south-west under French. This lasted until 1951, when Libya was granted independence as a constitutional monarchy under King Mohammed Idris. He was the head of the Senussi tribe of Cyrenaica, who had for many years resisted Italian domination. While Musaddiq railed against the British in Iran, Idris saw them differently, as liberators who had helped to drive out the Italians. He allowed the British to keep a large military base in Cyrenaica and the Americans to maintain a huge air base at Wheelus, near Tripoli. After independence, foreign advisers, mostly British, continued to assist the Libyan government, whose political and economic ties were with the West. Far from seeking to repel Western oil companies, Idris' regime wished to attract them.

Libya was certainly in need of economic assistance. When it became independent it was one of the poorest nations in the world. Apart from a few settlements around the oases in the Fezzan, its population of some one million Arabic-speaking Muslims lived in a narrow strip along the Mediterranean coast. Only a tiny fraction of Libya's land was cultivable, the rest being a barren desert where rain never fell. Some 90 per cent of Libyans were illiterate and lived at subsistence level.[47]

After the war, the Company and Shell jointly enquired about oil conces-sions, but the military administration was unwilling to grant them before Libya became independent. In 1951 the two companies were, though, per-mitted to carry out geological and geophysical surveys in Cyrenaica. Although they did not rate the oil prospects highly, they lodged applications for concessions over large areas of Cyrenaica and Tripolitania.[48] A fresh assessment of Libyan oil prospects was made the following year by Alwyne Thomas who, after being withdrawn from Iran in 1951, became the Company's regional geologist for Africa and the Mediterranean.[49] He argued that the Libyan territory bordering the Mediterranean was geologically similar to the Arabian coast of the Persian Gulf, which caused him to be opti-mistic about the possibilities of finding oil, especially in northern Cyrenaica.[50] There was still, however, no way of exploring the possibilities more closely, as there was no Libyan legislation under which exploration licences could be granted.

As a first step, the 1953 Minerals Law allowed prospectors to carry out surveys, but not to drill for oil. Several companies took out permits and made surveys, finding large anticlines which were thought to be particularly promising in two areas, northern Cyrenaica and north-western Tripolitania.[51] Meanwhile, the Libyan government, in consultation with the oil companies, drafted a comprehensive petroleum law to establish the terms on which concessions would be awarded.

The Libyan Petroleum Law, enacted in April 1955, divided the country into a myriad of comparatively small blocks which were offered to appli-cants on generous terms. The Law put an upper limit on the acreage which could be held by a single company and gave concession-holders only five years from the grant of a concession before they had to start relinquishing territory, which would then be offered to other companies. These terms were meant to promote competition and to encourage concession-holders to get on with exploration, for fear that they might otherwise relinquish land cov-ering valuable undetected oil reserves. Concessions were to be awarded to qualifying applicants on a first-come, first-served basis, though the Libyan Petroleum Commission, which was set up to administer the Law, could also take account of 'furtherance of the public interest' in deciding between appli-cants for the same areas. This effectively opened the door to competitive bidding by the offer of 'sweeteners' such as increased royalty payments, finance for non-oil projects and other inducements.[52]

The 1955 Law certainly achieved its purpose of encouraging competition. The first concessions were awarded in November 1955 and by the end of 1960 a total of eighty-nine concessions had been granted to twenty-two companies.[53] The US oil companies were used to competitive land acquisition, as this was the normal system in the USA. BP, on the other hand, was a novice at it and soon fell behind in the competitive race. Before the passage of the 1955 Law, BP and Shell tried to establish with the Libyan government that the priority of their earlier application for concession areas would be respected. But they did not succeed, and after the Law was passed the two companies, working independently, had to make new applications.[54] BP's main interest was in the large anticlines in northern Cyrenaica, but to its 'great shock and disappointment' the concessions covering the most desired areas were granted in late 1955 to one of the smallest and least experienced applicants, the Libyan American Oil Company. Feeling that it had been outmanoeuvred, BP submitted amended applications for acreage in Cyrenaica and Tripolitania. It was rewarded in January 1956 with four concessions (numbers 34 to 37), of which three were in northern Cyrenaica and one in Tripolitania (see map 4.3). BP commenced mine clearance operations and exploration surveys in these areas.[55]

Soon, however, the focus of exploration in Libya shifted to new areas. Libyan American's drilling in northern Cyrenaica failed to find oil and at the end of 1955 a substantial oil discovery was made in Algeria at Edjele, just across Libya's western border. The oil companies quickly turned from Cyrenaica to the Fezzan, where a competitive scramble for acreage soon developed. BP applied for acreage in concessions 63 (Tripolitania), 64 (the Fezzan) and 65 (southern Cyrenaica), but at the insistence of the Libyan government the application was scaled down so that it excluded concession 65.[56]

Drilling in the Fezzan produced disappointing results and the focus of exploration soon shifted once more. The new centre of attention was the Sirte basin, where a consortium of US independents, named Oasis, made two discoveries in the second half of 1958.[57] The real breakthrough, though, came in mid-1959, when Standard Oil (NJ) discovered the Zelten field, a giant of truly Middle East proportions. 'Libya', wrote BP's local exploration manager, 'is now a first-class prospect'.[58]

Oil companies rushed to grab stakes in the Libyan exploration boom, but BP was not one of the front-runners. It had earlier applied for acreage in concessions 80 and 81 to the south-east of the Sirte basin, but the concessions had not yet been granted and, after the Zelten strike, they became much more sought after. BP's chances of acquiring them were reduced by its poor record of exploration activity in Libya. It had drilled just three wells, in concessions 34, 63 and 64 respectively, with a combined footage that was near the bottom of the Libyan league (figure 4.2). The Petroleum Commission

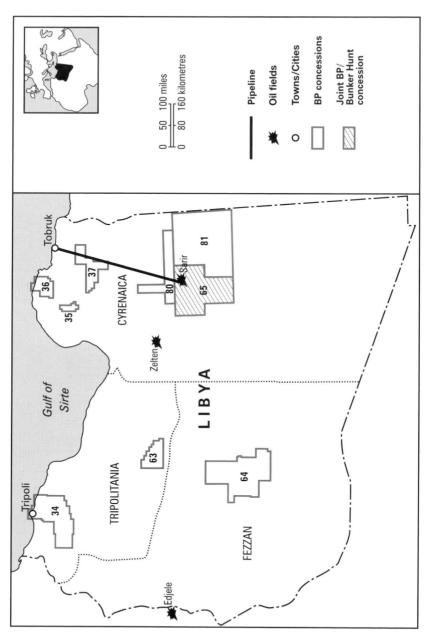

Map 4.3   Libya: BP's concessions and the Sarir oil field

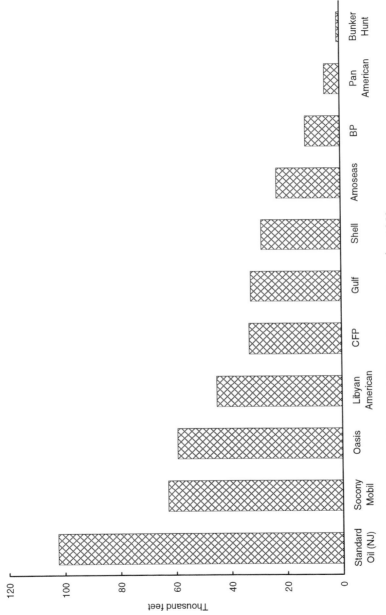

Figure 4.2 BP's and competitors' deep test drilling in Libya to September 1959

would not even consider BP's applications for concessions 80 and 81 unless BP committed itself to more active exploration. To recover from its backward position, BP had to raise its bid for concessions 80 and 81 by offering to forgo the depletion allowance (which reduced the tax on oil production) in all its Libyan concessions and to undertake a substantially increased programme of exploration. By offering these enhanced terms, BP secured concessions 80 and 81, which were awarded to it in September 1959.[59]

BP also raised its stake in Libyan exploration by a 'farm-in', industry jargon for an agreement to take the risks and costs of exploring for oil on another company's acreage in return for a share in the oil reserves, if any were discovered. In Libya, Standard Oil (NJ) had already come to a farm-in agreement with the US independent, Libyan American, which ran short of resources to continue exploring after drilling dry holes on the anticlines of northern Cyrenaica.[60] Another US independent with acreage in Libya was Nelson Bunker Hunt. He was one of the sons of Haroldson Lafayette Hunt, an eccentric and enormously wealthy US oil independent, 'a patron of right wing causes, a promoter of health foods, and an inveterate enemy of white flour and white sugar'.[61] Bunker Hunt, too, had interests outside the oil business and was an avid collector, principally of ancient coins. He also held two concessions in Libya, one of which was concession 65, for which BP had earlier unsuccessfully applied.

Hunt had undertaken very little exploration work on his concessions and his drilling footage was lower even than BP's (figure 4.2). He was a comparatively small operator and, even if he struck oil, he might not have the necessary capital to develop his find. By the end of August 1959 he had, according to BP's representative, an 'obvious desire' to consider a joint venture.[62]

In 1960 BP and Hunt reached a complicated agreement, about which there would later be much legal wrangling. Leaving the complications aside, it was agreed that BP would explore for oil in concession 65, initially paying all the costs, in return for a half-share in the concession.[63] BP started drilling and in November 1961 well C1–65, the fourth to be drilled, struck the giant Sarir oil field.[64]

Several other oil companies had already discovered oil in the Sirte basin and were developing fields for production. Standard Oil (NJ) was the first to commence oil exports in 1961, and others were exporting by 1967.[65] BP followed suit by developing the Sarir field and laying a pipeline 320 miles across the desert to an export terminal on the southern side of Tobruk harbour. The first cargo of Sarir crude left the terminal in January 1967.[66] Production rose rapidly and in 1970, when Libya was the world's fourth largest oil exporter, the Sarir field accounted for 12 per cent of Libyan oil production.[67]

After its farm-in to concession 65, BP did not acquire any more concessions in Libya and Sarir remained its only producing oil field there until the nationalisation of BP's share in concession 65 in December 1971. The concentration on a single giant field was in marked contrast to the position in Nigeria where BP's crude offtake was drawn from a multiplicity of smaller fields. That, however, was not the only contrast between exploration in Nigeria and Libya. In Nigeria, Shell-BP's pioneering exploration gave it a dominant position which endured throughout the 1960s. In Libya, BP and Shell tried to obtain the priority of a pioneer before the introduction of the 1955 Petroleum Law, but failed. Instead, they had to join in a competitive scramble in which BP trailed behind other companies until the Sarir discovery. Even after Sarir came on stream, BP-Hunt's 12 per cent share of Libyan production was far below Shell-BP's three-quarters share in Nigeria. The advantages of establishing an early lead over the competition were clear.

Although Sarir was a giant field, Libya's contribution to BP's worldwide reserves portfolio and production was quite small. In 1970, the field accounted for about 5 per cent of BP's total crude production and 1–2 per cent of its worldwide estimated recoverable reserves.[68] Its value to BP should not, however, be judged by volumes alone, for in the 1960s Libyan crude commanded a premium rating. Like Nigerian, it was low in sulphur, situated close to the oil-hungry markets of Western Europe, and in a country which posed few political worries while King Idris remained in power. The discovery of oil in Libya, as in Nigeria, therefore seemed to fit perfectly with BP's aim of finding secure supplies of low-sulphur crude outside the Middle East.

SOUTH AMERICA AND THE CARIBBEAN

In the prolific oil region of South America and the Caribbean, the Company had missed out on the pioneering phase of exploration in the first half of the twentieth century. The biggest oil-producing country in the region was Venezuela, where Shell had been the first large oil company to explore. After striking oil in the Maracaibo Basin in 1922, Shell established a leading position, which it later shared with Gulf and Standard Oil (NJ).[69] The Company was attracted by the oil prospects, but, being majority-owned by the British government, it held back because of uncertainties about its position under Venezuelan law, which forbade the granting of concessions to companies that were dependent on foreign governments.[70]

The neighbouring state of Colombia was a much smaller oil producer than Venezuela, but here, too, the Company failed to obtain first-phase exploration acreage. After Colombia was put on the world oil map by the discovery

19   King Idris of Libya at the official opening of the Tobruk oil terminal in February 1967

of the large La Cira-Infantas field in the Middle Magdalena Valley in the Andes in 1918, the ubiquitous Standard Oil (NJ) soon acquired control of the field and commenced production.[71] The Company carried out geological reconnaissance and rated Colombian oil prospects highly. In 1937 Cox was posted to Bogota to set up an office and investigate the possibilities more closely. But in 1938 he was recalled and the Bogota office was closed because Riza Shah objected to the Company engaging in exploration which might divert its attention from Iran.[72]

By 1950, the Company's only extant investment in South America and the Caribbean was in the British colony of Trinidad, where in 1938 the Company had purchased a one-third share in Trinidad Northern Areas Ltd with an eye primarily to the prospects of finding oil in marine concessions.[73] Even here, the Company was a late entrant. Commercial production had started in Trinidad in the 1900s and by 1950 more than 150 companies had been formed to explore for and develop Trinidad's oil, of which ten still survived.[74]

Starting from this backward position in 1950, BP made little progress in

South America and the Caribbean in the next quarter-century. After his stint in Papua, Frank Rickwood was appointed to investigate a number of oil prospects in Latin America. But active exploration was precluded by legislation in Venezuela and other Latin American countries, which discriminated against companies controlled by foreign governments.[75] In Trinidad, there was some success in 1955, when the large Soldado field was discovered in Trinidad Northern Areas' offshore concession in the Gulf of Paria.[76] Encouraged by this success, BP made a string of small acquisitions in Trinidad, including a majority interest in Trinidad Petroleum Development Company in 1957, the whole of Apex (Trinidad) Oilfields in 1960 and Kern Trinidad Oilfields, which came as part of the package of Kern oil interests in America, bought from Rio Tinto in 1961.[77]

The Soldado success was not, however, repeated. Trinidad's geology was fiendishly complicated, production was only small scale, and there were growing signs of economic and political trouble in the island after it was granted independence in 1962. The charismatic nationalist, Dr Eric Williams, dominated the government and tried to establish a regime which was moderate, democratic, capitalist and friendly towards foreign investors. But high unemployment fuelled labour discontent, which erupted in increasingly frequent disputes, not least between the oil companies and the powerful Oil Workers Trade Union under the radical leadership of George Weekes. Moreover, by the late 1960s the militant black power movement was gaining ground, accusing whites and other foreign minorities of perpetuating a neo-colonial system of exploitation.[78] Feeling that the prospects for profitable oil operations were poor, BP decided to withdraw and sold its interests to the Trinidad government in 1969.[79]

It was a timely withdrawal. In the next few years Trinidad was brought to the brink of economic and political collapse. Faced with a near-revolution from radical opposition, Williams resorted to authoritarian measures to assert his control. The economy, meanwhile, reached a nadir of virtual bankruptcy from which it was rescued by the Arab oil embargo and the quadrupling of oil prices by the OPEC countries in October–December 1973 (see chapter nineteen). The higher oil prices brought an economic bonanza to Trinidad, and in 1976 there was a return to multi-party politics. Although Trinidadian capitalism had survived, the state had greatly increased its penetration of the economy. Oil was a prime target for national ownership, as Trinidad joined in the great surge of oil nationalisations in the less developed oil-producing countries at that time.[80]

Meanwhile, BP had revived its interest in Colombia. Here, the Middle

Magdalena Valley in the Andean north-west of the country had been extensively developed since the discovery of the La Cira-Infantas field. Other oil fields had been found, roads and pipelines constructed and the area so heavily prospected that, although new discoveries would probably be made, there was, according to BP geologists who visited in 1955, little chance of a newcomer finding large oil reserves. In any case, they wrote, 'every acre of potential ground is leased'. To the south-east, however, lay the vast expanse of the Llanos plains, stretching eastwards from the Andes into the sparsely populated interior of the continent. This remote wilderness, accessible from settled Colombia only over the precipitous Andes mountains, had barely been touched by oil exploration (see map 4.4).[81]

Early in 1958 BP sent a representative to Bogota for talks with Ecopetrol, the national oil company, on the possibilities of co-operating on various projects, but before anything came of that, BP entered into a different alliance. This was with the Sinclair Oil Corporation, the seventh-largest US oil company, which possessed a large downstream network in the USA, but was short of crude oil. BP, on the other hand, had a surplus of Middle East oil, but no downstream outlets in the USA. It therefore seemed a perfect match for BP to supply Sinclair with Middle East crude. This arrangement, announced in October 1958, was soon frustrated by the USA's adoption of mandatory oil import quotas which restricted imports of Middle East oil to the USA. There was, however, another side to the BP/Sinclair alliance. This was that the two companies would set up a joint exploration company to explore for oil, primarily in Latin America. A new joint company, Sinclair and BP Explorations Inc., was duly formed in November 1958.[82]

An opportunity for the new BP/Sinclair partnership to explore in Colombia soon came when Intercol, Standard (NJ)'s Colombian subsidiary, offered to farm out (the opposite of farm in) some of its Colombian acreage to other companies. Warman and another geologist visited Colombia in January 1959 to investigate the possibilities and on their recommendation BP and Sinclair farmed into three Intercol blocks. After drilling a dry hole on one of them, the partners took a 50 per cent share (25 per cent each for BP and Sinclair) in another Intercol block in the Provincia concession in the Middle Magdalena Valley. BP/Sinclair's first well there discovered the medium-sized Provincia field in August 1960. Further success followed in 1963, when BP/Sinclair discovered the smaller Bonanza field, also in the Provincia concession.[83]

Outside the much explored Middle Magdalena, a BP/Sinclair geologist saw frontier exploration potential in the large, thrusted anticlines in the Llanos foothills of the eastern Andes, and picked out the Santiago de las

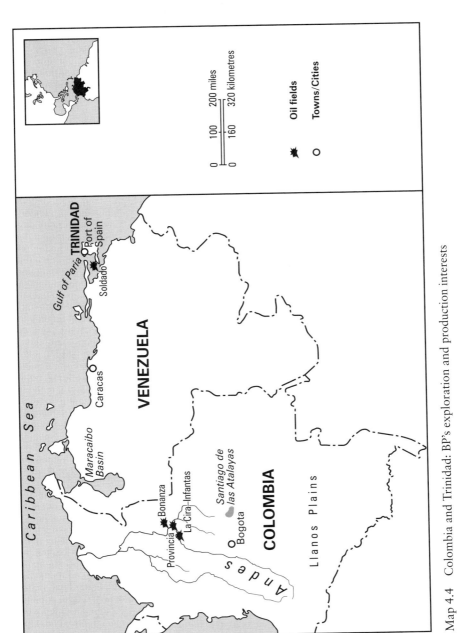

Map 4.4   Colombia and Trinidad: BP's exploration and production interests

20   Drilling rig at Provincia, Colombia, in 1962

Atalayas as an attractive prospect where no exploration had, so far as was known, been carried out. The westerly part of the Santiago de las Atalayas was in the foothills and BP favoured making geological and economic studies of oil prospects in the area. Sinclair, however, did not want to commit themselves to exploration in the Llanos and refused to support even a preliminary geological survey. The idea of exploring in the Santiago de las Atalayas was not, therefore, pursued at that time.[84]

In the meantime, the Provincia field came on stream in 1962 and for the next decade provided a tiny fraction, less than a third of 1 per cent, of BP's worldwide crude production. In 1968 Sinclair was taken over by Atlantic Richfield (ARCO), who in 1970 sold the ex-Sinclair 25 per cent share in the Provincia and Bonanza fields to Intercol. Previously, BP/Sinclair had managed operations in the Provincia block, but on purchasing the ex-Sinclair share, Intercol became the operator. BP had only a minority voting position and was unable to enforce its wishes in a series of disagreements with Intercol. Unhappy with this state of affairs, BP sold its 25 per cent share to Intercol in June 1973.[85] Having also pulled out of Trinidad in 1969, BP was left with no oil-producing interests in South America and the Caribbean.

After missing out on the earlier pioneering phase, BP had thus failed to establish a lasting presence in this major oil region. It had contemplated frontier exploration in the eastern foothills of the Andes, but found, not for the last time, that its US partner, Sinclair, lacked appetite for long-term frontier exploration prospects. A similar pattern was evident in Canada, where BP entered into an alliance with a smaller partner whose American shareholders preferred more certain, short-term returns to the more risky, but potentially more rewarding, longer-term frontier exploration prospects which BP favoured.

CANADA

In North as in South America, BP was a late starter in exploration. In the first fifty years of its existence the Company, alone of the oil majors, did not drill a single well in the USA, the world's largest oil-producing and oil-consuming nation. In Canada, the Company was only slightly more active. It drilled a few holes in the eastern maritime province of New Brunswick in 1919–20, but they turned out to be dry. After that, for more than twenty-five years the Company paid only sporadic attention to Canadian oil prospects, which were not inspiring.[86] Exploration by other companies resulted in only one significant oil discovery at Turner Valley, an anticline in the eastern foothills of the Rocky Mountains, near Calgary in the western province of Alberta, where oil was found in 1936 (gas had already been found there in 1914). The 1936 oil strike encouraged exploration activity in the Rockies foothills, but a decade of fruitless searching seemed to confirm that Alberta was just a parochial oil play. By 1947, Imperial Oil, Standard (NJ)'s Canadian subsidiary, had drilled 133 holes in western Canada without making a single commercial find.

However, as oil industry history had repeatedly shown, it needed only one large discovery to transform exploration backwaters into world-class prospects. In February 1947, Imperial, acting on geological interpretations which turned out to be erroneous, drilled a well at the village of Leduc, near

Edmonton in the Albertan plains to the east of the Rockies foothills (see map 4.5). The drill struck a prolific reservoir not in an anticline, but in a different type of geological formation, a stratigraphic reef.

The Leduc discovery precipitated a frenetic Albertan oil boom and more reef reservoirs were soon discovered in the Edmonton plains area. By 1951, while Musaddiq was nationalising the Company's Iranian oil fields, Alberta was emerging as a major oil province. 'It is a pity', wrote the *Daily Mail's* travelling correspondent, 'the Persians are not with me in Alberta. Perhaps if they were they would realise that one day they may have to peddle their oil for pennies, because this is where most of the Commonwealth's and Western World's oil is going to come from.'[87] The Company knew that this was a gross exaggeration of Canadian oil prospects,[88] but it nevertheless rated Canada highly and was keen to gain a stake in Canadian exploration.

Although the Leduc field and the other new discoveries were reef fields in the flat Albertan plains, the Company, influenced by its early years in Iran, was drawn more strongly to the anticlines in the Rockies foothills. It was convinced that these structures, 'comparable in size to the familiar whale-backs of Iran',[89] were outstanding oil prospects, at least as good as any others outside the Middle East. This opinion was shared by Cox, Bridgeman and main board directors including John Pattinson, the director responsible for exploration. 'It appears', remarked Pattinson to the chairman, Fraser, in 1953, 'most probable that some really large structures not entirely unlike some of the big limestone anticlines of south Persia and Iraq will be found in the deep foothill zone of the Rocky Mountains. (Cf. the Persian foothills in front of the Zagros range.)'[90]

While geological structures in western Canada seemed similar to those in Iran, the commercial conditions were entirely different. The Canadian oil industry had developed on the pattern of the USA. Oil leases covered small blocks, not vast concession areas, and there was strong competition from a myriad of oil companies ranging from the majors to a multiplicity of smaller independents.[91] Realising that it lacked both commercial experience and oil leases in Canada, the Company decided that it should enter Canada by acquiring a stake in an experienced local company. Accordingly, in 1953 it reached agreement to purchase an interest, initially of 23 per cent, in the Triad Oil Company, a small Canadian independent. Triad possessed the required attributes of good commercial management, well-chosen acreage and some existing oil production, providing dollar revenues which would help to finance exploration. Triad lacked technical skills in exploration, which the Company provided by seconding technical staff, as usual with previous experience in Iran, to the Canadian firm.[92]

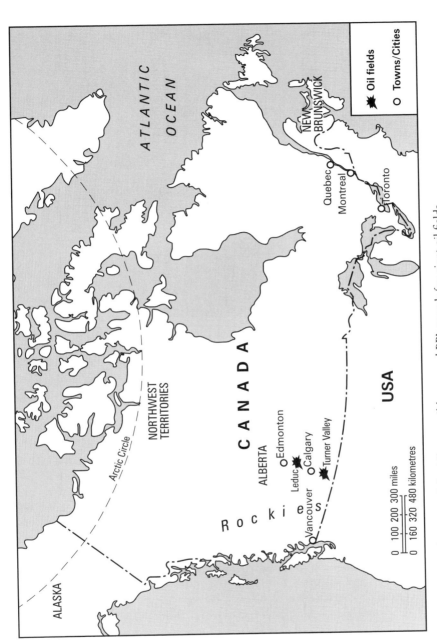

Map 4.5   Canada, scene of the Albertan oil boom and BP's search for giant oil fields

21   BP geologist examining exposed rock structure in Northern Alberta, Canada, in the 1950s

Alastair Down was posted to Calgary in January 1954 as the Company's representative in Canada and one of its two directors on Triad's board. He quickly found that it was not only the commercial conditions that differed from Iran. 'I must say', he wrote to Bridgeman, 'I find 40 to 50 degrees of frost fairly chilly'.[93] Property prices were, he found, 'astronomical'. Although he managed to purchase a house in a 'first class location', he added 'I would just like to say that it must not be assumed in London that we shall have accommodation on the Persian scale for all and sundry. You are aware of the servant problem here and my wife will be cook and housekeeper. We will be more than delighted to have guests, but I would like our position in this respect to be appreciated so that those guests that do come will realise the limitations that are imposed upon us.'[94]

After the Company and Triad formed their alliance, they concentrated most of their exploration effort on the foothills of the Rockies, as recommended by the Company. The results, however, were disappointing and after a few years without a significant discovery, differences began to emerge between the partners. Triad was a small company, ranking only about twentieth in Canadian oil production, and its Canadian shareholders expected quick returns. They were soon discouraged by the lack of success in frontier exploration and argued that Triad should give it up to concentrate on semi-proved acreage in easily accessible areas. Cox, on the other hand, argued in favour of a long-term approach to frontier exploration which, he insisted, would ultimately bring higher returns than the short-term approach.[95] He opposed the Canadian shareholders in Triad who, he wrote at the end of 1960, 'would like to bail out now'. 'Oil exploration', wrote Cox, 'is inevitably a gamble in which hazards are comparable with the racecourse . . . If your horse is well behind after the first lap, you shouldn't ask the bookie for your stake back.'[96]

To accommodate the divergent views of BP and the other shareholders, it was agreed that Triad would concentrate on short-term exploration prospects, while BP would take a direct stake in longer-term frontier exploration plays, most notably in the northern territories stretching up to the Arctic.[97] Meanwhile, in the 1960s BP raised its shareholding in Triad by stages to more than 62 per cent, partly in return for cash and partly for assets in Canada which BP acquired from Rio Tinto and Tidewater and transferred to Triad.[98]

In 1970 Triad, still without a major oil discovery to its name, was re-financed by a rights offer and its name was changed to BP Oil and Gas, in which BP held a 66 per cent stake.[99] By that time, BP had given up its earlier hopes of making major oil discoveries in the Rockies foothills and the focus of its north American exploration had moved to the new frontier of the Arctic, where BP's commitment to frontier exploration would produce a spectacular pay-off in the state of Alaska.

5

Finance and the British government

When BP emerged from the Iranian nationalisation crisis in an enviably strong financial position, there was still some unfinished business regarding dividends. In particular, the final dividend for 1954, due to be announced in May 1955, had still to be settled. As in previous years, the amount of the dividend was a matter in which the British government took a close interest. Now that the Iranian crisis had been settled it could no longer be argued that the level of the dividend had to be kept down for fear of upsetting the negotiations with Iran. However, dividend restraint was still seen as a necessary element in the government's domestic economic policy as it pursued that most desirable yet elusive of goals: a combination of full employment and low inflation. To that end, the government was trying to hold back wage rises and was well aware that nothing was more guaranteed to provoke large wage demands than a perception that the returns to capital were rising faster than those to labour. Moderation in dividends and wages was, a Treasury official noted, 'a doctrine which the Chancellor is constantly preaching in public to both sides of industry. We certainly could not refrain from making our views known to the Board of BP on any particular occasion out of an exaggerated desire not to interfere with the internal management of the Company.'[1]

Lord Strathalmond, however, had announced at the Company's extraordinary general meeting in December 1954 that the shareholders could expect more generous dividends in future. True to that statement, he informed the Treasury in May 1955 that he proposed to pay a final dividend for 1954 which, with the interim dividend already paid earlier, would result in a total dividend for the year of 20 per cent of the issued ordinary capital. As that capital had been increased fivefold by the bonus issue of 1954, the proposed dividend for the year was equivalent to 100 per cent on the old capital. This was more than twice the amount paid for 1953, when the dividend was equivalent to 42½ per cent of the issued capital.[2]

At the Treasury, Bridges acknowledged that Strathalmond's proposal was 'not as outrageous as it sounds, judged on commercial grounds'. BP's shares, he noted, had 'far the lowest yield of any of the active and important companies' and the balance sheet was 'extremely healthy'. 'I conclude', wrote Bridges, 'that, looked at simply on commercial grounds, there is nothing unreasonable in 15% or even 20%'. To which he added, 'I also conclude that from the point of view of Ministers, such a figure is quite out of the question.' Bridges had taken various soundings which supported this view. At the Bank of England, Lord Cobbold thought that a dividend rate of 12 or 13 per cent was nearer the mark. The Foreign Office 'doubt if they can object strongly to 15%, but would regard 20% as impossible'. Sir John Maud, top official at the Ministry of Fuel and Power regarded 12½ per cent as 'about right'. So did Bridges, who thought that Strathalmond had probably proposed 20 per cent in the expectation that he would have to compromise. 'But', Bridges presciently added, 'I think he will fight hard for 15%'. Bridges put these views to R. A. Butler, the Chancellor of the Exchequer, who said that 'he wd prefer 10 or 12 or 12½. But that 13% was his highest bearable figure.'[3]

Bridges passed that on to Strathalmond. 'I was quite clear', Bridges noted, 'that if he (Lord S) really pressed the Board to accept 13%, they would do so. And I urged him to take this line.' Strathalmond, however, refused to do so. He felt that the declaration of such a low figure would 'expose the Board to serious criticism'. He said that he would tell the board that 13 per cent was Butler's highest figure, but that he 'would not himself advocate this, or conceal his own views if pressed'.[4] Butler was furious that Strathalmond refused to be more co-operative. On seeing Bridges' record of the meeting with Strathalmond, he wrote on it: 'I am disgusted with attached.' And in a handwritten note he expressed his impatience with both Strathalmond and the government directors on BP's board, who he obviously thought were not doing enough to assert the government's views. In Butler's words:

> Lord S. should be informed that I shall take an opportunity of publicly repudiating a figure anywhere near as high as 20%.
> I wish to raise at Cabinet now in any case the position of the Gov^t Directors. It is impossible to go on with these <u>stooges</u> & I must review our association with this unpatriotic organisation . . .[5]

In this angry mood, Butler left instructions for an official to tell Strathalmond that if he exceeded 13 per cent and came in for public criticism he could 'count on no support' from Butler.[6]

This message was duly communicated to Strathalmond so that he would know what sort of a reception he would get when he visited the Treasury for a meeting with Butler. They met in early May, and after a long discussion agreed on a final dividend for 1954 which, when added to the interim dividend already paid, would bring the total for the year to 15 per cent. Significantly, the 'undesirability of disclosure of Government pressure' came up at the meeting.[7] After deducting income tax withheld by BP the net payment to the shareholders for 1954 would be 8·6p. per share. This was equivalent to 43p. per share before the 1954 bonus issue, and on this comparable basis was nearly double the net dividend of 23p. per share which had been paid for 1953.

This outcome of the negotiations between Strathalmond and Butler over the 1954 dividend was, of course, a compromise. But it was an uneven compromise in favour of Strathalmond, who had defied Butler face to face and come away with the sort of dividend that he in all likelihood expected in the first place. It is impossible to imagine that he ever hoped to get his starting figure of 20 per cent, which, as Bridges had suspected, can have been no more than an opening gambit. Butler, on the other hand, had committed himself to a maximum of 13 per cent, only to be forced higher. As in the Iranian crisis negotiations,[8] Strathalmond had shown how tough a negotiator he could be and how little he was intimidated by the most powerful politicians of the day. He had also again shown, not that further demonstration was needed, that he defined his duty entirely in terms of BP's commercial interests which for him took precedence over the matters of national policy with which governments were concerned.

What was it, though, that enabled Strathalmond in this instance to prevail over the outwardly more powerful figure of the Chancellor of the Exchequer? The answer, to a large degree at any rate, lay in their differing concerns about public disclosure. When Butler warned that Strathalmond could not count on support from the Treasury in the event of public criticism that the dividend was too high, he was missing the point. Strathalmond was more concerned that he would be criticised by BP's other shareholders (apart from the government) if he announced a dividend that was too low on commercial criteria. If he took such a step because of behind-the-scenes pressure from the government, he would be taking responsibility for a decision that was not really his, but the government's. That would have been anathema to Strathalmond and it may be assumed that when he and Butler met for their showdown on the dividend, the reference to the undesirability of the disclosure of government pressure was a veiled way of saying that Strathalmond would not agree to announce a lower dividend without it being disclosed that this was at the insistence of the government. That would have been awkward for the government, which publicly claimed that it did not interfere in the normal commercial affairs of BP and which had publicly rejected a policy of compulsory dividend limitation. In short, public disclosure would have been more embarrassing for the government than for Strathalmond.

This episode had a larger significance than might be apparent at first sight, for it was not the first – and nor was it to be the last – occasion which raised a fundamental question: could BP, as a commercial organisation, be expected to succumb to political pressure where this would put BP in the position of taking responsibility for unannounced government policy, against the interests of BP's shareholders *as a whole*? This issue had arisen in 1947 when Clement Attlee's Labour government tried to press the Company into reducing oil supplies to other countries in order to maintain supplies to Britain in a time of shortage.[9] And, as will be seen in chapter nineteen, it was to arise again in 1973. On each occasion BP took the view that if the government wished BP to act against its commercial interests, it was for the government to issue the necessary directive and take public responsibility for public policy.

Butler's efforts to limit BP's dividend in May 1955 were somewhat at odds with the incautious and economically injudicious national budget which he had introduced just a few weeks earlier. In it, he presented a rosy picture of the British economy and cut the standard rate of income tax from 45 to 42½ per cent.[10] With the balance of payments coming under pressure as consumption and imports exceeded the country's capacity to produce and export, the tax cut could hardly have been worse timed economically. It only fuelled the expansionary pressures on the economy, which was already showing signs of overheating after the strong recovery of the previous two years. The reduction in taxation was not, however, inspired by economic motives. Eden, having succeeded Winston Churchill as Prime Minister in April, had called a general election for 26 May and Butler was quite obviously trying to buy votes. He was justly criticised for doing so by the political opposition at the time, and by economic historians later.[11] But opposition notwithstanding, the Conservatives won the election with a comfortable overall majority of fifty-eight Members of Parliament.

It was not long, however, before Butler, who continued as Chancellor under Eden, had to resort to the emergency measure of a supplementary autumn budget to rein in the economy. On 26 October he announced that he was concerned about rising inflation and proposed to cut public spending, to increase Purchase Tax and to raise the Profits Tax that was levied on company dividends. At the same time, he appealed for dividend restraint.[12] It was therefore a cause of some embarrassment that the very next day BP, after consultations with the Treasury, announced an interim dividend for 1955 equal, tax-free, to 5 per cent of the issued capital. As a percentage, this was the same rate as the interim dividend which had been announced in autumn 1954. However, since then the issued capital had been increased five-fold by the bonus issue. Moreover, the interim rate of 5 per cent announced in 1954 had been a gross figure, from which standard rate income tax would be deducted at source by the Company. The 1955 rate of 5 per cent was net, with standard rate income tax already deducted. After taking account of these factors the interim dividend announced in October 1955 represented a payment to shareholders that was nearly eight times as much as the interim dividend of 1954. The enormity of the increase had been missed by the Treasury, which had noticed the change in tax treatment, but had overlooked the effective fivefold increase which resulted from the bonus issue. At the Treasury, Bridges was penitent for his oversight.[13]

But by then it was too late to forestall public criticism. The announcement that BP was increasing its interim dividend immediately after Butler's autumn budget urging dividend restraint gave rise to a series of questions by opposition MPs in the House of Commons.[14] In reply to one such question a government spokesman told the Commons that 'it has not been our practice to interfere with the commercial management of the Company'.[15] If Strathalmond read *Hansard*, in which the Parliamentary debates were published, he might with justification have regarded this statement as being at variance with his experience.

After Strathalmond retired at the end of March 1956, it was left to his successor, Basil Jackson, to consult the Treasury over the final dividend for 1955, which was to be announced at BP's annual general meeting in May 1956. Jackson and the board proposed to make a final dividend which, when added to the interim payment already made, would bring the total dividend for 1955 to 15 per cent of the issued capital. This was the same rate as had been paid for 1954, with the crucial difference that the 1955 dividend was tax-paid (at the standard rate of income tax), whereas the 1954 dividend was before tax. With the standard rate of income tax at 42½ per cent, the shareholders stood to get a much higher dividend for 1955 than they had for the previous year.

The Treasury had again to admit that on commercial grounds there was nothing excessive about the proposal to pay 15 per cent for 1955. Such a payment was well covered by BP's earnings and did not represent an unreasonable yield on the market price of the shares. As Bridges commented on

Table 5.1 *BP's dividends to ordinary shareholders, 1950–1960*

	Dividends, net of tax, per share (pence)	Bonus share issues	Dividends, net of tax, per share adjusted for bonus issues (pence)	Index of previous column adjusted for retail price inflation (1951 = 100)
1950	16	–	16	111
1951	16	–	16	100
1952	19	–	19	112
1953	23	–	23	133
1954	9	4 for 1	43	239
1955	15	–	75	399
1956	15	–	75	388
1957	15	–	75	374
1958	9	1 for 1	87	424
1959	11	–	108	522
1960	12	–	117	559

the proposed payment: 'On commercial and City grounds there is nothing wrong in this'. However, he thought that politically the proposed 1955 dividend would 'cause a howl'.[16] Bridges consulted Macmillan, who had succeeded Butler as Chancellor of the Exchequer. Macmillan said he would prefer it if the 1955 dividend was 12½ per cent, rather than the 15 per cent proposed by BP. Bridges passed this on to Jackson, telling him that 'although the City might think 12½% was too little, the Opposition would make a devil of a fuss about 15% free of tax: and this fuss would react on both of us'. Bridges' record of his discussion with Jackson on this matter showed none of the obvious acrimony of the earlier exchanges with Strathalmond. Yet Jackson was no more inclined than Strathalmond had been to yield to Treasury pressure. After seeing him, Bridges noted, 'on the dividend I got very little change'. All that Jackson agreed to do was to put Macmillan's view on the dividend to BP's board without arguing against it. He refused to argue in favour of it.[17] The board was as unmoved as Jackson and agreed on a final dividend which brought the total for 1955 to 15 per cent. This was equivalent to 15p. per share on the issued capital in 1955, or 75p. per share on the issued capital before the 1954 bonus issue, as shown in table 5.1. As the table indicates, the 1955 dividend brought to an end a five-year period of astonishing dividend growth. After adjusting for the bonus issue of December 1954, the dividend had risen from 16p. per share in 1951 to 75p. in 1955.

With price inflation also taken into account, the dividends rose fourfold between these years.

That rate of increase was not, however, sustained during the next five years. In consulting with the Treasury on the dividends for 1956 and 1957, Gass, who had succeeded Jackson as chairman, was given the usual Treasury view that 'it is very desirable that BP should not increase their dividend' because of the 'controversial effects which increases of dividends are likely to have in wage discussions'. Gass, concerned about the financial impact of the Suez Canal crisis on BP, readily concurred that there should be no increase.[18] The next year, the Treasury agreed to a dividend for 1958 which, after allowing for BP's one-for-one bonus share issue of that year, gave shareholders their first rise for three years.[19] There were further increases in the dividends for 1959 and 1960, but not on the lavish scale of five years earlier (see table 5.1). Thus by the end of the decade the spectacular increases in BP's dividends had abated, as had the storm they had caused in BP-government relations.

The main reason for the moderation in dividend growth in the second half of the 1950s was not political pressure, but the deterioration in BP's profitability. In 1955, the first full year of trading after the settlement of the Iranian nationalisation crisis, BP's post-tax profits were about £50 million, approximately double the level of 1954. That great leap was, however, very largely due to the fact that 1955 was the first year in which BP was able to claim significant relief from UK taxation in respect of overseas taxation paid under the 50:50 profit-sharing agreements with the governments of producing countries. Overseas tax relief in 1955 was nearly £60 million (see table 5.2), compared with less than £1 million in 1954 (see table 1.2 in chapter one).

In 1956 post-tax profits rose again, but the increase was only slight after making allowance for price inflation (see table 5.2). The profits that year would no doubt have been greater had it not been for the Suez Canal crisis, which disrupted oil supplies in November and December.[20] The crisis ran on into the spring of 1957, causing sales to fall and costs to rise.[21] Some financial relief was achieved through market price increases, but these were not enough to prevent a fall in BP's profits in 1957. The tight supply position in the Suez crisis also prompted the international oil companies to raise the published prices, known in the oil industry as posted prices, of their Middle East crude oils. In March 1957 BP, in line with other oil majors, raised the posted price of its Iraqi crude (36°) at the pipeline terminals on the Eastern Mediterranean seaboard from $2·46 to $2·69 per barrel. That was followed in May by increases of 13 cents per barrel in the posted prices of BP's crude oils in the Persian Gulf (equivalent to price rises of 6 to 8 per cent for different crudes).[22]

Table 5.2 *BP's profits, 1955–1960 (£millions)*

	Pre-tax profits	Deduction for overseas taxation	Profits after overseas taxation	Deduction for UK taxation before overseas tax relief	Overseas tax relief	Post-tax profits	Post-tax profits adjusted for price inflation (1950 prices)
1955	122	63	59	69	60	50	39
1956	136	64	72	79	63	56	41
1957	126	64	62	70	63	55	38
1958	138	81	57	74	80	63	43
1959	131	78	53	67	77	63	43
1960	145	92	53	77	86	62	42

After the Suez Canal crisis was over, BP's profits did not recover strongly. The principal reason was growing competition from new entrants to the international oil industry, who undermined the previously dominant oligopoly of the 'Seven Sisters', which included BP. The newcomers were generally either independent US oil companies seeking access to low-cost oil outside the USA, or state-backed companies from other oil-consuming nations which wished to break the dominance of the majors in their national markets.

The established majors and the newcomers succeeded in finding great quantities of oil both in traditional oil-producing countries and in new areas such as the Neutral Zone between Kuwait and Saudi Arabia (1953), Algeria (1955), Nigeria (1956) and Libya (1959). The situation soon developed where the industry's main problem was to prevent excess oil from swamping markets and undermining prices and profits. This problem was greatly exacerbated by protectionist measures in the USA, where the politically powerful lobby of domestic oil producers demanded protection against imports of cheap Middle East oil. An attempt to limit imports by voluntary restraint was tried from 1957, but it failed to achieve the desired results and in March 1959 President Eisenhower announced the imposition of compulsory import quotas. The new supplies of oil which were coming on stream outside the USA therefore had to find outlets elsewhere, mainly in Europe which was the largest market outside the USA. The problem of excess productive capacity was made still more acute by the resurgence of Soviet oil exports as the Soviet oil industry made huge increases in its output in the 1950s.[23]

The oil companies, faced with a glut of supply in international markets, resorted to unofficially discounting their published posted prices, which nevertheless continued to be used for calculating the profits on which the companies paid taxes (usually 50 per cent of profits) to the oil-producing countries. The posted prices thus became artificially high tax-reference prices, bearing no necessary relationship to the prices which the companies could realise in international markets. In effect, therefore, the profits of the international oil industry were redistributed in favour of the oil-producing countries, whose tax revenues from oil were not affected by the declining profitability of the industry. The 50:50 division of profits, which formed the basis of most Middle East concession agreements, became unbalanced, with the producing countries receiving substantially more than half, the companies substantially less.

BP was particularly badly affected by the international excess of oil supply over demand because its downstream operations were heavily concentrated in the increasingly competitive markets of Western Europe, and it lacked a significant presence in the protected, and therefore relatively profitable, US oil industry. This was reflected in BP's financial results: despite a 50 per cent increase in sales volumes, BP's profits remained static in the late 1950s, owing to falling profit margins, which fell from 23 to 15 per cent between 1956 and 1960.[24] In the same period, BP's return on capital employed fell from 18 to 11 per cent.[25]

At the same time, the burgeoning volume of BP's business called for heavy capital expenditure on the plant and equipment required to produce, transport, refine and market the growing flow of oil. With its profit margins falling and its capital expenditure rising, BP was unable to generate sufficient funds from its operations to cover its outgoings on capital expenditure and dividends. It was clear by 1957 that BP would have to raise new finance in the capital market by, it was thought, making a rights issue. That, however, raised some important questions: would the government, as majority shareholder, be prepared to spend public money in taking up its rights? If not, would it be prepared to lose its position as majority shareholder?

For several years, the advantages and disadvantages of the government retaining a majority shareholding had been kept under review and various consultations on the subject had taken place within and between the main government departments concerned. In sundry memoranda, it was pointed out that the government's shares had been an immensely profitable investment and would probably continue to be so in future. Moreover, the shareholding enabled the government to maintain close contact with one of the main players in an industry that was of national strategic importance. Against such factors, it was pointed out that the shareholding prevented BP from gaining concessions in some countries, most notably Venezuela, where petroleum legislation effectively excluded companies with foreign government shareholdings from participating in oil exploration and production. A stream of papers reiterated these (and other) points with different emphases, with the result that, for the time being at least, the government decided in favour of maintaining the status quo by keeping its majority shareholding, although BP would have preferred to see it reduced.[26]

However, while the government wished to retain its majority shareholding, it was also trying to pursue deflationary economic policies and for that reason had no desire to add to its spending by acquiring new BP shares in a rights issue. These potentially conflicting objectives were resolved after extensive consultation between BP and the Treasury.[27] It was decided that instead of making a rights issue, BP would issue convertible debentures to the public. The actual issue, made in December 1957, was of £41 million 6 per cent debentures (i.e. bonds), optionally convertible into ordinary shares in 1958–60.[28] If all the debentures were converted, the government's voting rights would fall from 55 per cent to just under 51 per cent, preserving its majority ownership. Lord Cobbold, the Governor of the Bank of England, was worried that, because the government was not subscribing, the issue would 'start with a bang and end with a whimper'.[29] However, it turned out to be a resounding success, being over-subscribed more than sixteen times.

Apart from issuing convertible debentures, BP also resorted to other ways of raising new finance in the late 1950s. One new source of funds was lease-back arrangements, whereby tankers built for BP were sold to the Clyde Charter Company and the Tanker Charter Company, financed by a combination of bank loans and debentures subscribed by institutions, and chartered back to BP. Other finance was raised by issues of loan stock in Switzerland and Canada, where the proceeds were applied towards the costs of constructing BP's new Montreal refinery.[30]

BP did not, however, rely solely on the raising of new capital to shore up its finances. By 1960 it was also curbing capital expenditure by deferring capital projects and reassessing the urgency of the need for new projects.[31] More importantly, BP, along with the other oil majors, cut the posted prices which were used to calculate the profits on which taxes were paid to the oil-producing countries. After the end of the Suez crisis the posted price of Iraqi crude (36°) at the Eastern Mediterranean pipeline terminals was reduced in stages to $2·49 per barrel in February 1958.[32] A year later, following a reduction in West Texas and Venezuelan crude prices, BP cut the posted prices of its Middle East crudes by 18 cents a barrel (about 9 per cent). The same cut was swiftly adopted by the other oil majors.[33] A further round of cuts, this time initiated by Standard Oil (NJ), would follow in August 1960 (see chapter six).

These cuts in posted prices had momentous consequences. They meant that the profits per barrel on which the companies paid taxes to the producing countries would fall, and that the producing countries would therefore receive smaller revenues per barrel. The producing countries saw this, not as a fair reflection of market conditions or as a reasonable attempt to redress the imbalance in the 50:50 division of profits, but simply as a revenue reduction imposed on them by the companies. In reaction, the leading producing countries would join forces in forming the Organisation of Petroleum Exporting Countries (OPEC), which would play a major part in the struggle between oil companies and producing countries for control of the world's largest industry: oil.

PART II
UNDER PRESSURE FROM THE PRODUCERS

= 6 =
The advent of OPEC

Despite rising competition and nationalism, the oil majors were still the most powerful players in the international oil industry in the late 1950s. Their dominant concessionary positions in the main oil-exporting states, their control over the international flow of oil, and their presumed ability to call on the power of their parent governments, all helped to give them an aura of unassailable power. The recent defeat of Musaddiq's attempt to take national control of oil operations in Iran, and the majors' ability to rearrange international oil movements during the Suez Canal crisis, were tangible examples of the majors' ability to deal with the volatile political environment in which they operated.

But although they still dominated the international oil industry, the majors faced growing problems in managing their relations with the oil-exporting states. For BP this was an absolutely crucial issue. With its near-total dependence on the Middle East for crude oil (see figure 6.1), BP was more exposed than any other major to the risks of political disturbances in a region that was in ferment. The rising tide of nationalism, the decline of the old European imperial powers, the Cold War between the superpowers of the USA and the USSR, revolutions in Middle East countries, the Arab-Israeli conflict, other rivalries between Middle East states – all these factors combined to make the Middle East highly unstable.

The end of the Suez crisis brought no respite. The humiliation of Britain and France revealed the extent to which European imperial power had already faded; the exhibition of the USA's overarching power emphasised its growing influence in the Middle East; and Nasser's triumph raised him to new heights as the hero of the Arab radicals, the flag-bearer of pan-Arab nationalism, whose ascendancy was confirmed in February 1958 by the union of Syria with Egypt in the United Arab Republic (UAR). Nasser's anti-Western

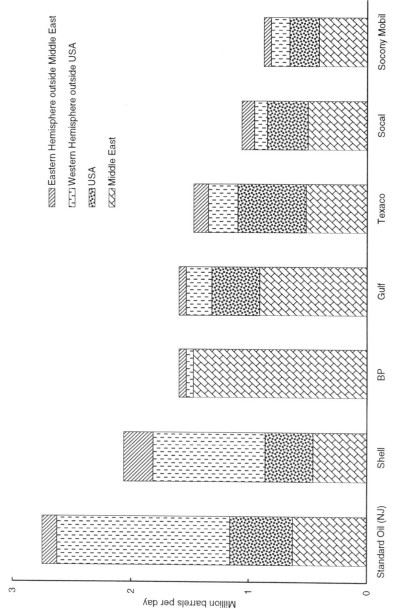

Figure 6.1 The majors' comparative crude oil production by area, 1961

onslaught, broadcast over the airwaves by Radio Cairo, put the conservative, pro-Western Arab monarchies of Iraq, Saudi Arabia, Kuwait, Jordan and Libya on the defensive against the seemingly unstoppable spread of Arab radicalism.[1]

In this atmosphere, none of the Middle East countries in which BP had oil-producing interests was stable and secure. In Iraq, the pro-British Hashemite monarchy and its leading politician, Nuri al-Said, were overthrown in July 1958 in a violent coup led by Brigadier Abdul Karim Qasim. The old monarchy was replaced by a republic, in which Qasim was Prime Minister. He rejected the pro-Western policies of the previous regime, signed trade and economic agreements with communist countries, and withdrew from the Baghdad Pact, which the previous regime had formed with Britain, Iran, Turkey and Pakistan as an anti-communist alliance. After Iraq's withdrawal, the Pact carried on with a reduced membership under the new name of the Central Treaty Organisation (CENTO).[2]

The overthrow of the Hashemites in Iraq dealt a further blow to Britain's waning influence in the Middle East and placed Iraq together with Egypt and Syria in the camp of the radical republics. Although they were fragmented by rivalry between Iraq and Egypt, and by the secession of Syria from the UAR in 1961, the radicals remained a potent force which permeated the Middle East to the discomfort of the conservative monarchies.

After the Hashemites were toppled in Iraq, leadership of the conservative Arab states passed to the rival Saud dynasty of Saudi Arabia, where the main oil concession was held by Aramco, the all-American consortium of Socal, the Texas Company, Standard Oil (NJ) and Socony Mobil. Their tax and royalty payments to the Saudi government served both their own and the US State Department's interests in strengthening the economic and political stability of the pro-American Saudi regime as a barrier against the spread of communism in the Middle East.[3]

Around the southern and eastern edges of Saudi Arabia, British influence still survived in the coastal belt that included a string of small British dependencies, a replica in miniature of the one-time British position in the Middle East as a whole. The most important from an oil-producing point of view were the tiny sheikhdoms on the Arabian littoral of the Persian Gulf. Most of them remained impoverished strips of desert and lagoon, but Bahrain and Qatar were established oil producers and some of the seven Trucial states, notably Abu Dhabi and later Dubai, were following in their footsteps.

But easily the richest of the mini-oil states, and certainly the most important to BP, was Kuwait. Here, the Al-Sabah dynasty ruled with support and protection from Britain over a population of less than a quarter of a million

Table 6.1 *Oil revenues of the main Middle East oil-producing states, 1958*

	Oil revenues (£millions)	Oil revenues per capita (£)
Iran	80	3·50
Iraq	79	12·50
Kuwait	125	555·00
Saudi Arabia	110	17·00

in an area no larger than an English county, but containing vast reserves of readily accessible oil which was the cheapest and easiest to produce in the world. Production by the joint concession-holders, BP and Gulf, had increased very quickly in the 1950s and since 1953 Kuwait had been the largest producer in the Middle East. It also had the highest oil revenues and its income per head was enormous compared to the other main Middle East producers (see table 6.1). There was, the British Foreign Office noted in 1958, a 'Midas touch' about Kuwait.

The situation of Kuwait and the other sheikhdoms on the Arabian side of the Persian Gulf was not conducive to regional stability. Britain's mantle, cast over them many years before to pacify the imperial sea route to India, was all that protected this fragmented collection of mini-states against larger, predatory neighbours. It was no secret that Iraq nursed claims to Kuwait, Saudi Arabia to the Buraimi oasis which lay partly in Abu Dhabi, Iran to Bahrain and other islands in the Persian Gulf. If Britain withdrew, there would be little to stop these claims being enforced.[4]

The non-Arab state of Iran occupied a vital strategic position between the USSR and the Persian Gulf. Its security was of such great concern to the USA that, after helping to overthrow Musaddiq, the USA turned Iran into a client state, providing military and economic assistance to fortify the Shah's regime as a buffer against the Soviet Union. The Shah, shedding the timidity and vacillation which he had shown during the Musaddiq crisis, aspired to be the strong man of the Middle East. He became increasingly vainglorious and autocratic as he tried to transform Iran from a US dependency into a major regional power. To that end, he pursued ambitious policies of social and economic modernisation, unaccompanied by political democratisation. His repressive rule was given a thin democratic veneer by a system of 'guided democracy', which was really a sham. Internal opposition was never far below the surface and sometimes broke through, before being put down by harsh counter-measures.[5]

The Shah looked to the Consortium to help provide the large revenues which he needed to finance his military and economic ambitions. He was determined to restore Iran to the position of top Middle East oil producer which it had held before the nationalisation of the Company in 1951.[6] Although Iran was aligned with the West as a member of CENTO, the Shah played on Western fears that if his demands for higher revenues were not met, he would turn to the communist bloc. But while superpower rivalry made it possible for the Shah to play off one against the other, his own position was not without dilemma. On the one hand, he wanted and needed Western support to strengthen his regime; on the other, he could not afford to look like a stooge of the West, which would undermine his position in the Middle East. The Shah thus trod a narrow path in his single-minded desire to raise Iran (and himself) to a power to be reckoned with at global level.

With such fragmentation and rivalry in the Middle East, the oil majors and the governments of Britain and the USA thought it unlikely that the oil-exporting countries would unite in a common front. A British official working party reported in 1958 that concerted action by the main exporting countries was improbable unless the Arab oil countries were reacting to Western measures which they regarded as provocative, or unless they were united by a single man or group.[7] BP and Shell were consulted about the report,[8] which was also sent to the US government for comment. It aroused little interest and would be forgettable, but for a tardy comment by the State Department in May 1960 that 'effective unified action on the part of the concessionary governments remains unlikely because of divisive political factors, mutual distrust and a natural inclination to profit from conflicts between the companies and other concessionary governments'.[9] Little did the State Department know that the ground for co-operation between concessionary governments was already being prepared.

THE FORMATION OF OPEC

The immediate, if unwitting, provocation came in February 1959, when the majors, reacting to a surplus of oil supply over demand, reduced their posted prices for oil in Venezuela and the Middle East. As taxes to producing country governments were paid on posted prices, the price cuts meant reduced revenues for the producing states.

Although the response of the producing countries to the price cut was muted on the surface, there was much discussion about it behind the scenes at the first Arab Petroleum Congress in April. Among nearly 600 people who attended were two who had never met before. One was Juan Pablo Perez

Alfonzo, Venezuela's Oil Minister, who was a strong advocate of using pro-rationing (rationing of production between producers, i.e. quotas) to stabil-ise output and prices in the international oil industry. Although John Loudon, the head of Royal Dutch-Shell, dismissed him at the time as a 'fourth-rate economist with ill-digested ideas',[10] Perez Alfonzo was to become one of the architects of co-operation between oil-exporting coun-tries. The second was Abdullah Tariki, the Director General (later Minister) of Petroleum and Mineral Affairs in Saudi Arabia. A geologist by training, he was a fervent Arab nationalist with Nasserite tendencies, described by Wanda Jablonski, doyenne of oil journalists, as 'a fanatic'.[11]

During the Congress, Perez Alfonzo and Tariki met in a chalet at the Maadi Yacht Club outside Cairo with Manucher Farmanfarmaian of Iran, Ahmed Said Omar of Kuwait, Mohammad Salman of Iraq and Salah Nessim of the United Arab Republic. These six individuals signed a document, the Maadi Pact, which was kept secret until 1961. It proposed the formation of a Petroleum Consultation Commission, to meet at least once a year to discuss matters of shared concern such as oil prices and production and the participation of producing countries in the oil industry.[12] The proposed Commission did not come into being and the Maadi Pact remained an infor-mal agreement between individuals, unsanctioned by their governments. Nevertheless, it was a harbinger of producer country co-operation.

After the Congress, Perez Alfonzo and Tariki kept in contact. In May 1960 they both attended the annual meeting of the Texas Independent Producers Organisation and saw at first hand the working of the prorationing system which had long been used to regulate Texan production. Tariki went on to Caracas, where he and Perez Alfonzo held a joint press conference at which Perez Alfonzo said that Venezuela would be prepared to join with the main Middle East oil-producing states to stabilise markets and defend oil prices.[13]

The oil majors also would have liked stable markets, but they could not wholly control the international oil surplus, which was swelled by growing Soviet exports and restrictions on access to the US market owing to the man-datory import quotas introduced in March 1959. Despite the persistent surplus of oil supply over demand, BP was against making another reduc-tion in posted prices, which BP believed would harm relations between oil companies and producing countries. In the spring of 1960, Harold Snow, one of BP's two deputy chairmen, warned Monroe ('Jack') Rathbone, chair-man of Standard Oil (NJ), that a reduction 'would cause a great deal of trouble'.[14] In July, the normally mild-mannered Snow repeated his concerns to Standard, expressing 'very strongly' the hope that Standard 'would not suddenly confront us with a fait accompli'.[15] Rathbone's background was

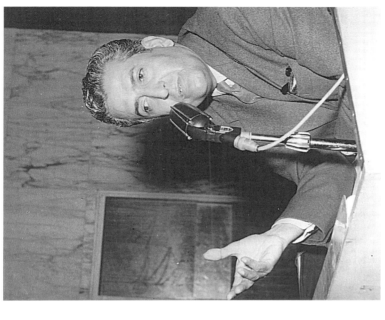

22, 23 OPEC's two founding fathers: Venezuelan oil minister, Juan Pablo Perez Alfonzo (centre left), and Saudi Arabia's first oil minister, Abdullah Tariki (right)

almost entirely in the domestic US oil industry and he had little experience of relations with foreign producing countries. He believed that posted prices should be reduced, against the advice of Howard Page, Standard's director for the Middle East, who opposed a reduction. But Rathbone overrode the opposition and on 9 August Standard announced that it was reducing its Middle East posted prices by up to 14 cents a barrel (about 7 per cent) with immediate effect.[16]

Snow cabled BP's New York office with a message for the press that BP had 'heard the news with regret' and would review the situation before taking any action.[17] On 13 August, Shell announced that it was following Standard's lead. BP still held back, but when news came through that CFP intended to make a cut, BP decided to follow suit, and on the 16th reduced its Middle East postings by up to 10 cents a barrel (about 5½ per cent).[18] Over the next week, the other majors also reduced their postings. Initially, some followed Standard with cuts of up to 14 cents a barrel, while others followed BP's smaller reductions.[19] Later, the group led by Standard aligned their prices with BP's.[20]

Although BP was soft-spoken about the price cuts in public, it was privately furious with Standard for initiating the cuts. Maurice Bridgeman, BP's chairman, complained to the Foreign Office that BP's 'hands had been tied' by Standard's isolated action, which was 'objectionable'.[21] The oil-producing countries had also not been consulted by Standard before the cuts. Provoked, Venezuela and the main Middle East oil-producing states arranged to meet for a conference in Baghdad on 10 September.[22] The day before the conference opened, a senior Foreign Office official noted BP's view 'that the price reductions precipitated by ESSO [Standard] in August, in such ham-handed fashion that they can scarcely bring themselves to talk about it, will be seen in perspective to have set in train a whole series of demands from Middle Eastern governments starting at the Baghdad meeting opening tomorrow, which will have far-reaching consequences in oil company/producing country relationships in the Middle East'.[23] This would prove to be a prophetic comment.

The meeting of the producer countries in Baghdad from 10 to 14 September 1960 was a seminal moment in oil industry history. The delegates resolved that posted prices should be restored to their pre-August levels and kept free from unnecessary fluctuations, and that the oil companies should consult with the producing countries before making price changes. They decided to study and formulate prorationing systems for the regulation of production. They agreed that if the oil companies used sanctions against any member to discourage the application of a unanimous conference decision, the other members should show solidarity and refuse beneficial treatment, such as higher production or prices, from the oil companies. And they decided to form the Organisation of Petroleum Exporting Countries (OPEC) to co-ordinate and unify the policies of its members. The five founding OPEC members were Venezuela, Iran, Iraq, Kuwait and Saudi Arabia.[24]

The formation of OPEC strengthened the bargaining power of the producing countries and would, in time, help to alter fundamentally the balance of power between the countries and the companies. A month after the Baghdad conference a Shell representative attended the second Arab Petroleum Congress in Beirut. He found that the great majority of Arabs were convinced that the cut in prices in August had been made in collusion by the oil companies 'whom they knew to be a cartel'. In reaction, the producing countries had formed OPEC 'as a cartel to face a cartel'.[25]

ROYALTY EXPENSING AND OPEC'S 'WEAK LINK'

Initially, the oil majors were not sure how to react to the formation of OPEC. BP favoured a 'wait and see' attitude. Standard Oil (NJ) regarded OPEC as 'a far more serious and immediate menace' than BP did, and thought about taking a more definite attitude of opposition or friendliness. Standard seemed to believe, wrote Bridgeman, that 'they must either murder it or make love to it', whereas he thought that OPEC might prove harmless 'provided it is not united by too much opposition from outside, or strengthened through too much love-making'. On this occasion, Standard apparently thought better of taking precipitate action and the majors adopted an attitude of studied neutrality in the hope that, without co-operation or hostility to feed on, OPEC would peter out in internal divisions. The oil majors and their Western governments thought that the most likely defector was Iran, which was regarded as the 'weak link' in OPEC.[26]

While OPEC was trying to establish itself in the first two years of its existence, its solidarity was indeed put to the test by internal divisions. The issue on which the Organisation looked most likely to split was prorationing. Perez Alfonzo and Tariki remained enthusiastic about the idea of producer countries controlling production by prorationing, but the Shah of Iran disagreed. He was determined to raise Iran's share of Middle East oil production and opposed the idea of limiting production. He aired his views in an interview which was published in *Petroleum Week* under the title 'Shah of Iran calls proration "impractical"'. The Shah was quoted (with his permission) as saying 'Iran must be restored to number one producer; international oil prorationing is nice in theory but unrealistic in practice.'[27] Another article on the same theme appeared in the same journal a week later, this time under the title 'OPEC won't be a "cartel", Iranians say'.[28] OPEC did not, however, push the matter to a division. Instead, the Organisation simply passed over it, making only perfunctory mention of prorationing at its Caracas conference in January 1961. In effect, the divisive issue of prorationing was shelved.[29]

A few months later the unity of OPEC was tested again. Kuwait, as has been seen, was a tiny state with enormous wealth. Defenceless on its own, it had traditionally relied on the guarantee of British protection for its security. In June 1961 the longstanding protective treaty between Britain and Kuwait was ended and Kuwait was declared to be independent. Within a week the Iraqi Prime Minister, Qasim, laid claim to Kuwait, announcing that it was part of Iraq. At Kuwait's request, British troops were rushed to its defence. Other Arab states publicly disapproved of Qasim's action and organised an Arab force, which soon replaced the British troops. Kuwait's independence was thus maintained, but the episode caused a rift in the Arab world. Iraq pulled out of the Arab League and recalled its ambassadors from countries that recognised Kuwait. OPEC, however, survived this rupture by a compromise whereby Iraq stopped sending representatives to OPEC meetings, but remained a member and endorsed OPEC's resolutions.[30] Iraq's nonattendance continued until Qasim himself fell victim to a coup in Iraq in February 1963.[31]

While surviving these divisive issues, OPEC followed a low-profile, moderate course, concerning itself primarily with matters of internal organisation and fact-finding studies on the oil industry.[32] It was not until its fourth conference in June 1962 that the Organisation passed three resolutions, numbers 32 to 34, which presented a serious challenge to the oil companies. Resolution 32 called on OPEC members to negotiate with the oil companies for the restoration of posted prices to their pre-August 1960 levels. Resolution 33 was concerned with royalties. Under the 50:50 agreements which were the concessionary norm in the Middle East, the producing countries received royalties, generally worth 12½ per cent of posted prices, plus income tax which, *taken together*, amounted to 50 per cent of the profits made by the concessionaire companies. In Resolution 33, OPEC demanded that royalties should be paid *separately* as an expense, after which profits should be split 50:50 between the concessionaires and the producing countries. This was known as royalty expensing. If adopted, it would add about 11 cents a barrel, or about 15 per cent, to the oil revenues of producer countries.[33] Resolution 34, the least important of the three, called for the elimination of the small marketing allowances of 1½ cents a barrel which concessionaires were allowed to deduct from posted prices before calculat-ing their taxable income.[34] OPEC requested Iran and Saudi Arabia to take the lead in negotiating with their respective concessionaire companies on the three resolutions.[35] They were to report back to OPEC at its fifth conference, scheduled to take place in November in Riyadh.

These three resolutions confirmed OPEC's arrival as a force to be reck-oned with. Its membership had by this time been increased from the five founding members to eight by the additions of Qatar, Indonesia and Libya. At the same time, OPEC's well-reasoned moderation had won it credibility. In the Foreign Office's view, OPEC had given the impression that 'it seeks to be regarded as a responsible organisation which is pursuing its aims with moderation'. Its conferences had been 'free from the fireworks which usually light up the gatherings of the Arab Oil Congress' and it had 'all the trappings of a respectable international body'.[36] OPEC, the Foreign Office realised, was 'here to stay'.[37] Sir Geoffrey Harrison, the British Ambassador to Tehran, noted that when the Organisation was founded in 1960, Bridgeman had 'dismissed it with the remark that, if left to itself, it would die a natural death or remain ineffectual . . . Time has proved him wrong.'[38] In the USA, an Interior Department official agreed that OPEC 'had turned out to be a much more solid structure than the US had expected'. 'Nobody', he went on, 'could deny that the present OPEC resolutions had created a potentially dan-gerous situation'.[39]

The oil companies, which earlier had hoped that OPEC might fall apart without their intervention, turned to a more active policy of trying to put a brake on OPEC by persuading the moderate members to slow down the rad-icals. To that end, the oil companies offered concessions to the moderates, trying to persuade them that they had more to gain from co-operation with the oil companies than they could win by confrontation through OPEC. BP played a prominent part in this essentially defensive campaign, and looked especially to Iran to split OPEC, whose constitution stipulated that resolu-tions could be passed only with the unanimous support of its members.

The opening moves were made in the summer of 1962 when, at Iran's request, the Consortium agreed to talks on Resolutions 32 to 34.[40] The two sides met in Paris at the beginning of October. In the space of four days the Iranians presented three memoranda, to which the Consortium responded with three 'rebuttal memoranda', to which the Iranians in turn responded with three 'counter rebuttal' memoranda. 'The whole meeting', wrote Geoffrey Stockwell, BP's regional co-ordinator for the Middle East, 'was rather like a pistol duel, each side firing its shots and then letting the other return the fire'.[41] The leader of the Iranian negotiators, Abdul Husayn

Behnia, used a negotiating tactic that would become an Iranian favourite: playing on the companies' fears, he warned that if the companies did not meet Iran's moderate demands, Iran would not restrain the radicals. 'The day of the big Companies ruling the world was past', he said, 'and it would be as well if the Companies understood this . . . OPEC was here to stay . . . Iran had had a beneficial and restraining influence on the Councils of OPEC, but if the Companies did not co-operate, the wild men would take over'.[42] At the end of the Paris talks the Consortium and Iran seemed to be set on a collision course.

Wishing to avoid a collision, Bridgeman, keeping the Foreign Office in the picture, went on a private diplomatic mission to Tehran at the end of October. He met the Iranian Prime Minister, Asadollah Alam, a man so quietly ruthless that it was said he 'could cut your throat with a feather'.[43] Alam said 'categorically that he was only concerned with Iran's interest, which was more money; he had no concern for OPEC qua OPEC . . . He made it clear, however, that if he was to drop OPEC, it would have to be made worth his while.'[44]

Bridgeman and Alam hatched what Alam called a 'gentlemen's agreement', in which Bridgeman offered financial inducements to secure Iran's promise to slow OPEC down on Resolutions 32 to 34. Though somewhat vague, the inducements in essence covered four points: first, that the Consortium's marketing allowance should be reduced by two thirds to half a cent per barrel; second, that if posted prices dropped in future, the Consortium would see that Iran did not suffer as a result; third, that Iran's oil production would be increased by 10 per cent annually; and fourth, he said that the Consortium would try to help Iran obtain finance for economic development projects. In return, Iran was to argue at the OPEC conference in Riyadh that more talks with the oil companies were needed before OPEC considered further action on Resolutions 32 to 34.[45] In other words, Iran would see to it that talks on the Resolutions were spun out and did not come to a head in Riyadh.

After his talks with Alam, Bridgeman went over the same ground with the Shah, who later confirmed that he had instructed the Iranian delegation at Riyadh to take the agreed line.[46] As a result, the OPEC members agreed to adjourn the conference, deferring consideration of Resolutions 32 to 34 until March 1963, when they would reconvene.[47] Bridgeman's mission to Tehran had accomplished its immediate purpose.

Indeed, Bridgeman was almost too successful in convincing Iran that it had more to gain from co-operating with the Consortium than it could win through OPEC. His purpose was to use Iran as a brake within OPEC, not to detach Iran from OPEC altogether. But a few days after OPEC's Riyadh meeting Sir Geoffrey Harrison reported that he had seen Alam, who 'rather electrified me by saying that he was anyhow intending to take Iran out of OPEC as soon as he could do so without loss of face . . . he saw no advantage in remaining in OPEC now that Iran's position was secure as a result of the assurances given by Bridgeman'.[48]

This news caused consternation in the Foreign Office, where officials were worried about what might happen if the Iranian government withdrew from OPEC on the strength of Bridgeman's vague assurances, and the Consortium subsequently failed to deliver what the Iranians expected. This would strengthen opposition to the Shah and reduce the political stability of the country. Moreover, if Iran left OPEC 'it would remove the one moderating influence in the Organisation'.[49] The Foreign Office, BP and Shell were agreed that 'it would be better for Iran to stay in and put a brake on it'.[50] A telegram was sent to Harrison notifying him of the official line. Harrison spoke to Alam, asking him to think carefully about the possible repercussions of leaving OPEC, and Alam came back with the reassuring message that the Shah was not in favour of such a move.[51]

In the early months of 1963, the oil companies had further reason to be pleased by the news that OPEC's fifth conference, already adjourned from November 1962 to March 1963, was now put off until November 1963. The reason given was, once again, that the negotiations with the oil companies had not reached a sufficiently decisive stage for OPEC usefully to discuss what action to take on Resolutions 32 to 34.[52]

The continued deferral of OPEC action appeared to confirm the ascendancy of the moderates in OPEC. Bridgeman thought that this had much to do with personalities. Perez Alfonzo and Tariki had provided the initial drive behind OPEC, but by this time they had both ceased to be involved in its affairs. Perez Alfonzo had become disillusioned with the Organisation and resigned as Venezuela's Oil Minister in 1963. Tariki had lost his post as Saudi Arabia's Oil Minister in March 1962 after injudiciously accusing Crown Prince Faisal of dishonest dealings in an oil contract.[53] He was succeeded by the more moderate Ahmed Zaki Yamani, a Harvard-educated lawyer who had previously been Faisal's legal adviser.[54] Bridgeman was glad to see the back of OPEC's two radical founders. 'With them', he said, 'had gone the driving force for the destruction of the international oil industry as it was now constituted'.[55] Stockwell, too, was in confident mood, saying 'that OPEC had shot its bolt, would perpetuate itself in the manner of all bureaucracies, but would certainly never again worry the oil companies'.[56]

Meanwhile, the Iranians were rewarded for their moderation by the fulfilment of some of the assurances which Bridgeman had given them. The oil companies' marketing allowances were reduced to half a cent a barrel, as he had promised. The Consortium also offered to send a three-man team to Iran to study development projects with which the Consortium might help financially.[57] The Shah, fearing accusations that he was an 'imperialist lackey' and a 'stooge', rejected the offer.[58] Alam suggested that instead the Consortium might discreetly appoint a continental banker to conduct a study.[59] A semi-retired Swiss banker, Hans Müller, was duly engaged and reported back with several schemes, including the construction of a harbour and the erection of a sugar refinery.[60] These ideas were received without enthusiasm by the Consortium, which prepared its own scheme for converting, at its expense, the crude oil terminal at Bandar Mashur into a products terminal, and re-routing crude exports from Bandar Mashur to Kharg Island through a new pipeline.[61] This project was accepted by the Shah.

OPEC, meanwhile, had authorised its Secretary General, the Iranian Fuad Rouhani, to negotiate on behalf of the member governments with the oil companies on the expensing of royalties (Resolution 33). The Consortium agreed to talks on the basis that Rouhani would be representing Iran, not OPEC. The Consortium appointed John Pattinson of BP, Howard Page of Standard Oil (NJ) and George Parkhurst of Socal – the '3Ps' – as their negotiating team.[62]

In the talks, held in September to October 1963, Rouhani made it clear that if OPEC's demands were not met, the Organisation would take unilateral action such as production restrictions and a tax on tankers exporting oil. On the other side, the '3Ps' made an offer which yielded no significant ground. This was that they would agree to the expensing of royalties, which would give the producing countries an extra 11 cents a barrel, in return for a royalty expensing allowance. This was a discount of 12½ per cent off posted prices, which would take 11 cents a barrel away from the producing countries. The net effect on the revenues of the producing countries would thus be zero.[63] Rouhani rejected this offer and demanded that the Consortium should produce a final offer which would satisfy Iran and OPEC in time for his next meeting with them.[64] This, after some rearrangement, was to be in New York on 4 November. Stockwell thought that it would be the 'make or break meeting'.[65]

At the New York meeting the Consortium representatives came up with an improved offer. They were now prepared to concede the expensing of royalties if Iran would agree to a royalty expensing allowance of 8½ per cent. This would have the overall effect of raising Iran's revenues by 3½ cents a barrel. Rouhani rejected it, saying 'I know this offer will be completely unacceptable to OPEC and the Iranian Government'.[66] His decision was unani-

24 OPEC's first Secretary General, Fuad Rouhani (centre), at a reception in Geneva in 1963

mously endorsed at an OPEC consultative meeting, and the Foreign Office and the oil companies received word that when OPEC met for its full conference in Riyadh (which had once more been rescheduled, this time to 24 December), the members would probably decide to pass simultaneous laws providing for the expensing of royalties. If the companies tried to retaliate by cutting posted prices, the OPEC countries would take unilateral control of prices and production.[67]

The prospect of an open confrontation between OPEC and the oil companies worried the Foreign Office. In the previous few months the Middle East had been even more unstable than usual. In Iraq, Qasim had been overthrown and killed by Baathist plotters in February 1963. The Deputy Prime Minister in the new Baath government was Ali Saleh al-Saadi, who controlled the newly formed National Guard, a fearsome paramilitary organisation. In November, army officers, concerned about the challenge which the National Guard posed to their authority, overthrew the Baathists and a new government was formed under Lieutenant-General Tahir Yahya.[68]

In Iran, also, 1963 had been a turbulent year. In January, the Shah had announced his 'White Revolution' of economic and social reforms, which aroused opposition in large sections of Iranian society, including the influential religious leaders. Throughout early 1963, Ayatollah Ruhollah Khomeini spoke out against the Shah's rule. On 3 June he delivered a fiery speech denouncing the Shah, for which he was arrested. Immediately, riots broke out across the country. The demonstrators were ruthlessly quelled by armed forces, who were sent into the streets with shoot-to-kill orders which they did not hesitate to carry out. Martial law was declared, though later lifted, followed by 'controlled' elections for the 21st Majlis (Iranian parliament) in September.[69]

Open conflict between the OPEC countries and the oil companies might, commented one Foreign Office official, 'involve a weakening of these countries' links with the West, the reappearance of the mob in Tehran and elsewhere, and further internal political upheavals in the Middle East countries concerned'.[70] Lord Carrington, Minister without Portfolio in the British government, was briefed to speak to the British Cabinet about the possibility of OPEC deciding to take unilateral action at the Riyadh conference. If that happened, 'oil producing governments and the international oil companies may well be set on a collision course'.[71] In Tehran, Sir Denis Wright, who had succeeded Harrison as British Ambassador, saw Prime Minister Alam, 'who was more worried than I have ever seen him'.[72] The Shah, Wright reported, 'seems to be as worried as Alam about what to do next'.[73]

In this crisis, the tactic adopted by the oil companies, with Foreign Office support, was largely a repeat of that used earlier, namely to persuade the Iranians to put the brakes on OPEC. To that end, Wright advised Alam that Iran 'would do herself great harm if she introduced legislation forcing the Consortium to expense royalties'. The Iranian delegate at OPEC's Riyadh conference should, Wright suggested, argue that the Consortium's offer was the best obtainable in the circumstances. He went on, 'the real issue for Iran was whether to risk being odd man out in OPEC or to fall in with the majority in OPEC against the oil companies. The former would lead to difficulties with the Arabs whereas the latter would lead to conflict with the oil companies and sooner or later with the British and American Governments . . . The Iranian Government could alone decide where their best interests lay in this dilemma.'[74]

The oil companies and the Foreign Office also tried to convince the Iranian government that Rouhani, who was meant to have been negotiating on behalf of Iran, had in fact been following a more radical OPEC agenda. By discrediting Rouhani, they hoped to persuade the Iranian government to go

25 Riza Fallah (on left) putting his feet up with Farhang Mehr (Governor of OPEC for Iran) while waiting for the banquet at the OPEC conference in Riyadh, Saudi Arabia, in December 1963

back on Rouhani's flat rejection of the Consortium's offer and to take a more moderate line.[75] The Consortium therefore responded enthusiastically when Alam suggested that the Iranians, acting 'for themselves and not (repeat not) for OPEC', should have further talks with the Consortium before OPEC's Riyadh conference. Better still for the Consortium, the Iranian representative was to be Dr Riza Fallah. He was known to be a moderate, opposed to Rouhani and OPEC, and so pro-BP that he was known in the OPEC secretariat as 'the BP delegate to the [OPEC] conferences'.[76]

Fallah visited London in mid-December for talks with the '3Ps'. He told them that the Shah no longer trusted Rouhani and did not like being a member of OPEC, but dared not leave it because 'he was afraid of being labelled an imperialist stooge by the Arab members'.[77] The '3Ps' insisted that the Consortium would not offer Iran an increase in oil revenues above the 3½ cents a barrel which had already been put forward. However, they were prepared to discuss the 'modalities', meaning the form in which the offer was presented. Various alternatives were discussed and Fallah returned to Iran, ready to make an 'all-out effort to persuade the Shah to accept the

Consortium offer'. The Shah was persuaded and word came back to London that he intended to stand firm against unilateral action at OPEC's Riyadh conference.[78] Meanwhile, talks also took place with Saudi Arabia and Kuwait, who also agreed to take a moderate line in Riyadh.[79]

These efforts to split the OPEC moderates from the radicals paid off. When OPEC finally held its fifth conference in Riyadh in the last week of December, the moderates, led by Iran, prevailed over the radicals, led by Iraq (the others were Venezuela and Indonesia). The members resolved that unilateral action was not called for. Instead, they nominated a three-man committee to continue negotiations with the companies.[80] Talks dragged on as OPEC's next conference, to be held in Geneva in July 1964, approached.[81] As the conference drew nearer the companies came up, as they had in similar situations before, with an improved offer: they would expense royalties and reduce the royalty expensing allowances by stages so that the revenues of producing countries would rise by 3½ cents a barrel in 1964, 4 cents a barrel in 1965 and 4½ cents a barrel in 1966; and they would consider a further reduction in the allowances after 1966, subject to market conditions at the time. At its Geneva conference, OPEC decided that this was a 'suitable basis' for agreement.[82]

The next few months were spent hammering out detailed agreements, in time for OPEC's seventh conference in Djakarta in November. At the conference the OPEC members split along the now-familiar lines of moderates versus radicals. The moderates, comprising Iran, Saudi Arabia, Kuwait, Qatar and Libya, expressed their intention of accepting the companies' offer. On the radical wing, Iraq rejected it and was supported by Venezuela and Indonesia, although these two countries were not directly affected by the royalty issue as their concession contracts already provided for royalty expensing. Faced with this division, the conference voted to leave acceptance or rejection of the companies' terms to the individual member countries.[83]

By the end of January 1965, Iran, Saudi Arabia and Qatar had ratified their respective royalty expensing agreements, though not before Iran had extracted some last-minute concessions from the Consortium (an interest-free loan of £6 million and a commitment to invest £2 million in an industrial investment corporation).[84] In Libya, ratification was delayed until January 1966, and in Kuwait until May 1967, owing to opposition from Arab nationalists in the National Assembly.[85] In Iraq, the royalty expensing terms were not ratified and joined the other issues on which the government and the IPC were in acrimonious dispute (see chapter seven).

In April 1966 OPEC resolved that each member should take steps towards the complete elimination of the royalty expensing allowances. Negotiations were started and were still in progress when the Arab–Israeli war of 1967 resulted in the closure of the Suez Canal. This put a premium on short-haul crudes from Libya and from the Mediterranean terminals of the pipelines carrying crudes from Saudi Arabia and Iraq. The companies therefore agreed to suspend the allowances on crudes shipped from these places. In January 1968 agreement was reached for the remaining allowances to be phased out so that they would cease entirely in 1972.[86] In the event, they were eliminated earlier than that under the Tehran agreement of 1971 (see chapter eighteen).[87]

With the cessation of the allowances, full royalty expensing was finally achieved after nine years of negotiations. By detaching the OPEC moderates, especially Iran, from the radicals, led by Iraq, the oil companies had achieved a large measure of success in slowing OPEC down and perpetuating the concessionary system.

7

The political balancing act

In the 1960s the oil majors were engaged in a political balancing act that was harder and harder to keep up. Every Middle East oil-producing state wanted higher national oil production (and thereby revenues), but they could not all be satisfied without the overall level of oil supply exceeding available demand. To balance supply with demand, the majors in effect prorationed production between the Middle East states.[1] This was done, not through a formal cartel or quota system, but by a process of political bargaining in which economic factors were of secondary concern. In essence, the competing demands of Middle East states were accommodated according to the political leverage which they could bring to bear on the companies, either directly or by going over the companies' heads to their parent governments.

The most important parent country by a long way was the USA, which sought to make up for the diminishing role of its close ally Britain in the Middle East by forging close relationships with Saudi Arabia and Iran. These two countries, especially Iran, became the pillars of the USA's policy of containing communism in the region. Although Saudi Arabia and Iran were both anti-communist, they were also rivals as regional powers. Iran was the closer ally of the USA and, being non-Arab, was less prone to tensions with the USA over the USA's support for Israel. The Shah never missed an opportunity to press for higher production on the grounds that Iran's capacity to resist communism depended on its economic and military strength, which in turn depended on oil revenues.

The other two main oil-producing states in the Middle East were Kuwait and Iraq. Kuwait was in the moderate, pro-Western camp, but was too small in area, population and military capacity to have much political weight. In any case, it was so rich that it had little need or use for higher oil production and revenues. Iraq was on the other side of the fence in the radical anti-

Western camp of Arab states. Its outright opposition to the West removed any possibility that it would or could adopt the Iranian tactic of appealing for Western support on the grounds that it was a bastion against communism. Instead, Iraq adopted the most directly confrontational attitude of the Middle East oil producers in its relations with the companies.

The companies came under considerable pressure from this mixture of forces. The USA and Iran squeezed them hard for higher production in Iran. Favouritism towards Iran, combined with US support for Israel, alienated the Saudis enough that the companies in Aramco feared for the security of their concession, which they valued above all others. In Iraq, the companies were locked in dispute with the government. Elsewhere, the rise of production in newer producing countries, most notably Libya and Abu Dhabi, added to their problems in balancing supply with demand.

The role of the companies in this period has often been likened to a buffer between the oil-producing and oil-consuming nations, preventing the nations from clashing directly over oil and so reducing political tensions.[2] This was not, by definition, a comfortable role, but if the pressure was at times intense, it was at least preferable to being squeezed out altogether.

THE IPC AND IRAQ

The overthrow of the Hashemite monarchy in Iraq in 1958 was followed not by the promised revolution, but by a series of coups. For the next decade, military governments came and went in quick succession, with the army never far from the centre of political life. Indeed, army officers became the new privileged class, eclipsing the merchants and great landowners in their status and influence. Politics consisted not of winning elections, but of intrigues and personal rivalries, as competing groups of communists, Nasserites, military politicians and Baathists sought power through armed force rather than the ballot box.[3]

In this unstable and unpredictable political atmosphere, relations between the IPC and the Iraqi government became a vicious circle of unending negotiations. For the Iraqi negotiators, there was much more political capital to be had from attacking the oil companies than from coming to terms with them, which smacked of compromise with imperialist exploiters. Whenever the possibility of reaching agreement came in sight, there was, therefore, a tendency for the Iraqis to step up their demands.

This helped to make the oil companies feel that any offer they made would be treated by the Iraqis not as the basis for a settlement, but as the starting point for further demands. Moreover, the companies could never be sure how long the government that they were negotiating with would stay in power, and whether successor regimes would be politically driven to make stiffer demands. In these circumstances, and as part of the normal bargaining process, the oil companies tended to hold back from offering all that they were prepared to concede. Instead, they made concessions by increments, and only when they were put under pressure by the Iraqis. This only encouraged the Iraqis to believe that a better offer could always be extracted from the companies by making new demands, which started another round of fruitless negotiations.

Talks between the IPC and the Iraqi government on revisions of the IPC's concessions were already in progress before the overthrow of the Hashemite monarchy. In early July 1958, Geoffrey Herridge, the managing director of the IPC, visited Baghdad. His talks with the government finished on 11 July with the IPC agreeing to relinquish some of its concessionary area of 168,500 square miles (virtually the whole of the country),[4] to increase crude oil production capacity to nearly 1·2 million bpd by the end of 1961 and to submit detailed proposals to the Iraqi government on other outstanding matters. After finishing the talks, Herridge flew home the day before Qasim seized power on 14 July.[5]

When Qasim came to power, the memory of Musaddiq's fate in Iran was still fresh, the power of the IPC was feared and Qasim needed oil revenues. These factors militated against confrontation, and, in his first public statement on oil, Qasim gave assurances that existing agreements would be honoured.[6] In August, Herridge visited Iraq, where he found that the IPC's operations were going on as normal. He met Iraqi ministers, confirmed the expansion of capacity to nearly 1·2 million bpd, and discussed other matters such as relinquishment. The tone of the talks was moderate and both sides seemed to be willing to carry on with the negotiations which had been started before Qasim came to power. The impression given was one of continuity, rather than revolution.[7]

Soon, however, tensions began to grow. Of the various issues under discussion, the two main ones were relinquishment and Iraqi demands for a 20 per cent shareholding in the IPC.[8] On the shareholding, the IPC was unyielding. At BP, Bridgeman was, according to a senior Foreign Office official, 'undoubtedly obsessed that a concession to Iraq would create a dangerous precedent elsewhere'.[9] According to Bridgeman, 'although the Arabs undoubtedly want a greater share in the business, their ideas and aspirations, when it comes down to detail, are remarkably uninformed and woolly . . . the whole approach to the problem in Arab countries is purely emotional and . . . lacks any vestige of practical or logical foundation'.[10]

On relinquishment, the IPC reiterated that it was not opposed in principle to surrendering some of its concessionary area. Indeed, the idea of another concessionaire operating in Iraq was welcomed, as this would make the IPC less vulnerable to accusations that it was an imperialist monopolist in Iraq.[11] However, the IPC and Iraq disagreed about the size of the area to be relinquished and the speed with which relinquishment should proceed. In autumn 1958 the IPC offered to surrender 50 per cent of its concessionary area phased over ten years, but the Iraqis wanted more and faster. By mid-1959, the IPC was offering to give up 54 per cent immediately, with a further review after five years. That was still not enough for Qasim, who demanded 60 per cent immediately and a say in selecting the areas to be surrendered.[12] By July 1959 the IPC felt that 'they must call a halt to the current process of bazaar bargaining'.[13] 'The time had come', thought Bridgeman, 'when the Company must dig in their toes'.[14] The IPC refused to concede that the Iraqis should have a say in choosing the areas to be relinquished and the talks broke down at the beginning of October.[15] Later that month, Qasim was wounded by a would-be assassin and was confined to hospital for nearly two months.[16]

By the time that negotiations restarted in August 1960, relations between the two sides had been soured by a dispute over a sudden and very large increase in Iraqi port dues at Basra and by the oil companies' cut in posted prices, also in August.[17] As before, the talks covered a number of issues, but relinquishment was the main topic of discussion. Under pressure from Qasim, the IPC was by the end of 1960 offering to relinquish 75 per cent of its concession area immediately, plus a further 15 per cent in nine years' time.[18]

After an adjournment of the talks, Herridge returned to Baghdad in March 1961 at the urgent insistence of the Iraqi government. The reason, it transpired, was that the government wanted the IPC to make its next quarterly tax and royalty payment, which was due in April, before the end of March (the close of the financial year). The IPC acquiesced to the tune of £20 million.[19] Meanwhile, also in March, serious unrest had broken out in Baghdad owing to discontent with Qasim's government. Though the disturbances were suppressed by the army, they demonstrated Qasim's political shakiness and brought about a hardening of his attitude towards the IPC. As the Foreign Office commented, 'Qasim has always been most demanding and inflexible as a negotiator when internally assailed, and the Iraqi Nationalists and Communists have for some time been accusing him of being an Imperialist puppet and unwilling to take a firm line with the oil companies.'[20] In these circumstances, it would have been political suicide for Qasim to come to terms with the IPC.

At the end of March, he made angry attacks on the oil companies, accusing them of fomenting the disturbances in Baghdad.[21] On 2 April, he saw Herridge and spent the first twenty minutes of a three-hour session reproaching the IPC for making heavy weather of the advance payment and for delaying the decision as a bargaining counter.[22] He accused the IPC of 'procrastination, unwillingness to reach settlement, unfairness and unfriendliness'.[23] Sir Humphrey Trevelyan, the British Ambassador in Baghdad, cabled, 'we shall do our best to keep negotiations going, but it is becoming almost impossible to negotiate with Qasim who is wholly unpredictable and takes any concession as an encouragement to ask for more'.[24] Herridge saw Qasim again on 6 April, but no progress was made and Qasim announced that the IPC must suspend all exploration activities and that the army would enforce this decision.[25] Herridge returned to England and the IPC stopped exploration work in Iraq.[26]

In August, a senior IPC delegation visited Baghdad to resume negotiations.[27] The main issues were participation, relinquishment and Qasim's demand for an increased share of the IPC's profits.[28] As usual, Qasim stepped up his demands. On relinquishment, the IPC had offered in June to surrender 75 per cent of its concession area immediately, followed by a further 15 per cent seven years later. Qasim, however, demanded 90 per cent immediately.[29] In Britain, Edward Heath (the Lord Privy Seal) informed the Prime Minister, Harold Macmillan, that 'this further instance of Qasim's technique of always asking for more had a very bad psychological effect on the companies'. On participation, Heath went on, the IPC was 'opposed to making concessions to the bad boy of the class, particularly when the likelihood was that Qasim would pocket any concession and ask for more'.[30]

By autumn, the prospects of a settlement were as remote as ever. Qasim was trying to put down a Kurdish revolt in northern Iraq, which he blamed on imperialists and the oil companies.[31] On 11 October, he presented IPC representatives with a final 'offer' for the immediate relinquishment of 90 per cent of the concession areas, Iraqi participation in the area left to the IPC, and a larger share of the IPC's profits. The IPC refused the offer and Qasim broke off the negotiations.[32] Two months later, on 12 December, the Iraqi government issued Law 80, which expropriated 99½ per cent of the IPC's concession area without compensation. The area of some 750 square miles left to the IPC included its producing wells. However, the highly prized North Rumaila field in southern Iraq, where oil had been discovered but was not yet in production, was expropriated.[33] This measure, Law 80, was the most extreme measure that had been taken in the Middle East against the oil companies since Iran's nationalisation of the Company's assets in 1951.

On 13 December, the IPC delivered letters to the Iraqi Oil Ministry pointing out that Law 80 was a breach of the IPC's concession agreements. The letters were returned as unacceptable. On the 17th, the IPC attempted to deliver formal letters of protest, but Oil Ministry officials refused to open them or to sign for their receipt.[34] In January 1962, the IPC requested that the dispute be submitted to arbitration, as provided for in the concession agreements. Qasim, however, ignored the arbitration request.[35] Sir Roger Allen, who had succeeded Trevelyan as British Ambassador in Baghdad, thought that it would be impossible to reach a satisfactory agreement with Qasim.[36] At BP, Stockwell held much the same view.[37] In effect, their policy was to wait out Qasim, hoping for a change of government.

That came in February 1963, when Qasim was killed in the Baathist coup which brought in Colonel Abdul Salam Arif as President (a newly created post), with Ahmad Hassan al-Bakr as Prime Minister. Army officers held other key posts in the new Cabinet, and the paramilitary National Guard was widely feared. However, the new Oil Minister, Abdul Aziz al-Wattari (a petroleum engineer with a degree from the University of Texas), was a moderate with whom the IPC could hold peaceable negotiations.[38] He managed to keep his post as Oil Minister despite the overthrow of the Baathist government in November 1963, when a new Cabinet was formed under Lieutenant-General Tahir Yahya as Prime Minister.[39]

Negotiations between the IPC and al-Wattari began soon after he became Oil Minister. After many meetings, al-Wattari announced in June 1965 that a draft agreement had been reached, subject to approval by the Iraqi Cabinet.[40] The agreement was in two parts. The first promised to restore to the IPC an area of about 750 square miles which had been expropriated under Law 80. This would double the area retained by the IPC. More important than the size of the area was that it included the highly prized North Rumaila field. Under part one of the agreement, other issues were also to be settled by a lump sum payment from the IPC. In part two, all the IPC participants except for Standard Oil (NJ) agreed to establish a joint venture with the state-owned Iraq National Oil Company (INOC), which had been formed in 1964. The joint venture, in which INOC was to have a one-third shareholding, was to explore for oil in an area of some 12,350 square miles.[41]

The al-Wattari agreement would thus have put the IPC in the strong position of sole concessionaire in an area of 1,500 square miles which contained its producing wells and North Rumaila, and majority shareholder in the joint venture with INOC. However, this favourable settlement for the IPC soon became, in Stockwell's words, a 'political football' in Iraq.[42]

When the al-Wattari agreement was presented to the Iraqi Cabinet in July, it was attacked by Nasserite ministers. This brought to a head deep divisions in the Cabinet on various issues, including oil, and six ministers resigned. A violent press campaign against the agreement followed.[43] Then came a new round of political intrigue and instability. On 6 September, Yahya resigned as Prime Minister. President Arif asked Brigadier Arif Abdul Razzaq, Commander of the Air Force, to form a new Cabinet. Al-Wattari was dropped from the Oil Ministry. Only ten days after the new Cabinet was formed, Razzaq led a coup to overthrow the President. The coup failed, Razzaq and nine accomplices fled to Cairo, and President Arif turned to Dr Abdul Rahman al-Bazzaz to be Prime Minister.

Al-Bazzaz was the first civilian Prime Minister since Nuri al-Said in the old days of Hashemite rule. As a former Secretary General of OPEC (he had succeeded Rouhani in 1964), he had some experience of oil matters. However, his first priority was to find a solution to the Kurdish problem, and consideration of the oil question was deferred. Meanwhile, the political turbulence continued. In April 1966 President Arif was killed in a helicopter crash. He was succeeded by his brother, Brigadier Abdul Rahman Arif, who promised to throw the al-Wattari agreement open to full public discussion. That had not been done by August, when Arif requested the resignation of al-Bazzaz and invited Brigadier Naji Talib to form a new Cabinet.[44] The government was once again in military hands, after a civilian interlude of less than eleven months.

On coming to power, Naji Talib set up an advisory committee to consider the Wattari oil agreement. It had made little progress before the oil agreement was overtaken by a fresh crisis originating not in Iraq, but in Syria. In August 1966 the Syrian government, which had recently come to power through a violent coup, requested negotiations with the IPC to increase transit payments for oil flowing through the pipeline from northern Iraq to the Mediterranean via Syria. Negotiations produced no result, and in December the Syrian government unilaterally raised transit and loading fees, and levied a new fee for alleged past underpayments by the IPC. The government told the IPC to stop pumping oil through Syria until it paid the amounts claimed. The IPC complied and the flow of oil was stopped until the two sides reached a settlement on transit payments in March 1967. In the meantime, Iraq, though not a party to the dispute, was deprived of about two thirds of its oil exports.[45]

The Iraqi government, concerned about its loss of revenues, demanded that the IPC's tax and royalty payment for the first quarter of 1967 should be based on normal exports. After negotiations lasting until May, the IPC agreed to make an advance of £14 million to Iraq to help cover the loss of revenues.[46] A week later, Talib fell from power and Arif assumed the posts of both President and Prime Minister.[47] Soon afterwards, any prospect of making progress with the al-Wattari agreement was dashed by the outbreak, for the third time in less than twenty years, of an Arab–Israeli war.

The build-up to the war was rapid. By May, a series of border clashes between Israel and Syria had raised the heat to the point where Israel was threatening military action against Syria if Arab terrorist activity did not stop. In the middle of the month, Nasser promised Syria that in the event of a war, Egypt 'would enter the battle from the first minute'. Two days later, he requested the withdrawal of the United Nations peacekeeping force which had occupied the Sinai since the conclusion of the 1956 Suez Canal crisis. UN Secretary General U Thant agreed to Nasser's request. Then Nasser, declaring that the Gulf of Aqaba was part of Egypt's territorial waters, forbade Israel access to the Red Sea through the Straits of Tiran. The USA and Britain supported Israel's cause and drew up a declaration stating Israel's right to free passage through the Straits. At the end of May, King Hussein of Jordan signed a defence pact with his long-time adversary, Nasser. On 4 June, Iraq adhered to the new Jordanian-Egyptian military alliance.

Israel, feeling increasingly threatened, decided to strike first. On 5 June a surprise attack by Israeli warplanes wiped out the Egyptian air force while it was still on the ground. In a six-day blitz before the war ended in a cease-fire, Israel seized and occupied territories three times the size of Israel itself in the Sinai (Egypt), the Golan Heights (Syria) and the West Bank (Jordan). It was a crushing military defeat for the Arabs.

In retaliation against the USA for supporting Israel, six Arab states (Egypt, Syria, Iraq, Yemen, Algeria and Sudan) broke off diplomatic relations with the USA. The Arabs also retaliated with the oil weapon. The Suez Canal (still a main oil artery from the Middle East to the West) was blocked, the flow of oil through Aramco's Tapline and the IPC's pipeline to the Mediterranean was stopped, and the major Arab oil producers called for an oil embargo against countries friendly to Israel. It was carried out with varying degrees of firmness against the USA, Britain and, to a lesser degree, West Germany, which had angered the Arabs by establishing diplomatic relations with Israel in 1965.[48]

On top of this Middle East crisis came another unrelated but simultaneous interruption in oil supplies from Nigeria. The federal system of Western-style multi-party democracy which was installed when Nigeria was decolonised by Britain in 1960 soon proved unable to accommodate the religious, tribal, regional and economic diversity of Nigerian society. After a series of alleged ballot frauds, and the boycotting of a general election by the principal party of opposition, a group of army officers staged a military coup in January 1966. Seven months later, a counter-coup brought Lieutenant-Colonel Yakubu Gowon to power as head of the federal government. Under the leadership of his rival, Lieutenant-Colonel Odumegwu Ojukwu, the eastern region seceded from the federation in May 1967, declaring itself to be the independent Republic of Biafra. In July, civil war broke out between federal and Biafran forces and onshore oil operations were halted by the hostilities. Production was resumed in 1968, while the civil war continued until January 1970, when Ojukwu fled the country and Biafra rejoined the federation. In the meantime, the loss of Nigerian production in mid-1967 added to the problems posed by the Arab oil embargo.[49]

A serious oil shortage did not, however, materialise. Although Western Europe and Japan imported over 80 per cent of their oil from the Middle East and Africa, the Arabs did not join in the embargo with enough unity to make it effective. Iraq led the hard-liners and called for a complete three-month ban on oil shipments to all customers to teach the West a lesson, but found no support among the moderate Arab states of Saudi Arabia, Kuwait and Libya. Meanwhile, increased exports from the non-Arab oil producers of Iran, Venezuela and Indonesia helped to make up for shortages. The most crucial factor, though, was that the USA was able to use its spare production capacity to produce a great surge in its oil exports, which ensured that other industrialised countries would not go seriously short of oil.

The rearrangement of normal oil-supply patterns required rapid action by the oil companies. As in the 1956 Suez crisis, they scrambled for tankers to carry oil from the Middle East to Western Europe around the Cape of Good Hope, while reprogramming their refineries to increase yields of the most needed products from crude oils that the refineries were in some cases not designed to process. At BP, the speed with which the mass of supply data had to be recalculated was too much for the computer models which were normally used, and the Supply and Development Department had to fall back on quick-fire human decisions (see chapter sixteen). These measures proved effective. By the time that the Arab countries met in Khartoum in late August there was no oil shortage and the last elements of the embargo were lifted, though the Suez Canal remained closed.[50]

The oil embargo was over, but the political fall-out of the Arab–Israeli war remained. The conflict radicalised Arab opinion against the USA and opened a period of Israeli ascendancy in Washington. The Arab–Israeli dispute became the dominant issue in the Middle East, where the USA supported Israel, Iran, Turkey and the conservative Arab states. Against them stood the radical Arab nations of Egypt, Syria and Iraq, backed by the USSR.[51]

The Six-Day War and the hardening of Iraq's anti-Western policy inevitably impacted on the IPC's negotiations with Iraq. Any remaining hopes of the 1965 al-Wattari agreement being approved were finally dashed. In August 1967, the Iraqi government issued Law 97, which barred the restoration of North Rumaila to the IPC, and gave INOC exclusive oil rights throughout Iraq, except for the area of 750 square miles allowed to the IPC under Law 80. This was followed, in September, by Law 123, which reorganised INOC and put it under more direct political control with the aim of making it a more effective instrument for the development of a national Iraqi oil industry. At the same time, the moderate head of INOC was replaced by Adib al-Jadir, one of the foremost opponents of the al-Wattari agreement.[52]

Laws 97 and 123 were quickly followed by new agreements between INOC and foreign contractors for oil exploration and development in areas outside the pocket of 750 square miles still held by the IPC. The first was in November, when the Iraqi government announced a twenty-year contract between INOC and France's state-owned Entreprise de Recherches et d'Activités Pétrolières (ERAP) for exploration in an area of some 4,200 square miles. This agreement was ratified in February 1968.[53] Meanwhile, in December 1967, al-Jadir had signed a letter of intent with the USSR for Soviet assistance to INOC.[54] Al-Jadir did not survive in power for long enough to see the letter of intent come to fruition. He lost his post in July 1968, when a Baathist military coup deposed President Arif.[55] The new President was Major-General Ahmad Hassan al-Bakr, who had been Prime Minister in the Baathist government of 1963. A Revolutionary Command Council was set up, consisting of army officers and one civilian, Saddam Hussein al-Takriti, who became Vice-President of the Council in November 1969. The main aims of al-Bakr and Saddam Hussein were to establish their supremacy in the army, the party and the state. Opposition was ruthlessly suppressed by strong-arm methods. Though the Baathist government roundly denounced the previous regime, it pursued broadly the same oil policy. In 1969 the letter of intent with the USSR was translated into definite agreements for Soviet financial and technical assistance in developing the North Rumaila field, which was brought on stream in April 1972.[56] Meanwhile the IPC continued to produce most of Iraq's oil exports from the tiny fraction (0·5 per cent) of its original concession areas which it still retained.

THE PIVOTAL ROLE OF IRAN

Iraq's expropriation of nearly all the IPC's concession area reduced both the scope and the enthusiasm of the IPC's members to raise production in Iraq

26   Saddam Hussein, Vice-President of Iraq's Revolutionary Command Council, speaking at a rally in a stadium in Baghdad in 1970

at the same rate as in other, more friendly Middle East producing countries. As a result, Iraq's oil output grew slowly by Middle East standards in the 1960s. In Kuwait the growth rate, though higher than in Iraq, was also fairly flat by the standards of the region (see figure 7.1). There was some concern that this might annoy the Ruler who gave numerous warnings that his desire to be co-operative with the oil companies should not be regarded as a sign of weakness.[57] But his need for protection against Iraq made his weakness obvious to all, and his warnings cannot have been taken too seriously. Certainly they carried much less weight than those of Iran and Saudi Arabia, the pro-Western heavyweights of the region. The competition for higher production between these two countries was intense. BP, with its 40 per cent share in the Consortium and no stake at all in Aramco, was directly interested only in Iran. But the four US majors in Aramco also held shares in the Consortium, and their interlocking interests ensured that the majors' dealings with Saudi Arabia and Iran were intertwined.

When the Consortium restarted the Iranian oil industry in 1954 after more than three years of shutdown owing to the nationalisation crisis, the immediate problem was to find room for Iranian oil in markets where Iran's posi-

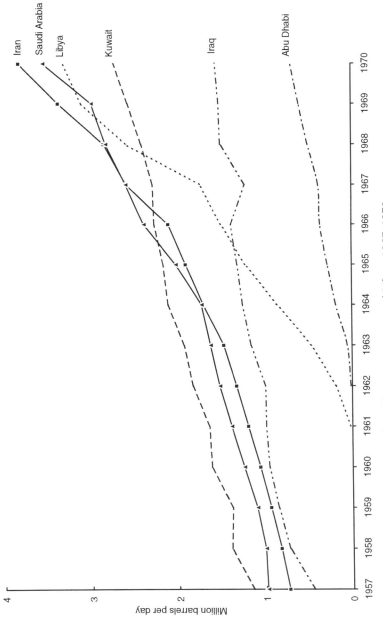

Figure 7.1 Crude oil production of Middle East countries and Libya, 1957–1970

Million barrels per day

Iran
Saudi Arabia
Libya
Kuwait
Iraq
Abu Dhabi

tion had been taken by other producers. To guarantee Iran's position the Consortium pledged that it would raise its production by stages to at least 603,000 bpd in 1957, and that after 1957 it would adjust production in Iran to 'reasonably reflect the trend of supply and demand for Middle East crude oil'.[58] This vague second pledge was generally taken to mean that the Consortium would raise Iran's production in line with the growth in output of the Middle East as a whole. Over the next decade, these targets were achieved. The Consortium's liftings in 1957 were 672,000 bpd, and between 1957 and 1965 they grew faster than overall production in the Middle East, comfortably outstripping the growth rate achieved by Aramco.[59]

The Shah, however, was not wholly satisfied. Although Iran's oil output was growing faster than Saudi Arabia's or Kuwait's, Iran was still in third place behind them in the Middle East production rankings (see figure 7.1). The Shah thought that Iran, as an ally of the West with a population of over twenty-five million, vast undeveloped natural resources and a stable government, should be restored to its pre-1951 position of number one oil producer in the region. That other countries which he regarded as underpopulated and barren sand wastes should be receiving more oil revenues than they could spend seemed to him to show 'an incomprehensible lack of judgement on the part of the oil companies'.[60]

In early 1966 the Shah was more than usually dissatisfied. The Consortium's performance in the previous year had been disappointing, an increase in production of only 9 per cent, compared with 18 per cent by Aramco and 10 per cent for the whole Middle East. Abu Dhabi, still a newcomer to oil production, had enjoyed a 51 per cent rise in output.[61] In an address to the Majlis, the Shah declared that the Consortium's 9 per cent increase was not enough.[62] He complained to the Consortium about the 'monstrous' rise in Abu Dhabi, stressed his concerns about the defence of the Persian Gulf area and pointedly compared the high prices which the USA was asking for F-111 aircraft with the much lower Soviet prices for MIG fighters. The Consortium, he said, must present him with a forecast of future oil liftings so that he would know exactly how far he could count on oil to finance his economic and defence plans.[63]

He also made representations to the British and US governments. The British Foreign Office in turn took up Iran's case with the State Department and suggested that they should urge the companies to raise their production in Iran. The State Department, after talking with the US majors, was satisfied that they were doing their best and took the view that the production dispute was something for the Consortium to settle with Iran without interference.[64]

To that end, a top-level Consortium delegation, including Bridgeman for BP, met the Shah and his main ministers in Tehran in July. The Iranians stressed that a foreign exchange deficit could only be avoided by a 17½ per cent increase in oil revenues, without which Iran's national development would be thwarted.[65] The Shah, as usual, emphasised his defence worries and the need for a militarily strong Iran. He stressed that if the West could not provide enough money for his military and economic plans, he would have to turn to the Soviet bloc. In fact, he had already decided to obtain less sensitive items such as anti-aircraft guns from the East, but wanted to continue purchasing advanced equipment, such as planes and tanks, from the West. This worried the British and Americans, who did not like the idea of Soviet technicians in Iran 'fooling around' with sensitive American and British equipment.[66]

In the autumn, the Iranians presented new demands: a 12 per cent annual production increase, the relinquishment by the Consortium of part of its concession area, and the supply of oil at cost price to the National Iranian Oil Company (NIOC) to sell for its own account in Eastern Europe, where it would not compete with the Consortium companies.[67] If the demands were not met, Iran would seize part of the concession area. 'It seems', wrote a Consortium executive, 'we have a pistol held at our head'.[68] The British press raised the tension by talking of a showdown and comparing the situation to the Iranian nationalisation crisis of the early 1950s.[69]

The US majors conferred again with the State Department, who thought that the Shah's needs for more money were largely of his own making, resulting from his penchant for state-of-the-art military equipment. 'It is not our idea', noted a State Department official, 'that he should have all the most expensive equipment'.[70] Although the British government thought that pressure should be put on the Consortium to go as far as it could to meet Iran's demands, the US government still favoured a more hands-off policy of leaving the companies to make their own decisions.[71] The US government therefore held back while a Consortium team, including David Steel who was by this time a managing director of BP, flew to Tehran and negotiated a settlement.[72] Under its terms, announced in December, the Consortium agreed that its production in 1966 would be 12 per cent higher than in 1965, and that there would be further increases in 1967 and 1968. The Consortium was also to supply the NIOC with 15 million barrels of crude at discounted prices in 1967, rising to 45 million barrels in 1971, for barter transactions in Eastern Europe. In addition, the Consortium was to relinquish 25 per cent of its unexploited land within three months.[73]

For the moment, at least, the Shah was happy with the Consortium.[74] The events of the next few months helped to sustain his good mood. Iran, being

27   The coronation of the Shah as Emperor of Iran on his forty-eighth birthday in October 1967

non-Arab, did not join in the Arab oil embargo that was imposed against supporters of Israel after the outbreak of the Six-Day Arab–Israeli War in June 1967. To help make up for the loss of Arab supplies, the Consortium's liftings in Iran were rapidly increased, so that in 1967 the Consortium's production was 22 per cent higher than in 1966.[75] This unforecast increase helped to boost the Shah's oil revenues beyond expectations.

At the same time, however, the changing balance of power in the Persian Gulf region fuelled the Shah's demands for higher revenues. The Six-Day War brought the USA and Israel much closer together in a special relationship which, by turning the Arab world against America, limited the USA's ability to exert a positive influence in the Persian Gulf. At the same time, Britain continued to withdraw from her old imperial role in the region. In November 1967 the last British troops withdrew from the colony of Aden, which was swallowed up by the Marxist regime of the People's Republic of South Yemen. In January 1968, Harold Wilson's Labour government, beset by economic problems and the urgent need to cut government spending, announced that British forces stationed in the Persian Gulf would be withdrawn by the end of 1971. This withdrawal would be the end of the last important vestige of the *Pax Britannica* which had kept the peace in the Gulf since the nineteenth century.

The West, worried that the Soviet Union would try to fill the power vacuum left by Britain, looked to Iran and Saudi Arabia to act as guardians of stability in the region.[76] The Shah relished the role and lost no time in pointing out to Bridgeman that it would place an additional defence burden on Iran. The Consortium, he said, 'would clearly have to make a contribution to this additional expense'.[77] The Consortium's forecast oil offtake of 915–925 million barrels in 1968 was 'totally unacceptable'.[78] To meet the requirements of Iran's Fourth Five-Year Plan (1968–72), the Consortium would have to produce revenues for Iran of $865 million in 1968, which would require substantially higher production than that forecast.[79] The Shah warned that if his requirements were not met, 'he would be forced to take "very unpleasant steps" which could include the expropriation of a proved oilfield'.[80]

The threats continued. In March 1968, the Iranian Finance Minister, Jamshid Amouzegar, told Consortium representatives that if they stuck to their forward estimates on a 'take it or leave it basis', it would be like 'cornering someone'. He said that in negotiations he personally always avoided doing that because a cornered person would 'use his claws to get out of the corner'.[81] The Shah gave the Consortium until 20 April to come up with higher production estimates.[82]

The State Department viewed the Consortium's problems with the Shah as a threat to Middle East stability and became more interventionist. Under-Secretary of State Eugene Rostow told top executives of US members of the Consortium that he was worried about the situation in the Middle East, particularly the growing Soviet influence in the region and the continuing Arab–Israeli skirmishes. He thought that the situation was even more serious than it had been when the Six-Day War broke out. He feared that there might be 'another blow-up and an oil boycott'. Iran was the strongest state in the area and was very important to the USA. The State Department, he revealed, had already discussed the Shah's demands for higher oil revenues with Iran and Britain. Despite warnings from the oilmen that favouritism towards Iran would endanger relations with Arab states, Rostow asked them to do whatever was possible to avoid a confrontation with the Shah. He said that the State Department would tell the Shah that the companies were trying to solve the problem and that the Consortium's production was 'likely to be nearer the Shah's demands than it now appeared'.[83] The Consortium, in other words, was now being put under pressure by the State Department to satisfy the Shah for strategic reasons.

Within a few days, countervailing efforts to dissuade the Shah from taking drastic action against the Consortium were made by an influential private emissary to Iran. He was John McCloy, one of the USA's top political elite, who had very strong connections in Washington, Tehran and with the oil companies, many of whom he represented as a lawyer.[84] Starting in 1962, the heads of the oil majors, including BP and Shell, had met twice-yearly in his presence with the permission of the US Attorney General, who had no objections as long as the meetings avoided cartel-like collusion which would violate the US antitrust laws. At first, the meetings were concerned mainly with Soviet oil exports, which were seen as part of a general Soviet offensive against the economic and political systems of the West. By the mid-1960s, however, the political situation in the Middle East had become the main item on the agenda of the McCloy group meetings.[85]

McCloy frequently used his exceptionally strong links with the political establishment in Washington to lobby on behalf of the oil companies. He could also obtain direct access to the Shah and his senior ministers, whom he visited at the end of March 1968, just a few days after Rostow had met with members of the Consortium. In conversations with Iran's Prime Minister, Minister of Court and the Shah, McCloy discussed the threat of a Soviet build-up in the Middle East, the importance of Iran as a guarantor of stability in the area, and the Consortium's forthcoming meeting with the Shah on 20 April. McCloy emphasised that 'it was particularly important to Iran that there be no "rupture" . . . with the west, meaning the oil companies'.[86]

Despite McCloy's visit, the Shah took a hard line when he met a Consortium team led by David Steel on 20 April. Steel reported that the Consortium hoped to be able to close the gap between its forward estimates and Iran's requirements by $20–30 million, or about one third, in 1968. The Shah replied that this 'still left a gap which it would need an inter-continental ballistic missile to cover while he had been thinking in terms of a pistol shot. The Shah said that nothing could stop Iran's development. Iran must have the revenue required for the Fourth Plan.' If the Consortium could not sell enough oil to produce the required revenues, he would 'give them a choice of weapons with which to shoot themselves'. These consisted of three alternatives: first, that Iran would seize the Consortium's Marun oil field; second, that Iran would set up a company to become a member of the Consortium; or, third, that Iran would make up the 'revenue gap' by itself selling oil which the Consortium would supply at a discounted price. At the beginning of his meeting with the Consortium team, the Shah had quipped that although he was wearing military uniform, 'it is not battle-dress'. He ended the meeting by saying 'that next time he might be wearing his battle-dress'.[87]

Afterwards, the Iranians said that they would 'call it a day' if the Consortium could provide the required revenue of $865 million in the Iranian year 1347. By switching from the Gregorian calendar (1 January to 31 December 1968) to the Iranian calendar (21 March 1968 to 20 March 1969) the Consortium would gain three months on the upward production trend, which would enable them to meet the Shah's target. The Consortium agreed to this clever expedient and the offtake dispute with Iran was resolved, for the time being.[88]

Looking ahead, though, the oil companies were deeply worried that the Shah's rapacity would make it extremely difficult, perhaps impossible, for them to keep up the political balancing act. As David Steel commented, 'the companies cannot meet the Shah's demands for increased offtake without creating tremendous problems elsewhere. If Iran should receive a preference, pressures for similar treatment would inevitably be exerted by Saudi Arabia and probably other producing countries, and it would be impossible for the companies to meet all demands.'[89] The US oil majors with interests in Arab countries had even more to worry about than BP. They were becoming increasingly concerned that America's pro-Iranian and pro-Israeli foreign policy in the Middle East might provoke an Arab backlash which would endanger US oil interests in Arab countries. They were displeased that the Shah had succeeded in putting pressure on the State Department which, via Rostow, had in effect told the companies to meet the Shah's demands. Saudi Arabia had already warned the companies that if they yielded to the Shah's threats, the Saudis would use the same tactics as Iran.[90]

If they entertained hopes that US policy might change after Richard Nixon won the US presidential elections in November 1968, they were soon disappointed. When Nixon assumed the Presidency, he inherited a difficult foreign policy legacy from his predecessor, Lyndon Johnson. The war in Vietnam was still unresolved, the 1967 ceasefire between Israel and Egypt was being constantly violated, six Arab states had broken diplomatic relations with the USA, and a military vacuum was in the making in the Persian Gulf.[91] Far from deserting the Shah, Nixon grew close to him. The two leaders developed a warm personal relationship, frequently exchanging greetings and sugared messages of admiration. Extravagant gifts of caviar, Persian carpets and other Iranian artefacts regularly arrived at the White House from the Shah and his most senior ministers. At the same time, Nixon and the Shah kept their nations in close alliance. Nixon saw the Shah as the strongest upholder of peace and stability in the Middle East. He admired the Shah's 'White Revolution' of economic and social reforms, and he was aware that the Shah's ambitious plans depended on higher revenues, not least from oil.[92] Henry Kissinger, Nixon's influential national security adviser, was no less enthusiastic in his support for Iran under the Shah. It was, he later wrote, 'one of America's best, most important, and most loyal friends in the world'.[93]

In January 1969, the same month that Nixon was formally inaugurated as President, a Consortium team led by David Steel met the Shah in Zurich, where he had gone for a medical check. After a congenial start, the meeting turned sour when Steel said that the Consortium's estimated exports in the Iranian year 1348 (year ending in March 1970) would produce revenues for Iran of $900 million. This was $100 million short of the amount required from the Consortium for Iran's Fourth Plan. Steel added that it also now looked as though the Consortium's payments to Iran in the year 1347 would be $10 million below the required amount of $865 million. The Shah was livid. He raged: 'You promised $865 million and you have broken your promise. I will remember this . . . It is unthinkable that my plans should have to be altered because you sit on my oil wealth and refuse to produce the oil. No doubt we will have to tighten our belts, but we will see that you have to tighten yours also.' With the Shah's fury unabated, the meeting ended abruptly 'in a very sombre and threatening atmosphere . . . no smile relieved the gloom'.[94]

Afterwards, the Shah sent a message that the Consortium must pay $865 million for 1347 and $1,000 million for 1348. He threatened that if this were not agreed by the time of the Consortium representatives' next visit to Tehran in March, he would 'take away 50% of the Consortium' or take oil from it at cost.[95] At the March meeting, the matter was put off until May, but the outlook remained grim. In late March to early April the Shah visited Washington at Nixon's invitation for the funeral of former US President, Dwight D. Eisenhower, who had authorised US covert support for the coup that overthrew Musaddiq and restored the Shah's authority in 1953. While in Washington, the Shah had private meetings with Nixon and others, including the US Secretary of State, William Rogers. The Shah said that he was unhappy with the performance of the Consortium and attacked the oil companies for providing unnecessarily high revenues for Libya, Kuwait and Abu Dhabi.[96] He warned Nixon and Rogers that if the Consortium did not meet his demands for 1348 he would take retaliatory action. The State Department was convinced that the Shah was 'completely serious'. 'The

28 President Richard Nixon (on right at lectern) and the Shah of Iran (wearing medals) standing among others before the coffin of the former US president, General Eisenhower, in Washington in April 1969

"clincher" to them (if one was required) was that the Shah told this to Nixon and Rogers and won't back down'.[97]

In Tehran, the situation was regarded no less seriously. Tim O'Brien, a senior Consortium executive, told Alam, now Minister of Court, that the revenue shortfall in 1347 was now estimated to be $19 million. Alam 'was greatly taken aback, and said "His Majesty will be very upset at this – very upset indeed. This is very serious."'.[98] Three days before the May meeting, the Shah was described by O'Brien as 'implacable'. O'Brien feared that there would be 'an outright explosion' if the Consortium could not come up with the $1,000 million wanted for 1348.[99] The British press raised the temperature with headlines like 'Showdown looms in Shah's oil row with BP' and 'Oilmen and Shah fight over $1000m'.[100] On the day, Steel, again leading the Consortium team, put forward an estimate of $925–930 million for the Consortium payments from oil exports in 1348. He also suggested bringing forward one month's payment from 1349 into 1348. This would add an estimated $75–80 million to the $925–930 million, which would be enough to

29   Muammar al-Qaddafi, head of Libya's Revolutionary Command Council and Prime Minister of Libya, in 1970

meet Iran's target of $1,000 million. The Shah accepted these proposals.[101] Once again, a crisis had been averted.

The underlying problem, however, remained. The Shah, determined to make Iran the number one power in the Middle East, continued, according to a State Department report in September 1969, 'to let everybody know that Iran, as the major riparian power, intends to play a leading role in the area after the British withdrawal'.[102]

Meanwhile, developments in Libya heightened the oil companies' concerns about the radicalisation of the Arab world. On 1 September the moderate, pro-Western Libyan regime of King Idris was overthrown by a coup of radical young officers led by Muammar al-Qaddafi. Inspired by the example of Nasser, Qaddafi picked up the mantle of revolutionary Arab leader, breathing new energy into the anti-Zionist, anti-imperialist cause. At about the same time, the USA began deliveries of new Phantom aircraft to Israel, helping to fan the flames of Arab radicalism.

These developments so alarmed the US oil companies that they took up a vigorous campaign to change US foreign policy in the Middle East. They took every opportunity they could to put across to the US government their views that the USA's pro-Iranian and pro-Israeli policies were radicalising the moderate Arab states, especially Saudi Arabia. They lobbied the State Department themselves and through John McCloy.[103] In September, they presented their views to the US Ambassador-designate to Tehran, Douglas MacArthur II (son of the famous general).[104] In November, Rawleigh Warner and J. Kenneth Jamieson, the heads of Mobil and Standard Oil (NJ) respectively, wrote to Nixon requesting the opportunity to put their views to him in person. 'Our long experience and extensive contacts in the Arab world indicate', they wrote, 'that America's political, strategic and economic interests throughout the area are in jeopardy. Anti-American sentiment has intensified to an unprecedented level. Unless present trends are reversed, the United States is faced with the loss of the Arab Middle East.'[105] Nixon saw them on 9 December, but their meeting was overshadowed by another event.[106] On that same day, Secretary of State Rogers made a public speech which became famous as the 'Rogers Plan' to achieve an Arab–Israeli peace. One of the features of the Plan was that the Israelis would have to withdraw from the territories which they had taken from the Arabs in the 1967 Six-Day War.[107] That was anathema to the Israelis, who unleashed a storm of protest. The White House was deluged with a flood of letters and petitions from American Jews attacking the Rogers Plan as a betrayal of Israel. Some of the protesters saw the Plan as a sop to oil interests. 'Let America not be drugged by the power of oil', ran one petition; 'oil interests should not dictate fate of Israel', ran another.[108] The American Jewish protesters need not have worried. They were a far more powerful lobby than the international oil companies. The visit of Warner and Jamieson made no apparent difference to US policies towards Israel or Iran. Far from modifying his pro-Iranian stance, Nixon instructed Peter Flanigan, his assistant on international economic matters, to keep up the pressure on the Consortium to raise production so that Iran's revenues would be increased.[109]

While Flanigan applied pressure on the Consortium in Washington, Ambassador MacArthur did the same in Tehran. Ignoring the oil companies' earlier attempts to influence him, in March 1970 he treated the Consortium members to a one-and-a-half hour monologue on the importance of Iran and said that Iran should be given 'a strong preference' for oil production. When the Consortium responded that the stability of the Middle East would be endangered if Iran was given preference, MacArthur 'did not appear to listen'.[110]

As for the Shah, he, of course, could be relied on to be relentless in his demands. In April the Consortium suggested at meetings with Amouzegar and others that Consortium operations in 1349 would raise revenues of $1,010 million for Iran. This was well below the $1,155 million which was required from the Consortium to meet the Fourth Plan. The Shah telephoned Amouzegar during one of the meetings and reportedly 'tore . . . [him] to ribbons'. Amouzegar was 'visibly shaken'. The Shah was incensed by the Consortium's estimate of $1,010 million and insisted that there could be no compromise on $1,155 million.[111] After further talks, a *procès verbal* was agreed in May, meeting the Shah's demands.[112] Flanigan reported this result to Nixon.[113] The Shah, once again, had got his way.

The *procès verbal* of May 1970 came at a turning point in relations between oil companies and producing countries. During the 1960s, the producing countries, especially Iran, had sought to increase their oil revenues by putting pressure on the oil companies to increase production. As a result of this pressure, which was known in the industry as 'volume push', the oil companies were faced with the almost constant problem of disposing of ever-increasing supplies of oil. The ways in which BP found outlets for its burgeoning production are described in chapters nine and ten.

However, even as the *procès verbal* was being signed, events were taking place in Libya which signalled a change in the producing countries' approach to raising their oil revenues. After inconclusive negotiations for increased oil prices with Libya's concessionaire companies, Qaddafi took action in May which turned 'volume push' on its head. Instead of demanding increased production, he did the opposite and restricted the amount of oil that the companies were allowed to produce. The threat of an oil surplus was replaced by the threat of shortage. Libya's enforced production cuts were intensified in the summer and proved effective. In the autumn, the oil companies conceded the higher prices and taxes which Libya demanded.[114] This set in motion a train of events which brought about the dissolution of the concessionary system which had been the traditional basis of the international oil industry.

=== 8 ===

The 'Holy Grail' of exploration

While the rise of Qaddafi intensified anti-Western radicalism in the Arab world, BP's explorers were making major oil discoveries in politically more secure areas. In a bid to diversify its sources of crude oil away from the politically volatile Middle East, BP had launched a great burst of international exploration in the 1950s (see chapter four). The results by the early 1960s were mixed. New reserves had been discovered in Trinidad, Colombia and Canada, but only on a modest scale by international standards at that time. In Nigeria, where Shell was the operator in the Shell-BP partnership, substantial new reserves had been added in a multitude of comparatively small oil fields. In Libya, BP had discovered the giant Sarir field. Overall, BP's estimated recoverable reserves outside the Middle East had increased from a mere six million barrels in 1950 to more than two billion barrels after the Sarir discovery was made in 1961. In the next decade, that total would be increased more than fivefold, mainly by new successes in frontier exploration in Alaska and the North Sea.[1]

ALASKA

Lying at the north-western extremity of the North American continent, Alaska was real frontier territory. It had been purchased by the USA from Russia in 1867 for a price equivalent to less than two cents an acre. This would later seem a cheap price to pay for the ejection of Russia from the American continent and for Alaskan oil, but at the time many Americans regarded the purchase as an expensive folly. Indeed, Congress was prevailed upon only with great difficulty to appropriate the funds required to complete the purchase.[2] What, Americans asked, was the value of this wilderness, which was cut off from the Lower 48 states of the USA and largely uninhabitable?

185

Alaska's quarter of a million people inhabited an area bigger than France, Germany, Britain and Italy combined, giving an average population density of one person to two square miles, compared with 862 persons to each square mile in England. Three quarters of the population lived in the three cities of Anchorage, Fairbanks and Juneau, south of the mountain barrier of the Brooks Range, which extends unbroken from the Canadian frontier in the east to the Bering Sea in the west. To the north of the Brooks Range lies the Arctic North Slope, a vast plain of treeless, frozen tundra sloping down from the Brooks foothills to the Arctic Ocean, where land and sea merge for much of the year in one white, flat wilderness of ice and snow. Home to Eskimos, caribou, wolves, polar bears, seals and many species of birds, the North Slope is forbiddingly cold for most other forms of life. For most of the year, the temperature stays below freezing and in the heart of winter there is constant darkness, as the sun never rises above the horizon. With the coming of summer all is light. The sun never sets and the temperature rises slightly above freezing, melting the top few inches of soil, turning it into a marshland carpeted with millions of tiny flowers of dazzling colours, laced with thousands of streams and lakes. Beneath the top surface, however, the ground remains permanently frozen through the short summer, before the winter again closes in.

Oil seepages had been known in Alaska for many years and there had been some small-scale production and refining in southern Alaska from 1907 until 1933, when the refinery was destroyed by fire. Meanwhile, in northern Alaska the US government, wishing to safeguard future oil supplies for the US navy, had set aside some 35,000 square miles in the western half of the North Slope as a Naval Petroleum Reserve. Subsequent exploration carried out by the US Geological Survey for the US navy resulted in the discovery of several oil and gas fields. The largest were the Gubik gas field and the Umiat oil field on the Colville River near the Naval Reserve's eastern boundary (see map 8.1). These fields were, however, too small and remote to justify development.

In 1958 federal lands on the North Slope between the Naval Reserve in the west and the Arctic Wildlife Reserve in the east were opened for leasing. The next year, Alaska achieved statehood as the 49th state of the USA and proceeded to exercise its right to select federal acreage as state land, to which it was entitled under the provisions of statehood. The area of the North Slope selected by the state was a strip on the shore of the Arctic Ocean north of 70°N latitude.

The main excitement in the late 1950s, though, was in southern Alaska, where the Richfield Corporation made a commercial discovery on federal

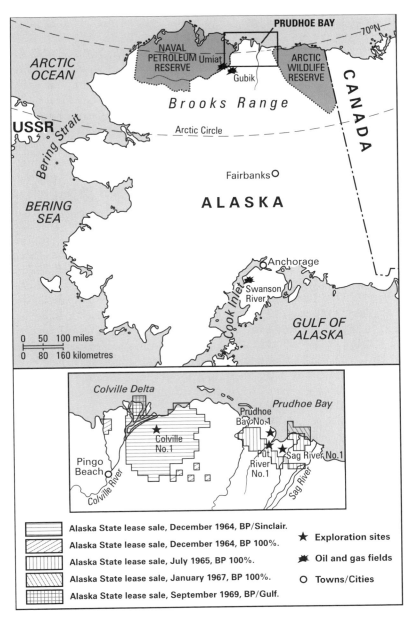

Map 8.1   Alaska: locating the Prudhoe Bay oil field

land at Swanson River in 1957. The US federal leasing system virtually guaranteed that this would result in a competitive frenzy for acreage. It allowed individuals or companies to apply for leases on blocks of 2,560 acres (about four square miles) of federal land for 25 cents an acre on a first-come, first-served basis, with simultaneous applications being decided by drawing cards from a box. Designed to protect the rights of the individual to a stake in the USA's mineral resources, the system was in practice a speculator's paradise. It resulted in exploration acreage being divided into a multitude of individual holdings, often held by dealers and speculators seeking a fast profit from a resale. The only way that a major company could gain significant continuous tracts of prospective land was by wheeling and dealing in land options, subject to the overall limitation that no single company was allowed a land holding of more than 300,000 acres of federal land in Alaska.

The Swanson River discovery set off just such a leasing boom as oil companies flocked to explore in southern Alaska, whose ice-free ports were conveniently situated to supply oil consumers on the US west coast. The search for oil soon moved offshore into the Cook Inlet, where several oil fields were discovered and brought into production. By international standards, these were comparatively small, high-cost fields, but that did not put off US oil companies. They were used to wheeling and dealing for leases in the USA and to working small fields, whose profitability was underpinned by the pro-rationing of domestic production and quota restrictions on imports. BP, on the other hand, was not used to commercial land dealing and small oil fields. Most of its experience was in discovering giant oil fields in vast concessions obtained directly from governments in the Middle East.[3]

The southern Alaska oil boom was in full swing when BP and Sinclair made their 1958 alliance, providing for joint exploration. The two companies agreed to join forces in Alaska, where neither company was already involved, as well as Colombia. BP examined the prospects and decided that southern Alaska, though highly ranked by other companies, was of limited interest because it had already been intensively leased and was unlikely to contain very large individual fields. BP was much more interested in the North Slope and in particular, true to BP tradition, in the giant anticlines in the northern foothills of the Brooks Range. 'Alaska', wrote BP geologist Peter Kent, 'is probably the world's last major oil province available for exploration . . . The areas of sedimentary basins are very large. That of the Arctic Slope measures 105,000 square miles – larger than our entire Persian concession . . . It contains a wealth of drillable anticlines on the Persian scale, with lengths of the order of 20 miles . . . the prospects are of a high order'. Harry Warman, another BP geologist, agreed, arguing that the most attractive area in the North Slope was the Brooks foothills, which contained simple folding 'on a scale unique in North America'. Moreover, this area had not been extensively leased because other companies had used up most or all their permitted federal land holdings in the Cook Inlet area. With most of its competitors hamstrung by the Alaskan acreage limitation, BP had an opportunity to establish a strong land position on the North Slope.[4]

Peter Cox, who was by this time managing director of the BP Exploration Company, was receptive to these arguments. He recommended to Maurice Bridgeman, at that time BP's main board director of exploration, that BP should enter into joint exploration with Sinclair in Alaska. In support of his recommendation, he set out one of the classic arguments for frontier exploration: 'Unless our entry into this field is secured at this relatively early stage in the country's exploration the cost of entry later will be high if commercial production on a substantial scale proves feasible within the next ten years.' Bridgeman and his board colleagues agreed.[5]

If BP and Sinclair had set up a joint company to explore in Alaska, their joint acreage might have been restricted to the single maximum of 300,000 acres. They therefore decided to set up independent subsidiaries, though in practice they would co-operate on the basis of equal shares. BP duly set up a new subsidiary, the formal proceedings being carried out in a twenty-minute meeting in New York in, one participant later recalled, 'a mood of great hilarity – most unusual for such a major move'.[6]

The most important immediate objective was to acquire acreage, a task which involved precise procedures which had to be closely followed if the title was to be secure. Any movement by a major company was bound to attract attention and might have introduced competition. BP, needing both land-dealing know-how and secrecy, engaged George Jenkinson, an American with oil leasing experience, as land consultant. He was given power of attorney and detailed instructions on BP's acreage requirements, focusing on a group of anticlines in the Brooks foothills near those where oil and gas had been discovered at Umiat and Gubik. Jenkinson prepared about seventy separate applications for lease blocks of 2,560 acres each and filed them all simultaneously at the land office in Fairbanks, so that by the end of February 1960 BP held leases or options to lease on about 200,000 acres on the North Slope. A much smaller acreage was also acquired in southern Alaska, where BP participated in some exploration plays, but its concentration was focused mainly on the northern Brooks foothills.[7]

In the summer of 1960, BP put its first geological survey party in the field to begin mapping the foothills and identifying the most promising structures, which could be done only in the short summer season when the snow melted and the rocks could be seen. By 1963 geological and seismic surveys had located the anticlines thought to be most promising, and in the summer of that year a drilling rig was brought from Canada on a barge down the Mackenzie River to the Arctic Ocean, then westwards along the coast of the North Slope and up the Colville River to a landing point forty miles upriver at Pingo Beach. There the barge was unloaded, the rig assembled and, after the winter freeze-up had hardened the ground, the rig was hauled sixty miles across country to a drilling location in the Umiat area of the Brooks foot-hills.

By that time the only company to have drilled on the North Slope was Colorado Oil and Gas, who used a small rig that the US Geological Survey had left behind to drill a well near the Gubik gas field in 1963. BP and Sinclair paid 90 per cent of the drilling costs in return for a 50 per cent share in Colorado's Gubik acreage. The well, however, turned out to be dry.

Between the end of 1963 and the spring of 1965 BP and Sinclair used the rig brought from Canada to drill six wells in the Brooks foothills. Working in blizzards and temperatures far below zero was slow and hazardous. Metal became brittle and snapped, lubricants solidified, drilling mud froze and simple matters, like having to wear mittens, hampered straightforward oper-ations like tightening and loosening bolts. Worst of all, the results were poor. Although some oil shows were encountered, they were in 'abysmally bad reservoirs' of low porosity and permeability. The Brooks foothills, on which such high hopes had been pinned, turned out to be a complete disappoint-ment.[8]

As in Canada, BP had been influenced by its longstanding preference for exploring in areas with a geological likeness to the Zagros foothills in Iran. In 1959 the US Geological Survey opened all the information obtained on behalf of the US navy in the earlier exploration of the Naval Reserve on the North Slope. Wilf Bischoff, a BP geologist who subsequently joined the US oil company Sohio, later recalled that the problems of poor reservoir quality in the Brooks foothills were fully documented, 'but BP did not dig deeply enough due to its preoccupation with the simple structural likeness to Iran'. Another geologist who was there at the time was also critical of BP's over-rating of the Brooks foothills. This was Alwyne Thomas, who was appointed executive vice-president of BP's Alaskan subsidiary in 1963. In 1964 he wrote: 'It is unfortunate that most of the unit appraisals on our files here were written in superlatives, creating the general impression that the anticli-nal belt between Shale Wall and Gubik had outstanding prospects. The evi-dence of poor reservoir development throughout the Cretaceous section was ignored, and an optimistic view was pressed with enthusiasm and energy by those concerned with exploration in BP Alaska. This has led to a series of wells all investigating the same basic problems in a very restricted area. We must accept the probability that this part of the programme will be abortive.'[9]

In exploration, however, the process of repositioning as understanding grew could play as big a part as initial commitment. Nowhere in BP's exploration history was this more clearly demonstrated than in Alaska, where if BP had not initially seen potential in the exposed anticlines of the Brooks foothills it might never have ventured into the coastal plain further north.[10]

While BP and Sinclair were drilling unsuccessfully on their first choice acreage in the foothills, they also turned their attention to the state lands in the flat coastal plain. Here, there was no scope for surface geological surveys as there were no visible outcrops on the featureless landscape. Relying on the evidence of seismic surveys, Jim Spence, chief geologist of BP Alaska, identified a huge coastal arch extending from Colville in the west to Prudhoe Bay in the east beneath the state lands. The arch had two crests at Colville and Prudhoe Bay respectively, joined by a saddle in the middle. Spence was convinced that 'this vast basement of the Coastal Arch must be rated as an extremely attractive oil prospect. Comparison with other oil producing areas shows that this is the almost classical location for large oil fields or even of oil provinces ... There are closed highs [crests] with a total enclosed area of more than 200,000 acres ... in view of the vast size of these closed structures and other favourable factors, it is recommended that this whole structural province is worth a major exploration effort'. This would prove to be a brilliant appraisal. BP and Sinclair, Spence pointed out, had a commanding lead over competitors because they were the first to have made seismic surveys of the area 'which must surely be the largest such unexplored prospect in the United States'.[11]

BP and Sinclair seemed to be in the near-perfect frontier exploration situation of having identified a major new prospect in previously unexplored territory ahead of the competition. Yet they still had no acreage, for the state lands on the North Slope had not been opened for leasing. After an approach from BP/Sinclair, the state decided to open the western half of the area, covering the Colville crest, for competitive bidding in December 1964. Under this procedure leases were awarded to the highest bidder, unlike the system of first-come, first-served on open federal lands. Alwyne Thomas was in complete agreement with Spence's appraisal and formulated BP's bidding strategy accordingly. At the state sale BP and Sinclair purchased the major part of the lands offered, including 317,934 acres covering the Colville crest at an average price of $12·47 an acre.[12]

They had still not drilled on the Colville structure when the state held its second sale of North Slope acreage in July 1965. This time, the acreage on offer covered the Prudhoe Bay structure to the east of Colville. In Spence's view, based on his study of the tilting of the two structures over time, Prudhoe Bay was a better prospect than Colville. Sinclair, disappointed by the failure of the drilling in the foothills, was not convinced by Spence's analysis and decided against participating in the bidding. BP therefore decided to go ahead on its own. By this time, however, BP/Sinclair's early lead over competitors was diminishing. Other companies had become interested in the North Slope and had made seismic surveys of their own. They included the Richfield Corporation and Humble Oil, a subsidiary of Standard Oil (NJ). These two had formed an alliance which could muster much greater financial resources than BP on its own. BP therefore knew that it would face strong competition for the most sought-after acreage, which would be on the structure's crest, the high point in which oil was most likely to be trapped as a result of its upward migration. By analysing the history of the structure's tilting in another superb piece of geological interpretation, Spence calculated how far the oil-filled section of the reservoir (if there was one) extended down the flanks of the structure as it dropped away from the crest. On the basis of his appraisal, BP submitted bids not only for the crest, but also down the flanks. When the results of the sale were announced BP had, predictably, been easily outbid by Richfield/Humble on the crest acreage, but had succeeded in acquiring 81,137 acres on the flanks at an average price of $17·80 an acre. This would prove to be 'unquestionably the lowest per barrel acquisition cost in any US competitive lease sale'.[13]

That, however, could not have been foretold at the time. Indeed, the more immediate news was discouraging. In late 1965 to early 1966 BP and Sinclair drilled their first well on the Colville crest, but it encountered only small, uncommercial quantities of oil. This was their seventh unsuccessful well on the North Slope. Disappointed, they came to a farm-out agreement with Union Oil, under which Union would drill another well in return for a share in some of BP/Sinclair's Colville acreage.[14]

Despite the Colville setback, BP had still not given up hopes of finding oil at Prudhoe Bay. Admittedly, the Colville crest was higher than its Prudhoe Bay counterpart, which, on the principle that oil migrates upwards, would have suggested that if Colville was dry, so was Prudhoe. Even Spence had to admit that the prospects of the whole North Slope had to be downgraded after the Colville failure. Yet, by retracing the evolution of the coastal arch through geological time, he argued that Prudhoe had been higher than Colville at the time of oil migration. This increased the chance of finding oil at Prudhoe Bay.[15]

In this situation, BP decided to hang on to its Prudhoe Bay acreage and to wait for others to take up the running in sinking wells. BP and Sinclair were the only companies to have drilled on the North Slope since Colorado Oil and Gas's initial well in 1963. Now that Atlantic Richfield (ARCO, newly formed by the merger of Atlantic Refining and Richfield) and Humble had put up $2½ million for Prudhoe Bay acreage the feeling in BP was that 'whatever they say they must drill it. Consequently we are in an excellent position to play a waiting game.' An offer from ARCO/Humble to farm in to BP's acreage by drilling a well on it was therefore turned down. While waiting for ARCO/Humble to drill, Monty Pennell, general manager of BP's exploration department, authorised Frank Rickwood, by this time regional manager for exploration in the Western Hemisphere, to bid up to $250,000 in the third state lease sale of North Slope acreage, held in January 1967. In the event, Rickwood acquired 15,259 acres at Prudhoe Bay at an average price of $17·25 an acre, a total expenditure of $263,000.[16]

But the drilling news continued to be disappointing. ARCO/Humble's first well, Susie No. 1, was drilled on federal land in the Brooks foothills, a long way south of Prudhoe Bay, and was completed as a dry hole in January 1967. A few weeks later, Union Oil's farm-in well at Colville was also completed as a dry hole. BP, its confidence in the North Slope diminishing, began to think of withdrawal. Its drilling rig was stacked at Pingo Beach, ready for shipment out; the Los Angeles office, a managerial base for directing Alaskan exploration, was closed and its staff disbanded; and informal feelers were put out to other companies for the sale of BP's North Slope acreage.[17]

ARCO/Humble decided, however, to drill one more well. Their Susie rig was hauled to Prudhoe Bay and started drilling Prudhoe Bay No. 1 well in April 1967. Drilling was suspended in the summer, when it was difficult to operate without damaging the tundra, but resumed in the winter. In January 1968 rumours of a possible discovery were circulating. They were confirmed in March when ARCO/Humble announced that they had struck oil and intended to drill a second well at Sag River, seven miles south-east of Prudhoe Bay No. 1. The Sag River No. 1 well, which was drilled using the rig which BP had laid up at Pingo Beach, struck oil in June, confirming that Prudhoe Bay was a major oil field.[18]

The discovery precipitated a frenzy of activity as oil companies flocked to the North Slope in a classic American oil rush. There was still considerable unleased acreage around Prudhoe Bay and the announcement by the State of Alaska that it would hold another lease sale in September 1969 raised the exploration temperature to fever pitch. Rejecting offers by

ARCO to purchase its Prudhoe Bay acreage, BP hastened to resume drilling operations. There were many rigs available in southern Alaska and one of these was hauled through the forest, dismantled and loaded onto barges in the Cook Inlet, together with equipment and supplies. The small convoy set sail on the long northward voyage around the Aleutian Islands, on through the Bering Straits and then eastwards through the Arctic Ocean, eventually arriving at the depot on Foggy Island, a few miles from Prudhoe Bay, in mid-August, just in time to unload and get back to the Pacific before the winter ice-pack re-formed. The rig was held at Foggy Island until the winter freeze-up hardened the ground so that the rig and equipment could be hauled to the drilling site on the banks of the Putuligayak River, three miles south of ARCO/Humble's original discovery well. Drilling of the Put River No. 1 well began in November 1968 and in March 1969 the well struck oil. BP knew that it had a large share of the Prudhoe Bay field, but wishing to preserve secrecy in the run-up to the lease sale it issued only a tight-lipped statement that it had found oil, giving no indication of the estimated quantity.

Meanwhile, the North Slope was being transformed by the exploration boom. A jetty was built on the shores of Prudhoe Bay for unloading summer shipments of supplies. Landing strips were laid for aircraft. Companies providing services and supplies for the oil industry set up workshops and warehouses stacked with steel pipes and casings, bags of cement and other equipment. Gravel roads snaked in all directions, linking base camps to drilling sites. There were vehicles of all kinds, from small pick-ups to caterpillar tractors and massive overland transporters, which came up the new 330-mile road laid over the Brooks Mountains to connect the Arctic Slope to Fairbanks, which became a boom town after years of economic eclipse. Oil companies brought in rigs by sea or flew them in from Fairbanks in Hercules freighters. New wells were drilled, while geophysical parties swarmed over the North Slope, carrying out seismic surveys. By the time of the lease sale Anchorage airport was home to at least a dozen corporate aircraft, as executives gathered for the sale, which was held in public with sealed bids for individual blocks being opened before an audience of oilmen, bankers and journalists.

The sale proved to be a bonanza for the State of Alaska. The three previous competitive lease sales of North Slope acreage had raised a combined total of some $12 million at an average price of about $12 an acre. The fourth sale in September 1969 raised more than $900 million at an average of $2,180 an acre. In the bidding, BP, in partnership with Gulf, bid $47·2 million for the most promising block of 2,560 acres, but they were easily outbid by others, including Amerada/Getty who acquired the block for $72·3 million, or $28,233 an acre. However, Gulf, who rated the Colville delta more highly than BP did, put up most of the money to acquire 15,360 acres in the delta, some twenty miles to the west of Prudhoe Bay. Later drilling there did not find oil.[19]

After the sale, BP announced that the estimated reserves in its share (about half) of the Prudhoe Bay field were nearly five billion barrels. The field was the biggest ever discovered in the USA and was of truly Middle East proportions. BP's estimated finding costs at Prudhoe Bay were a mere 0·30 cents a barrel, virtually the same as BP's Middle East average of 0·28 cents.[20] And that was before allowing for the unquantified premium which BP attached to finding new oil reserves outside the Middle East. Although BP had initially overrated the Brooks foothills, its record of exploration in Alaska had been in many respects a model of frontier exploration, a paradigm of the long-term search for giant oil fields in remote locations where BP gained the first entry and took an early lead over competitors. It was not, though, the only area where BP was engaged in frontier exploration. Closer to home, BP was also deeply involved in searching for oil and gas in the North Sea.

BRITAIN AND THE NORTH SEA

By the 1950s the Company was well established as the only significant explorer and producer of oil on land in the UK. It had started exploration in Britain in the 1930s at the instigation of Lees, who had noticed, typically, that the anticlines of southern England were similar to those in Iran and that surface oil seepages were common. The Company's first wells were drilled near the south coast, but encountered only minor, uncommercial shows of oil and gas. The search moved northwards and minor discoveries were made, the largest being in the Eakring anticline, near Nottingham in the east Midlands.

During World War II, when Britain desperately needed secure oil supplies, further drilling in the Eakring area led to the discovery of new fields at Duke's Wood, Caunton and Kelham Hills, all of which were intensively developed for maximum wartime production. When peace was restored and the national emergency was over, the tempo of the Company's UK exploration was reduced. But it picked up again in the 1950s and early 1960s, when improved seismic reflection methods were used to help discover about a dozen further fields. Most of them were in the east Midlands, though two, at Kimmeridge (1959) and Wareham (1964) were near the south coast in Dorset. Encouraged by the success at Kimmeridge, BP ventured offshore and

30 Fitter inspecting rock-drilling bits at Eakring in the UK in 1955

drilled England's first offshore test well at Lulworth Banks, an offshore anti-cline, in 1963, without finding commercial oil.

Though numerous, BP's onshore UK oil fields were extremely small, high-cost units by international standards. They were made economically viable as producers only by tariff protection against imported oil, which ceased at the end of 1964, when Britain entered the European Free Trade Area (EFTA). BP thereupon reduced its UK production and virtually ceased onshore exploration until the 1970s, when higher oil prices led to renewed interest in Britain's onshore oil prospects. In the meantime, however, offshore exploration in the North Sea had yielded major oil discoveries.[21]

31 Diver being prepared for descent to the seabed at Lulworth Bay, Dorset, UK, in 1963

BP's experience in UK land exploration caused it to be initially cautious about prospects in the North Sea. By inference and geological extrapolation, the North Sea was regarded as having similar prospects to onshore Britain, where the oil and gas fields that had been discovered were too small to be economically viable in the more expensive offshore environment. For that reason, George Lees gave the North Sea a low exploration rating.[22] In 1959, however, his prognosis was made obsolete overnight by the discovery of a giant gas field in the Gröningen province on the coast of the Netherlands (see map 8.2) by Shell and Standard Oil (NJ), acting in partnership. This immediately awakened interest in the southern basin of the North Sea because the geological conditions which had yielded gas at Gröningen were thought to extend under the sea.[23] Shell and Standard (NJ) agreed to conduct a joint offshore geophysical survey, which they invited BP to join. Norman Falcon, BP's chief geologist at the time, was in favour, as was Cox, who was attracted by the gas prospects. As for oil, Cox still adhered to Lees' assessment. 'We would not', he wrote in 1962, 'recommend any expenditure in search of oil alone, on the grounds that the chances of finding a field of sufficient size and capacity per well are too small'.[24]

Terms with Shell/Standard (NJ) were agreed and joint survey operations

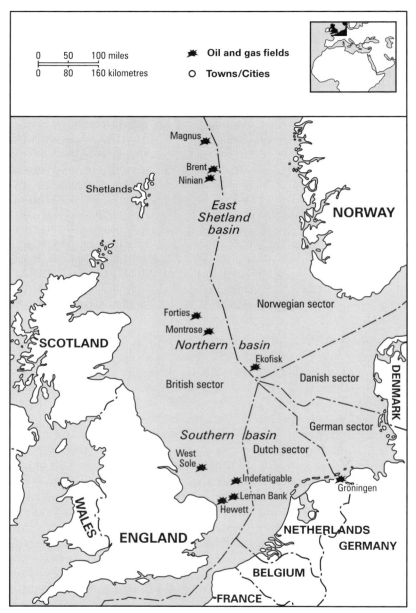

Map 8.2   The North Sea: opening up a new oil province

were carried out, though BP chafed under restrictions imposed by Shell, which limited the area covered and the freedom of the participants to exchange information with other companies. The joint survey agreement was therefore terminated in December 1963, leaving BP free to co-operate in surveys and data exchanges with other companies. By these means, BP avoided falling behind in the surveying of the North Sea, though it did not, as on the North Slope of Alaska, have a lead over competitors.[25]

As territorial rights over the seas and seabed still ended at the traditional three-mile limit, exploration farther offshore was limited to surveys, carried out on the basis of non-exclusive reconnaissance permits. The grant of licences giving the holders exclusive rights to drill for and produce petroleum outside the three-mile limit could not be made until the legal basis of their tenure was established. To that end, in May 1964 Britain ratified the UN Continental Shelf Convention, which gave coastal states sovereignty over continental shelves out to a water depth of 200 metres. In seas where coastal states had overlapping claims, the boundary was to be the median line, equidistant from their shores. Although negotiations on the median line continued between North Sea states, the British ratification of the UN Convention at least enabled the British government to issue licences over areas of the North Sea which definitely fell within British jurisdiction.

The terms on which licences would be issued were governed by the 1964 British Continental Shelf Act. This provided for the British sector of the North Sea to be divided into licence blocks of 250 square kilometres (96 square miles). Licences, valid initially for six years, would confer exclusive rights on the holder to drill for and produce petroleum. Unlike the State of Alaska's competitive lease sales, licences in the British North Sea would be awarded not to the highest bidders, but at the discretion of the Ministry of Power on the basis of several criteria. One of these was the extent to which the applicant had already made or was making a contribution to Britain's fuel economy.[26]

This gave BP good reason to hope that its record of onshore exploration and production would help it to secure North Sea licences.[27] It could also expect its position as a wholly British company, majority-owned by the British government, to count in its favour in licence awards. One BP manager even thought that BP had a 'moral right' to favourable treatment.[28] For whichever reason, Angus Beckett, the Permanent Under-Secretary at the Ministry of Power, later remembered ensuring that BP received the most sought-after blocks in the face of stiff competition from other companies.[29] This is borne out by the results of the first licensing round in September 1964, when BP received twenty-two of the thirty-five blocks for which it had applied, including block 48/6 in the southern basin, which was the most desired block of all.[30]

This was where BP decided to drill its first North Sea well. First, however, BP had to obtain a suitable drilling rig. There was already a long history of drilling in shallow, calm, inland waters such as the Caspian Sea in Asia, Lake Maracaibo in Venezuela and the Louisiana swamps in North America. Offshore drilling in open seas was a newer development, which had started in the Gulf of Mexico in the 1940s. In the mid-1960s, the technology was still quite immature and there was far to go in pushing back the limits to make offshore drilling possible in greater water depths, wave heights and other environmental extremes.

BP had already gained some offshore experience in the shallow waters of the Gulf of Paria off Trinidad and the Persian Gulf off Abu Dhabi. The waters in the southern basin of the North Sea were also quite shallow, with a depth of some eighty feet at BP's chosen drilling location in block 48/6, about forty miles off England's east coast. This fell comfortably within the range of existing offshore drilling technology and could be drilled by a self-elevating rig of the same type as the *ADMA Enterprise* which had been used off Abu Dhabi. This type of rig consisted, basically, of a drilling rig mounted on a platform, with legs that could be raised and lowered. When the legs were lowered to rest firmly on the seabed, the platform could be jacked up out of the water to a height where it was safely out of reach of waves. Drilling could then take place. Once a well had been completed, the platform could be jacked down so that it floated on the water, the legs raised, and the structure, resembling an upturned table with its legs in the air, towed to another location.

To drill on block 48/6, BP chartered a barge from Wimpey and had it converted into a self-elevating drilling rig by, among other things, the installation of second-hand drilling equipment from Trinidad. This rig, named the *Sea Gem*, was towed to block 48/6 in June 1965. Meanwhile, BP also ordered a new, much larger rig of a different design, known as a semi-submersible, which would be more suitable for drilling in deeper, rougher waters. The semi-submersible was a comparatively new type of rig which had been developed in the USA. It consisted, basically, of a drilling platform with legs, to which flotation chambers were attached at the bottom. These enabled semi-submersibles to float, not on the rough surface of the water, but in stiller water beneath the surface, making them comparatively stable. BP's semi-submersible, which would be named the *Sea Quest*, was ordered in the summer of 1964. Weighing 15,000 tons and standing 320 feet high, it had a triangular-shaped platform with three legs, each 35 feet in diameter and 160 feet

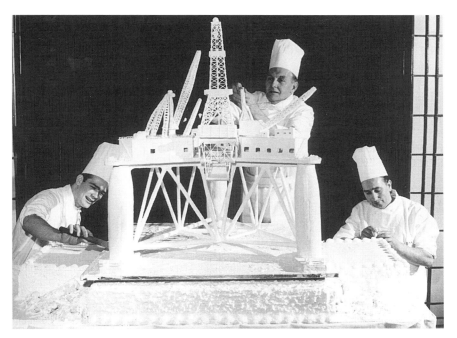

32   'An engineering job in pastry': Christmas cake, weighing over 4 cwt, of the
Sea Quest, in December 1966

long. To emphasise its enormous size, one press release showed an artist's
impression of the structure towering over Piccadilly Circus in London. It was
because the construction of such a leviathan would take some time that the
self-elevating *Sea Gem* was used to drill BP's first North Sea well.[31]

By the time that the *Sea Gem* began drilling on block 48/6 in June 1965,
some of BP's North Sea competitors (Amoseas, Shell/Standard (NJ) and
Gulf) had already started drilling. The *Sea Gem*, though, was the first to discover petroleum, striking a gas field, to be named West Sole, with its first
well in September. After further drilling, the field was confirmed as a commercial discovery in December. The celebrations, though, were short-lived.
After completing the discovery well, the *Sea Gem* was due to be moved to
another location. On 27 December, during preparations for lowering the
platform to the water, part of the structure collapsed, the platform lurched
uncontrollably, fell into the water, turned over and sank. Thirteen of the
thirty-two crew members died.[32]

In the shadow of this tragedy, the West Sole discovery was quickly followed by other, larger and more productive gas finds by other oil companies. These included Shell/Standard (NJ)'s giant Leman Bank field, British Gas/Amoco's Indefatigable field and Arpet/Phillips' Hewett field – all discovered in 1966. Soon, however, the oil companies' enthusiasm for further exploration for gas in the British sector of the North Sea diminished. BP developed the West Sole field, which came on stream in May 1967, but was disappointed by the prices it obtained from British Gas, the state monopoly gas company, which was the only buyer, as exports were prohibited. Other companies, too, thought that the price was too low to make gas exploration and production worthwhile. By mid-1968 exploration in the British sector of the North Sea had slumped.[33]

As had happened time and again in oil exploration history it took just one major discovery to transform a dull prospect into an exciting one. In the northern basin of the North Sea, the breakthrough came in December 1969, when Phillips Petroleum discovered the giant Ekofisk oil field in Norwegian waters. In the same month, Amoco discovered the smaller Montrose field in British waters, also in the northern basin, about 135 miles east of Aberdeen. Here, north of latitude 56°N, the waters were much deeper and rougher than in the southern basin where the gas fields had been discovered. Water more than 400 feet deep, winds of 125 miles an hour, waves 100 feet high, freezing temperatures and blankets of fog presented new challenges to the oil explorer. This, like northern Alaska, was frontier territory.[34]

BP, though, had not previously been strongly attracted to it. In the second North Sea licensing round held in the autumn of 1965 BP had applied for some blocks in the northern basin, as well as further south. BP's application provided for only seismic surveys and one test well on the northern blocks. Beckett, at the Ministry of Power, thought that BP should commit to drilling two wells instead of just one. BP was reluctant, admitting that it had 'little enthusiasm' for the northern blocks. Falcon warned that 'by expanding north of Lat.56°N we are following quite different hypothetical objectives which bear no relation to the reasons which caused us to go into the British North Sea'. He argued that in view of the remoteness and unknowns of the northern basin, BP should have no hesitation in withdrawing from the area if work commitments further south necessitated it.[35] BP came close to withdrawing its application for northern blocks because of Beckett's insistence on a commitment to drill two wells. Eventually, they compromised. BP undertook to drill two wells unless information at the time made this technically unjustifiable, in which case BP would be absolved from drilling in this 'depressing' area. Thus in November 1965 BP was awarded some blocks in the northern basin.[36]

It was a sign of BP's lack of enthusiasm for the oil prospects in the northern basin that for the next four-and-a-half years, BP did not drill on its northern basin acreage. Seismic surveys were carried out with results that interested Peter Kent enough for him to suggest in 1967 that block 21/10 might be a promising prospect. He suggested that BP's hitherto rather dismissive attitude towards the northern basin be reviewed. A study was carried out, leading to the negative conclusion that there was little chance of finding a field in the northern basin large enough to be economically viable. Block 21/10 and the other northern basin blocks were put back on ice.[37] By the time that Phillips discovered the Ekofisk field, the expiry of BP's northern basin licences was less than two years away and no wells had been drilled on them.

That soon changed after the Ekofisk discovery, when oil companies rushed to move drilling rigs to the northern basin. One of them was BP's *Sea Quest*, which began drilling in 250–300 feet of water in block 30/1, about fifty miles north-west of Ekofisk, in May 1970. The well yielded only a small show of oil and was abandoned in August. The *Sea Quest* was immediately moved further north-west to block 21/10, about 110 miles offshore from Aberdeen, and began drilling in about 400 feet of water. At the beginning of October, BP's head office received news that the well had struck oil, but attempts to send samples ashore for testing were thwarted by a sudden and violent storm. For nearly a week, the *Sea Quest* was cut off. The only physical contact with the outside world came when a Dutch helicopter ran into trouble in the storm and had to make an emergency landing on the rig. After the storm lifted, BP announced that it had discovered the Forties field, the first giant oil field in the British sector of the North Sea.[38]

'A SECOND-CLASS POWER'

The Prudhoe Bay and Forties discoveries in 1969–70 were the crowning glories of BP's long-term policy of seeking new oil reserves outside the Middle East. These two fields added an estimated 6½ billion barrels to BP's recoverable oil reserves and raised BP's reserves outside the Middle East to an estimated 11 billion barrels.[39] Only the world's two biggest oil companies, Standard Oil (NJ) and Shell, held larger reserves outside the Middle East.[40]

BP's overall reserves, including the Middle East, were even more impressive. They had doubled from 38 billion barrels in 1950 to 75 billion barrels at the end of 1970, when only one company, Standard Oil (NJ), had more reserves than BP (see figure 8.1).[41] BP's estimated share of the world's reserves had fallen from 23 per cent in 1950 to 13 per cent in 1970 (see figure 8.2), owing to the loss of reserves in Iran as a result of the nationalisation crisis, the

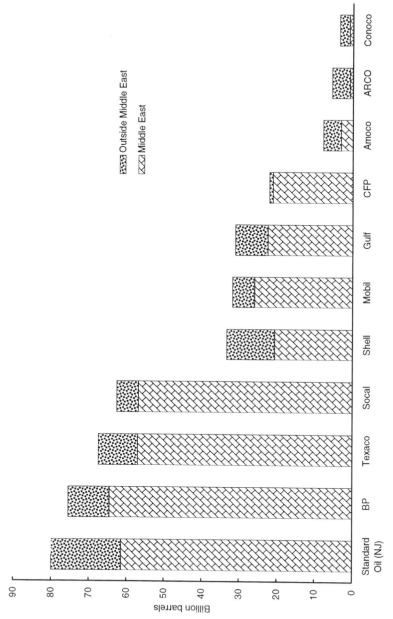

Figure 8.1 BP's and competitors' estimated recoverable oil reserves, 1971

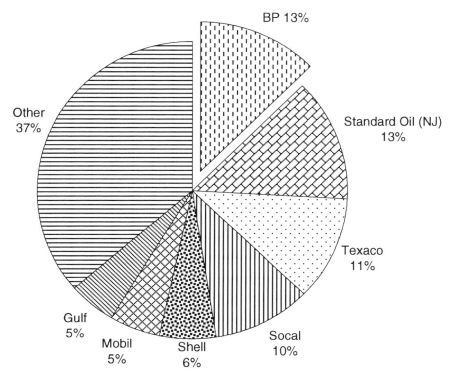

Figure 8.2   The majors' shares of world oil reserves, 1971

increased estimates of Saudi Arabia's oil reserves in which BP had no share, and the growth of competition from new entrants to the international oil industry. None of these factors detracted from BP's exploration record, which was the best in the industry. In the years for which comparative data are available, from 1963 to 1970, BP discovered more than twice as much oil at a lower finding cost per barrel than any comparable company (see table 8.1).

BP's exploration record owed much to an exploration strategy which was based on four key elements, learned from the Company's early experience as a first mover in Iran. These were early entry, a frontier focus on new oil provinces, an emphasis on giant fields, and an attraction to exposed anticlines. The first three of these factors helped BP to achieve a successful focus in the hunt for oil. The fourth, the preference for exposed anticlines, led BP astray. In Libya BP had been most attracted to the anticlines of northern Cyrenaica, in Canada to the folded foothills of the Rockies, in Alaska to the exposed

Table 8.1 *BP's and competitors' estimated crude oil finding costs (excluding USA), 1963–1970*

	Exploration expenditure ($millions)	Estimated oil reserves discovered (billion barrels)	Estimated expenditure per barrel found (cents)
Amoco	329·7	4·3	7·7
ARCO	127·4	0·6	21·2
BP	354·5	30·9	1·1
CFP	115·3	10·0	1·2
Conoco	210·2	1·0	21·0
Gulf	386·4	7·2	5·4
Mobil	389·4	6·8	5·7
Shell	900·5	13·4	6·7
Socal	236·7	11·2	2·1
Standard Oil (NJ)	652·6	15·4	4·2
Texaco	na	14·1	na

anticlines of the Brooks foothills and in Britain to the onshore anticlines. None of these prospects was found to contain a giant oil field. But BP, unsuccessful in its first moves, went on to find giant oil fields in, or offshore, three of these four areas.

It would be easy to put that down to a few strokes of good fortune, reinforcing the folklore of the oil industry, in which lucky wildcatters abound, and geoscience gets little mention. But more thoughtful study suggests that BP's high-risk, high-reward frontier exploration strategy was not such a random process. In any activity, the higher the risks taken and the higher the rewards, the greater the likelihood that a successful outcome will be interpreted as uncalculated good luck, rather than a measurable probability. But BP's exploration was based on exactly that, the continuous analysis of probabilities, and the progression of knowledge derived not only from successes, but also from failures which yielded improved understanding of the oil prospects of new provinces. Indeed, one of the advantages of being a first mover in a new province was that it enhanced the scope for learning, enabling frontier explorers to reposition, to take a second chance, if they were initially unsuccessful.

BP's exceptionally low finding costs per barrel also owed much to its continuing exploration in the Middle East, where giant oil fields were much more common than elsewhere. Between 1950 and 1970 BP had a share in no less than twenty-four newly discovered giant oil fields, each with estimated recoverable reserves of more than 500 million barrels, in Iran, Iraq and Kuwait. In Abu Dhabi, where the onshore concession was held by the IPC group (BP 23¾ per cent), a further five giant oil fields were discovered on land.[42] Offshore, the concession was held by Abu Dhabi Marine Areas (ADMA), owned two thirds by BP, one third by CFP. When ADMA discovered the giant Umm Shaif field in 1958, BP already had such an excess of Middle East oil that the news of Umm Shaif was greeted with 'general gloom' in BP's London headquarters at Britannic House.[43] Five years later, ADMA's discovery of the still larger Zakum field, a true super-giant, only added to the surplus.

The same partnership of BP and CFP also constituted Dubai Marine Areas (DUMA), which obtained the Dubai offshore concession in 1954. DUMA dragged its feet over searching for oil, but drilling commenced in 1964 after the more enthusiastic Continental Oil Company had acquired a half-share in the concession.[44] They discovered the giant Fateh field in the waters of the Persian Gulf in 1966, but BP had no use for still more Middle East oil and three years later sold its share in DUMA to CFP, 'ridding ourselves of the embarrassment of an additional source of Middle East oil'.[45] In 1970, BP disposed of more of its Middle East reserves by selling half of its two-thirds share in the El Bunduq field which ADMA had discovered in the waters off Abu Dhabi, to a Japanese consortium of energy and industrial interests.[46] Even after these disposals, 86 per cent of BP's worldwide estimated recoverable reserves were in the Middle East.[47]

The sale of BP's interests in DUMA and the El Bunduq field came at a turning point in perceptions about the sufficiency of the world's oil supplies. From the mid-1940s to the mid-1960s the rate at which the world was discovering new oil reserves rose to new peaks (see figure 8.3). As more oil was discovered, estimates by geological experts of the world's ultimately recoverable reserves rose steadily from a range of 400–600 billion barrels in the mid-1940s to about 2,000 billion barrels in the mid-1960s.[48] Lulled by these trends and estimates, producers and consumers alike came to take it for granted that there was a surplus of oil supply over demand.

The wave of new discoveries reached, however, its high point in the mid-1960s and then fell sharply. Estimates of the world's ultimate reserves (of discovered and undiscovered oil) also stopped rising and settled on a level around the 2,000 billion barrel mark. By 1970 the falling discovery rate had come to look like a confirmed trend, to which producers and consumers suddenly awoke. Fears that the world would soon run short of oil began to take

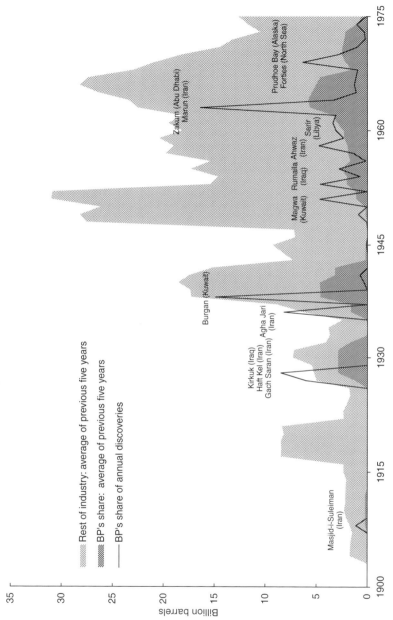

Figure 8.3   Oil discoveries by BP and the rest of the oil industry, 1900–1975

hold. One of those who raised the alarm was Harry Warman, who became BP's chief geologist in 1971. He published articles which suggested, by extrapolation of past demand trends, that the world's proved oil reserves would be exhausted in the 1980s and that the world's undiscovered reserves would not be enough to meet demand (rising at historic rates) beyond the end of the century.[49] He was accused of being a 'prophet of doom forecasting the imminent end of our oil supplies'.[50] Yet he was positively optimistic compared to Jim Spence, who succeeded Warman as BP's chief geologist in 1973. Whereas Warman had estimated in 1973 that there were still more than 1,000 billion barrels of recoverable oil to be discovered, Spence's estimate the following year was little more than a quarter of Warman's figure.[51] The mood concerning the future availability of oil had turned unmistakably to deep pessimism.

The seeming irreversibility of the trend towards oil exhaustion could be seen in the world's declining reserves:production ratio. This is a simple calculation of the number of years for which discovered recoverable reserves can sustain prevailing rates of production. The historical data, plotted in figure 8.4, showed that the world's reserves:production ratio had been on a downward trend since about 1950 and seemed to be heading inexorably towards the accepted minimum of 10:1, below which production rates would have to be reduced.

Although BP's reserves:production ratio was higher than the industry average, it too was on a downward trend. Even BP, with its traditional surplus of reserves, would, it was forecast, have to reduce production in the 1980s unless it made colossal new discoveries. 'To be in a comfortable position', wrote Kent and Warman, 'our finding rate would need to be at least doubled'.[52] Yet the prospects of finding new giant fields in large numbers seemed remote. 'We have perhaps been searching', Kent remarked, 'for a mythical Holy Grail, and the evidence becomes stronger that it does not exist, that there is only one super-oil province in the free world'.[53] 'There can be little doubt', wrote a colleague, 'that we are rapidly passing from the position of a reserves-rich company to a reserves-poor one'.[54]

BP's explorers blamed BP's top management for a lack of commitment to exploration. From 1970 Kent repeatedly complained that BP's exploration had been underfunded and its explorers actively discouraged from finding oil by higher management.[55] A report emanating from the Exploration Department in 1971 noted that only in very recent years had 'the discovery of a new field raised in BP more than the reaction "what on earth are we going to do with all this oil"'. Many BP geologists had, according to the report, spent their careers working 'against opposition as ignorant as it was

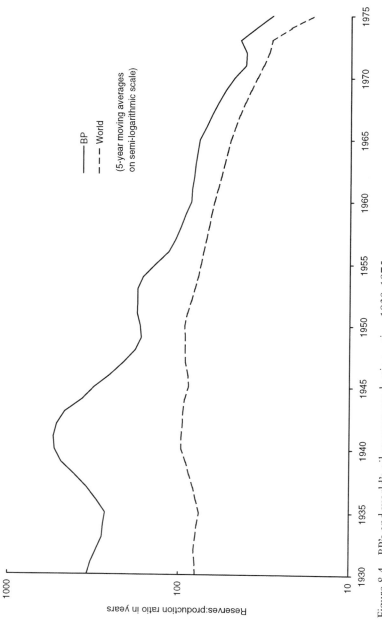

Figure 8.4   BP's and world's oil reserves:production ratios, 1930–1975

powerful'.[56] Bridgeman, Kent admitted, was the most exploration-minded chairman since Cadman (1927–41). But throughout Bridgeman's term as chairman (1961–9), most of BP's capital expenditure had gone on building up BP's downstream position to match more closely its surplus upstream capacity. After the burst of exploration expenditure in 1952–9, there had been a decline from the peak expenditure in 1959, which was not surpassed until 1969, by which time inflation had eroded its value substantially (see figure 8.5). Moreover, other oil companies were then spending proportionately more on exploration than BP (see figure 8.6). In December 1972 Kent wrote despairingly, 'with great regret I have seen our exploration effort progressively lose momentum since the time of the Alaska discovery, so that we have declined from our proud position of the most successful oil finding organisation in the world to a decidedly second-class power'.[57]

After strong representations from the explorers the level of exploration expenditure was very substantially raised in 1974 (see figure 8.5).[58] Most of the increase went on the North Sea, where BP's Forties discovery had been followed by a spate of discoveries by other companies much further north in the East Shetland Basin, sparked by Shell/Standard Oil (NJ)'s discovery of the giant Brent field in 1971. BP itself struck the giant Ninian and Magnus fields in 1974, adding about a billion barrels to its recoverable reserves, as then estimated.[59]

But while these discoveries were being made BP's oil reserves elsewhere were falling fast as OPEC countries took ownership and control of their oil resources in a wave of expropriations and participations, and as a result of BP's sale of 45 per cent of its share in ADMA to a consortium of Japanese companies in 1973. By the end of 1975 the reserves which BP had gained from its original pioneering exploration in the Middle East were mostly gone (see figure 8.7). Outside the OPEC countries, BP's reserves portfolio was based mainly on its discoveries in Alaska and the North Sea, which masked underlying weaknesses in BP's exploration position by the mid-1970s. Between 1973 and 1977 BP's gross additions to reserves were the lowest of the oil majors apart from Gulf. BP was still in this period spending less on worldwide exploration than any of the other majors, including Gulf. Perhaps most importantly, BP's exploration acreage was also the smallest of the majors by 1976. In the words of an internal report, 'notwithstanding BP's success in Alaska and the North Sea, our low level of exploration expenditure over many years outside the areas which are now OPEC have [sic] left us in an uncompetitive position relative to our major competitors, in terms of current discovery rate and holdings of prospective acreage on which to explore'.[60]

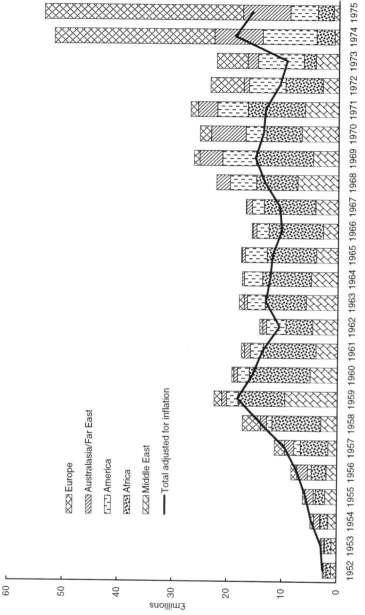

Figure 8.5 BP's exploration expenditure, 1952–1975

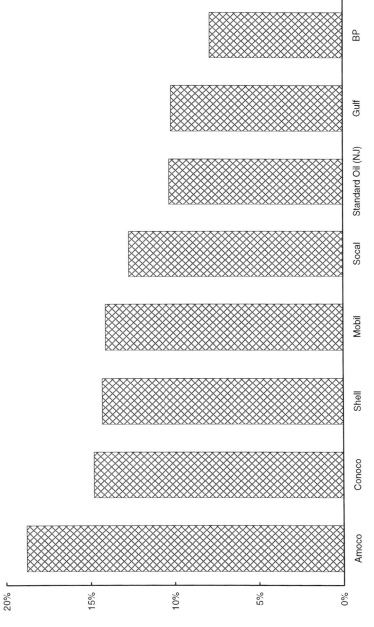

Figure 8.6 BP's and competitors' exploration expenditures as percentages of their total capital expenditures, 1970

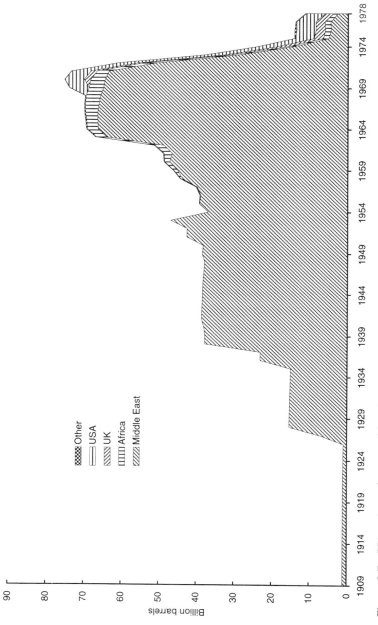

Figure 8.7   BP's estimated recoverable oil reserves, 1909–1978

For most of its history, BP's whole business had been based on success in exploration. The sequence of events and the order of priorities had been the same: firstly, to find giant oil fields, secondly to find outlets for the oil. The finding and development of giant oil fields had been the core of BP's business and the source of its competitive advantage. By 1975, however, most of the fruits of BP's early frontier exploration in the Middle East had been nationalised; BP (like virtually everyone else) believed that there was not much oil left to be discovered outside the Middle East; and, perhaps most importantly of all, BP seemed to have lost its lead in the activity which it traditionally did best.

= 9 =
The push for outlets . . .

In the 1950s and 1960s there seemed to be no end to the rising consumption of oil. Supplanting coal as the world's most popular fuel, oil rose inexorably to take a dominant share of a fast-growing energy market.[1] The sustained expansion of the non-communist world's major economies depended upon it as a cheap and convenient fuel. Consumers relied on it for heat, light and transport, and as a raw material for a host of items made from plastics, resins, synthetic rubbers and fibres. Its uses in road-surfacing, car tyres and upholstery, as well as in greases, lubricating oils and gasoline, ensured that motorists used oil not only to lubricate and propel their vehicles; they also quite literally drove on it. In the home and the workplace, oil found its way into paints and carpets, waxes and polishes, detergents and insecticides, furniture, packaging and innumerable appliances. In air travel, oil fuelled the new generations of jet aircraft, which carried a growing number of passengers on business and holidays. Oil was ubiquitous. The world's consumption of it grew at an average rate of more than 7½ per cent each year. In the USA, easily the world's largest and most mature oil market, the growth rate was comparatively subdued, at only 4 per cent a year. But in the other major industrial economies the growth was spectacular. In Western Europe, oil fuelled a golden age of economic growth, during which oil consumption grew at an average annual rate of more than 12 per cent. In Japan, the demand for oil grew still faster, at an average annual rate of more than 24 per cent, as the country rose rapidly to become a major industrial power.[2]

But although oil markets outside the USA were booming, they could not absorb all the oil that could be produced from the prolific oil fields of the Middle East and from new producing countries elsewhere, such as Algeria, Libya and Nigeria. In the international oil industry (the world outside the USA and the USSR) there was as a result a surplus of supply over demand. In the 1950s this was masked by the disruption of normal supplies in the Iranian

216

You get such a lot out of oil.

33 'You get such a lot out of oil.' BP advertisement in 1972

nationalisation crisis of 1951–4 and in the Suez Canal crisis in 1956–7, but after the end of the Suez crisis the supply surplus came to the fore.

'OUR MARKETING POLICY HAS BEEN OUR CONCESSIONS POLICY'

Among the oil majors, BP was the most extreme case of having the capacity to produce more oil than it could sell.[3] This imbalance went back to BP's origins as a company which was founded in 1909 to develop the giant oil field at Masjid-i-Suleiman in Iran without at the time having any market outlets. That beginning set the pattern for the Company's subsequent development, in which the discovery and development of giant oil fields came first, and the rest (shipping, refining and marketing) followed, essentially as a means of disposing of crude oil production.

The Company's marketing never caught up from its backward start. Its concession in Iran gave it exclusive access to a virgin exploration territory of vast potential. In exploiting the opportunities, the Company developed early prowess in exploration and production. But the formation of a comparable marketing capability was held back by a much more restricted set of opportunities. By the time that the Company was formed, international oil markets were dominated by Royal Dutch-Shell and Rockefeller's Standard Oil group. A late starter, the Company initially resorted to selling crude oil and products under contract to Shell, and it was not until the decade 1917–27 that the Company developed a significant marketing organisation of its own, primarily in Britain and continental Europe.[4] Even so, the Company remained a marketing minnow compared with its larger rivals, Shell and Standard Oil (NJ).

From the late 1920s, further expansion in marketing was stultified by restrictive agreements, designed to limit competition at a time when world oil markets were glutted with oil. In 1928, the Company agreed not to enter the Indian oil market in return for the right to supply the joint Indian marketing subsidiary of Burmah Oil and Shell with some of its oil requirements.[5] In the same year, the Company and Shell formed the Consolidated Petroleum Company, a joint company which was to be the sole vehicle for all their marketing operations in a vast area including the whole of eastern and southern Africa. The two parents shared equally the ownership and profits of Consolidated, and they also had the right and obligation to supply in equal shares the crude oil and products sold by Consolidated. But on the crucial question of who should manage Consolidated, it was agreed that Shell should be solely responsible for running and staffing the business.[6] Not surprisingly, Shell-branded products continued to hold the lion's share of sales in the Consolidated area, with sales of products carrying the BP trademark lagging far behind.[7] The essential feature of the Burmah-Shell and Consolidated agreements was, therefore, that in both cases the Company forfeited the freedom to establish and run its own marketing organisation, to promote its own brands and to attract its own customers in exchange for the right to supply a guaranteed share of the sales achieved by others.

Another, more famous, restrictive agreement into which the Company entered was the 1928 Achnacarry Agreement to share out international oil markets by a system of quotas. Though it was only partly successful, the cartel reduced both the incentive and the need for the Company to develop competitive marketing capabilities.[8]

In 1932 the Company went still further in reducing its scope for independent, competitive marketing by merging its UK marketing subsidiary, British Petroleum, with Shell-Mex, the joint UK marketing company of Shell and Mexican Eagle (controlled, though only minority-owned, by Shell). Ownership of the new joint company, Shell-Mex and BP, was split 40:40:20 between the Company, Shell and Eagle respectively. Their supplies of products to SMBP were fixed by quotas on the basis of their market shares at the time of the merger. Managerial control of SMBP (unlike Consolidated) was shared equally by the Company and Shell.[9] The formation of SMBP confirmed that the main thrust of the Company's marketing policy in the 1930s was to avoid or restrict competition, rather than to engage in it.

As the end of World War II approached, the Company's backward marketing position was a matter of much concern. Iran was in political ferment, and the Company's chairman, William Fraser, knew from experience that failure to deliver rising revenues to Iran could well precipitate concessionary problems, as had happened in 1932 and, to a lesser degree, in 1938–40. As Iran's oil revenues were based on the tonnage of oil produced, Fraser decreed that the Company should aim to increase production in Iran at a rate which would keep pace with the growth in world oil consumption. The increase in production was itself easy to achieve, but the task of finding markets was much harder. By 1950 the Company had increased its share of continental European markets to 13 per cent (from 8 per cent in 1938), but that was not nearly enough to absorb the growth in crude oil production. The Company therefore had to resort to selling unrefined crude oil, which grew from a mere 1 per cent of total sales tonnage in 1946 to 22 per cent in 1950.[10]

After the interlude of the Iranian nationalisation crisis, BP's old problem of surplus production capacity soon reasserted itself. Its oil liftings from Kuwait, Iraq and Qatar had been rapidly increased to cover the loss of Iranian production, and were enough to meet BP's requirements. But after the settlement of the crisis, BP also had to find outlets for its large share of the Consortium's production in Iran. In addition, large-scale production would soon come on stream in Nigeria, Libya and offshore Abu Dhabi.

As the pressure from the producing countries to raise output intensified in the 1960s, BP struggled to bring its upstream and downstream operations into better balance. One possibility was for BP to reduce its supply surplus by disposing of some of its concessionary interests. That possibility was considered in the mid-1960s,[11] but not pursued except in particular circumstances (see later in this chapter). BP's general aim was in fact the opposite: namely, *to hold on to its concessions*, just as it had tried to do before and during the Iranian nationalisation dispute. To that end, BP followed a policy of forced growth in its shipping, refining and marketing operations in an attempt to find outlets for the increases in production demanded by the producer countries. As one BP executive summed it up: 'Our Marketing policy has been our Concessions policy.'[12]

Despite all its efforts to expand its downstream operations, BP's perennial problem of having more production than markets would not go away. Its crude oil production rose rapidly from the nadir of 1951–2, and by 1970 BP was firmly re-established as one of the top three oil-producing companies in the world, in the same league as Shell and Standard Oil (NJ) (see figure 9.1). But in refining and marketing, Shell and Standard Oil (NJ) were in a class of their own, with BP far behind as one of the pack which also included Gulf, Mobil, Socal and Texaco (see figures 9.2 and 9.3).

In its quest for new markets, BP was to some degree hampered by trends in the international trading environment. The economies of the main oil-consuming countries were becoming as dependent on oil as those of the oil-producing states. Knowing that oil was a strategic commodity which they could not do without, and uneasy about their dependence on the oil majors for supplies, oil-consuming nations turned increasingly to nationalistic oil policies, aimed at securing control over their own oil supplies. This trend was apparent both in industrialised countries and in the less developed world, where the process of decolonisation left the vast expanses of the old European empires criss-crossed with national boundaries. Many of the new nation states equated political independence with economic autonomy and reacted against multinationals, whose association with the former colonial powers, employment of expatriates in senior positions and alien cultural values all contributed to the suspicion with which they were regarded.

With the rise of economic nationalism came the widespread adoption of discriminatory policies (often directed specifically against the majors), the erection of barriers to free international trade in oil and the formation of state-owned oil companies to provide nationally controlled supplies of oil, while enjoying privileged positions in national markets. The result was a fracturing of international oil markets into a growing number of national fragments. This did not suit the oil majors, whose integrated international supply systems were geared to moving oil across national boundaries without impediment.

While the majors generally argued in favour of a liberal international economy, they were themselves partly responsible for economic nationalism.

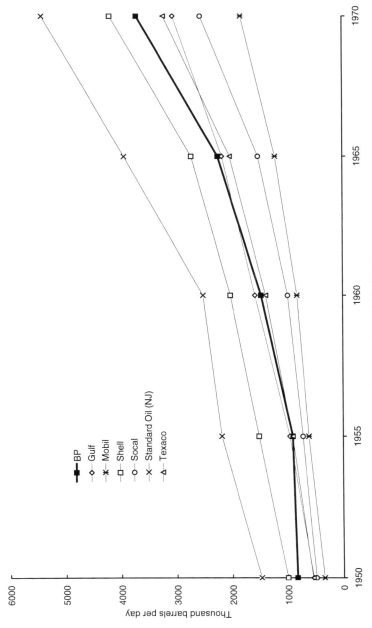

Figure 9.1   The majors' comparative crude oil production, 1950–1970

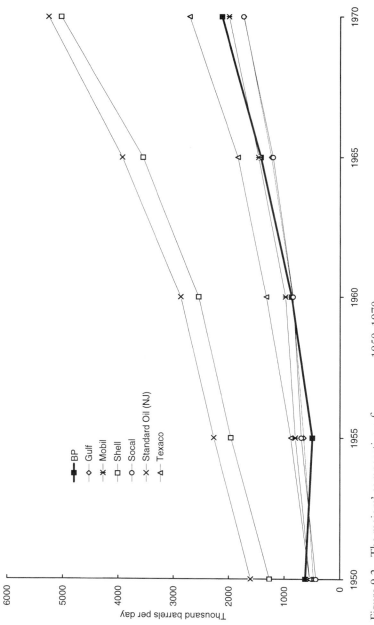

Figure 9.2   The majors' comparative refinery runs, 1950–1970

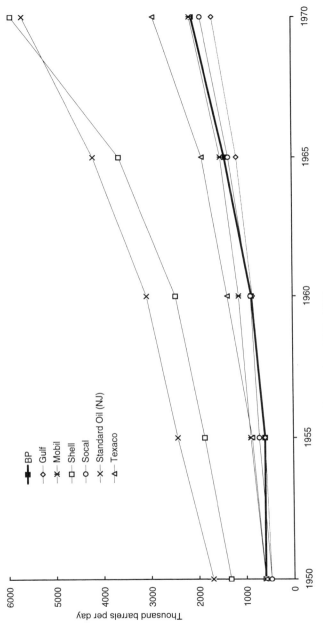

Figure 9.3 The majors' comparative products sales, 1950–1970

Through the international petroleum cartel and other restrictive agreements, they had carved up international oil markets, acquiring a reputation for overriding national sovereignty, which tended to fuel nationalistic reactions against them.

In BP's case, the association with national self-interest was more direct. In 1914, it had become in many respects the prototype national oil company when, financially overstretched and fearing takeover by Shell, it had turned for help to the British government, which backed the Company with new capital and became its majority shareholder. The motives of the British government then were similar to those of other governments which later formed their own national oil companies; namely, to secure nationally controlled supplies of a strategic commodity and to avoid becoming dependent on foreign multinationals.[13] Although the British government continued to honour the pledge it gave in 1914 that it would not interfere in the management of the Company, BP could not credibly claim to be wholly independent of the British government as long as the government remained a major shareholder.

But the most important national barrier to an open oil economy in the non-communist world came, ironically, from the country with least cause to feel oppressed by the power of multinationals. The USA, while outwardly promoting a liberal international economy as its multinationals spread across the globe, protected its domestic oil industry against foreign competition by imposing mandatory oil import quotas in 1959. US protectionism had far-reaching effects, which were felt acutely by BP. In the 1960s the five US majors all produced more than 20 per cent of their crude oil and sold more than 30 per cent of their refined products in the USA. The comparable percentages for Shell were not much less.[14] BP, however, failed to gain entry before the erection of the protective wall. It was therefore excluded from the largest and most lucrative oil market in the world.

US protectionism also had unwelcome knock-on effects for BP. In the second half of the 1950s a growing number of US independent oil companies with little or no international business began to search for low-cost oil abroad. Some, such as Occidental, Marathon and Continental, found oil in large quantities. Unable to ship it to the USA because of the import quotas, these and other companies crowded into Western European markets, intensifying the competition for sales.[15] BP was, and remained, far more dependent on West European markets than any other major (see figure 9.4) and was therefore the most susceptible to the growth of competition in this region.

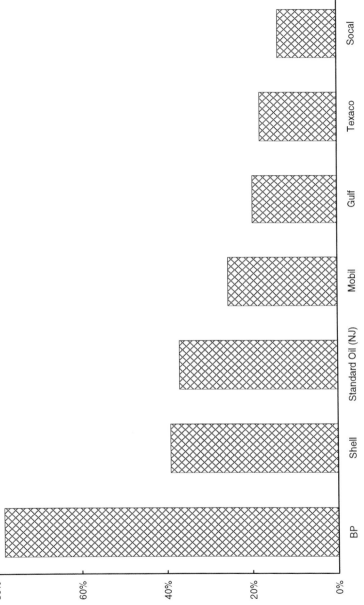

Figure 9.4   The majors' comparative products sales in Western Europe as percentages of their worldwide sales, 1969

EUROPEAN MARKETS

In marketing almost as much as in crude oil production, BP was not a global company. Its crude production was very heavily concentrated in the Middle East, which provided more than 99 per cent of BP's crude in 1955, falling to 84 per cent in 1970, primarily because of new production in Nigeria and Libya. In the marketing of refined products, the focus was on Western Europe. BP's European subsidiaries and associates accounted for 69 per cent of the sales of all BP's marketing subsidiaries and associates in 1955, rising to 83 per cent in 1970.[16] BP was, to a large degree, a company of two regions: the Middle East upstream; Western Europe downstream.

Across Western Europe as a whole, BP was the third largest marketer of oil products. Its market share of 12–13 per cent put it well below the market leaders, Shell and Standard Oil (NJ), but well above the other majors (Mobil, Texaco, Gulf and Socal), whose market shares were each less than 5 per cent (see figure 9.5). Collectively, the majors held a market share of 64 per cent in 1963, falling to 60 per cent in 1970 as their dominance was slowly eroded by state-controlled companies and private independents.

Western Europe was not, however, a single market. The creation of the Common Market under the 1957 Treaty of Rome did not lead to the adoption of a common oil policy by the member states, and Western Europe remained a disparate collection of nations, which made up, broadly speaking, four oil markets.[17] One was the UK, the most highly concentrated market in Western Europe, with the majors holding 90 per cent of the market in 1963, falling to 83 per cent in 1970. Nearly all of that share was held by an oligopoly of Shell, Esso (the marketing arm of Standard Oil (NJ)) and BP, who dominated the market with little interference from the government. Here, as elsewhere in Western Europe, the market share of the BP brand was third-placed, well behind Shell and Esso, but well ahead of the rest of the majors.

A second market was France, which since the 1920s had adopted highly *dirigiste* oil policies, limiting the role of the majors by a complex system of controls and the creation of powerful state oil companies. As a result, the majors held a smaller share of the market in France than in any other of Western Europe's main markets. Indeed, their combined share of 48 per cent in 1963, falling to 46 per cent in 1970, was not much more than half their share of the UK market. Unusually, only four of the majors – Shell, Esso, BP and Mobil – had a presence. The others – Gulf, Socal and Texaco – were absent. BP, once again, came third, after Shell and Esso.

A third market, grouped around the artery of the River Rhine, comprised Germany, the Netherlands, Belgium, Austria and Switzerland which, together

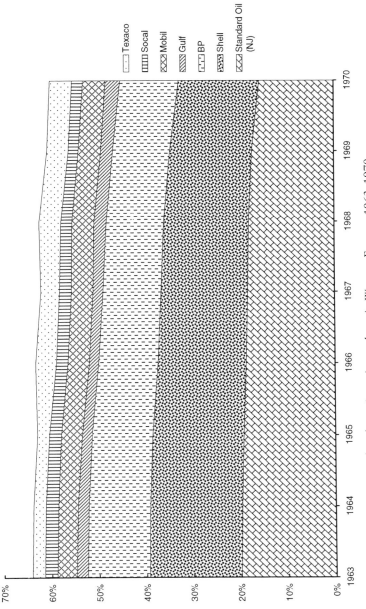

Figure 9.5   The majors' market shares in main products in Western Europe, 1963–1970

with Sweden and Denmark, followed liberal, *laissez-faire* oil policies. Germany was by far the largest national market in this group. Here, the majors held a combined market share which fell between their dominance in the UK and their weaker position in France. They held 55 per cent of the market in 1963 rising to 57 per cent in 1970. The increase was entirely due to Texaco's acquisition of one of the German independents, Deutsche Erdöl AG, in 1966. BP, once again, was the third-largest marketer, after Shell and Esso.

Finally, a fourth market was made up of the Mediterranean countries, including Italy, Greece, Turkey and Spain. Their crude oil supplies came mainly from nearby Libya and from the Mediterranean terminals of the trans-desert pipelines carrying crude oil from Iraq and Saudi Arabia. Italy was easily the largest national market in the Mediterranean area and discriminated in favour of its state oil company, ENI, which held a particularly strong position in the gasoline market. Otherwise, however, Italy was fairly liberal in its oil policies, and interfered much less than France with the operations of the majors. Their combined share of the Italian market, falling from 50 per cent in 1963 to 45 per cent in 1970, was comparable with France, but lower than in the UK and Germany. The ranking of the majors followed the usual pattern, with Shell and Esso at the top, and BP well behind in third position.

Across this spectrum of markets, there were some common features in BP's position. One, already noted, was its middling position as the smallest of the big three, and the largest of the rest. This had drawbacks. On one hand, BP had a large and costly marketing organisation which was similar to Shell's and Esso's, but which failed to attain the volume of business necessary to reap the economies of scale which they enjoyed. On the other, BP lacked the nimbleness and the lower costs to compete effectively with smaller, faster-moving, more entrepreneurial competitors, especially the independents. BP thus got, as one executive put it, 'the worst of both worlds'.[18]

Secondly, compared with its main competitors, BP's sales mix contained a disproportionately high percentage of low-value, heavy products, most notably fuel oil, which were sold wholesale to commercial customers; and a correspondingly low percentage of higher-value products, such as gasoline, retailed to the general public.[19] As the general manager of BP's marketing commented in 1968, BP needed 'to increase the volume of our sales of the more remunerative products, which demand a much higher degree of marketing skill, and where our performance in most markets has traditionally lagged far behind that of our main competitors'.[20]

BP's failure to secure a larger share of high-value business was a reflection of its marketing culture. In an industry where the products of separate companies were much the same, success in luring customers required imaginative marketing, powerful brands and trademarks with popular appeal. These were not BP's strong points. The problem was recognised, but the actions taken failed, on the whole, to dispel BP's image of being stodgy, old-fashioned and (in some views, synonymously) quintessentially British. The redesign of BP's trademark in 1957–8 is illustrative. In 1957, BP engaged the Compagnie de l'Esthetique Industrielle, founded by Raymond Loewy, and hailed as 'without doubt, the greatest design company in the whole of Western Europe', to redesign BP's symbolic trademark, the BP Shield. The design consultants were keen to use the colour red, in addition to the BP colours of yellow and green, and presented alternative designs accordingly.[21] But traditionalists in BP considered that red was, among other things, 'too revolutionary and vulgar', and opted instead for the retention of a modified green and yellow shield, which was duly launched as BP's 'New Look'.[22] The green was more vivid, the yellow more luminous, and the shape 'less heraldic in a medieval way, more firm-shouldered, modern and definite' than before.[23] But the co-ordinator of the project in BP was disappointed that BP's symbol had 'not been changed as radically as some of us would have wished', and later reported that it was not 'sufficiently forceful with the present competition'.[24] It was still in use in 1971 (and long afterwards), when a consultant remarked that it was 'defiantly ethnocentric'.[25]

The Britishness of the modified Shield may have served a purpose when it was introduced, for at that time the UK was BP's predominant market, far ahead of the two largest continental markets of France and Germany. But BP's sales in the comparatively slow-growing UK economy were not rising as quickly as in the faster-growing French and German economies. By 1970 sales in Germany were very nearly as high as in the UK, and sales in France were not far behind those in Germany. These three markets accounted for about two thirds of the sales tonnage of all BP's European subsidiaries and associates.[26]

THE UK

By 1950 motorists in the UK had long been used to restrictions on their consumption of gasoline. Rationing and the suspension of competition between brands, introduced as wartime stringencies, were still in force. Motorists could buy only limited quantities of one grade of gasoline, which was sold, whoever the supplier, as 'Pool' gasoline of 70–72 octane. In May 1950 this restrictive regime was relaxed by the ending of rationing. Motorists could now buy whatever quantities of gasoline they wanted, but by government direction 'Pool' gasoline was still the only one available.

By late 1952, however, it was clear that supplies of higher-grade gasoline would soon be readily available from UK refineries equipped with new catalytic 'cracking' units for the up-grading of products. The Company had catalytic crackers due to come on stream at its Llandarcy and Grangemouth refineries in 1953, and construction of its new Kent refinery (also with a catalytic cracker) was nearing completion. Shell had expanded its Stanlow refinery, whose catalytic cracker came into operation in 1952. Esso's big refinery at Fawley had been running since 1951, and Mobil's new refinery at Coryton was scheduled for completion in 1953.

At the same time, motoring organisations were calling for higher-grade gasoline. The Society of Motor Manufacturers and Traders complained that British motor manufacturers were seeking to produce high-performance cars for export markets, but could not sell them at home because the octane rating of 'Pool' gasoline was too low for them. The government's acceptance of these arguments was confirmed in October 1952, when the Ministry of Fuel and Power announced that the oil companies would be permitted to re-introduce their brands, and to sell higher grades, from the beginning of February 1953.[27]

From that date, motorists were faced with an unaccustomed choice. Shell-Mex and BP the largest gasoline marketer, alone sold four brands, namely Shell, BP and the two minor brands of Power and Dominion; and would soon (in 1956) acquire the National brand as well. Esso, the second-largest marketer, sold both its own brand and that of its subsidiary, Cleveland. Third came Regent (later acquired by Texaco). In addition to these brands, all resurrected from prewar days, were two new entrants: Fina (later Petrofina) and Mobil.

Each of these ten brands (Shell, BP, Power, Dominion, National, Esso, Cleveland, Regent, Fina and Mobil) came in a range of two to three grades, which were soon extended as the companies competed on quality by introducing higher-octane fuels. For example, at first BP 'Ordinary' (74 octane) and 'BP Super' (90–92 octane) were the only two grades sold under the BP brand. By 1956 these two grades had moved up to 76 and 95 octane respectively. In July that year BP launched a third grade, 'BP Super Plus', with 100 octane. It was soon followed by 'Esso Golden Extra', 'Regent 100' and 'Super Fina Plus', all of similar quality.[28]

While the oil companies tried to lure customers by quality competition, they also tried to secure their positions in the market by another method, which in the space of a few years transformed the structure of UK gasoline retailing. Before the war, most gasoline stations were owned and operated by independent garage owners, who generally sold a variety of competing brands from separate pumps at their filling stations. In the 1950s, this system virtually disappeared. It was replaced by a retailing method known as 'solus' trading, whereby the large gasoline suppliers extended their control down the distribution channel by securing 'tied' filling stations, which sold only one supplier's brand or brands of gasoline. This was much the same as the method adopted in the brewing industry, where large brewers sold their beers through tied public houses.

Esso was the first company to introduce solus trading in the UK when in 1950, anticipating the ending of wartime controls, it offered incentives to retailers with their own filling stations to buy their 'Pool' gasoline exclusively from Esso. SMBP, Regent, Mobil and Fina quickly followed with their own solus schemes. They spread so quickly that by 1953, when brands were re-introduced, at least 80 per cent of all filling stations had entered into exclusive agreements with one or other of the five major suppliers, who shared almost all the UK retail gasoline market.

The solus schemes of different suppliers varied in detail, but followed the same basic principle of exclusive brand representation. At first, SMBP's solus agreements provided some flexibility by allowing retailers to sell more than one of SMBP's brands. In 1956, however, SMBP changed to a single-brand system, which required tied outlets to sell either Shell or BP, or a combination of National (for higher gasoline grades) and the lesser Power brand (for the standard grade). The minor Dominion brand had by this time virtually disappeared.

Solus trading evolved into a regime which gave the main suppliers considerable control over previously independent retailers through a complex system of incentives and restrictions. These included financial rebates, free staff training, advice on the layout of service stations, free maintenance of pumps and other equipment, and loans for the purchase and improvement of premises. Solus agreements could last for up to twenty years; and they could also restrict the retailer's freedom to sell competing brands of lubricating oils and greases.

It was not only by solus agreements with independently owned filling stations that the leading suppliers secured greater control over the retail market. They also purchased or constructed their own filling stations, usually letting them to tenants on condition that they sold only the brand designated by the landlord company. Oil company ownership of filling stations, which had been negligible before the war, spread rapidly in the 1950s. By the end of 1964, stations owned by SMBP, Esso, Regent, Mobil and Petrofina accounted for nearly a quarter of the retail gasoline market. In line with this trend, SMBP's sales of gasoline through company-owned stations by then accounted for almost a quarter of its total retail gasoline sales. Overall, more than 95 per cent of filling stations were by this time tied to selling the brands of a single supplier.[29]

SMBP and Esso no doubt hoped that the solus system would help to keep out competition and perpetuate their dominance of the UK market. Until 1960 they seemed to be achieving those aims. SMBP's share of the market was roughly stable at about 50 per cent, and so was Esso's at about 30 per cent. There was virtually no competition, except from the 'second league' of Regent, Mobil and Petrofina.

From 1960, however, the dominance of the market leaders was challenged by smaller competitors who succeeded in entering the market despite the solus system. Some of the new entrants, most notably Jet, were independents who undercut the prices of the market leaders and gained a growing share of the market. Others, such as Total, Murco, AGIP and Amoco, were subsidiaries of international oil companies, who broadly adhered to the prices set by the market leaders. In 1961, Jet became a subsidiary of the Continental Oil Company, one of the US companies which had ventured abroad and found large quantities of oil in Libya, for which it was seeking outlets in European markets.[30] Under Continental's ownership, Jet kept up its price-cutting tactics and continued to increase its market share. Faced with growing competition, SMBP's share of the UK retail gasoline market fell from 49 to 39 per cent between 1960 and 1970, while Esso's fell from 30 to 22 per cent. Jet and other new entrants meanwhile raised their combined share from a mere 2 per cent to 22 per cent (see figure 9.6).

Meanwhile, the solus system came under attack from a group of specialist manufacturers of lubricants and accessories, who felt that they were being denied access to the market. In 1955, they formed the Motor Accessories Manufacturers' Association, which campaigned for an investigation into the solus system by the Monopolies Commission, set up under the Monopolies and Restrictive Practices Act of 1948. The driving force in the Association was C. C. Wakefield, whose Castrol brand was the best-selling motor oil on the market. Castrol's market share was, however, falling because, Wakefield claimed, the big oil companies were insisting that tied outlets should sell only their own lubricants.[31] BP, which sold a wide range of single-grade oils under the brand name Energol, as well as its multigrade oil, BP Energol Visco-static (launched in 1954), was accused by Wakefield of being the worst offender.[32]

The Wakefield-led campaign against the solus system achieved a breakthrough in 1960, when the Board of Trade (the government department responsible for identifying monopoly cases for investigation) referred the solus system to the Monopolies Commission. The Commission's report,

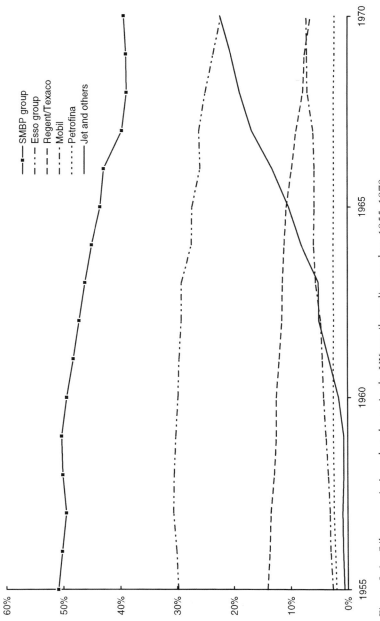

Figure 9.6 Oil companies' market shares in the UK retail gasoline market, 1955–1970

issued in 1965, made a number of recommendations to limit solus trading. The principal ones were that the duration of solus agreements between suppliers and tied retailers should not exceed five years; that large suppliers who did more than 15 per cent of their retail gasoline business through their own filling stations should not be allowed to build or acquire any more stations of their own; and, to the satisfaction of Castrol (Wakefield changed its name to Castrol in 1960) and its fellow campaigners, that solus agreements should not prevent retailers from selling competing brands of lubricants, kerosenes, antifreeze, tyres, batteries and accessories.[33] In sum, the recommendations were quite mild insofar as they affected gasoline sales, but sought to establish an open, competitive market in lubricants and accessories.

The Board of Trade accepted the Commission's recommendations with one significant relaxation; instead of banning large suppliers from opening new filling stations of their own, the Board of Trade agreed to allow them to open new stations as long as they disposed of an equivalent number of old ones. This would enable them to replace low-volume sales outlets with new higher-volume ones, so that they would be able to raise the market shares, though not the number, of their own outlets. SMBP and the other companies gave voluntary undertakings to comply with the recommendations, as relaxed by the Board of Trade. Although the overall impact on their operations was slight, it did reduce their ability to control the market through solus trading.[34]

By the time of the Monopolies Commission's report, both Shell and BP were concerned about SMBP's falling market share.[35] They did not, however, approach the problem from the same standpoint. As figure 9.7 shows, the Shell brand, along with Esso, was a market leader. It held about 25 per cent of the market in 1960, and was capable of standing alone against competition. The BP brand was much less popular, with a market share which was less than half of Shell's. It would have been even lower without SMBP, which resorted to a variety of measures to support the BP brand in the hope that BP's share of the overall market could be held up without Shell's coming down.

The problem of balancing the market shares of Shell and BP was a constant headache for SMBP, and became more acute after SMBP decided to convert from multiple-brand to single-brand solus trading in 1956. The programme to separate Shell and BP filling stations soon ran into objections from retailers who did not wish to convert to the weaker BP brand. Their resistance was difficult to overcome and delayed the conversion programme. Indeed, there were allegations that some SMBP retailers were coerced into accepting the BP brand.[36] Over the next decade, SMBP ran down the

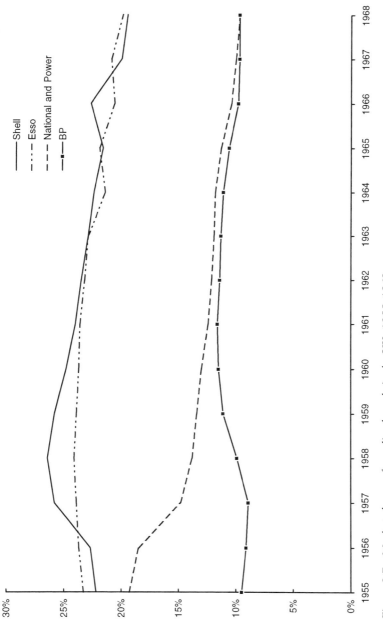

Figure 9.7 Market shares of gasoline brands in the UK, 1955–1968

National brand to strengthen both the BP and Shell brand positions. But it was the BP brand which needed most support, which was provided by concentrating SMBP-owned filling stations on BP.[37] It was recognised in 1964 that if SMBP had instead followed a policy of maximising sales, the 'Shell brand would today be in incontestable supremacy; National Benzole would have steadily deteriorated; and BP would virtually have not emerged from its postwar insignificance'.[38] This assessment was supported by statistics, which showed that BP filling stations had substantially lower average sales than Shell stations.[39] In essence, therefore, the BP brand was being supported by SMBP at the expense of Shell. This put the two parent companies in contrary positions: whereas it was in BP's interest to perpetuate SMBP, it was in Shell's to end it.

At the heart of the problem lay the lacklustre image of the BP brand, attested to by a series of opinion surveys in the early to mid-1960s. One found that the BP brand was seen as 'old fashioned, conservative and stodgy', while BP as a company was seen as 'dull, characterless and lacking in distinction'.[40] Another, that if seven gasoline brands were available, 30 per cent of motorists would buy Shell or Esso, 12 per cent National, 7 per cent Regent, and only 6 per cent BP. Yet another, that motorists who bought BP gasoline were 'notably lower class and economy minded' and had smaller and older cars than buyers of Shell and Esso.[41]

A BP-commissioned survey by the Tavistock Institute of Human Relations came out in 1965 with similar findings. It reported that one in three motorists and more than half of the general public did not spontaneously mention BP when asked what oil companies they knew. The recall of BP advertising was less than a quarter of the recall for Shell and Esso, although expenditure on BP advertising was about two thirds of that by Shell and Esso. When asked to name the best all-round company between Shell, Esso, BP, Regent and Mobil, 47 per cent answered Shell, 22 per cent Esso, and only 5 per cent BP. On the other hand, BP was seen by a wide margin as more old-fashioned and as more British than either Shell or Esso. When asked to give their general impressions of BP, more than half of the motorists and two thirds of the general public in the survey sample answered 'Don't know', 'No strong impressions', or 'Never dealt with them'. It was beyond doubt that BP lacked a positive image.[42]

BP accepted the validity of the Tavistock survey, and some BP executives recognised that action was needed.[43] As one put it, 'BP needs more than any other major to establish its identity and to establish itself as a lively company'.[44] He was not the only one to recognise the need for BP to get lively. Esso was conducting a memorable advertising campaign based on the theme that Esso was 'the tiger in your tank'. BP's response, described in the media as 'the British answer to the American tiger', was a BP Holiday Insurance scheme.[45] A BP executive commented ruefully that 'Esso puts "a tiger in your tank", "get-away people buy Super National" but all we can do is to worry about having a breakdown in one's car on holiday and offer the security of insurance'.[46] No wonder that an article in the London *Evening Standard* described BP as 'a sort of dull man's Shell. Family motoring. An old lady's petrol'; whereas Esso had an image of 'cheap and cheerful happy motoring . . . with a very strong image, that tiger'.[47]

BP's chairman, Maurice Bridgeman, so at home in the corridors of Whitehall and the world of international oil diplomacy, was, however, less attuned to mass marketing, certainly of the cheap and cheerful variety. He was content for BP to take its 40 per cent share of SMBP's profits 'regardless of the brand', and did 'not consider it to be his job to use BP shareholders money to improve the brand image of BP products in the UK'. He preferred to leave that task in the hands of SMBP.[48]

It was indicative of this reluctance to engage with the consumer that BP sought to increase its market share in the UK less by popularising the BP brand than by trying to increase its quota of SMBP's business by negotiation with Shell. In the mid-1950s, negotiations had taken place which resulted in the adjustment of quotas to bring them into line with the 40:40:20 percentage shareholdings in SMBP held by BP, Shell and Eagle respectively.[49] BP raised the question of increasing its quota of SMBP's trade from 40 to 50 per cent, but Shell's unequivocal response in 1954 was that 'such a horse must be regarded as dead before reaching the starting post'.[50] The next year, BP raised the idea again, but Shell still thought that it was a non-starter, pointing out, significantly, that 'the products in which BP were strong were on the reducing side and the Shell products appeared to be on the growing side'.[51]

BP tried again in 1960, shortly after Shell had raised its shareholding (and quota) in SMBP to 60 per cent by taking over Eagle. Shell, however, was convinced that 'the strength of any oil company depended primarily on its hold over outlets', and refused to reduce its 60 per cent quota of SMBP's trade.[52] In late 1964 the matter came up once more, this time on the basis that BP might exchange a share in one of its Middle East oil-producing interests for an increased share in SMBP.[53] This idea had obvious attractions for BP, as it would at a stroke reduce BP's crude oil surplus while simultaneously increasing its outlets. It would thus help to remedy the continuing problem, noted by a British government official in early 1966, that 'the obvious weakness of BP's marketing position could have the most serious consequences for the Company's ability to hold its Middle Eastern concessions'.[54]

However, the idea of exchanging production for markets came to nothing, primarily because Shell decided that its crude oil requirements were largely covered and was reluctant, as ever, to give up profitable market outlets.[55] Instead of granting BP a larger share in SMBP, Shell was turning to another, more radical idea. Impatient with SMBP, and nagged by the thought that the BP brand was being supported at its expense, Shell suggested in 1966 that SMBP should be split up.[56]

BP, the weaker partner, did not welcome the idea. A split would leave BP in an uncompetitive position not only because of the relative weakness of the BP brand, but also because BP would have higher distribution costs than Shell. The reasons for this were quite simple. If SMBP were split in line with shareholdings and quotas, both Shell and BP would have national marketing networks, but Shell would inherit 60 per cent of SMBP's trade, compared with BP's 40 per cent. BP would therefore have a substantially lower density of business than Shell with consequently higher distribution costs per unit of sales. Moreover, while Shell's refineries, connected to the main Thames-Mersey pipeline, were well placed to serve the inland market, BP's were situated more to serve export markets. BP would thus find itself with a comparatively high-cost distribution system in support of a weaker brand than Shell.

Knowing that Shell was keen on a split, which could not be achieved without BP's agreement, BP held out for favourable terms. It pressed in particular for either financial compensation or for BP's market share of SMBP's trade in main products to be brought up to 40 per cent before the split became effective; and for supply arrangements which would enable each company to make use of the other's refineries, depots and terminals after the split. In addition, BP took the opportunity to press for changes in another joint Shell-BP marketing area. Specifically, BP had for years been nettled by the fact that it had virtually no say in the marketing of BP products in the Consolidated area, where sales of both Shell and BP products were managed by Shell. This, BP was convinced, had resulted in the BP brand's share of the market being depressed. BP therefore demanded, as a *quid pro quo* for its agreement to break up SMBP, that Shell should agree to the break-up of Consolidated in South Africa, the biggest and most profitable market in the Consolidated area.[57]

While discussions on these issues continued, new initiatives were being taken to strengthen the BP brand in the UK retail gasoline market. The most important was the introduction in 1966 of the Superblend pump. Before its introduction, BP filling stations were offering three grades of BP gasoline (Regular, Super and Superplus) from three separate pumps. The Superblend, however, was a single mixer pump, offering Regular, Super and three intermediate blends of gasoline. BP's top grade, Superplus, continued to be offered from a separate pump. BP mixer pumps had earlier been introduced in Germany, Switzerland, the Netherlands and Austria with favourable results. It was hoped that the wider choice of prices and octane ratings would similarly enhance BP's competitive position in the UK, where no other company had introduced mixers on a national scale.[58] Other initiatives in the UK included the spread of self-service at BP forecourts, the introduction of BP Autocare (a diagnostic and minor repair service) and the launch of the BP Autoshop for the sale of antifreeze, accessories and various car care products.[59]

In lubricants, meanwhile, BP had in 1963 introduced a new oil, BP Energol Visco-static Longlife, alongside its existing multigrade, BP Energol Visco-static. Both oils had the same 10W/30 viscosity bracket, but the difference was that with Longlife the oil needed to be changed much less frequently. Longlife was not, however, a success. It was expensive, garage owners did not favour it because it reduced the frequency of customer visits, and motor manufacturers were reluctant to recommend extending intervals between oil changes.[60] It did not, therefore, catch on, and by 1966 BP was worried about its declining market share in retail motor oils which had fallen from 13 per cent in 1962 to little more than 10 per cent in 1965.[61]

There was not much prospect that market share in lubricants could be regained with BP's existing product range. Shell had recently introduced with great success Shell Super Motor Oil with a viscosity bracket of 10W/40, whereas BP's multigrades were only 10W/30. Moreover, fierce competition was expected from Castrol now that the lubricants market had been opened up as a result of the Monopolies Commission report.[62] To bolster its position, therefore, BP launched a new multigrade, BP Super Visco-static, in 1966. This oil, advertised as 'Super V', had a 10W/40 viscosity bracket and replaced Visco-static and Visco-static Longlife.[63]

But 'Super V' failed to stem the decline in BP's market share, which by 1969 had fallen to 7½ per cent. In contrast was the meteoric rise of the specialist lubricants firm of Alexander Duckham and Company, whose high-quality, keenly priced multigrades attracted a strong following among motor enthusiasts, especially in the fast-growing do-it-yourself sector. By 1969 Duckhams' market share had risen to 26 per cent (from a mere 1 per cent in 1960), putting it close on the heels of Castrol, which had been acquired by Burmah Oil. With Burmah's backing, Castrol was making a 'determined assault' on Duckhams, which felt that its position would be strengthened by association with a major company. BP, keen to increase its share of the retail lubricants market, stepped in and acquired Duckhams in 1970. BP did not, however, transfer Duckhams to SMBP. It retained ownership, but allowed Duckhams to operate at arm's length, selling its own brands.[64]

BP's retention of Duckhams outside SMBP was a forerunner of much more sweeping changes in its joint UK marketing arrangements with Shell. In April 1971, after protracted negotiations, the two companies were ready to announce that they had reached agreement on the break-up of SMBP.[65] BP, as 'the unwilling party in the divorce',[66] succeeded in securing its main demands, including the end of Shell's management (through Consolidated) of BP's marketing operations in South Africa.[67] In the UK, SMBP's trade and supporting operations were to be split over a five-year period so that Shell and BP would inherit 60 and 40 per cent respectively of SMBP's main products' markets and assets. SMBP's subsidiary, the National Benzole Company, and its National brand name would go to BP, but some National outlets would be rebranded and passed to Shell so that SMBP's retail gasoline business would, overall, be split in the 60:40 ratio. During the implementation period, which would end on 31 December 1975, all proceeds and costs would continue to be shared by Shell and BP through SMBP, as before. Thereafter, the two parents of SMBP would market independently in the UK, but, to preserve the logistical benefits of SMBP's distribution network, a system of 'supply support' would operate for at least twelve years. This would involve BP and Shell having 'guest' rights at each other's terminals, and exchanging products much as they had done under SMBP. The break-up was duly implemented on these lines, and from 1 January 1976 a BP subsidiary, BP Oil, took over BP's marketing operations. For the first time in forty-five years BP was marketing on its own account in the UK.[68]

FRANCE

While the British government acquiesced in the majors' dominance of the UK market, the situation in France was quite different. The clear objective of state policy in France was to achieve national independence in oil by creating a powerful state-owned oil sector and by closely controlling the operations of other oil companies.

The earliest French state oil company was the 35 per cent state-owned Compagnie Française des Pétroles (CFP), formed in 1924 to hold France's stake in the Iraq Petroleum Company.[69] The state reserved a 25 per cent share of the French refining industry for CFP's refining subsidiary, the Compagnie Française de Raffinage. After World War II, CFP also moved into marketing, adopting Total as its brand name and acquiring various independent marketers. One of the largest, Desmaris Frères, was acquired in 1966, raising CFP's share of the French market to 22 per cent.[70]

34   '. . . but to most people, of course, BP stands for . . .' BP advertisement in 1973

In the meantime, the French government, inspired by Gaullist distrust of the Anglo-Saxon oil majors, stepped up the search for French-controlled sources of crude. To that end, in 1945 it set up the Bureau de Recherches de Pétrole (BRP) to sponsor oil exploration in France's empire. Intensive exploration followed, primarily in Algeria and Equatorial Africa, resulting in several finds of which the most significant by far was the discovery of very

large oil and gas reserves in Algeria in 1956. After Algerian production
started in 1959, a new state-owned company, the Union Générale des
Pétroles (UGP), was created to refine and market the oil which BRP-backed
companies had found in the French Franc Zone. The majors were thus faced
with another new state-owned competitor, which began to look aggressively
for market share in France. UGP's acquisition of the French refining and
marketing operations of Caltex (jointly owned by Socal and Texaco) confirmed that it was a force to be reckoned with.

In 1966 the BRP, UGP and sundry other state-owned oil interests were
merged into a new state-owned entity, the Entreprise de Recherches et
d'Activités Pétrolières (ERAP), which adopted a new trademark, Elf. In 1970
Elf-ERAP acquired Antar, the biggest of the dwindling band of French independents. This increased Elf's share of the oil products market in France to
23 per cent. The two vertically integrated state companies, CFP and Elf-
ERAP, by then held about half of the French market in oil products.[71]

At the same time, the French government used the powers it had established under petroleum legislation in 1928 to extend its control over other
oil companies. In shipping, oil companies were required to use French flag
vessels for 66 per cent of their non-franc crude imports, and for 100 per cent
of their imports from the Franc Zone. After Algerian oil came on stream,
devoir national obligations were imposed on refiners and marketers, requiring them to use Franc Zone crude for a large proportion of their crude supplies. At the same time, a complex system of licences and permits enabled
the state virtually to control the market share of every company. All this
added up to an elaborate and comprehensive system of state control which
was more far-reaching than in any other major industrialised nation. A con-
solation for the companies was that the French government allowed prices
to be set at high levels which virtually guaranteed that the companies would
make profits from refining and marketing in France.[72]

Within this highly regulated market, BP's 70 per cent owned French subsidiary, Société Française des Pétroles BP (SFPBP), suffered from some of the
same weaknesses evident in BP's marketing elsewhere. SFPBP's sales mix was
weighted towards low profitability products, particularly heavy fuel oil, and
was relatively weak in more profitable products such as gasoline and lubricants.[73] French opinion surveys indicated that BP was thought to be more
old-fashioned than either of its main rivals, Shell and Esso, and much less
likely to have gasoline available where it was needed. BP was also thought
to be less interested in the motorist and a less good all-round company than
Shell or Esso.[74]

Although SFPBP benefited from the virtual guarantee of profitability in the

French market, it was nevertheless irked by the government's controls and discrimination in favour of the state-owned companies. Elf-ERAP received a large proportion of the Fonds de Soutien aux Hydrocarbures (an excise tax on petroleum products) and was, in effect, a subsidised competitor. The state-owned companies also received preferential treatment in the allocation of refining and marketing licences, enabling them to expand, while SFPBP was held back. But the most irksome of the French regulations for SFPBP was the obligation to purchase Franc Zone crude to cover more than 50 per cent of SFPBP's crude requirements. The primary function of BP's downstream operations in the 1960s was, as noted, to serve as an outlet for the growing volume of crude oil production demanded by the producing countries. BP was not a producer of franc crude, and the displacement of over half of its French outlet was a serious blow to its volume-driven downstream strategy.[75]

WEST GERMANY

Free on the one hand from the exceptionally concentrated oligopoly in the UK, and on the other from French-style state controls, the West German oil market was one of the most competitive in Europe. The West German government adopted liberal economic policies, and few sectors of the economy were closed to foreign participation.[76] In oil, there was some customs protection for products refined from West Germany's small-scale indigenous crude production, but it was phased out from 1963. At the same time the government, seeking to reduce Germany's reliance on foreign oil companies, began to offer loans on favourable terms to encourage German oil companies to explore for oil abroad. Apart from that, the oil industry in West Germany was left largely to itself, with little state intervention by the mid-1960s.[77]

West Germany was the fastest-growing oil market in Western Europe, expanding at an average annual rate of 19 per cent between 1960 and 1965, more than twice as fast as the UK market.[78] BP's West German subsidiary, BP Benzin und Petroleum AG, had a market share of about 13 per cent, and sold a larger volume of products than any other BP marketing company outside the UK. West Germany was, therefore, a vitally important outlet for BP in the quest for volume-driven growth. It was utterly essential for BP that an outlet of this magnitude and potential should be kept open.

In 1966, and for several years after, maintaining open access to the West German market was a matter of great concern to BP. In discussions on energy policy by the six countries in the Common Market (France, West Germany,

Italy, the Netherlands, Belgium and Luxembourg), most members were voicing concern about foreign domination of their oil industries. Calls for intervention came not only from government representatives, but also from a group of ten companies, including the French state-owned ERAP, the Italian state-owned ENI, and the main West German independents. They petitioned the Common Market Commission, demanding discrimination in favour of 'Community companies' and against non-Community multinationals. One of the measures they suggested was the granting of preferential market access to companies which were domiciled and controlled within the Community. As Britain was not a member of the Community, BP would not qualify as a Community company.

Of the member states, France was the most aggressive advocate of discrimination against foreign oil multinationals, and was keen to see its *dirigiste* system of state controls adopted as a model for the whole Common Market. BP, anxious to contain the spread of economic nationalism, looked upon West Germany as a pivotal country. If West Germany was won over to the French view, the whole Common Market would in all likelihood follow suit and BP would lose free access to its most important refining and marketing area. It was already shut out of the US market and simply could not afford to be shut out of most of Western Europe as well. It was, therefore, essential to do everything possible to avert a Franco-German axis on oil.[79]

In West Germany, the pressure for discrimination against the foreign multinationals was growing in the mid-1960s. Despite the favourable loans which were on offer from the government, German oil companies had not been notably successful in foreign exploration. The only German company with large-scale overseas oil reserves was Gelsenberg, which had acquired a minority share in Mobil's Libyan concessions, from which production had commenced in 1963.[80]

German insecurity about foreign domination of the German oil industry was heightened by Texaco's takeover of Deutsche Erdöl AG in 1966. As the West German Economics Minister later told BP, 'had there been any further moves after the Texaco/DEA take-over the Government would have been forced to introduce immediate legislation to prevent such take-overs and to restrict the freedom of the major international companies in Germany, as greater domination by the international companies would have been politically quite unacceptable'.[81]

By 1967, with West Germany's oil policy on a knife-edge between liberalism and nationalism, BP's fears that Germany might align itself with France were growing. A senior BP political analyst urged BP to consider 'whether there is anything we could do which might divert them from entering the

French camp lock, stock and barrel'. BP's top-level Planning Committee, under deputy chairman Eric Drake, agreed that the need to take action was urgent.[82]

In 1968 the lines on which the West German government was thinking became clearer when the Ministry of Economics indicated that while it wished to retain a liberal oil policy, it also wished to strengthen the position of the German independents. This, the Ministry hoped, could be achieved by encouraging the fragmented independents to merge, by safeguarding their 25 per cent share of the West German oil market and by encouraging them to form a government-backed joint exploration company to seek access to foreign crude supplies. At the same time, West Germany's resolve to prevent further foreign takeovers of the independents was confirmed when CFP attempted, with the support of the French government, to acquire the Dresdner Bank's 30 per cent shareholding in Gelsenberg. The acquisition was blocked by the West German government, to the fury of the French, whose objections created a diplomatic rumpus between the two countries.[83]

BP, meanwhile, argued before the European Parliamentary Committee on Energy in favour of a liberal international economy, free from discrimination against the oil multinationals, whose integrated international supply systems were, BP claimed, the best guarantee of security and cheapness in oil supplies.[84] In putting this view to Professor Schiller, the West German Economics Minister, Bridgeman argued that BP was a natural ally for the German independents insofar as BP had plentiful supplies of crude, which they lacked. BP, he wrote, would welcome opportunities to co-operate with them and to consider what it could do to meet their needs.[85] In December 1968 Bryan Dummett, one of BP's two deputy chairmen (the other being Drake), went on a mission to Bonn, the seat of the West German government, to meet Schiller and to confirm that BP saw itself as the ideal supply partner for the crude-short German independents.[86]

In 1969 the idea of an alliance between BP and the German independents was taken further. At the instigation of the West German government the leading eight German independents formed a joint company, Deminex, primarily to explore for oil outside Germany and to acquire shares in existing sources, with state financial backing. The idea was to increase direct German access to overseas oil reserves through what was, in essence, a private sector consortium, with the West German government performing a role akin to that of a bank, while preserving its liberal oil policies.[87] Deminex, in short, was an alternative to French-style *dirigisme* and was designed largely to head off calls for a more Francophile German oil policy.

BP, keen to support this liberal alternative to French policy, was willing to part with a share in one of its concessions to help uphold West Germany's liberal oil regime. It offered to sell Deminex a 5 per cent share in the Iranian Consortium from its own 40 per cent share, or to give up a pro rata share with the other members of the Consortium if they preferred the Deminex share to be spread between them. As Bridgeman told Rawleigh Warner, the chairman of Mobil, 'it was very important to offer the Germans some chance of a private source of crude oil rather than force them into the policies of French dirigisme'.[88]

BP's proposal to admit Deminex to the Iranian Consortium soon, however, ran into difficulties. The unanimous consent of the Consortium members was required, but CFP bluntly opposed the idea because it wanted to take any additional share in the Consortium that was going rather than let it pass to the Germans. The four US majors who were parents of Aramco as well as members of the Consortium also opposed the idea because they were resisting Saudi pressure for participation in Aramco, and their case against Saudi participation would be weakened if they allowed Germany to participate in the Consortium. The opposition to Deminex's participation in the Consortium could not be overcome, and in February 1970 BP told the West German Ministry of Economics that nothing could be done on that front.[89]

As an alternative, BP and Deminex turned to the possibility of Deminex acquiring a share in the BP subsidiary which held BP's two-thirds share in Abu Dhabi Marine Areas (ADMA). BP and Deminex could not, however, reach quick agreement on various points, including the extent to which Deminex would be allowed to participate as a full partner in ADMA, the price to be paid for Deminex's proposed share, the extent of ADMA's crude oil reserves, the production potential of the oil fields and the costs of their future development.

As a result of the long time spent in negotiations on these and other issues, no final agreement had been reached by the autumn of 1971, when the OPEC countries began more actively to seek to participate in the concessions held by the oil companies. The West German government was not prepared to underwrite the risk of Deminex buying an asset which would soon be diluted by producer-country participation. Deminex therefore lowered its offer price so that, in effect, BP would bear the risk of participation. BP found this unacceptable, the differences between BP and Deminex could not be resolved, and the negotiations were eventually called off in July 1972.[90] A few months later, BP sold 45 per cent of its share in ADMA to a consortium of Japanese companies who were seeking foreign oil supplies on much the same basis as Deminex. For that reason, the sale to the Japanese was known in BP as the 'Japinex' deal (see next chapter).

In the meantime, BP's brand image in West Germany had improved considerably in the first half of the 1960s, when it was boosted by the adoption of the 'New Look' at service stations, and the introduction of Super-Mix blending pumps, self-service and Visco-static Longlife oil. In the second half of the 1960s, however, BP's image in West Germany went into decline. In 1969, a German brand image study found that BP was weaker than its main competitors in 'all positive brand attributes'. BP, the study found, was 'particularly associated with being a reserved and anti-dynamic brand'. In lubricants, Longlife oil had been more successful in West Germany than in Britain, and after its introduction in the German market BP had become the market leader in lubricants. But between 1965 and 1969, a period which saw the introduction of Super Visco-static, BP's lubricants lost popularity. At the same time, West German motorists considered BP gasoline stations to be less spacious, clean and bright than those of its main competitors. Staff at BP stations were thought to be less responsive, slower in service and less technically qualified than at Shell and Esso stations.[91]

Other indicators also pointed to weaknesses in the competitive position of BP in Germany. For example, much of BP's trade was in short-term wholesale or bulk business, which was highly vulnerable to cut-throat price competition. The proportion of BP's sales which was sold under the BP brand direct to the final consumer through retail outlets was smaller than that of its main competitors. This retail business was the more desirable trade, with higher margins and greater opportunities to build customer loyalty to the brand than in the wholesale trade. A key measure of performance in the retail market was the volume of sales per gasoline station. BP's average volumes were markedly lower than those of its main competitors, except for Texaco.[92]

Similar weaknesses were evident in BP's refining and marketing operations in other countries in the area of the Rhine. For example, in Belgium BP's gasoline outlets had lower average sales volumes than those of Shell and Esso, and BP's overheads were disproportionately high.[93] Belgians thought of BP as being more old-fashioned, less interested in the motorist and less likely to have gasoline where needed than Shell and Esso.[94]

In the Netherlands, it was much the same. BP's share of the retail market was substantially lower than its overall market share because most of BP's trade was in bulk business. BP gasoline outlets generally had lower sales volumes than those of Shell and Esso. BP's overheads were disproportionately high. 'There seemed to be', went a familiar BP refrain, 'a disproportion between our organisation and our sales structure. We had similar organisation to the market leaders with much lower quantities, particularly of direct [retail] trade. On the other hand, we had to compete with the minor market-ers who had smaller organisations and therefore a lower cost profile.'[95] The Dutch, echoing the Belgians, thought of BP as more old-fashioned, less inter-ested in the motorist and less likely to have gasoline where needed than Shell and Esso.[96]

The only country in a 1971 BP corporate image study where BP was con-sidered to be less old-fashioned than Shell and Esso was Italy.[97] Here, BP evi-dently had an image, utterly incongruous with BP's own culture, but consonant with the image of swinging London in the 1960s, of being a 'hippy company'.[98]

ITALY

In the 1920s, the Company had formed a small marketing organisation in north-eastern Italy and had entered into negotiations to purchase a 30 per cent interest in the marketing subsidiary of the state-owned oil company, AGIP. The foray into Italian marketing was soon, however, ended when, in accordance with the market-sharing principles of the Achnacarry Agreement, the Company broke off negotiations with AGIP, and agreed to give up direct marketing in Italy except for a very small continuing presence in the Trieste area. In return, it received undertakings from Shell and Standard Oil (NJ) that they would purchase 20 per cent of their main prod-ucts requirements in Italy from the Company.[99]

After World War II, with the Achnacarry Agreement no longer extant, the Company looked again at the Italian market. A subsidiary, Britannica Petroli, was formed in 1948 to look after the Company's interests in the Trieste area. In the same year, the Company acquired a 49 per cent interest in AGIP's Marghera refinery at Venice, with the right to supply the refinery with its crude oil requirements. Products from the Marghera refinery were marketed by AGIP. In 1958 BP made a larger re-entry into the Italian market by purchasing two independent marketers, Sarom 99 and Italiana Carburanti, both of which marketed in northern Italy. In 1959 the name of Sarom 99 was changed to BP Italiana, with which Italiana Carburanti was subsequently merged. In 1961 BP purchased AGIP's share in Rifaer, a joint BP/AGIP aviation refuelling company, and it too was merged into BP Italiana, together with Britannica Petroli. BP Italiana thus embraced all BP's interests in Italy apart from its 49 per cent share in the Marghera refinery.[100]

BP's re-entry into the Italian market soon ran into problems. BP Italiana was originally expected to secure 10 per cent of the market throughout the whole of Italy.[101] It was, however, a late entrant facing the entrenched market leaders,

Shell, Esso and AGIP.[102] Moreover, although Italy was second only to Japan in its dependence on imported crude oil, it exploited its strategic position on the tanker route from Suez to develop, with government encouragement, the largest refining industry in Western Europe. The resultant surplus of Italian refining capacity promoted stiff competition for outlets. Price-cutting was exacerbated in the 1960s by the surge of crude oil exports from nearby Libya, which gave the independents ready access to low-cost raw material.[103]

In this difficult environment, BP Italiana failed to reach its target market share of 10 per cent and consequently was saddled with a marketing organisation which could support a much larger volume of trade than it achieved.[104] 'We have', ran the minutes of a BP meeting, 'the costs of a market leader without the high market share of Shell and Esso, and we lack the freedom of manoeuvre of the smaller competitors'.[105] BP Italiana also suffered from other weaknesses which echoed BP's problems in other European markets. In Italy, the 'undeniably poor BP network' had lower sales volumes per retail outlet than Shell or Esso.[106] And BP Italiana's sales mix had a preponderance of low-value products.[107]

In addition to these shortcomings, BP Italiana faced problems peculiar to Italy. One was the favoured position of the state-owned company, ENI, which was formed in 1953 to hold the state's oil and gas interests, including AGIP. ENI was notoriously hostile towards the oil majors, especially under its first president, Enrico Mattei. One of the most powerful men in Italy, he engaged in an aggressive vendetta against the majors. Disappointed that ENI was excluded from the Iranian Consortium, he undercut the majors' standard 50:50 profit-sharing terms in seeking overseas concessions. After his death in 1962, his less charismatic successors toned down ENI's crusade against the majors, but ENI retained a privileged position in Italy and remained a thorn in the majors' sides.

For example, in theory all companies bidding for refining licences ranked equally, but BP found that in practice no grant of a licence would be considered by the government unless the proposal had ENI's agreement. Because of its privileged position ENI could, according to a BP report in 1969, 'ride roughshod' over regulations controlling, for example, rights of way for pipelines. Indeed, such was ENI's influence that it could even 'exercise power over appointments of both civil servants and ministers in ministries relevant to ENI's activities'. It could also influence the government's energy policy 'to a point where government policies are likely to mirror precisely ENI's own attitudes'.[108]

A second problem for BP in Italy came when its short-haul Mediterranean crudes were nationalised by radical Arab states in the early 1970s. First came the nationalisation of BP's operations in Libya in December 1971; followed in June 1972 by Iraq's nationalisation of the IPC's production in northern Iraq, which was piped to the Mediterranean through the IPC's trans-desert pipeline (see chapter nineteen). The loss of these short-haul crudes put BP in an uncompetitive supply position in Italy. Whereas the US majors in Aramco and other oil companies operating in Libya retained (for the time being) their access to short-haul supplies, BP had to supply Italy with long-haul crudes transported by tanker all the way from the Persian Gulf.[109]

Another, but by no means lesser, problem was the endemic corruption of Italy's political and bureaucratic machine, which required constant lubrication with *omaggi* (gifts), backhanders, contributions and commissions. In 1972 BP's senior Executive Committee noted that 'it seemed hard for law-abiding foreign oil companies to make a profit in this country'.[110] Another BP report remarked that 'in Italy, the financial and administrative preference given to ENI, the surplus refining capacity deliberately created by the Italian Government and the availability of short haul crudes in the hands of new-comers have combined to create a market in which we could not operate profitably, even if we could compete with Italian methods of state intervention and political skulduggery'.[111]

What the authors of these remarks in all likelihood did not precisely know at the time was that BP was itself involved in making political contributions to the centre-right coalition government headed by Giulio Andreotti in return for favours. This would later become public knowledge, widely reported in the British media.[112] BP and other oil companies channelled their payments through the Unione Petrolifera, an association of oil companies operating in Italy. The favours they sought included the continuation of an allowance for increased transportation costs after the Suez Canal was closed in the 1967 Arab–Israeli war; the deferral of excise tax payments on favourable credit terms; a system of tax relief known as 'defiscalisation', which alleviated the tax burden on the companies; and the continued use of oil for electricity generation by ENEL, the state electricity utility. BP's contributions amounted to £800,000 in the years 1969–73, and were paid through a Swiss bank account designated by the Unione Petrolifera.[113] The top managers of BP Italiana argued before an Italian Parliamentary Commission of enquiry that BP's payments were normal, lawful political contributions, made to support democratic government in Italy, as opposed to bribes to secure particular favours.[114] But few believed these claims, and BP had to admit to the US Securities Exchange Commission that the payments had been made for specific purposes.[115] It remained, however, a moot point whether the companies were guilty of bribery, or the bureaucrats and politicians of extortion.

Certainly, Italian government controls on the oil industry seemed to be designed to make it impossible for the oil companies to operate at a profit. At the heart of the matter was the system of government price controls on oil products, which pegged prices at unremunerative levels. In 1972 BP Italiana and virtually every other oil company operating in Italy, including Shell Italiana and Esso Italiana, made losses in the Italian market.[116] Unhappy about their losses, they began to consider withdrawing from the Italian market. BP was the first to go. In May 1973 it received rival offers for its Italian refining and marketing operations from AGIP, and from a consortium organised by the Italian entrepreneur, Attilio Monti, whose business interests spanned oil refining and marketing, sugar and alcohol production, real estate and farming. Monti's offer was the higher, and agreement was swiftly reached for the sale of BP Italiana to his consortium, the Oil Chemicals and Transport Finance Corporation. BP's 49 per cent share in the Marghera refinery was sold separately to AGIP.[117] Later in the year, Shell reached agreement to sell its Italian refining and marketing operations to AGIP with effect from the beginning of 1974. Meanwhile, Esso and Gulf were trying to dispose of some of their Italian refining and marketing operations.[118]

BP's withdrawal from Italy was a turning point in its approach to refining and marketing. By 1973 BP's strategy of trying to protect its concessions by raising crude oil production and seeking outlets merely to dispose of the increase was manifestly obsolete. The concessionary system was collapsing, and the OPEC countries were taking control of crude oil production and prices. It would not be long before the dramatic rise in crude oil prices in the last quarter of 1973 would precipitate a world economic recession and, unthinkably for those conditioned by the buoyancy of the 1960s, a downturn in oil demand. But before that happened it was already becoming painfully clear that BP's earlier volume-driven growth in refining and marketing had saddled it with a vast investment in unprofitable downstream assets. BP urgently needed to stem the losses and to focus not on sales volume, but on the profitability of its refining and marketing operations. In this new situation, BP Italiana was an easy first choice for divestment. Its sale was the first major move in a campaign of downstream rationalisation and restructuring which would prove to be a very long-term undertaking as BP tried to shed the legacy of its great push for outlets.

$$===\quad \mathrm{IO} \quad ===$$

. . . And more outlets

Outside Western Europe the Company's markets before World War II reflected a largely imperial design. The Company's extra-European sales were made not in the industrial economies of Japan and the USA, but mainly in the less economically dynamic territories of the vast British Empire and at bunkering ports on the sea lanes of the Eastern Hemisphere, where Britain was the supreme maritime power.

The Company's connection with Britain's maritime power was long-standing. Before World War I, when Britain was engaged in a tense naval race with imperial Germany, it was the Company's ability to supply the Royal Navy with British-controlled supplies of fuel oil which had provided the strategic rationale for the British government's acquisition of a majority shareholding. Thereafter, the Company developed an extremely strong position in the marine oil bunkering trade of the Eastern Hemisphere, where it held a market share of 37 per cent by mid-century.[1] Fitting the imperial connection, the British colony of Aden, one of the strategic port-bases on the imperial sea route to India, became the Company's largest bunkering station.[2]

Turning to inland markets, in Australia the Company and the dominion government were partners in Commonwealth Oil Refineries, formed in 1920 to refine and market oil products in Australia. Oil markets in most of the rest of the British Empire were shared out in 1928 in an imperial carve-up between the three main British or part-British oil companies, the Company, Burmah Oil and Royal Dutch-Shell. The carve-up created two vast 'joint areas' in which the Company's sales were governed by restrictive agreements with Burmah and/or Shell.

One of the joint areas was India, the jewel in the British Empire. Here, a set of agreements between the Company, Burmah and Shell laid down that the Burmah-Shell Oil Storage and Distributing Company of India, jointly owned by Burmah and Shell, would be the sole vehicle for the marketing of the three companies' products in India. The Company, which had no equity stake in the Burmah-Shell company, agreed not to establish its own marketing outlets in India, and received the right to supply some of Burmah-Shell's oil requirements. Specifically, Burmah would have a pre-emptive right to use its indigenous oil supplies in India and Burma to supply Burmah-Shell. The remainder of Burmah-Shell's requirements would be met by imports, supplied equally by the Company and Shell. Burmah, with a guaranteed local outlet for its production (it had no production outside India and Burma), agreed not to market directly in other parts of the world, and to consign any oil it came to possess elsewhere to the Company and Shell in equal quantities.[3]

The second 'joint area', even larger in extent than India, was called the Consolidated area. It covered an expanse, roughly triangular in shape, between Cyprus, South Africa and Ceylon, taking in a great swathe of British imperial territory, primarily in East, Central and Southern Africa. The area was named after the Consolidated Petroleum Company, which was owned by the Company and Shell in equal shares, and which was the exclusive vehicle for the ownership and management of all their marketing operations in the Consolidated area. Shell and the Company were entitled and bound to supply in equal shares the crude oil and products sold in the Consolidated area, and the profits of Consolidated were split equally between the two parents.

That, however, was as far as the equality went. On the vital matter of who should manage Consolidated, it was agreed in the main Consolidated Agreement of 1928 that Shell should be responsible for the management on terms acceptable to Consolidated. A management agreement between Shell and Consolidated was duly reached in 1931, providing for Shell to manage the entire commercial business of Consolidated and its subsidiaries.[4] This meant that Shell would be responsible for the promotion of the BP brand, the use of the BP trademark and sales of BP products in the whole Consolidated area.

In retrospect, it seems extraordinary that the Company should have been prepared to cede control over its brand and trademark to a rival company in this way. But at the time, it was not so strange. When the Consolidated Agreement was signed in 1928, the air was thick with restrictive international pacts between the oil majors, such as the Red Line Agreement covering concessions in the whole area of the old Ottoman Empire, the Achnacarry Agreement and, of course, the Burmah-Shell Agreements. The

Company, a latecomer in marketing, could not hope to negotiate on equal terms with Shell, and accepted a subsidiary role in both the Burmah-Shell and Consolidated areas. Later, however, BP would regret the restrictions which it accepted in 1928.

At the end of World War II, the Burmah-Shell and Consolidated Agreements were still extant. So too, despite the enormous stresses and strains of the war, was the British Empire. The seismic changes in international relations wrought by the war were not yet wholly apparent. The USA and USSR had emerged as the two superpowers, but their influence did not spread immediately into every corner of the globe, and Britain remained the predominant foreign power in the vast area stretching from the Middle East and Africa across the 'British lake' of the Indian Ocean to Malaysia and Australasia. Here, in a region relatively remote from the new superpowers, the British Empire was still whole, an apparently secure field for British settlement and investment (see map 10.1).

However, the map of the British Empire flattered to deceive. Undermined by the impact of the war, the imperial order of the old European powers could not long survive the challenges of growing nationalism in the colonies and the rising hegemony of the superpowers. In the space of thirty years the European empires all but disappeared as the European countries gave up their political sovereignty over the peoples of Asia and Africa in a great wave of decolonisations. The old empires were replaced by a 'world of nations', the 'colonial world' became, broadly speaking, the 'Third World', and the USA and the USSR became the new arbiters of world affairs.

Britain, emerging victorious from World War II with a larger overseas empire than any other country, suffered a greater diminution of power than any other nation over the next quarter-century. Lacking the resources to sustain its far-flung possessions, Britain liquidated most of its empire in two great convulsive movements. The first centred in Asia in 1947–8, when India, Pakistan, Ceylon and Burma received independence; the second focused on Africa, where Britain's imperial position in West, East and Central Africa was undone in a flurry of decolonisations, most of them in the first half of the 1960s.

These convulsions fractured the great imperial expanses of the Burmah-Shell and Consolidated areas into new nation states, free from British rule. In most cases, their emancipation went further than Britain had intended. In the prelude to independence, Britain generally tried to ensure that new states would have moderate governments which would maintain close political, economic, social and cultural ties with the old mother country through membership of a multiracial Commonwealth. The new Asian and African

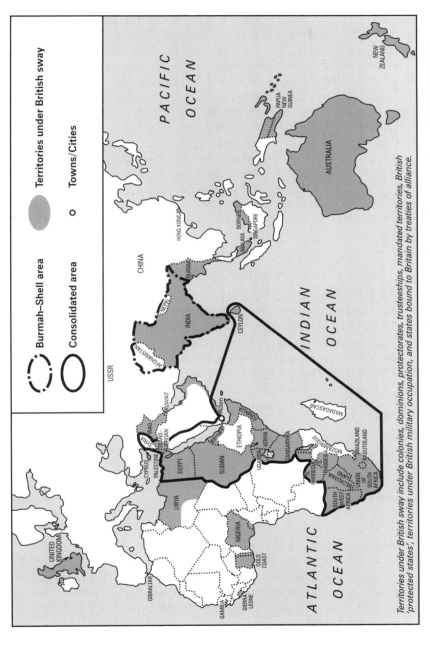

Map 10.1   The Consolidated and Burmah-Shell areas, and territories under British sway at the end of World War II

Territories under British sway include colonies, dominions, protectorates, trusteeships, mandated territories, British 'protected states', territories under British military occupation, and states bound to Britain by treaties of alliance.

nations would, it was hoped, remain 'open' economies, as favourable to British trade and investment as they had been in the imperial era. But these hopes rarely came to pass. The attainment of political independence was in most cases followed by a weakening of the economic links with the old mother country. As ex-colonies strove to achieve economic liberation consonant with political independence, the 'open' colonial economies of the imperial era were widely supplanted by 'closed' economies, designed to limit the exposure of the ex-colonies to external economic forces. Restrictions on foreign trade and investment, controls on foreign companies and the extension of state control over the 'commanding heights' of the economy were the typical economic hallmarks of the emerging nations. Foreign companies had, therefore, to adapt to new environments in which they were less secure than hitherto. These winds of change blew through the Burmah-Shell and Consolidated areas to the discomfort of BP and other oil multinationals.[5]

INDIA AND PAKISTAN

On the eve of independence, India was a large, diverse, poor country, predominantly agricultural, with the great majority of the population living in rural villages. A large proportion of energy consumption was provided by fuels such as wood, dry leaves and cow dung. Per capita oil consumption was only a small fraction of that in the more advanced industrial nations, and the main demand was for kerosene, used for lighting in rural areas which lacked electricity. Indigenous crude oil production and refining were small scale, and the oil market was dominated by foreign companies which supplied most of India's demand with imports of refined products, mainly from the Middle East. Burmah-Shell was the market leader, with about half the market, followed by Standard-Vacuum (an alliance of Standard Oil (NJ) and Socony-Vacuum), and Caltex (an alliance of Socal and the Texas Company). Their dominance would not, however, continue unchallenged after Indian independence, which resulted in the political and economic fragmentation of British India and the growth of state control over oil and other key economic sectors.

The British, who had taken pride in having established the political unity of India, undid their achievement at the end of their rule. Under the Independence of India Act of 1947, British India was partitioned between the two newly independent states of India and Pakistan, with West Pakistan and East Pakistan (later Bangladesh) separated by more than 1,000 miles of Indian territory. Mirroring this partition, the Burmah-Shell area was split between the old Burmah-Shell Oil Storage and Distributing Company of India and a newly formed sister company, the Burmah-Shell Oil Storage and Distributing Company of Pakistan.

In India, a substantially larger oil market than Pakistan, the government pursued nationalist economic policies of self-reliance and inward-looking industrialisation, based on import substitution and the Indianisation of industry. The state came to dominate the 'commanding heights' of the economy, including basic strategic industries such as transport, energy and steel, while in the private sector Indian business houses such as Tata and Birla dislodged the British managing agencies of the old colonial era.

Initially, state intervention in the oil industry was comparatively muted. The government, seeking to save some of the foreign exchange spent on imported oil products, encouraged the oil companies to construct refineries in India. The companies, with plentiful supplies of products available from the nearby Abadan refinery, at first resisted the idea of refining in India. But the nationalisation and shutdown of the Abadan refinery in 1951 altered the picture, and in 1951–3 Burmah-Shell, Stanvac and Caltex each agreed to build a new refinery in India. Stanvac's and Burmah-Shell's refineries were both at Bombay (Mumbai), and came on stream in 1954 and 1955 respectively, while Caltex's refinery at Vizagapatnam on India's east coast came on stream in 1957. Burmah-Shell's refinery was the largest (bigger than the other two combined), and supplied 45 per cent of India's total requirements of oil products after it came on stream. With these new refineries, the three dominant foreign companies of Burmah-Shell, Stanvac and Caltex continued to dominate the Indian market for the time being.

From the late 1950s, however, the left-wing Oil Minister, Keshiva Deva Malaviya, greatly extended the state's role in all sectors of the Indian oil industry. In the upstream sector, the state-owned Oil and Natural Gas Commission was set up in 1956 to search for oil in India with financial and technical assistance from the communist bloc, and soon made a series of discoveries. The state also participated in Oil India Ltd, formed in 1959 to exploit oil fields which Burmah had earlier discovered in Assam in the northeast of India. A state-owned refining company, Indian Refineries Ltd, was formed in 1958 to construct new refineries with Russian and Rumanian assistance. In marketing, the state-owned Indian Oil Company was formed in 1959, initially to supply oil to state enterprises. Later, however, its role was expanded and it became the sole distributor for products from the state refineries, and from other new refineries which the state owned in partnership with new private entrants to the Indian oil industry. In 1964 the Indian Oil Company and Indian Refineries Ltd were merged into a single state-owned refining and marketing company, the Indian Oil Corporation, which was given a monopoly on the import of oil products into India. With these developments, leadership of the Indian market passed from Burmah-Shell to the Indian Oil Corporation, which established a dominant position in the Indian market (see figure 10.1).[6]

In Pakistan, relations between the established foreign oil companies and the government were less adversarial than in India, but the same tendency for restrictions to be placed on the foreign companies was at work. By 1961 the government of Pakistan was making it clear that it expected majority national participation in Burmah-Shell. When the two parent companies appeared to be dragging their feet, the government set up the state-owned Pakistan National Oil Ltd in 1962 to import and market oil products. Three years later, Shell agreed to sell its half-share in Burmah-Shell's assets in East Pakistan to Burmah, and the following year a new marketing company, Burmah Eastern Ltd, was formed to take the place of Burmah-Shell, with the majority of its shares offered to and taken up by Pakistanis. In West Pakistan, Shell maintained its interest in Burmah-Shell, but majority local participation was conceded in 1969 when a new company, Pakistan Burmah-Shell, was set up. It was 51 per cent Pakistani-owned, with the government's National Investment Trust taking up part of the controlling interest.[7] By the end of the 1960s, therefore, the Burmah-Shell company, which had dominated the oil market of British India before independence, had suffered a huge loss of market share, and was breaking up.

Although BP had no equity stake in Burmah-Shell, it was tied to Burmah-Shell as the Indian outlet for its oil by the restrictive conditions of the 1928 agreements. In the 1950s and 1960s the original arrangements became immensely complicated as various further agreements, running into several hundred pages, were drawn up to cater for the fragmentation of Burmah-Shell's marketing area, the importation and processing of crude oil at the Bombay refinery, and the diversification of Burmah Oil into new areas outside India and Burma. The effect of the revised agreements was, on the whole, to make the Burmah-Shell arrangements still more restrictive for BP.

From the beginning of 1962, BP's share of the Burmah-Shell outlet was reduced by a new system of supply quotas whereby all Burmah's preferential supplies of indigenous oil to Burmah-Shell were deducted from BP's supply quota, and not from BP's and Shell's equally, as was previously the case. At the same time, BP continued to be precluded from supplying oil to other outlets in India, except in special circumstances, and even when they arose BP would have to grant 50 per cent of any new outlet to Shell. In addition, BP entered into a new 'right-through-sharing' arrangement with

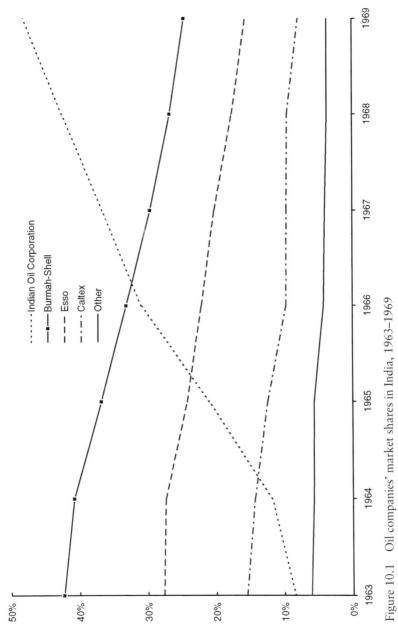

Figure 10.1 Oil companies' market shares in India, 1963–1969

Burmah whereby the two companies shared the profits on the whole chain of operations – including production, refining and marketing – of the oil which BP supplied to Burmah for sale through Burmah-Shell. However, outside the scope of right-through-sharing, BP agreed to purchase and export some of the products, especially gasoline, produced at the Bombay refinery which were surplus to Indian requirements.[8] As BP already had a surplus of gasoline in its international supply system, and could in any case produce gasoline more cheaply at other refineries, this was an onerous obligation. The fundamental problem, however, was that BP remained shackled to Burmah-Shell, whose ability to provide BP with an expanding, profitable outlet in India was curtailed by the precipitous fall in its market share and the downward pressure on oil prices from the Indian government. Not surprisingly, BP came to view the Burmah-Shell agreements with little enthusiasm.[9] Indeed, it came as something of a release when the Indian government nationalised Burmah-Shell at the end of 1975, and the Burmah-Shell agreements were terminated.[10]

THE CONSOLIDATED AREA

In the Consolidated area, as in the Burmah-Shell area, the imperial design of the market-sharing agreements of 1928 was broken up by decolonisation, market fragmentation, and growing state intervention in the third quarter of the twentieth century. At the same time, in another echo of the Burmah-Shell area, BP grew increasingly restive with the restrictions on its marketing to which it had agreed in 1928.

By the 1950s, BP had become deeply unhappy with its almost complete exclusion from any say in the running of Consolidated. There was, BP complained, little effort made by Shell to promote the BP brand, and little or no consultation by Shell with BP about the way that BP products were sold. The Consolidated area was, in effect, closed to BP, which had very little knowledge of the area, which Shell jealously protected as its preserve.[11]

Worse still, BP seemed powerless to break Shell's hold. BP could not, under the terms agreed in 1928, terminate the Consolidated Agreement without Shell's consent. BP could withdraw by selling its shares in Consolidated to Shell, but in that event BP would not be allowed to trade in the area for 50 years. The 1931 management agreement between Shell and Consolidated, giving Shell complete managerial charge of Consolidated, could be terminated by Shell or by Consolidated, but not by BP. In a nutshell, BP was shackled to the Consolidated company under Shell's management. The logic of these arrangements, as a Company executive explained it in 1928, was that 'the underlying principle of the Agreements being a marriage, no provisions have been made for divorce'.[12]

The part of the Consolidated area in which BP most wanted to manage its own marketing was southern Africa. BP acquired a marketing organisation in this area in 1954 when it purchased the Atlantic Refining Company of Philadelphia's refining and marketing interests in the Eastern Hemisphere. These included the Atlantic Refining Company of Africa, which marketed Atlantic products across southern Africa.[13]

The plum in this large area was the British self-governing dominion of South Africa, the largest and most profitable oil market in the Consolidated area. Here, the white National Party was firmly entrenched in power, which it sought to perpetuate by adopting apartheid policies of white leadership and racial separation. Though the black majority was comparatively impoverished, the sizeable white community enjoyed a high standard of living in an economy which was growing rapidly, providing a large energy market. South Africa's plentiful supplies of cheap coal met most of its requirements for heavy fuels, but gasoline was much in demand for transport.[14]

After BP acquired the Atlantic Refining Company of Africa, it was immediately brought under the ownership and management of Consolidated, in keeping with the 1928 agreement, which stipulated that all BP/Shell marketing operations in southern Africa would be conducted by Consolidated. However, the Atlantic company continued to trade under the Atlantic name, selling Atlantic brand products until 1959. In that year, the rebranding of products from Atlantic to BP was completed and the name of the company changed to BP Southern Africa (Pty) Ltd.[15] In the run-up to these changes, BP proposed that it should second a senior manager to Consolidated to take over the management of BP Southern Africa, but Shell, protecting its managerial preserve, rejected the proposal.[16] BP was thus shut out of the management of BP Southern Africa, which remained a subsidiary of Consolidated under Shell's management, holding about 13 per cent of the South African market for oil products (see figure 10.2).

BP remained powerless to alter this situation until the late 1960s when, as was seen in the last chapter, Shell became keen to break up Shell-Mex and BP, the joint Shell/BP marketing company in the UK. BP did not welcome the idea of splitting up SMBP, and agreed to it only on the condition that the two parent companies would also split their operations in South Africa. Shell reluctantly agreed to this bargain, and in the early 1970s a Market

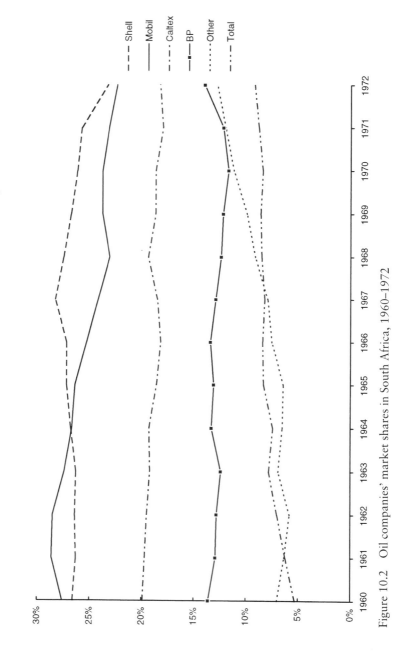

Figure 10.2   Oil companies' market shares in South Africa, 1960–1972

35 BP filling station
at Misty Mount,
South Africa, in 1963

Equalisation Group was set up to split Consolidated's operations equally between Shell and BP in South Africa, South West Africa (Namibia), Botswana, Lesotho and Swaziland. As the BP brand's market share was at the outset substantially lower than Shell's, the split involved the transfer of 182 gasoline stations from Shell to BP. This operation took time, and it was not until 25 June 1975 that the split became effective. From that date the Shell and BP marketing companies in southern Africa ceased to be governed by the Consolidated agreement, and ceased to be jointly owned. They became, instead, wholly owned subsidiaries of their respective parent companies, operating independently of each other. At last, BP owned and managed its own marketing company in southern Africa.[17]

Meanwhile, elsewhere in the Consolidated area British decolonisation had proceeded apace. In East and East-Central Africa, Tanganyika, Uganda,

36   BP road tanker delivering BP products to farmers in Natal, South Africa, in the 1960s

Kenya, Northern Rhodesia (Zambia) and Nyasaland (Malawi) all received independence in the first half of the 1960s. As the Consolidated area fractured into these and other new nations, a growing number of independent governments nationalised or participated in Consolidated's subsidiaries. By mid-1970 Egypt, Syria, Ceylon (Sri Lanka), South Yemen and Somalia had all nationalised Consolidated's local marketing operations, while Tanzania (formerly Tanganyika and Zanzibar), Uganda and Zambia had acquired shareholdings of 50 per cent or more.[18]

But of all the African nationalists, none caused the British more trouble than the white settler nationalists of Southern Rhodesia (Zimbabwe). Here the white settler minority had enjoyed self-government, though not full independence, since 1923. In the early 1960s, as black Africans in other countries achieved majority rule, white Rhodesians grew increasingly determined to perpetuate their rule by achieving independence on their terms before their dominance in Rhodesia's economic and political life was eroded. The

37   BP service station at Salisbury (Harare) Airport, Rhodesia (Zimbabwe), in the 1960s

decisive moment came in November 1965, when the Rhodesian Prime Minister, Ian Smith, made a Unilateral Declaration of Independence (UDI) from Britain. The British government immediately declared that Rhodesia was still a British dominion, and economic sanctions were imposed on Rhodesia by Britain and the United Nations. The sanctions proved, however, to be largely ineffective, and the intransigent white Rhodesian regime was able to hold out for a decade and a half before Rhodesia eventually attained independence as Zimbabwe under black majority rule in 1980.

The local management of Consolidated in southern Africa, the parent companies of Shell and BP in London and the British government could not escape some of the responsibility for the ineffectiveness of sanctions. Rhodesia was a landlocked country with no indigenous sources of crude oil, but it had plentiful supplies of cheap coal and hydroelectric power, and was therefore much less dependent on oil than most comparably industrialised nations. Nevertheless, Rhodesia needed oil for transport and industrial purposes. The

38   Road tankers at Umtali refinery, Rhodesia (Zimbabwe), in October 1965, before the imposition of international sanctions against Ian Smith's white settler regime. The refinery, in which BP held a 21 per cent shareholding, was shut down in January 1966, after the sanctions were imposed

denial of oil supplies was therefore an important component of the sanctions policy.

But large supplies of oil continued to reach Rhodesia from South Africa, whose white community passionately supported the cause of the white settlers in Rhodesia, and went to great lengths to organise an 'oil lift' to their besieged neighbours. They did so with the blessing of South Africa's white National government, which refused to join in the sanctions, and insisted that companies registered in South Africa (including BP Southern Africa and Shell South Africa) should continue to supply customers without discrimination, although it was obvious that many South Africans were purchasing oil for transport to Rhodesia. Some supplies reached Rhodesia directly from South Africa, while others were routed via neighbouring Mozambique, whose Portugese colonial rulers also refused to apply the sanctions.

Shell South Africa and BP Southern Africa provided a large share of Rhodesia's oil requirements, mainly through an intermediary South African forwarding company. When this became known to the London managements of BP and Shell in 1968, they discussed it with the British government, who accepted that sanctions against Rhodesia could not be effective unless they were also applied to South Africa. The British government was, however, unwilling to provoke a confrontation with South Africa, and acquiesced in a complicated exchange agreement whereby the South African subsidiary of CFP would supply oil to Rhodesia via Mozambique in return for supplies in South Africa from the local Shell and BP marketing companies. This was no more than a cosmetic device, which did nothing to prevent oil reaching Rhodesia, but which allowed the British government to state publicly that British companies were not involved in sanctions-busting. The exchange arrangement continued for about three years, after which the South African Shell and BP marketing companies resumed making supplies to Rhodesia via Mozambique, through the same intermediary as beforehand. The trade to Rhodesia was divided between Shell and BP when they split their South African marketing operations in the early 1970s, and continued until 1975–6, when the newly independent Mozambique closed its borders with Rhodesia, and the South African government-owned oil company, SASOL, took over the business of supplying Rhodesia.[19]

By the time that these facts were laid bare in 1978 in the Bingham Report (the result of an official investigation requested by the British Foreign Secretary), the remains of joint BP/Shell marketing in the Consolidated area were looking decidedly ragged. The two companies had split their marketing operations in southern Africa; the rest of the Consolidated area was crisscrossed with national boundaries within which many independent governments had nationalised or participated in Consolidated's operations; and BP's and Shell's reputations in black Africa had been badly damaged by the revelation that they had broken the sanctions on Rhodesia. At the end of 1981 Shell and BP separated their interests in most of the remaining Consolidated area, leaving only Kenya and Zimbabwe as joint marketing areas.[20] The last traces of the restrictive Consolidated agreement of 1928, based on imperial boundaries, had all but disappeared, along with the British Empire.

THE FAR EAST

In the Far East, as in India and Africa, the Company's marketing interests before 1950 reflected both the imperial emphasis of its extra-European marketing and the Company's tendency to make restrictive agreements, giving up its freedom to establish its own market outlets in return for supply rights to other oil companies.

In keeping with the imperial emphasis, the only countries in the Far East where the Company had its own market outlets were the British dominions of Australia and New Zealand.[21] In each of these countries the dominion government held a 51 per cent shareholding in the Company's local associate, mirroring the British government's majority shareholding in the parent

Company in London. In the 1950s, however, both dominion governments sold their shareholdings. In Australia, in 1952 the Company purchased the Australian government's shareholding in Commonwealth Oil Refineries, which was renamed BP Australia in 1957;[22] and in New Zealand BP purchased the government shareholding in The British Petroleum Company of New Zealand in 1955.[23]

While Australia and New Zealand continued to provide steadily growing market outlets for BP, there were other countries in the Far East which offered far more spectacular growth prospects. The most important by far was Japan.

Before World War II Japan had not been a major oil market. The Japanese economy was fuelled mainly by coal, and oil accounted for only a small proportion of energy consumption. There was very little indigenous oil production, and the small demand for oil was met mainly by imports. The oil market was dominated by Shell, which had entered the Japanese market in 1899, and Standard-Vacuum, which had entered the market in 1900. Both of these companies imported refined products and sold them through their own market outlets in Japan. If the Company had any designs on the Japanese market in these prewar years, no record of them has been unearthed.

Devastated by World War II, Japan in the immediate postwar years was under the administration of the US occupying forces commanded by General Douglas MacArthur. The USA, seeking to remove Japan's capacity to make war, ordered the closure of Japan's oil refineries, apart from those processing the small quantities of locally produced crude. The aim was to prevent Japan restoring its Pacific coast refineries, which had run on imported crude, and to ensure that Japan would have to rely on imports of refined products for most of its oil needs. These were not expected to be all that high, as coal was still regarded as Japan's basic energy industry.

In 1949, however, the USA reversed its previous policy and decided to allow the rehabilitation of Japan's Pacific coast refineries. For a brief period from 1949 to 1952 Japan welcomed foreign investment to help rebuild the Japanese refining industry. In these few years there was a flurry of foreign investment as international oil companies entered into partnerships with Japanese refining and marketing companies. The partnerships usually took the form of an international company – Shell, Standard-Vacuum and Caltex were the main participants – agreeing to supply crude oil, capital and refining technology in exchange for an equity stake in the Japanese partner.[24]

The Company, however, was debarred from taking part in the Japanese oil industry by one of the restrictive agreements which had become the hallmark of its downstream relationships. Specifically, in the late 1940s the Company negotiated the Far Eastern Agreement with Shell, which took effect from the beginning of 1948. For a period of at least twenty years, the Company was to supply Shell with products from the Abadan refinery equal to 50 per cent of Shell's sales in the Far Eastern markets of Malaya, Indo-China, Siam, China, Hong Kong, the Philippines, Japan, Australia and New Zealand. In return for its supply rights, the Company undertook not to establish its own outlets in these markets, except in Australia and New Zealand, where it was already established.[25]

The Far Eastern Agreement did not, however, run its full course. Deliveries of oil under the Agreement continued until June 1951, when they were suspended because of the imminent shutdown of the Abadan refinery after its nationalisation by Iran. In February 1955 BP and Shell agreed that the Agreement had been frustrated by BP's loss of its exclusive possession and control of its Iranian oil fields and the Abadan refinery. With the Agreement at an end, BP was free to establish its own sales outlets throughout the Far East.[26]

For BP, under mounting pressure from the producing countries to raise production from its vast oil reserves in the Middle East, the Japanese market had obvious attractions. In 1950, on the eve of the Iranian nationalisation, oil had accounted for only 7 per cent of Japan's energy requirements, compared with coal's 60 per cent share. The next two decades witnessed a dramatic transformation as Japan turned increasingly to oil to fuel its spectacular economic growth. By 1970 oil accounted for about three quarters of Japan's energy needs, and Japan ranked third in the world, behind only the USA and the USSR, in oil consumption. Lacking significant indigenous production, Japan was the world's largest importer of crude oil, and its economy was more dependent on imported energy than that of any other major industrial nation.[27]

In Japan, resistant at the best of times to foreign control of Japanese industry, the rising dependency on foreign oil supplies and companies aroused the same insecurity that was felt in other oil-consuming countries, and provoked a similarly nationalistic response. The welcome which the Japanese had given to foreign companies to help rebuild the Japanese refining industry in the early 1950s was soon replaced by measures to discourage or prevent further inroads by the foreign oil majors.[28] BP, having missed the opportunity to get into the Japanese market while the door was open, found, after being released from the restrictions of the Far Eastern Agreement in 1955, that the door was closed.

At intervals over the next fifteen years BP reviewed the possibilities of entering the Japanese market on its own or, more feasibly, in partnership with a Japanese refiner/marketer. Repeatedly, however, BP came to the conclusion that Japanese concerns about foreign domination of the oil industry were so strong that there was, in the words of one report, 'no possibility whatsoever of obtaining a substantial interest in a Japanese oil company'.[29]

While failing to establish a direct presence in the Japanese market, BP did, however, succeed in making very large sales of crude oil to independent Japanese refiner/marketers. By 1972 BP's crude sales to Japan were so large that they accounted for 8½ per cent of the Japanese market.[30] In some cases the sales were conventional exchanges of crude oil for cash, but in others they were linked to loans which BP made to Japanese firms for the construction of refineries, to BP orders for new tankers from Japanese shipbuilders and, in 1970, to Japanese participation in BP's share of the El Bunduq oil field in the waters off Abu Dhabi and Qatar.[31]

Japanese participation in the El Bunduq field reflected the priorities of Japanese oil policy, which had obvious parallels with West Germany's (see previous chapter). In both countries, the government sought to limit the penetration of foreign companies into the national market, while encouraging national companies to search for oil abroad. In Japan's case, the first significant entrant to overseas exploration and production was the Arabian Oil Company, a Japanese consortium which obtained concessions in the Neutral Zone between Saudi Arabia and Kuwait in 1958, and discovered the Khafji field in the Neutral Zone in 1960. The Japanese government gave preference to imports of Khafji crude and aimed to raise the nationally owned share of Japan's crude imports to 30 per cent. Imports of Khafji crude came nowhere near this target.[32] To stimulate the search for nationally controlled crude, the Japan Petroleum Development Corporation (JPDC) was established in 1967 to provide finance for overseas exploration by Japanese companies.[33] The president of the JPDC personally supported Japanese participation in the El Bunduq field, and was keen for Japanese companies to enter into further upstream co-operation with BP.[34] In late 1971, the JPDC approached BP, expressing interest in securing Japanese participation in Abu Dhabi Marine Areas (ADMA). BP was at the time negotiating with Deminex for the sale of part of BP's share in ADMA to the Germans and kept the Japanese on the sidelines while proceeding with the Deminex negotiations. When these broke down, BP turned to the eagerly waiting Japanese, who in December 1972 agreed terms for the purchase of a 45 per cent share in BP's two-thirds interest in ADMA. The agreement, which was known as the 'Japinex' deal, was completed in February 1973.[35]

While BP remained without any refining or marketing interests in Japan, it had established a small refining and marketing foothold elsewhere in the

Far East by purchasing the Maruzen Toyo Company's Singapore refinery and entering the Malaysian market in 1964. After Singapore separated from Malaysia in 1965, BP split its marketing operations between BP Singapore and BP Malaysia, and also considered expanding into other countries such as Indonesia, the Philippines, Thailand, Hong Kong, South Korea and Taiwan.[36] No further action had, however, been taken by the mid-1970s, when BP's refining and marketing operations in the dynamic tiger economies of South-East Asia were still restricted to Singapore and Malaysia.

ENTRY INTO AMERICA

Although Japan was the fastest-growing national oil market in the world in the 1950s and 1960s, its oil consumption at the end of these decades was still dwarfed by that of the USA, which was far and away the world's largest oil consumer. Protected by import quotas, the US oil industry was also highly profitable. This combination of size and profitability, coupled with political stability and a private enterprise culture, made the USA the most desirable country in the world for investment by the oil majors.

Up to the late 1960s, BP, alone of the majors, remained outside the enticing US market. In neighbouring Canada, BP entered the market in 1957, when it began to build up a substantial refining and marketing business in Quebec and Ontario, the main oil-consuming provinces of the country.[37] After acquiring the Canadian refining and marketing assets of Cities Service Oil Company in 1964, BP became the fifth-largest refiner and marketer in Canada, behind Standard Oil (NJ), Shell, Gulf and Texaco.[38] In 1971, BP further expanded its marketing outlets in Quebec and Ontario by purchasing a controlling interest in an independent Canadian marketing company, Supertest Petroleum.[39] But these moves in Canada were no substitute for a presence in the USA.

In keeping with the time-honoured pattern of BP's development, it was the discovery of a giant oil field which paved the way for BP's entry into refining and marketing in the USA. After Standard Oil (NJ) and ARCO struck oil at Prudhoe in the first half of 1968, BP not only hastened to drill on its own Prudhoe Bay acreage (see chapter eight), but also began to address the enormous opportunities and challenges which the Prudhoe discovery had opened up for BP in the US market. Two matters, in particular, stood out. One was how crude oil could be transported to markets from the icy wilderness of the Alaskan North Slope; the other, how BP should deal with the commercial exploitation of its prospective Alaskan oil reserves.

On transportation, in the second half of 1968 BP, ARCO and Standard Oil (NJ) jointly formed the Trans Alaska Pipeline System (TAPS) and prepared a feasibility study for a trans-Alaska pipeline. In February 1969 the partners in TAPS announced plans for the construction of a 48-inch diameter pipeline with an initial capacity of about 500,000 bpd, traversing a distance of some 800 miles from the North Slope to a terminal in the Gulf of Alaska. It was planned that the pipeline would be brought into operation in 1972, at an estimated cost of approximately $900 million. In the event, both the timescale and the costs would greatly exceed these early estimates, primarily because of the time and money expended on native land claims and environmental concerns, unforeseen in the early stages of the project.[40]

The question of how BP should commercialise Alaskan oil was also considered in the second half of 1968. The idea that BP might become solely a crude oil seller in the US was not seriously entertained because BP, steeped in the tradition of vertical integration, wanted to have the security of its own outlets for Alaskan crude.[41] There were, however, strong financial, managerial and political arguments against BP trying to go it alone in establishing an integrated US operation from scratch.

Financially, it was clear that very large expenditures would be required to develop BP's Alaskan reserves, to construct transportation facilities and to establish downstream operations on the necessary scale. BP could not afford to finance these projects on its own and there was, in any case, the problem that the UK Treasury would not have permitted BP to remit the necessary funds from the UK.[42] With regard to management, BP lacked an established US managerial organisation capable of undertaking a large-scale entry into the US market, and it was therefore felt that BP should acquire American management for its US operations.[43] Politically, there was the danger of arousing American antagonism towards BP on the grounds that US oil reserves were falling into the hands of a foreign company which, worse still for Americans, was majority-owned by a foreign government.[44] For these reasons, BP concluded that it should seek a US partner which possessed efficient downstream operations, financial strength, good management and a widespread American shareholding which would afford some political protection in the USA.[45]

In November 1968 BP held exploratory discussions with a number of US oil companies, including Cities Service, Union Oil and Sinclair, which was at the time the target of a hostile bid from Gulf and Western. Nothing had come of these discussions when ARCO stepped in with an agreed offer for Sinclair. The ARCO-Sinclair amalgamation would be the largest merger in the history of the oil industry, creating the seventh-largest US oil company, with assets of $3·7 billion. But before it could come about a way had to be found around the US Justice Department's antitrust concern that the amalgamation would have too large a share of the USA's eastern markets. To circumvent this

39   BP signs being prepared for shipment to the US for the rebranding of the
Sinclair stations which BP acquired in 1969

problem ARCO offered to divest Sinclair's downstream assets on the US
eastern seaboard to BP, a willing buyer. The outcome was that the ARCO-
Sinclair merger went ahead, and BP acquired a chain of nearly 10,000 Sinclair
gasoline stations (about double the number of BP filling stations in Britain)
stretching down the eastern seaboard from Maine to Florida (see map 10.2).
Also part of the package were two refineries, one at Marcus Hook in
Pennsylvania and the other at Port Arthur, Texas, plus some pipeline interests.
The purchase of these assets was completed on 1 April 1969 at a price of $400
million (£166 million) payable over five years starting at the end of 1972 or
six months after the first commercial shipment of crude from BP's Prudhoe
Bay reserves, whichever occurred first. BP immediately started rebranding the
former Sinclair gasoline stations, replacing the soon-to-be-extinct Sinclair
trade mark (somewhat aptly a diplodocus dinosaur) with the BP Shield, which
made its debut on US forecourts amid much publicity in the spring of 1969.[46]

The rebranding programme could not fully mask the low quality of the
assets that BP had acquired. The old Sinclair stations were thinly spread over
a large marketing area, with low market shares, low sales per outlet and

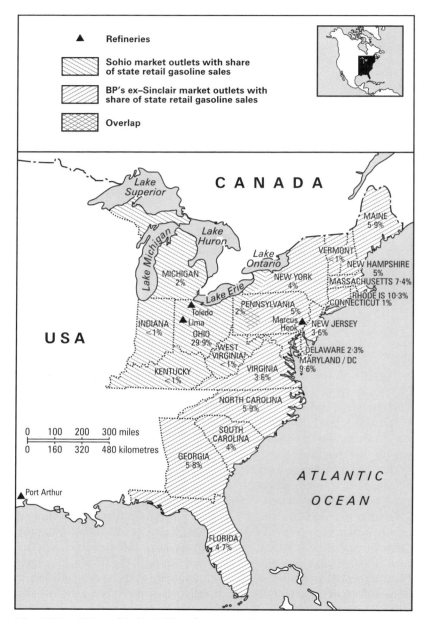

Map 10.2   BP's and Sohio's US refineries and markets, 1970

long, costly supply lines.[47] The refineries at Marcus Hook and Port Arthur were, a BP director would before long admit, 'pretty out of date', with high operating costs.[48]

Nevertheless, BP's entry into the US market was widely reported as a major corporate coup. Since World War II US multinationals venturing abroad had dominated global flows of foreign direct investment, spreading American management, capital and technology throughout the world. By comparison, non-US companies were at that time making only small inroads into the USA.[49] BP's incursion was a riposte to the established trend, and its purchase of the ex-Sinclair assets made it the largest single all-British investor in the USA.[50]

More, however, was to come. Although the acquisition of the ex-Sinclair assets provided BP with a potential outlet for Alaskan crude, it did not provide the financial, managerial and political vehicle which BP had been seeking for its entry into integrated US operations. The purchase of assets was not, therefore, a substitute for a US partner. It was in the search for such a partner that, contemporaneously with its purchase of downstream assets from ARCO, BP held its first discussions with Sohio.[51]

Sohio had been founded in 1870, when it was incorporated under the name Standard Oil as the original company of John D. Rockefeller, who subsequently expanded his oil interests into the huge empire which dominated the American oil industry until 1911 when the US Supreme Court ordered its dismemberment for antitrust reasons. On the break-up of Rockefeller's oil trust, Standard Oil of Ohio (Sohio) was left without its own crude production, and became a regional refining and marketing company operating solely within the state of Ohio. In later years Sohio expanded into petrochemicals, coal production and other diversified activities such as motels, restaurants, vending machines, uranium exploration and shale oil. But in its core oil business Sohio remained essentially a one-state refiner and marketer, purchasing most of its crude supplies.[52]

By the mid-1960s Sohio, worried about its lack of crude oil, was seeking an alliance with a crude-rich company. BP, with its large share of the Prudhoe Bay field, was an obvious candidate.[53] While BP possessed the crude which Sohio was seeking, Sohio had the qualities which BP was looking for. With two refineries (at Toledo and Lima) and 3,800 filling stations, Sohio held the dominant share (about 30 per cent) of the gasoline market in the state of Ohio, the fourth-largest gasoline market in the USA, with heavy urban concentrations in Cleveland, and along Lake Erie in Akron, Columbus, Dayton and Cincinnati. Sohio's marketing management under chief executive officer, Charles Spahr, was thought to be one of the best in the whole of the USA. In addition, Sohio had widespread shareholders, numbering 49,000 at the end of 1968, and was financially strong, with little debt.[54] In sum, BP and Sohio possessed complementary attributes, and seemed natural partners.

For all the mutual attraction there were some significant differences between them, especially in their corporate cultures. BP was unmistakably British, well-practised in international oil diplomacy, long-established as a major player in the international oil industry, with an excellent record in exploration and production, but limited in its marketing capabilities. Sohio, on the other hand, was quintessentially American, highly market-orientated, and much more geographically focused than BP. Its headquarters in Cleveland, Ohio, seemed small scale and parochial compared with BP's head office in cosmopolitan London. But although Sohio was much the smaller, more local company, it had a tradition of managerial independence which it did not wish to give up.[55]

Negotiations between the two companies commenced early in December 1968 and resulted in a Memorandum of Intent, signed in June 1969, followed by the Principal Agreement in August. Under its terms BP was to transfer to Sohio its acreage at Prudhoe Bay, partial interests in other BP leases in Alaska, its minor production interests in the lower forty-eight states and the ex-Sinclair downstream assets which BP had acquired from ARCO. In return, BP was to receive 1,000 shares of newly created special shares in Sohio, which carried the same rights as Sohio's ordinary shares except that they were not entitled to dividends until sustainable production from Sohio's Prudhoe Bay reserves reached 200,000 net bpd or until 1 January 1975, whichever occurred earlier.

Initially, each Sohio special share was to be equivalent to 4,466 ordinary shares, so that BP would start out with a 25 per cent equity stake in Sohio. Thereafter, BP's shareholding would escalate in line with the growth in Sohio's crude production from Prudhoe Bay, until the point was reached where Sohio's production was 600,000 bpd, and BP's equity stake was 54 per cent. That was the maximum interest in Sohio to which BP was entitled by virtue of its ownership of special shares. If Sohio's Prudhoe Bay production exceeded 600,000 net bpd BP would benefit, not by a further increase in its stake in Sohio, but by what was known as its Net Profits Royalty Interest (NPRI). This entitled BP to 75 per cent of the net profits from Sohio's Prudhoe Bay production of between 600,000 and 1,050,000 net bpd. BP could choose to take its NPRI in crude oil, rather than cash. If, as was thought unlikely, production exceeded 1,050,000 net bpd, Sohio was to be the sole beneficiary of the excess production.[56]

1 Sir Neville Gass, who succeeded Basil Jackson as BP's chairman and chief
executive in 1957, at a lunch given by the Ruler of Abu Dhabi in his palace in 1958

2 BP service station in
Aden in the 1950s

3 Preliminary examination of rock samples in a BP palaeontological laboratory in the 1950s

8 BP service station
at Wursthorn,
Germany, with the
'New Look'

9 BP service station at Pont de l'Isère, France

10, 11 BP tankers: the 28,000-ton *British Adventure* in 1958 (above) and BP's first 100,000-tonner, the *British Admiral,* which was delivered in 1965 (below)

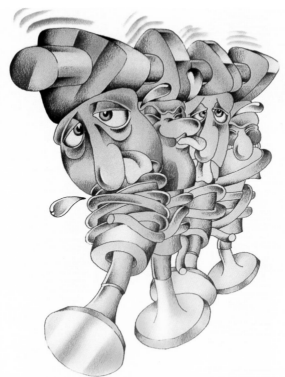

Have mercy on your tappets.

BP Super Visco-static. The Greater Lubricator

12 'Have mercy on your tappets.' Advertisement for BP Super Visco-static in the 1960s

13 Flaring of gas from BP's jack-up drilling rig, the *Sea Gem*, after it discovered the West Sole gas field in the southern basin of the North Sea in 1965

14 BP survey party in the Libyan desert in 1966

15 Atlas computer at London University in 1966

16 BP's old and new head offices: the old Britannic House (curved building, bottom right) in Finsbury Circus, London, and the new Britannic House (skyscraper, centre) in Moor Lane in 1967

17 The boardroom in the old Britannic House

18 The boardroom in the new Britannic House in 1969

19 Cellulose acetate
being processed at BP
Chemicals' Stroud
works in 1967

20, 21 Naphtachimie's chemical works at Lavera, France, in 1967 (above) and
Erdölchemie's chemical works at Dormagen, Germany, in 1967 (below)

22 Models in PVC uniforms at opening of BP's all-plastic service station at Baldock, UK, in 1968

23 Lounge of BP show-home at Great Burgh, Surrey, UK, incorporating plastic building components, fittings and furnishings, in 1969

24 Aerial view of Yukon River and Brooks Range, Alaska

25 Put River No 1 camp at Prudhoe Bay, Alaska, where BP struck oil in 1969

26 Man in freezing blizzard at Put River, Alaska, 1969

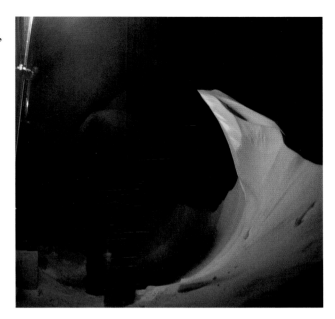

27 BP geologists (from right to left) Alwyne Thomas, Harry Warman and Peter Kent, who received the 1970 MacRobert Award for Engineering and Technology for the oil exploration in Alaska that resulted in the discovery of Prudhoe Bay

28 One of the Sinclair gasoline stations on the eastern seaboard of the USA acquired by BP in 1969 following ARCO's takeover of Sinclair. The Sinclair trademark, a diplodocus dinosaur, would soon become extinct

29 Unveiled: the first BP sign to appear on a service station in the USA at Atlanta, Georgia, on 29 April 1969

30 A Sohio service station in the 1960s

31 The ex-Sinclair
Marcus Hook
refinery in
Pennsylvania, USA,
in 1975, after
modernisation

32 BP's semi-submersible drilling rig, the *Sea Quest*, which discovered the Forties
oil field in the northern basin of the North Sea in 1970

33 Taking sterile samples at the pilot proteins plant at BP's Grangemouth refinery, Scotland, in 1970

34 Whisky and liquid Toprina: Scotland's oldest and newest fermentation products in 1970

35 'Scappa con Superissima!' BP Italiana advertisement in 1970

36 One of the BP Tanker Company's new VLCCs, the 215,000-ton *British Scientist*, off the coast of Shikoku, Japan, in 1971

Turning to other matters, BP was to pay the $400 million due to ARCO for the ex-Sinclair downstream assets, while Sohio was to take over responsibility for capital expenditures and operating costs on these downstream assets, on BP's interest in TAPS and on the development of the Prudhoe Bay reserves with effect from the beginning of April 1969.[57] On management, BP was entitled to nominate at least two members of Sohio's board of directors, and BP Alaska Inc., a wholly owned BP subsidiary, was to continue as operator of the Alaskan leases and therefore to be responsible for the management of all but an insignificant part of Sohio's upstream operations.[58] Finally, there was an unwritten agreement between the two companies that BP would conduct its US operations through Sohio as its 'chosen instrument', and that Sohio would keep out of international operations. This was to prevent the two companies from overlapping and competing.[59]

The only significant blot on this happy wedding was that the Department of Justice (the antitrust watchdog of the US government) objected to the BP/Sohio alliance on antitrust grounds. After lengthy negotiations, the Department of Justice filed a Consent Decree under which Sohio was required to divest itself of 400 million gallons of annual taxable motor fuel sales in the state of Ohio. Of much less importance, in volume terms, was the additional requirement that Sohio should divest itself of service stations in western Pennsylvania, where ex-Sinclair stations overlapped with some of Sohio's existing outlets marketing under the brand name 'Boron'. This latter divestiture was to be achieved by Sohio ridding itself of either the ex-Sinclair stations or the Boron stations.[60]

For Sohio, the Consent Decree was undoubtedly a blemish on its alliance with BP. The volume of 400 million gallons represented about 30 per cent of Sohio's annual taxable motor fuel sales in Ohio, and motor fuel sales were the most profitable part of Sohio's business.[61] BP's willingness to accept the terms of the Consent Decree was unpopular with Spahr who would have preferred BP to be less yielding in the negotiations with the Department of Justice.[62]

Notwithstanding Spahr's feelings on this issue, both BP and Sohio stood to realise their strategic objectives when, on 1 January 1970, the Consent Decree became final and the merger of interests between BP and Sohio was consummated.[63] Sohio thereby received BP's share of more than 50 per cent in the Prudhoe Bay field, the largest oil field ever discovered in the USA and was positioned to transform itself into a fully integrated oil company. With the prospect of a controlling interest in Sohio and the establishment of a major presence in the USA, BP was poised to pull off the most important development in its history since the settlement of the Iranian dispute in 1954. After more than sixty years of exclusion from the world's largest oil market it was on the verge of achieving a radical change in geographical emphasis, which would give it major interests in both hemispheres of the globe for the first time.

II

Refining and shipping

While BP was achieving the major strategic triumph of entry into the USA, its downstream expansion in Western Europe continued unabated. Here, after two golden decades of uninterrupted economic growth, memories of the depressed 1930s had faded and the general assumption, shared by BP, was that there was no end in sight to the continuing rise in oil consumption. Expansion was the unquestioned norm, oil was the cheapest and most versatile fuel, and by the early 1970s BP, the third-largest oil marketer in Europe, was selling gasoline through more than 26,000 filling stations in Western Europe, from Scandinavia in the north to Greece and Turkey in the south.[1]

Meanwhile, great changes were taking place in refining and shipping. Before World War II it had not been viable to construct major 'base-load' refineries in the oil-consuming countries of Western Europe because local oil consumption was not big enough to absorb the full output of a refinery that was large enough to be economic. That was why international oil companies had generally constructed their major refineries close to their sources of crude production, and shipped the refined products to their various markets in large numbers of comparatively small tankers. But as demand in the oil-consuming countries grew to a size where they could absorb the full output of a substantial refinery, the disadvantages of market-located refining diminished. At the same time, scale economies in transport helped to tip the balance in favour of market-based refining. It was much cheaper to ship bulk cargoes of crude from sources of production to market-located refineries using big tankers, than to ship lighter cargoes of individual refined products in smaller vessels.

THE EUROPEAN REFINING BOOM

The incentives to locate new refining capacity in oil-consuming countries, combined with the fast growth in oil demand, created a European refining boom in the 1950s and 1960s. In these decades, Western Europe's refining capacity grew at an average annual rate of 13 to 14 per cent, surpassing even the rapid rate of growth in the region's oil consumption.[2]

BP was a major participant in the expansion of European refining. Before the Iranian nationalisation crisis, the Company was already increasing the number and size of its European refineries, and by the end of 1950 it had seven wholly or partly owned refineries in Western Europe, with a combined capacity of 150,000 bpd.[3] The nationalisation of the Abadan refinery in 1951 accelerated the movement to Western Europe, which quickly became the main location for BP's refining operations (see figure 11.1).

By 1973, BP had twenty wholly and partly owned refineries in Europe. They were outnumbered by BP's wholly and partly owned refineries elsewhere, details of which are shown in table 11.1. But in terms of capacity, Western Europe was predominant, accounting for three quarters of BP's worldwide refining capacity. At more than 2,000,000 bpd, BP's European refining capacity was fifteen times as large as it had been at the end of 1950.[4] This represented an average annual growth rate of 13 per cent between the end of 1950 and the beginning of 1973.

Up to the mid-1960s much of the expansion was in the UK, where the prewar refineries at Llandarcy and Grangemouth were substantially expanded; the Kent refinery, commissioned in 1953, grew to become BP's largest wholly owned refinery, with a capacity in excess of 200,000 bpd; and a new wholly owned refinery was commissioned at Belfast in 1964. Continental Europe was not, however, neglected. In France, the refineries at Lavera and Dunkirk were expanded, and BP had a minority shareholding in a new refinery at Strasbourg. In West Germany, a new wholly owned refinery at Dinslaken, in the Ruhr, came on stream in 1960, and BP's Hamburg refinery was increased in size. In Belgium, the large Antwerp refinery in which BP and Petrofina held equal shares, was expanded; and in Italy the Venice refinery, 49 per cent owned by BP, was also enlarged.

From the mid-1960s to 1973 the expansion continued unabated, but with a shift in geographical emphasis from the UK to continental Europe. In the UK, no new refineries were built by BP, but the Grangemouth refinery had its capacity doubled in an expansion scheme which was completed in 1970.[5] On the continent, five new grass-roots refineries were brought on stream in as many years. They were at Rotterdam in the Netherlands (1967), Gothenburg in Sweden (1967), Vohburg in Bavaria (1968), Vernon near Paris in France (1969), and Volpiano near Turin in Italy (1971). Elsewhere, BP's refineries at Lavera and Dinslaken were more than doubled in capacity. The expansion of the new Rotterdam refinery was, however, the most

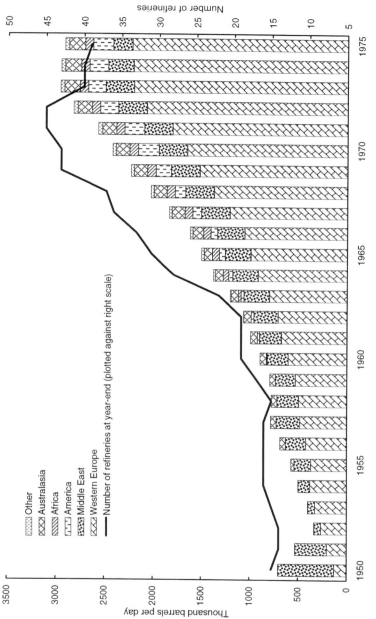

Figure 11.1   BP's refining capacity at wholly and partly owned refineries, 1950–1975

Table 11.1 *BP's wholly and partly owned refineries, 1950–1975*

Region	Country	Refinery	Year acquired or commissioned	BP share-holding (%)	Comment
Africa	Algeria	Algiers	1964	10	BP share sold with effect from 1 January 1967
	Gabon	Port Gentil	1967	3	
	Ivory Coast	Abidjan	1965	10	
	Kenya	Mombasa	1963	50	BP share reduced to 25% in 1964, and to 13% in 1970
	Madagascar (Malagasy Republic)	Tamatave	1966	7	BP share reduced to 6% in 1967
	Nigeria	Port Harcourt	1965	25	
	Rhodesia (Zimbabwe)	Umtali	1965	21	Shut down in January 1966 owing to oil sanctions on Rhodesia
	Senegal	Dakar	1963	12	
	Sierra Leone	Freetown	1969	7	
	South Africa	Durban	1963	50	
	Sudan	Port Sudan	1964	50	
America	Canada	Montreal	1960	100	BP share reduced to 80% in 1971, and to 66% in 1972
	Canada	Trafalgar	1964	100	BP share reduced to 80% in 1971, and to 66% in 1972
	USA	Marcus Hook	1969	100	Transferred to Sohio (BP shareholding 25%) in 1970

	USA	Port Arthur	1969	100	Transferred to Sohio (BP shareholding 25%) in 1970
	USA	Lima	1970	25	
	USA	Toledo	1970	25	
Australasia	Australia	Laverton	1924	100	Closed in 1955
	Australia	Kwinana	1955	100	
	Australia	Westernport	1966	100	
	New Zealand	Whangerei	1964	15	
Europe	Belgium	Antwerp (SIBP)	1951	50	
	Belgium	Antwerp (T&T)	1959	100	Ceased crude processing in 1962
	Cyprus	Larnaca	1971	26	BP share reduced to 15% in 1972
	Eire	Whitegate	1959	16	
	France	Lavera	1933	70	
	France	Dunkirk	1950	70	
	France	Strasbourg	1963	23	
	France	Vernon	1969	70	
	Germany	Hamburg	1948	100	
	Germany	Hamburg (Schindler)	1951	97	
	Germany	Dinslaken	1960	100	
	Germany	Vohburg	1968	100	
	Italy	Venice	1948	49	BP share sold in 1973
	Italy	Genoa	1965	20	BP share sold in 1973
	Italy	Volpiano	1971	100	Sold in 1973
	Netherlands	Rotterdam	1967	100	

Table 11.1 (cont.)

Region	Country	Refinery	Year acquired or commissioned	BP share-holding (%)	Comment
Europe (cont.)	Sweden	Gothenburg	1967	100	
	Switzerland	Aigle	1966	23	BP share raised to 24% in 1967
	Turkey	Mersin	1962	17	
	UK	Pumpherston	1919	100	Ceased crude processing in 1965
	UK	Llandarcy	1921	100	
	UK	Grangemouth	1924	100	
	UK	Kent	1953	100	
	UK	Belfast	1964	100	
Far East	Singapore	Singapore	1964	100	
Middle East	Aden (Yemen)	Aden	1954	100	
	Egypt	Suez	1928	31	BP share halved in 1961. Nationalised in 1964
	Iran	Abadan	1912	100	Nationalised in 1951
	Iran	Kirmanshah	1935	100	Nationalised in 1951
	Iraq	Alwand	1927	100	Sold to Iraq government in 1951
	Israel	Haifa	1939	50	BP share sold in 1958
	Kuwait	Kuwait	1949	50	Kuwait obtained 60% participation in 1974, and 100% in 1975
	Lebanon	Tripoli	1946	24	Nationalised in 1973

40 Inside one of the oil storage tanks at BP's refinery at Gothenburg in the 1960s

spectacular. Rotterdam was the world's largest port, a great entrepôt close to the main industrial areas of north-west Europe, and the site of a vast refining and petrochemicals complex. When BP's new refinery opened here in 1967, it had a capacity of 100,000 bpd. Five years later, on completion of a two-stage expansion scheme, Rotterdam became BP's biggest refinery, with a capacity of 470,000 bpd.[6]

Astonishingly, even after this expansion the Rotterdam refinery was not as large in crude oil processing capacity as the great, sprawling Abadan refinery had been in 1950. The Rotterdam refinery was, however, vastly more efficient. Although smaller than Abadan in overall capacity, the Rotterdam refinery exploited much greater economies of scale by utilising much larger plant. For example, the two crude oil distillation units (central equipment in refining) which were installed at Rotterdam in the early 1970s had a capacity of 200,000 bpd each. In 1950, the largest unit at Abadan was about half that size.[7]

Technologically, also, BP's refineries had come a long way since the days when the Company owned Abadan. When the Company originally designed and constructed the Abadan refinery in 1909–12, it relied, like other refiners at that time, on the simple distillation of crude oil to yield consumable

41   New reactor being erected at SFPBP's Lavera refinery in 1973

products. This basic refining process proved difficult for the Company to master using Iranian crude oil whose unique characteristics were then unknown. More intractably, the simple distillation process suffered the fundamental problem that the yield of products, such as fuel oil, kerosene and gasoline, obtained from a given crude was virtually fixed in its proportions. Different crudes produced variations, but for the individual refiner using a given crude there was virtually no scope for adjusting the yield of products to match market demands that were in constant and often unpredictable flux, depending on seasons, cycles, trends and all manner of local conditions. A refiner with only simple distillation capabilities who wished to supply extra quantities of one product would have no option but to increase his output of all products, some of them unwanted.

The vital breakthrough out of this situation came in America at just the time that the Company was planning and erecting the Abadan refinery. In 1909–12 William Burton, a scientist with Standard Oil of Indiana (later Amoco), developed and patented the first commercial 'cracking' process. By subjecting the larger hydrocarbon molecules of heavy oil products such as fuel oil to high pressure and high temperature, Burton succeeded in cracking them, so that they broke down into smaller molecules, allowing increased output of lighter products such as gasoline, which was in growing demand.

42   New crude oil distillation unit for BP's Rotterdam refinery in 1971

The introduction of Burton's process of thermal cracking meant that the oil industry could make relatively more gasoline from a barrel of crude oil. The refiner's yield of products was no longer arbitrarily bound by the atmospheric distillation temperatures of the different components of crude oil; now the refiner could manipulate the molecules to increase the yield of the most desired products.

The technique of thermal cracking was quickly taken up in the USA, which led the world in the adoption of road transport and, therefore, in the demand for gasoline. The Company, however, concentrated on simple distillation techniques until 1926–7, when two US-designed thermal crackers were installed at the Llandarcy refinery in south Wales. These were the first full-scale cracking plants to be installed by the Company at any of its refineries. Soon afterwards, a continuous cracking process, developed at the Company's Sunbury research centre, was brought into operation at Llandarcy, and at the Grangemouth refinery in Scotland. But these plants were dwarfed by the four US-designed thermal crackers which the Company decided to install at Abadan in 1929. These came on stream in 1931, and revolutionised the Company's gasoline output, both in quality and quantity.

A major advance in cracking technology was made in the USA in the mid-1930s, when a process was developed to enhance the cracking process by the use of a catalyst. The process was widely adopted in the USA during World War II to meet the wartime demand for high-octane fuels, and soon after the war the Company decided to install its first catalytic cracker at Abadan. However, the US-designed plant took more than four years to complete, and by the time it was commissioned the Company was engulfed by the Iranian nationalisation crisis. The stoppage of the Abadan refinery during the crisis prompted the Company to push on with its plans for catalytic cracking in Britain, and in 1953 three identical catalytic crackers, designed and installed by the US firm of Kelloggs, came on stream at the Llandarcy, Grangemouth and Kent refineries respectively. In 1955 a new catalytic cracker of different design (but also by Kelloggs) was commissioned at BP's Kwinana refinery in Australia, and in the same year another US design, this time by Lummus, was chosen for a new cracking installation at the Antwerp refinery, which BP shared with Petrofina. And in 1960 another US design, this time by Universal Oil Products (UOP), was chosen for a sixth catalytic cracker at BP's Montreal refinery in Canada. By 1970 the number of catalytic crackers in use at BP refineries had increased to nine, and BP made little use of the earlier process of thermal cracking.

Another breakthrough in refining technology came in 1949 when UOP launched the process of catalytic reforming (platforming), a new technique which could upgrade low-octane products to higher octanes. The introduction of catalytic reforming marked the beginning of a new era in refinery design, and was soon adopted by the Company, which signed a licensing agreement for use of the process with UOP in 1951. In the next two decades platformers became increasingly standard equipment in BP's refineries.[8]

These advances were not, however, enough to place BP in the top rank of refiners. Benchmark studies carried out by BP in the early 1970s contrasted BP's high performance in exploration, in which it was ranked first among the majors, with its low rating as a refiner/marketer, in which it was ranked bottom (see table 11.2). In terms of simple volumes, BP had broadly speaking kept up with the industry as a whole in refining and marketing in the 1950s and 1960s, but its refineries were comparatively inflexible in their ability to upgrade low-value products to higher-value ones, and the profitability of BP's product mix was low.

In this period, however, the profitability of BP's refining and marketing operations was of secondary concern. BP's higher priority was to hold on to its oil concessions by satisfying the demands of oil-producing countries for increased crude oil production. Production required outlets, preferably guaranteed ones, which BP largely provided for itself by refining its own crude, and marketing its own products. In this integrated operation, refining and marketing were not free-standing operations, required to be

Table 11.2 *The majors' comparative strengths and weaknesses, 1970–1972 (rankings)*

	Exploration skill	Refining flexibility	Profitability of product mix
BP	1	7	6
Gulf	6	1	1
Mobil	4	4	2
Shell	5	2	4
Socal	2	6	3
Standard Oil (NJ)	7	5	5
Texaco	3	3	na

competitive and make profits in their own right. Indeed BP, wedded to integration, continued to measure only its overall profits, and had no means of knowing what contributions were made by different sectors, such as refining and marketing.

What mattered more to BP was simply the expansion of its refining and marketing capacities so that they could absorb increased production. The timing of refinery expansions was, however, difficult to get exactly right. It took several years to plan and construct a major new refinery, whose size and design were based on forecasts of the future market demand for oil products. In the 1960s, when demand was rising fast, the penalties for over-expanding were quickly dissipated as rising demand soon absorbed any over-capacity. Under-expansion, on the other hand, could result in loss of trade, the very opposite of BP's primary goal. There was, therefore, an inbuilt tendency to err on the side of over-expansion.

More fundamentally, BP and other oil companies became so accustomed to fast-growing demand that its continuance came to be assumed. In 1970, BP was forecasting that Europe's demand for oil over the next five years would rise by 7 per cent a year, which a BP director publicly stated 'could be an under-estimate'.[9] Similarly, BP's forward plan for 1972–6 was based on a forecast growth rate of more than 7 per cent a year in products sales. In January 1971 several senior BP executives said they 'feared that this might be an under-estimate'.[10] It was on the basis of forecasts such as these that BP, not knowing what lay ahead, was continuing to expand its refining capacity in the early 1970s (see figure 11.1).

SHIPPING

In shipping, the expansion of BP's tanker fleet was propelled by the same forces that lay behind the growth in refining capacity. Pressure from the oil-producing states for higher production, the lure of economies of scale, and expectations of ever rising demand combined to push and pull BP along a path of rapid expansion.

In the case of shipping, economies of scale loomed still larger than in refining, and prompted quite phenomenal increases in tanker size in the third quarter of the twentieth century. This era of the growth of the giant tanker was without precedent in shipping history. Previously, tanker sizes had shown little tendency to increase. The *Murex*, the first tanker to carry oil in bulk through the Suez Canal, was launched in 1892 with a capacity of 5,010 deadweight tons. Owned by Marcus Samuel (founder of Shell Transport and Trading), it was an average-sized vessel for its time.[11] Nearly a quarter-century later, in 1915, the Company ordered its first seven tankers, of which two were in the 5,000-ton class, and five were 10,000 tonners.[12] By the 1920s the Company and Shell had adopted the 10,000 tonner as their standard size tanker, which was not superseded until the introduction of a new class of 12,000 tonners in the 1930s.[13]

The longevity of the 12,000 tonner as a standard size was remarkable. After World War II the Company, seeking to rebuild its war-ravaged fleet, continued to see the 12,000 tonner as the most useful size. It took delivery of fifty-seven of them between the end of the war and 24 December 1951 when its last vessel in this class, the *British Maple*, was delivered. By that time no fewer than ninety-three ships of this size had been built for or acquired by the Company.[14] They remained the backbone of the fleet, whose standard size of vessel had for more than thirty years been in the 10–12,000-ton bracket.

One of the advantages of these vessels was their versatility. Unlike later generations of tankers they were general purpose ships, which could be used to haul crude oil from the Middle East via the Suez Canal to the Company's European refineries or, more commonly, to ship refined products from Abadan to Western Europe, also through the Suez Canal.

Life on these vessels was a far cry from the onshore jobs in which most Company staff were employed. The crew of sailors and firemen lived in the cramped forecastle, and followed an immutable routine of tank-cleaning as they passed, on a typical outward bound voyage, through the English Channel, across the Bay of Biscay, through the Strait of Gibraltar and into the Mediterranean. The tank-cleaning had to be finished by the time they arrived at Port Said, so that the captain could declare a clean, gas-free ship for the Suez Canal transit. In the Canal, busy with trade, the Red Ensign of the British merchant fleet was ubiquitous, and the waterway was crowded with the funnels and house flags of eastern-trading British shipping companies such as P&O, Orient, the City Line, Clan, Blue Funnel, Glen, the Ben Line and, of course, the oil tankers of Anglo-Saxon (Shell's shipping subsidiary) and the British Tanker Company (the Company's shipping subsidiary).

East of Suez it was a different world, as the voyage continued down the Red Sea, stopping at Aden to take on bunkers before turning east along the southern coast of Arabia, into the Gulf of Oman, through the narrow Strait of Hormuz, and up the Persian Gulf. At its head was the mouth of the Shatt, confluence of the Tigris and Euphrates Rivers, a landscape of featureless mud flats edging into low desert. Towering over it was Abadan, the world's largest refinery, 'an incongruous 20th century intrusion upon a deserted wilderness', as it struck a junior officer at his first sighting of the refinery's great complex of towers, retorts, tanks, pipelines and jetties. A hundred ships could be at anchor here, predominantly tankers, high out of the water with a minimum of ballast on board as they waited for a loading berth. While the first three months of the year were bearable, the sweltering heat of the summer and autumn meant spending the day under cover except for the early dawn and late dusk. Canvas awnings rigged over the forecastle were the only way of reducing the temperature of the metal deck. Eventually each tanker went alongside, pumping the last of its water ballast into the Shatt-al-Arab waterway, and rising even higher in the water, its bow projecting upwards to expose the curved forefoot at its base. But once loaded, all that could be seen of the vessel was a low silhouette, with most of the hull submerged for the return passage.[15]

This pattern of trade would not, however, last. In the late 1940s the shifting balance of advantage between production- and market-located refining was closely examined by the Company, which concluded that market-located refining would be competitive if 28,000-ton tankers, which could transit the Suez Canal fully loaded, were used to haul crude oil to market-located refineries.[16] Accordingly, in 1949 orders were placed for the first of six 28,000-ton 'supertankers', which were delivered in 1950–2.[17] Other oil companies were making similar moves. Shell Transport and Trading, for example, also took delivery of several new 28,000 tonners in the early 1950s.[18] The entry into service of these vessels was a watershed, marking the beginning of the end of the general purpose tanker and the ascendancy of specialised crude carriers of ever larger sizes.

43   A busy scene at the cargo jetty of the giant Abadan refinery before it was nationalised in 1951

In BP's case, the last of its general purpose tankers was delivered in 1957.[19] Meanwhile, more and larger crude carriers were being ordered by BP and other international oil companies. In 1951, the British Tanker Company ordered six 32,000 tonners for delivery between 1953 and 1955. During the turmoil of the Iranian crisis in 1951–4, eight more tankers in this class were ordered for delivery in 1957–8. In 1955, after the settlement of the Iranian crisis had dispelled uncertainty, a still larger order for more than twenty new supertankers was placed.[20]

The expansion of the tanker fleets of BP and Shell provided a bright spot, virtually the only one, in the otherwise beleaguered British shipping and shipbuilding industries. The tanker sector was the fastest-growing sector in world shipping, but British shipowners and builders, who had once led the world, were generally failing to capitalise on the shifting pattern of trade. For the most part, British shipowners clung to their traditional businesses, operating liners in the declining sector of passenger transport (facing growing competition from air travel) and tramps carrying exports of

44   Chief engineer's cabin on the BP Tanker Company's 32,000-ton oil tanker, *British Sailor*, in the 1950s

Britain's fading old industries such as cotton goods and coal. Although the oil companies offered a ready market for tanker charters, few British ship-owners went into the tanker business. They were even said to have 'a dis-taste for tankers which they scarcely regarded as ships at all'. As a result, while Japanese, Norwegian and Greek shipowners expanded their super-tanker fleets, there was a shortage of independently owned British super-tankers available for charter to the oil companies. Their chartered tonnage was very substantial, accounting for between one and two thirds of their total requirements, depending on conditions.[21]

BP and other oil companies also faced growing problems ordering new tankers from British shipbuilders, whose share of world shipbuilding contin-ued to fall as they became increasingly uncompetitive. They were less prepared than foreign builders to quote fixed prices, and their preference for cost-plus, rather than fixed-price, contracts meant that owners could not be sure how much new orders would eventually cost. In addition, delivery times from British shipyards were generally far longer than those of foreign competitors, and British shipbuilders were developing an unenviable reputation for failing to meet even their distant delivery dates.

Increasingly, the uncompetitiveness of British shipbuilders overcame the long-established tradition for British shipowners to have their ships built at home. BP was a case in point. Until 1955 every tanker order placed by the British Tanker Company was with a British builder. But in 1955, when the British Tanker Company placed orders for more than twenty new supertankers, six of the orders were placed in Italy. Ironically, these ships were bedevilled by electrical problems, and earned the dubious soubriquet that they were 'more handsome but less reliable' than the British-built equivalents.[22] The decision to order abroad was, however, a sign of the times. In 1956 the shipping tonnage launched from Japanese yards was higher than that from British yards for the first time.[23]

That news was, however, overshadowed by the much more dramatic impact on the tanker market of the Suez Canal crisis. Nasser's blocking of the Canal in November 1956, in retaliation against the ill-advised Anglo-French–Israeli invasion of Egypt, precipitated a spectacular boom in the world tanker market. As tankers were re-routed to make the long voyage around Africa (see chapter three) there was a worldwide shortage of shipping. Freight rates soared, the scrapping of obsolete vessels was deferred, and orders for new tonnage were accelerated. The boom, however, was short-lived. After the Canal was reopened in the spring of 1957 tanker owners the world over were left with a large surplus of tonnage. To reduce its surplus, the BP Tanker Company (as the British Tanker Company was renamed in 1956) laid up some of its tankers, transferred others to the grain trade, negotiated deferred delivery of new orders and scrapped many of its prewar general purpose 12,000 tonners. While the number of tankers owned by the BP Tanker Company and BP's other shipping associates fell rapidly from its peak of more than 170 ships in 1957 (see figure 11.2), crude carriers which were already on order were increased in size. The effect of the post-Suez slump in the tanker market was, therefore, to accelerate the trend towards larger tankers in BP's fleet. At the end of 1956 tankers of 28,000 deadweight tons or more had accounted for only 9 per cent by number and 20 per cent by tonnage of the BP fleet. By the end of 1963 these percentages had risen to 42 and 66 per cent respectively.[24]

In the meantime, new supertankers of ever larger sizes came into service. BP's first 50,000 tonner, the *British Queen*, was delivered in 1959.[25] In 1960 Shell's 71,250-ton *Serenia* was launched, followed in 1963 by BP's first 70,000 tonner.[26] With the introduction of these larger vessels came improvements in comfort for crews, such as air-conditioning, swimming pools and

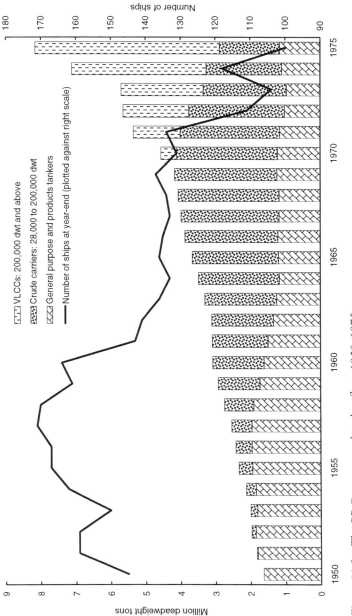

Figure 11.2   The BP Group owned tanker fleet, 1950–1975

45   Cricket on the foredeck of the *British Queen*, the BP Tanker Company's first 50,000-ton tanker, delivered in 1959

more generous cabins; and as the standards improved, more wives began to join their husbands at sea.[27] The relentless increase in the size of tankers still, however, had far to go. In 1965 BP's first 100,000 tonner, the *British Admiral*, was delivered, followed quickly by another, the *British Argosy*, and in 1967 Shell's largest tanker to date, the 117,000-ton *Naticina*, was completed.[28] Despite progressive enlargements of the Suez Canal, these vessels drew too deep a draught to navigate the Canal fully loaded. They could, however, pass through the Canal in ballast and return part-loaded, topping up in the Eastern Mediterranean.[29]

The Canal would soon, however, become irrelevant. In June 1967, as a result of the Arab–Israeli Six-Day War, it was closed again. As in 1956, tankers were re-routed round Africa, there was a scramble for tonnage, and freight rates rose dramatically. This time, though, the closure would be much more long-lasting, and the Canal would remain closed not for a few months as in 1956–7, but for eight years.

Already, before the Canal was closed in mid-1967, orders had been placed for a new, still-larger class of tankers which would in any case have made the Suez Canal partly redundant unless it was substantially enlarged. With a capacity of 200,000 deadweight tons or more, these tankers were designed to navigate the Canal in ballast, but to sail around Africa when loaded. They represented such a quantum leap over earlier generations of supertankers that a new term, the Very Large Crude Carrier (VLCC) was coined to describe them. Such were the economies of scale of these vessels that even sailing on the longer voyage around Africa they could deliver oil more cheaply than Canal-routed supertankers could.[30]

Ahead of the closure of the Canal, BP had in 1966 ordered a number of VLCCs to enter service beginning in 1969 on long-term charters of ten to fifteen years.[31] Soon afterwards, in May 1967, BP announced its first order for VLCCs to join its owned tanker fleet. The order was for two VLCCs to be built by Mitsubishi in Japan. BP, richer in crude oil than in cash, and as ever looking for new outlets for oil, was to pay for the VLCCs in crude oil.[32]

The Mitsubishi order was just the beginning. The closure of the Canal, combined with the scale economies of operating VLCCs, led to a great boom in orders for new VLCCs throughout the world. BP did not hold back. In 1967–8 it agreed two more crude-for-ships deals with other Japanese firms. One was for two VLCCs with Mitsui; the other for a single VLCC with Kawasaki. By mid-1968 the opportunities for further orders against payment in crude oil had apparently dried up, but after taking price indications worldwide, BP found that Japanese yards offered the most competitive combination of price, credit terms and quality. An order for two more VLCCs was therefore placed with Mitsubishi. With this order, BP had a total of seven VLCCs on order in Japan for delivery between 1970 and 1972.[33]

The ordering boom continued in the early 1970s, when BP ordered thirteen more VLCCs. Eight of the orders were with Japanese shipbuilders, four with the Dutch shipbuilder, Verolme, and one with the French firm, Chantiers d'Atlantique. All were for delivery between 1973 and 1976. The three British shipbuilders with the capacity to build VLCCs (Swan Hunter, Lithgows and Harland and Wolff) were given opportunities to bid for the contracts, but could not match their foreign competitors.[34] Although the names of the BP Tanker Company's tankers continued to have the prefix 'British', the days when BP ordered its new tankers from British shipbuilders were past.

In addition to the twenty VLCCs ordered by the BP Tanker Company, a further seven were ordered by BP's French shipping associate, Société Maritime des Pétroles BP, all for delivery between 1970 and 1976.[35]

The first BP-owned VLCC to be launched was the *British Explorer*, built by Mitsubishi, which joined the BP fleet in 1970. Others quickly followed,

46   Main deck of BP's first
VLCC, the 215,000-ton
British Explorer, rolling in
Atlantic swell in 1970

and by the end of 1973 BP had taken delivery of twelve VLCCs, with more
than that number under construction or on order. This great expansion in
BP's fleet, shown in figure 11.2, could be no more quickly reversed than the
growth in refining capacity. A great amount of investment in ships and refineries was hung on the forecasts of continuing growth in the world's demand
for oil.

Financial strains

Behind the majority of BP's actions, and vital to understanding them, lay a strategy of fundamental simplicity and continuity, enduring from the Edwardian era before World War I to the 1970s. The essence of the strategy was to uphold the concessionary regime which provided BP, and other oil companies, with the licence to operate in the less developed oil-exporting countries, whose crude oil was the cheapest to produce in the world.

Of course, the strategy sometimes had to be adapted to changing circumstances. In the first half of the century, BP's uppermost concern was to hold on to its exclusive concession in Iran, the jewel of BP's business. When that concession was terminated in the Iranian nationalisation crisis of 1951–4, BP diversified its operations and reviewed its long-term policy, deciding to intensify its search for oil outside the politically volatile Middle East (see chapter one). But although exploration in new areas was stepped up (chapters four and eight), nearly all BP's crude production continued to come from concessions in less developed countries in the Middle East, and in Africa. The retention of these concessions remained BP's top priority.

This was the purpose motivating BP as it strove to restrain OPEC by splitting the moderate members from the radicals (chapter six), to balance the demands of the producing countries for more production (chapter seven) and to expand its refining and marketing capacity so that it would have guaranteed outlets for more crude oil (chapters nine to eleven). BP's far-reaching activities, potentially bewildering in their complexity, thus fell into place as components of the essentially simple, but not easy, strategy of upholding the concessionary system.

In the 1960s and early 1970s the financial consequences of this strategy became of growing concern. Reflecting the importance which BP attached to increasing production to meet the oil-producing countries' demands, BP planned its operations on the basis of meeting volume rather than profit targets. The emphasis on maximising volume resulted in a substantial physical expansion of sales, which rose from just under a million barrels per day in 1955 to nearly four million barrels per day in 1970. This fourfold increase represented an average annual growth rate of 10 per cent.

While the volume of BP's sales rose rapidly, profits stagnated. After adjusting for price inflation they remained broadly level from 1955 to 1970. There were, to be sure, some fluctuations, but the broad trends (see figure 12.1) were clear. In 1970, BP's net profits per barrel, adjusted for price inflation, were little more than a quarter of what they had been fifteen years before.

BP's shareholders, having made spectacular financial gains in the last three years of Strathalmond's chairmanship in the mid-1950s (chapters one and five), had to get used to lower returns. After the peak of 1956 ten years passed before the price of BP shares, adjusted for bonus issues and price inflation, recovered its 1956 level. In 1968–9 the share price peaked again, on news of the Prudhoe Bay discovery in Alaska, but quickly fell back in 1970. Meanwhile real dividends per share, having climbed spectacularly in the mid-1950s, increased more slowly up to the mid-1960s and then went into decline (see figure 12.2).

BP's poor profits performance mostly reflected general conditions in the international oil industry. With oil in plentiful supply, and competition from recent entrants, there was almost constant downward pressure on oil prices in international markets. The brunt of weak prices was borne by the international oil companies. The producing countries were for the most part unaffected because their tax receipts from the companies in most cases represented their agreed shares of profits based on inflexibly high posted prices, which the companies dared not cut after the formation of OPEC in 1960 (see chapter six). As a result, the share of the producing countries in the overall profits of the international oil industry increased, while that of the companies declined.

Although all the majors felt the effects of weak international prices, BP was easily the worst hit, mainly because it alone of the majors lacked a presence in the sheltered US market, where the introduction of import quotas in the late 1950s helped to protect domestic oil producers from international competition. The figures speak for themselves: in 1955 BP was the most profitable oil major, with a return on capital employed of more than 15 per cent; by 1970 BP was the least profitable, with a return on capital employed of less than 5 per cent (see figure 12.3).

BP's poor profits were part of the reason for its growing reliance on borrowing to finance the expansion of its business. Here, also, BP compared

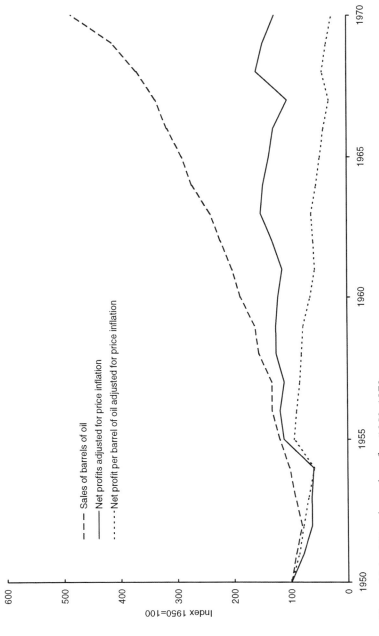

Figure 12.1   BP's sales and profits, 1950–1970

Index 1950=100

Sales of barrels of oil

Net profits adjusted for price inflation

Net profit per barrel of oil adjusted for price inflation

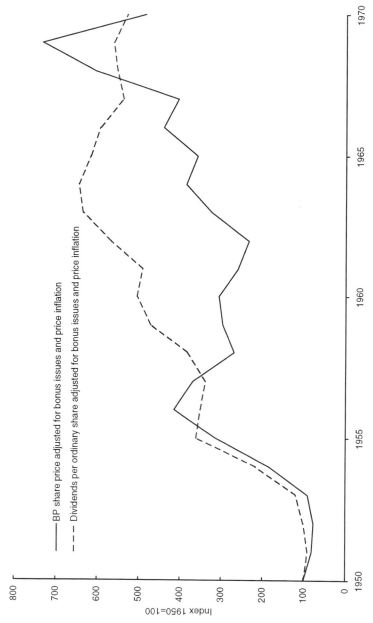

Figure 12.2 BP's share price and dividends, 1950–1970

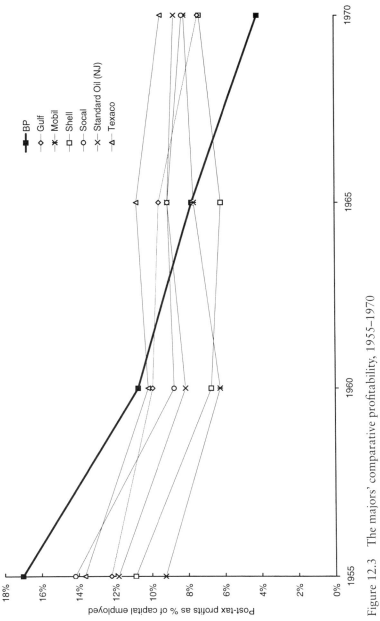

Figure 12.3   The majors' comparative profitability, 1955–1970

unfavourably with the other majors. In 1955 BP emerged from the Iranian nationalisation crisis with the lowest reliance on long-term debt of all the majors, apart from Socal. But over the next fifteen years BP became more reliant on borrowing than any other major apart from Gulf (see figure 12.4).

BP's inability to finance its expansion from retained profits reflected not only its poor profit performance, but also the heavy investments that were made to support its volume-maximising strategy. In the early 1960s this was not an immediate cause for concern. Between 1960 and 1963 annual capital expenditure rose from £109 million to £129 million. Most of the increase was in refining and marketing, whose combined share of BP's capital expenditure rose from 52 to 58 per cent in these years. The increase in capital spending was, however, mostly covered by funds generated from trading, with little need for borrowing. At the end of 1963 BP's debt ratio (debt as a percentage of debt-plus-equity, the standard measure of a company's reliance on borrowing) was 13 per cent, with which BP was comfortable.[1]

But from 1963 to 1965 BP's financial position deteriorated rapidly as annual capital expenditure shot up from £129 million to £207 million. Refining and marketing continued to be the most capital-hungry sectors, accounting for most (61 per cent by 1966) of BP's capital spending. As BP's capital expenditure exceeded its retained income by a widening margin, the gap was filled by new borrowing. As a result, BP's debt ratio climbed to 16 per cent in 1964, and to 20 per cent in 1965.[2] This was getting uncomfortably close to the limit of 25 per cent which BP felt was the maximum that a company like itself, with a large part of its operations in politically unstable areas, should prudently allow.[3] By February 1965 Alastair Down, who had become the BP managing director responsible for finance, was voicing his anxiety to Bridgeman about BP's ability to finance the scale of capital expenditure that was envisaged for the next few years. He called for immediate and urgent action to deal with the situation.[4]

A series of measures was put in motion. One was a review of BP's planned capital expenditure, which was cut back.[5] Another was the decision to raise new capital from BP's shareholders through a rights issue. This was carried out in 1966, when £60 million was raised, primarily from Harold Wilson's Labour government, which held nearly 52 per cent of BP's shares, and from the Burmah Oil Company, which held close to 25 per cent. This was BP's first issue of ordinary shares for new capital since 1923.[6]

Although the rights issue strengthened BP's finances, it was clear that BP could not afford to continue expanding its capital-intensive downstream operations fast enough to keep up with the forecast general increase in Middle East oil production. This forced BP to reconsider one of its most fundamental

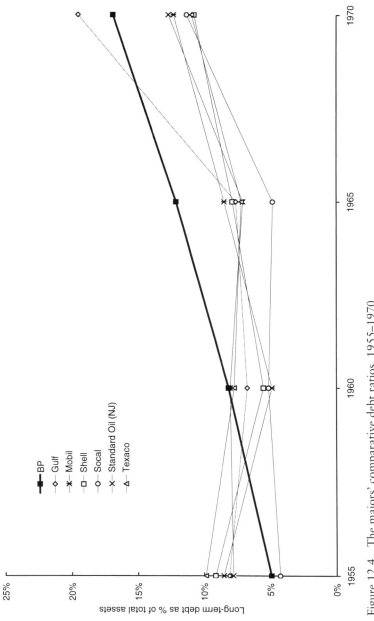

Figure 12.4   The majors' comparative debt ratios, 1955–1970

tenets. In its early years as Anglo-Persian, the Company had been convinced that to achieve security of outlet for its production, it must become a vertically integrated organisation, possessing enough refining and marketing capacity to provide a guaranteed means of bringing its production to markets. That was what Charles Greenway, chairman and chief executive from 1914 to 1927, meant when he spoke about his aim to create 'an absolutely self-contained organisation'.[7]

This policy was not held to unerringly by his successors, in particular by William Fraser (Lord Strathalmond), chairman and chief executive from 1941 to 1956. Under his aegis the Company, short of refining and marketing capacity after World War II, resorted to selling crude oil in great quantities, hoping that the increased offtake from Iran would help to make its Iranian concession more secure. Long-term, high-volume contracts for the sale of crude to two other majors (Standard Oil (NJ) and Socony-Vacuum) were entered into, and crude oil sales became a substantial and growing part of BP's trade in the late 1940s and 1950s.[8]

Nevertheless, BP remained wedded to the idea of vertical integration. In 1962 Bridgeman explained to shareholders that 'if oil in the ground is to acquire a value, those who have access to it must also have outlets for its sale. An integrated company must therefore not only be sure of its supply of crude oil, but must have refineries, ocean-going tankers, storage installations and marketing networks.'[9] The same theme was reiterated by BP's Central Planning Department in 1966: 'the ultimate consumer wants products, and security of outlet in a competitive situation must be through a vertically integrated operation'.[10]

These statements were backed up by action, and in the first half of the 1960s BP succeeded in increasing its sales of refined products through its own marketing outlets faster than its sales of crude.[11] But by 1966 capital constraints ruled out the continuance of this policy, and Bridgeman resorted to the same solution as Fraser had adopted after World War II. Insisting that BP must not fall behind in raising its crude oil offtake from the Middle East, and recognising that the scope for increased products sales was limited by capital constraints, Bridgeman stipulated that the additional growth would have to come from crude oil sales, which would not require downstream capital investment.[12] Long-term policy was duly set on the basis that 'the strategic plan is taken to be the achievement of a sales target – mainly through selling more crude oil – which will give a satisfactory offtake from all our concessions'.[13]

Over the next five years BP's sales of crude grew more than twice as fast as products sales.[14] The long-term contracts with Standard Oil (NJ) and

Socony-Vacuum (by this time renamed Mobil) were extended for another fifteen years in 1966, and remained the backbone of BP's sales of crude to other majors. Petrofina was another large-scale buyer, purchasing all its crude requirements outside the USA from BP. At the same time, BP tried to avoid supplying crude oil to refiners and marketers where they would compete with BP's own downstream operations. BP therefore concentrated its crude oil selling on areas where it had no downstream presence, most importantly Japan (see chapter ten), but also South America and Eastern Europe.[15] Crude oil sales grew so rapidly that by 1970 they accounted for nearly half of BP's total sales volume.[16]

In the meantime, BP had started questioning another aspect of its integrated operations. When Bridgeman addressed BP's shareholders in April 1962, he went to great lengths to explain the indivisibility of BP's overall profits, stating:

> We believe that it is impossible to assess in isolation the true profitability of any one of the several functions which are involved, such as the actual production or transport or refining or marketing of the oil, without making assumptions which are largely arbitrary as to the part which each plays in determining the price at which the product is sold. To say that an integrated oil company makes a profit at the producing stage is like saying that the profit of the village baker is derived from mixing his dough. The baker cannot make a profit until he has bought an oven, and probably a van as well, and has baked and delivered his bread and been paid for it. The only figure that really matters in the case of an integrated oil company is the net income after tax remaining with the group when the results of all its constituent elements have been set one against the other.[17]

Like so many statements in the international oil industry, this one was partly politically motivated. The US consulting firm Arthur D. Little had recently completed a study, commissioned by OPEC, in which it was calculated that the oil companies had made an enormous return of 64 per cent on their net assets in OPEC countries in the Middle East in 1956–60. The study was due to be presented to OPEC's fourth conference in April 1962.[18] Bridgeman's statement, claiming that it was meaningless to single out one element in BP's overall results, was designed as a counter to OPEC claims that the oil companies were making excessive profits from their producing operations in OPEC countries.[19]

An equally politically motivated, but differently slanted, commentary on oil company profits was published in a series of studies undertaken by the First National City Bank (FNCB) of New York at the request of the oil majors. Their purpose was to rebut claims by Arab nationalists in the late

1950s that Arab governments should be entitled to share the large profits which, it was claimed, the international oil companies were making on their Eastern Hemisphere downstream operations. The first FNCB study, timed to appear in time for the Arab Petroleum Congress in Cairo in 1959, sought to demonstrate that the companies' downstream activities in the Eastern Hemisphere were in fact loss-making.[20] Subsequently, further FNCB studies appeared on the same theme.[21] They were refuted by Francisco Parra, the Chief of OPEC's Economics Department, in a searching article which attempted to probe the opaque indivisibility of the oil companies' profits.[22]

The central problem for everyone in breaking down the oil companies' profits was the setting of transfer prices at which crude oil passed from the companies' producing operations to their downstream subsidiaries and associates. The FNCB studies were fundamentally flawed insofar as they used the official posted prices of crude as the transfer prices. In reality, of course, posted prices were simply tax reference points for the calculation of payments to the producing countries; they bore no necessary relationship to costs or market prices, and were an essentially arbitrary basis for economic calculations, apart from tax. But if posted prices were virtually meaningless as transfer prices, what were the alternatives? Most of the oil in international trade moved through the integrated channels of the oil majors. To be sure, large quantities of crude oil changed hands, but mainly under private contracts between the integrated companies. There was nothing approaching an impersonal, transparent market, reflecting arm's length transactions between competing firms.

Outside observers may have found it difficult to believe, but actually BP was as baffled as anyone else by the problems of breaking down its profits. This problem came to the fore in the mid-1960s, when BP began to take a more active interest in its downstream profitability, not for political reasons, but because capital stringency made it necessary to channel capital expenditure more selectively into profitable operations.

The task of measuring the comparative profitability of BP's downstream operations was assigned to BP's Central Planning Department in 1966, and soon ran into difficulties, primarily because different departments took opposing views of the matter. The aim, put simply, was to compare the profitability of BP's downstream subsidiaries in different countries by deducting from their sales revenues the costs of crude oil production, transportation, refining, distribution and selling which were attributable to them. This was much easier said than done. The costs incurred by BP in the whole of its integrated operations were known, but it seemed impossible to arrive at an agreed set of transfer values at which different grades of crude and products from BP's many oil fields and refineries were supplied to individual down-stream subsidiaries.

The Finance and Accounts Department argued that the right approach was simply to take the actual costs incurred in producing, shipping and refin-ing the oil supplied to each individual market. Suppose, for example, that this approach was applied to the production and transportation of Libyan crude for refining and marketing in the Netherlands. In this example, the accounting method of calculating profitability would be to deduct from sales revenues in the Netherlands the sum of the actual costs incurred in selling, distribution and refining in the Netherlands; in freight from Tripoli to Rotterdam; and in producing the Libyan crude. This, the accountants argued, would reflect the 'real world' of actual costs incurred.

The Central Planning Department disagreed with this method, arguing that while it was historically accurate as an accounting record, it was mean-ingless as a basis for decision-taking. Transfer values should not, the plan-ners insisted, be simply the actual costs of supplying individual markets as isolated units. Rather, they should reflect the impact of decisions in individ-ual markets on the costs of BP as a whole. If, for example, BP Holland were to increase its sales of products refined from Libyan crude, it did not follow that BP would necessarily increase its Libyan crude production. It might just as easily reallocate Libyan crude from other refining/marketing areas and, perhaps for political reasons, lift additional supplies of, say, Iranian crude from the Persian Gulf. In that case, the cost of supplying more Libyan crude to the Netherlands should be the costs of producing Iranian crude and trans-porting it to other refining/marketing areas.

By 1970, after several years of arguing over this knotty problem, the accountants and the planners were nowhere near agreement on the transfer values to be adopted in measuring BP's comparative profitability in different markets.[23]

In the meantime, the increase in crude oil sales was helping to reduce the need for capital expenditure on refining and marketing, which levelled off in the late 1960s. But even so, refining and marketing continued to account for a high proportion (more than half) of BP's capital expenditure in the second half of the decade.[24] More tellingly, during the 1960s as a whole BP spent a higher proportion of its capital expenditure on refining and marketing than any other oil major,[25] but still, at the end of it, ranked bottom of the majors in refining and marketing competitiveness (see table 11.2 in chapter eleven). Over the same period, BP, whose main competitive strength was in explora-tion and production, spent proportionately less on these activities than any other major.[26]

BP's apparently contrarian policy of investing most heavily in its least competitive activities was easily explicable. The other oil majors were investing heavily in domestic oil production in the USA, where BP had no oil fields, apart from the minor Kern oil fields in California (chapter four), until it struck oil in its Prudhoe Bay leases in 1969. In the Eastern Hemisphere, where BP's operations were concentrated, the industry as a whole was allocating its investment between the upstream and downstream sectors in much the same proportions as BP.[27] The logic, in BP's case, was simple. BP already had the capability to produce much more crude oil than it could sell and was under pressure from the producing countries to increase production; the last thing it needed was still more production capacity. More pressing was the need to secure outlets to meet its volume-maximising targets.

The ironic result was that by 1968–9 BP had a greater concentration of its fixed assets (book value) in the downstream sector than any other major, although it was the least competitive refiner and marketer.[28] The downstream profitability studies were held up by the problems of measuring profitability, but there was little doubt that, however measured, most of BP's downstream operations in Western Europe (BP's main refining and marketing area) were making losses. Much of BP's capital expenditure in the previous decade had, to BP's growing but not yet urgent consternation, been invested in unprofitable activities.

At the end of the 1960s, BP also had other financial worries. One was that a substantial proportion (about a fifth) of the funds required to cover its capital spending came from non-trading items which would soon disappear, or at least diminish. BP's large non-trading income came from three main sources. One was the other participants in the Consortium, established in 1954 to operate the nationalised assets in southern Iran which had previously been exclusively owned and operated by BP. In 1954, as a consideration for their shares in the Consortium, the other participating companies agreed to pay BP a lump sum of £32·4 million in the first year of operations, plus a further payment of 10 cents per barrel on the Consortium's exports of crude and products from Iran until a total of $510 million had been paid. By the end of the 1960s most of that sum had been paid, and it was clear that BP's income from this source would end in 1971.[29]

A second source of non-trading income arose from changes in company taxation introduced by the Labour Chancellor of the Exchequer, Jim Callaghan, in 1965. Previously, company profits in the UK had been subject to a profits tax of 15 per cent plus income tax of 41 per cent. Companies withheld income tax on dividends to shareholders, and in most cases passed the withheld tax to the Inland Revenue as an advance payment of the shareholders' income tax liability. However, companies with overseas operations were allowed to deduct their foreign taxes from their UK tax liabilities, including the withheld income tax on dividends. This was called double taxation relief and was designed to relieve companies from having to pay tax twice on the same profits (overseas and in the UK). It allowed companies like BP, whose foreign taxes exceeded their UK tax liabilities, completely to avoid paying UK tax on profits.

Under the system introduced by Callaghan, companies would in future pay a new Corporation Tax instead of the previous taxes on profits, while continuing to withhold income tax on dividends as before. They would, however, be allowed to claim double tax relief only against Corporation Tax, and not, as before, against the withheld income tax on dividends. Strong protests from BP and other similarly affected companies failed to prevent the introduction of this new system. However, to soften the loss of double tax relief, BP and other companies in the same position would be allowed to claim transitional relief. This would compensate them for the loss of double tax relief on withheld income tax for three years (1966–8), and provide decreasing relief over a further four years, ending in 1972.[30] As it turned out, transitional relief would in 1972 be extended for a further five years by Edward Heath's Conservative government, which succeeded Wilson's Labour government in 1970.[31] But in the late 1960s this extension of transitional relief lay in the future, and BP seemed to be faced with the disappearance of both the Consortium payments and transitional relief in the early 1970s.

To add to the prospective cessation of these sources of income, BP was also faced with the termination of its third main source of non-trading income when, in 1970 Heath's government decided to replace investment grants, which had been introduced in 1966 to subsidise corporate investment in the UK, with tax allowances on qualifying expenditure.[32] BP, which had received £83 million in investment grants in 1966–70, stood to lose by this change because, on account of double tax relief, it paid no Corporation Tax against which it could claim the new investment tax allowances. The immediate impact of the loss of investment grants was reduced by the government's decision to continue paying grants on contracts made prior to the changeover to tax allowances; but this palliative was not enough to stop BP being concerned about the phasing-out of the grants, and the diminution of its non-trading income.

By the time that the Heath government decided to put an end to investment grants, it was already clear that BP would have to turn again to its shareholders for new capital. For the previous four to five years capital expenditure on BP's organic business (i.e. excluding acquisitions), although rising less fast than in 1964–6, was still substantially higher than the retained funds generated from trading.[33] And on top of capital expenditure on organic business came two major acquisitions.

The first, BP's purchase of Distillers' chemicals and plastics interests in 1967, was largely paid for by the issue of new BP shares with a value of £61 million to Distillers (chapter fifteen). This method of payment had two obvious advantages for BP. One was that the issue of new shares to Distillers diluted the British government's BP shareholding, which was reduced to less than 50 per cent for the first time since 1914. This was welcomed by BP, which had come to see the government's large shareholding as a handicap, limiting BP's freedom to operate in countries like oil-rich Venezuela which did not allow foreign state-owned companies to hold oil concessions. The dilution of the government's stake in BP by the issue of new shares to Distillers was only slight, a reduction from 51·6 to 48·9 per cent,[34] but by bringing the government's stake under the psychologically important 50 per cent mark it had a symbolic importance.

During the negotiations over the issue Callaghan was consulted about this and raised no objection.[35] The dilution of the government's shareholding did not affect the government's rights to appoint two of BP's directors and to veto board decisions. These were written into BP's articles of association, which could not be changed without the approval of shareholders controlling at least 75 per cent of BP's ordinary shares. The British government's powers over BP were not, therefore, diminished in practice by the loss of its majority shareholder status.

In acquiescing in the state becoming only a minority shareholder in BP, Callaghan was putting pragmatism before ideology, as previous British governments had mostly done in their relations with BP. But on the ideological surface, it was something of an irony that BP, having been semi-nationalised in 1914 by a Liberal government with an essentially free market ideology, was partially denationalised in 1967 by a Labour government committed to national ownership of the economy's commanding heights.

The second obvious advantage of financing the acquisition from Distillers mainly by the issue of new BP shares was that it reduced the need to find cash to pay for the purchase. BP's finances were not, therefore, put under strain by the acquisition.

The same could not be said of BP's purchase of the ARCO/Sinclair refining and marketing assets on the eastern seaboard of the USA in 1969 (chapter ten). Although BP transferred these assets to Sohio at the beginning of 1970, BP remained liable for repayment of the $400 million purchase price, plus capitalised interest. This new debt raised BP's debt ratio to 33 per cent, too high for comfort.[36]

To strengthen its finances, BP approached its main shareholder, the British government, for an injection of new capital through a rights issue. Heath's Conservative government, pledged to reducing the state's role in the economy, did not as a matter of policy favour adding to the state's investment in British business. There was also another problem. BP's other large shareholder, the Burmah Oil Company, was negotiating with the Continental Oil Company (Conoco) on the possibility of a merger. One of the elements in their merger talks was the cancellation of Burmah's 23 per cent shareholding in BP in exchange for some of BP's upstream assets. Until the uncertainty about Burmah's shareholding in BP was settled, little progress could be made with BP's proposals for a rights issue.

In 1971 these obstacles to the proposed rights issue melted away. After top-level lobbying by BP that government support was in the national interest, Heath's anti-interventionist government, in no less an ideological twist than its Liberal and Labour predecessors, agreed to back BP with new capital from the public purse. Meanwhile, the Burmah-Conoco merger did not materialise, and Burmah agreed that while it could not afford to invest new capital in BP on the scale required, it could pass on its rights to buy new BP shares to Burmah shareholders, who would be given the opportunity to invest directly in BP.[37]

On this basis, BP raised £120 million of new capital through a rights issue in September 1971. The government, whose shareholding in BP had been further diluted to 48·6 per cent by the issue of new BP shares for the purchase of Duckhams in 1970 (chapter nine), took up its full entitlement. At the same time, BP gained 50,000 new shareholders as a result of Burmah shareholders taking up their rights to invest in new BP shares.[38]

The receipt of £120 million from the rights issue was not, however, enough to stop the relentless rise in BP's debt ratio. In the early 1970s BP's capital expenditure increased sharply in all sectors of its business. The modernisation of BP's service station network, refinery expansions, the construction of new giant tankers, expansion in petrochemicals (chapter fifteen), appraisal drilling of the recently discovered Forties field in the North Sea, all contributed to a rise in capital expenditure from £244 million in 1969 to £375 million in 1971.[39] To help relieve the pressure, in September 1972 BP signed an agreement with a consortium of banks for the financing of Forties' development by the method, adapted from the USA, of making advance payments for future production from the field.[40] But nevertheless, BP's debt ratio at the end of 1972 stood at 38 per cent (see figure 12.5).

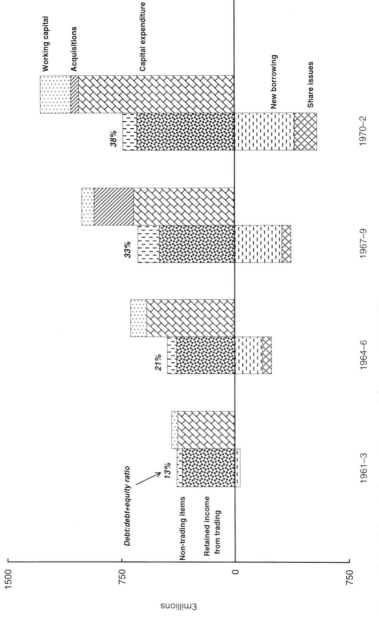

Figure 12.5 BP's capital expenditure and finance, 1961–1972

This would have been uncomfortably high at the best of times, let alone in the early 1970s when BP and the other international oil companies were facing a more powerful and concerted challenge to the concessionary system from the OPEC countries than they had ever faced before (chapters eighteen and nineteen). In 1951, on the eve of the Iranian nationalisation crisis, the Company had been in an exceptionally strong financial position, which had helped it to weather the crisis largely unscathed (chapter one). At the end of 1972, on the eve of the seismic climax of the OPEC crisis, BP was in a much weaker financial position as it faced the greatest upheaval in its history.

=== 13 ===
The managerial hierarchy

THE BOARD OF DIRECTORS

Maurice Bridgeman's appointment as chairman and chief executive in 1961 ended the succession problems which had unsettled the board for several years in the wake of Lord Strathalmond's dominant rule as chairman and chief executive from 1941 to 1956. Under Bridgeman from 1961 to 1969, the board enjoyed a period of calm stability, in which a collegiate atmosphere prevailed, untroubled by rivalries and contests for the leadership. As one director put it, commenting privately on BP's management: 'The only reason it works is that there are seven friendly gentlemen at the top – one self-seeker and the system would fall down.'[1] In the words of another: 'If you got bitter in-fighting at Board level, I, for one, wouldn't stay.'[2] And yet another: 'We know each other and work well together . . . It's very rewarding to work for BP, first of all to belong to a marvellous club which I don't believe any other oil company can achieve.'[3]

Remarks such as these, made shortly after Bridgeman retired in 1969, reflected the emphasis he had consciously placed on building a board that functioned as a team, as opposed to a collection of individuals. His first board appointments were made in 1962, when one of BP's two deputy chairmen, Sir Harold Snow (he had been knighted in 1961), retired and two new managing directors were appointed. One was Lord Strathalmond's son, William (Billy) Fraser, who had been one of Bridgeman's protégés in BP's Exploration Department in the mid-1950s (see chapter two). He had spent the late 1950s and early 1960s as BP's chief representative in New York, before being appointed managing director of the Kuwait Oil Company, through which BP and Gulf jointly held the Kuwait oil concession. More tactful than his father, Billy Fraser was well-liked by his colleagues, and by BP's partners in associated companies such as the KOC.

The second new managing director appointed in 1962 was Alastair Down, a qualified accountant, who had been financial adviser on exploration and production under Bridgeman in the early 1950s, before being posted to Canada in 1954 to take charge of BP's entry into exploration and production, and later refining and marketing, in that country. From Canada, Down kept in friendly touch with Bridgeman in London. Possessing a strong grasp of finance and a powerful strategic focus, Down could be tough, but he was not abrasive and fitted comfortably into the collegiate atmosphere fostered by Bridgeman. On his appointment to the board, Down took responsibility for BP's finances, and for its growing petrochemical interests (see chapters fourteen and fifteen).

The epitome of the Bridgeman school was, however, David Steel, who joined the board as a managing director in 1965, when John Pattinson retired after forty-three years with the Company. Like Billy Fraser and Down, Steel had worked in the Exploration Department under Bridgeman in the 1950s, after which he spent three years in BP's New York office. From there, he returned to London in 1961 for a spell as regional co-ordinator for the Western Hemisphere before succeeding Fraser as managing director of the KOC in 1962. Described by Billy Fraser as 'quite the most conscientious person I have ever met',[4] Steel stood out for his modesty, warmth towards people from all walks of life and patience. He played, above all, a diplomatic role for BP, leading several Consortium delegations to Iran for difficult negotiations with the demanding Shah (see chapter seven).

After the arrival of these three newcomers in the boardroom, the executive element on the board consisted of the chairman/chief executive (Bridgeman), two deputy chairmen (Drake and Banks) and four managing directors (Dummett, Billy Fraser, Down and Steel). In terms of pre-BP qualifications, these seven 'friendly gentlemen' included one natural scientist manqué (Bridgeman) as chairman and chief executive; four professionals in law and/or accountancy (Drake, Fraser, Down and Steel), who were responsible for exploration and production, commercial, concessionary and financial affairs; one linguist (Dummett) in charge of marketing; and one scientist (Banks) in charge of refining and research and development. All the executive directors had been to British universities – four (Bridgeman, Drake, Dummett and Fraser) to Cambridge; one (Steel) to Oxford; one (Down) to Edinburgh; and one, the only science graduate (Banks), to Manchester.

The tendency for BP's science-qualified technical directors to come from Britain's provincial universities was confirmed when Banks retired in 1967, and was replaced on the board by Denis Barker. A chemistry graduate from the University of Sheffield, Barker had worked his way up BP through a

47 David Steel (on left), managing director of the Kuwait Oil Company, with King Hussein of Jordan (centre) and HH Shaikh Abdullah, Ruler of Kuwait (right), in 1964. Steel joined the BP board as a managing director in 1965, and would later become BP's chairman and chief executive

series of posts in refining in France, Israel and Australia, as well as in the London head office. He had also, like most of BP's other executive directors of the time, had a spell in BP's New York office, reflecting the growing tendency for BP to look to the West, as much as to the East, as a training ground for top management. Barker was the fourth new managing director appointed by Bridgeman, and was for a time the only science graduate among the executive directors.

Broadly speaking, therefore, the executive directors under Bridgeman were predominantly professionals trained in law or accountancy, and were mostly graduates from Cambridge and Oxford (especially Cambridge). BP's technical graduates from other universities, though large in number (see chapter two) had not, with the exceptions of Banks and later Barker, found their way to the top. But if the board that Bridgeman built was short on science and technology, it did at least have much broader international experience, and a much less Anglo-Iranian emphasis, than its predecessors (see table 13.1).

When Sir Maurice Bridgeman (he had been knighted in 1964) stepped down there was no doubt but that Eric Drake would succeed him as BP's new chairman and chief executive. Drake was by far the most assertive of BP's executive directors. In his early days working for the Company in Iran he had made no secret of his ambition to rise to the top of the head office in London (see chapter one). He did not court popularity, and was criticised for being too aggressive towards colleagues and for showing a lack of sympathy for 'Iranianisation', i.e. the training and promotion of Iranians engaged in the Company's operations in Iran. In 1944 the Company's staff manager, having taken soundings about Drake, wrote that:

> I was informed that Drake had fallen foul of the Management and Staff at every centre at which he had worked. His ability was admitted, but his total incapacity to co-operate with people both on and off the job greatly restricted his usefulness; I verified that he is unpopular, and has a rather contemptuous attitude to those he regards as his social inferiors.[5]

Not everyone agreed with this assessment[6] and Drake survived the criticisms, rising to become general manager in Iran and Iraq, in charge of the Company's operations in Iran, just six months before the Iranian nationalisation crisis. Drake's handling of the nationalisation dispute, in particular his resistance to the Iranians' demands and his departure from Iran while other British expatriate staff stayed behind, made him, once more, a focus of controversy. The Iranians accused him of sabotage, some of the British staff on the spot felt that he had left them in the lurch, and on returning to London he clashed with Sir William Fraser, the Company's autocratic chairman and chief executive.[7] But once again, Drake managed to live down the criticisms. After a spell in the Company's New York office he came back to London in 1954 to head the newly formed, and increasingly powerful, Supply and Development Division, which soon became a fully fledged Department. He went on to join BP's board, by then under Sir Neville Gass, in 1958 (see chapter two).

After Bridgeman became chairman Drake was soon marked out as Bridgeman's most likely successor, and in 1962 Drake succeeded Snow as a deputy chairman, alongside John Pattinson. When Pattinson retired in 1965, Banks stepped up to become a deputy chairman alongside Drake, but there was no contest between them. Drake, it was clear, was favoured for the chairmanship by Bridgeman, who told the British government, watchful as

Table 13.1 *BP's executive directors, 1956–1975 (selected years)*

							Other	**Continental**
	Age	University/college	Degree/qualification	UK	Iran	Middle East	America	Europe	Australasia

At the end of Lord Strathalmond's chairmanship, 31 March 1956

	Age	University/college	Degree/qualification	UK	Iran	Other Middle East	America	Continental Europe	Australasia
Lord Strathalmond	67	Glasgow	Chemistry	37	–	–	–	–	–
Basil Jackson	63	–	–	20	–	–	14	–	–
Edward Elkington	65	–	–	19	16	–	–	–	–
Neville Gass	62	McGill	Railway work	22	15	–	–	–	–
John Pattinson	56	Cambridge	Engineering	11	23	–	–	–	–
Harold Snow	58	Cambridge	Mathematics	35	–	–	–	–	–

At the beginning of Bridgeman's chairmanship, 1 July 1960

	Age	University/college	Degree/qualification	UK	Iran	Other Middle East	America	Continental Europe	Australasia
Maurice Bridgeman	56	Cambridge	–	28	3	–	3	–	–
Harold Snow	62	Cambridge	Mathematics	38	1	–	–	–	–
John Pattinson	60	Cambridge	Engineering	15	23	–	–	–	–
Maurice Banks	58	Manchester	Science	33	3	–	–	–	–
Eric Drake	49	Cambridge	Law and accountancy	9	13	1	1	–	–
Bryan Dummett	47	Cambridge	Languages	10	–	–	–	10	4

At the beginning of Drake's chairmanship, 25 January 1969

Eric Drake	58	Cambridge	Law and accountancy	18	13	1	1	–	–
Alastair Down	54	Edinburgh	Accountancy	14	1	2	8	–	–
Bryan Dummett	56	Cambridge	Languages	19	–	–	–	10	4
George Ashford	57	Cambridge	Natural sciences and law	1	–	–	–	–	–
Denis Barker	60	Sheffield	Chemistry	16	–	6	2	11	2
William Fraser	52	Cambridge	Law	13	–	3	2	–	–
David Steel	52	Oxford	Law	13	–	3	3	–	–

ever over BP's succession plans, that Drake was the 'obvious choice'. Sir Dennis Proctor, then Permanent Under-Secretary at the Ministry of Power, had some reservations about this, but by the autumn of 1966 the Ministry of Power believed that Drake had 'strengthened and mellowed' since their earlier assessments of him. He was, they believed, the 'only internal possibility' as a successor to Bridgeman.[8] Over the next few years, no new rivals appeared. When Banks retired in 1967, Bryan Dummett took his place as the second deputy chairman alongside Drake, but Drake was obviously the senior of the two in all but name. It was, therefore, without contest that Drake took over as chairman and chief executive when Bridgeman retired in January 1969.

Although the succession was smooth, there were marked contrasts in style between Bridgeman and Drake as chairmen. Bridgeman was not outspoken, did not court controversy and did not seek to stand out from his colleagues. He brought onto the board people like Billy Fraser, Down and Steel, whose careers he had personally nurtured, and who formed a close-knit circle. Drake, on the other hand, was domineering, widely feared in BP, and more adversarial (though no more likely to prevail) than Bridgeman had been in external relations. For example, whereas Bridgeman had without fuss kept on good terms with the British government, Drake would clash with the Prime Minister, Edward Heath, in 1973 (see chapter nineteen). Similarly, whereas Bridgeman had been careful not to take a stand against OPEC (see chapters six and seven), Drake would favour taking a harder line (see chapters eighteen and nineteen). The difference between Bridgeman and Drake was encapsulated in their attitudes towards colleagues. Bridgeman had seen his fellow executive directors as parts of a team; Drake described them as his 'alter egos'.[9]

Although Drake brought a substantial change in leadership style to the boardroom, he did not introduce much new blood to the executive directorate in his first three years at the top. When Drake became chairman, Down succeeded him as a deputy chairman (alongside Dummett), and George Ashford was appointed a managing director. Ashford was unusual in this role, insofar as he had not, like all the other executive directors, worked his way up through BP's managerial hierarchy. A graduate in natural sciences from Cambridge University, and a qualified lawyer, Ashford had spent his main career at Distillers, where he rose to become a director of the main board, and chairman of Distillers' Chemicals and Plastics Group. When BP acquired Distillers' chemical and plastics interests in 1967 (see chapter fifteen), Ashford joined BP to become managing director of BP's chemical subsidiary (BP Chemicals Ltd) and a member of BP's main board, but without the status of a managing director until the rearrangement occasioned by Bridgeman's retirement.

After that, there were no changes in BP's executive directorate until 1972–3, when Dummett, Barker and Ashford retired, and Billy Fraser, who had inherited the title Lord Strathalmond on his father's death in 1970, had to relinquish his executive duties owing to ill health. For the same reason, he left the BP board altogether in 1974. These departures called for a substantial rearrangement of the board. Steel moved up to become the second deputy chairman, alongside Down, and three new managing directors were appointed.

The first was Monty Pennell, a physics graduate from Liverpool University, who had been recruited by the Company in 1946. He was noted for his ability to form good-humoured relationships with almost everyone he met, and for his technical ability. A keen rugby player, he claimed that he had been recruited to fill the vacant position of fly-half in the Company's Fields team in Iran, rather than as a physicist. Under the patronage of Bridgeman, and later Steel, he rose through a series of postings in exploration and production, working not only in the UK, but also in the Middle East, Sicily, East Africa, Libya and the USA.[10]

Pennell was the last of the Bridgeman school to join BP's board. The other two managing directors appointed in 1972–3 came from different moulds. One was Christophor Laidlaw, a languages graduate from Cambridge University who had worked his way up BP primarily in marketing. He was extremely bright, independently minded, frank and self-confident with, it was noted in 1958, a capacity for 'mordant critical analysis'. In contrast to the typical member of the Bridgeman school, he was impatient, at times brutally outspoken, and made, it was noted early in his career, 'insufficient allowance for his seniors and elders being somewhat slower than he is'.[11] Laidlaw liked to shake things up and certainly did not conform to the image of the 'friendly gentlemen' who had populated the executive directorate under Bridgeman.

The same could be said, albeit for different reasons, of Peter Walters, who became a managing director in 1973. At forty-one years old, he was easily the youngest of BP's executive directors (the others were all over fifty), and to some extent he represented a new generation of BP management. He came from what he described as an intelligent working-class background, was educated at a grammar school and took a degree in Commerce at Birmingham University. He joined BP in 1954 and worked his way up mainly in the powerful Supply and Development Department, interspersed with two spells in the USA in 1960–2 and 1965–7. Conscious that he had not come from a privileged background, he believed in 'an elitism based on merit', and frowned on old-boy networks and systems of personal patronage of the sort that Bridgeman had operated at BP.[12]

While the arrival of Pennell, Laidlaw and Walters reinvigorated the board, the choice of a successor to Sir Eric Drake (knighted in 1970) as chairman and chief executive remained shrouded. The two deputy chairmen, Down and Steel, were realistically the only internal candidates, but it was not until January 1975 that it became obvious which one of them would take over from Drake. That month, having presumably learned that he would not be Drake's successor, Down resigned from BP to become chairman of the Burmah Oil Company. Overburdened with debt after several years of over-ambitious expansion, Burmah was in a crisis. The Bank of England had tried to bail it out by guaranteeing some of its debt and then, just days before Down became chairman, by buying Burmah's 20·15 per cent unpledged shareholding in BP.

Down's departure from BP left a vacancy on the board for a new managing director, which was filled by Robin Adam, a qualified accountant whose BP career had been spent mainly in finance and planning. More importantly, it was clear to all that Steel was the chosen successor to Drake. The handover took place on Drake's sixty-fifth birthday, 29 November 1975. By a coincidence, it was also Steel's birthday, his fifty-ninth.

Apart from the executive directors, there were also a large number of non-executive directors who came and/or went from BP's board between 1960 and 1975, when Bridgeman and then Drake were chairmen. Some of the non-executives were government directors, appointed by the British government as BP's main shareholder. Others came from Burmah Oil, which continued until 1975 to hold more than 20 per cent of BP's shares. Still others were ordinary non-executive directors, appointed to balance the board between executives and non-executives, and to bring relevant knowledge and experience to BP's board. Some information on who they were, what they did and the time they spent on BP's board is given in table 13.2.

HUMAN RESOURCES AND ORGANISATION

A fundamental issue for Bridgeman and Drake, as for all leaders of big business, was the use and organisation of human resources. The challenge, in essence, was to recruit and promote the right people, and to develop an organisation which could effectively co-ordinate the international flow, not only of oil, but also of management skills, best practices, information and technology.

Table 13.2 *BP's non-executive directors, 1960–1975*

	Titles and peerages	Primary occupation	Dates on BP board
Government directors			
Frederic Harmer	Knighted 1968	Shipping	1953–70
Ronald Weeks	Knighted 1943; cr. peer 1956	Manufacturing industry	1957–60
Lionel Robbins	Cr. peer 1959	Economics	1961–68
John Stevens	Knighted 1967	Finance	1969–73
Robin Brook	Knighted 1974	Finance	1970–73
Denis Greenhill	Knighted 1967; cr. peer 1974	Diplomatic service	1973–78
Tom Jackson	–	Trade union	1975–84
Burmah directors			
William Eadie	–	Oil industry	1955–65
Robert Smith	–	Oil industry	1957–71
The Earl of Inchcape	Hereditary peer	Shipping	1965–83
Ordinary non-executive directors			
William Keswick	Knighted 1972	Trade	1949–73
Desmond Abel Smith	–	Manufacturing industry	1950–62
Cameron Cobbold	Cr. peer 1968	Finance	1962–74
Humphrey Trevelyan	Knighted 1955; cr. peer 1968	Diplomatic service	1965–75
Samuel Elworthy	Knighted 1961; cr. peer 1972	Armed forces	1971–78
Val Duncan	Knighted 1968	Minerals industry	1974–75
Michael Verey	–	Finance	1974–82
(John) Lindsay Alexander	Knighted 1975	Shipping	1975–91

Until the nationalisation of its operations in Iran in 1951, the Company, aptly named Anglo-Iranian, had concentrated on a simple two-way flow of British managers and staff between the head office in London and the oil fields and Abadan refinery in Iran. This system was brought to an end by the nationalisation crisis, when the Company's staff in Iran were repatriated to England, where many of them were made redundant (see chapter one). After the immediate crisis was over, BP engaged in a recruitment drive from 1953 to about 1960 to rebuild its managerial resources at the same time as it internationalised its business across a wider geographical spread. But BP retained, as seen in chapter two, a centralised, Anglocentric management system in which large numbers of British managers were sent out from the centre to the periphery of BP's operations, but few foreign nationals came the other way, from the periphery to the centre.

By the time that Bridgeman became chairman and chief executive in 1961, the head office staff headcount had been built up to about 3,000,[13] and concerns that BP was short of staff were turning into worries that the growth in numbers needed to be checked. For the rest of Bridgeman's chairmanship, as BP's profitability declined (see chapter twelve), concerns about the inexorable growth in the number of head office staff were often expressed, but with little result. The upward trend in headcount (see figure 13.1) continued without interruption.

This was not for lack of attention to what in those days was called 'manpower planning'. BP's head office Staff Department, a keen proponent of the then-fashionable social sciences, issued a profusion of studies, papers, reports and recommendations on manpower planning, 'organisation and methods', organisational behaviour and the like.[14] But these made little impact on BP's board of directors. As Alastair Down would comment in 1969: 'Central Staff Department is too preoccupied in producing bright ideas which have absolutely no point at all as far as the Company is concerned.'[15] Billy Fraser separately confirmed that there had been 'no results from work in social sciences'.[16] Most important of all, Bridgeman, preoccupied with the problems of how BP could hold on to its oil concessions, showed no apparent interest in manpower planning.

Nor did he seem greatly interested in changing BP's organisation. After BP's system of regional co-ordination was introduced in 1960, he quashed the suggestion that the regional co-ordinators should be given more power in relation to the dominant functional departments. Putting failings in the system down to the personalities involved, he made it clear that he saw no need to change the balance between the regional and functional elements (see chapter two). The functional departments therefore remained the core of

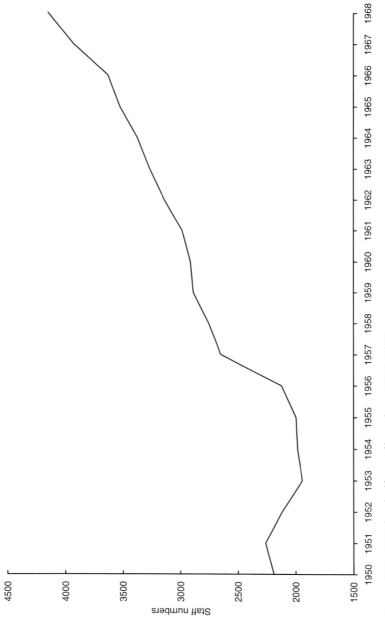

Figure 13.1 BP's head office staff numbers, 1950–1968

BP's head office organisation. The most powerful of them, especially the Exploration, Supply and Development and Marketing Departments, were largely independent baronies, internalising their own supporting services and controlling the development and promotion of staff through closely guarded internal hierarchies. The departments converged only at the apex of BP's organisational pyramid, the board, where each executive director had specific functional responsibilities as well as wider strategic issues to consider.

But it was not only within the head office that the functional departments held sway. They also exercised highly centralised control over BP's international operations. For example, the flow of oil between BP's oil fields, refineries and market outlets was centrally programmed by the Supply and Development Department, using ever larger computer models. This system did not allow BP's numerous national refining and marketing subsidiaries and associates to choose their own crude oil supplies, to decide on their own products mix or, consequently, to maximise their own profits. With so many crucial decisions lying outside their control, they could not be held fully accountable for their performance. They operated in what was, in effect, a centrally planned BP economy.

Nor could local national managers hope, whatever their abilities, to rise to the top of BP, whose articles of association continued to stipulate that the main board directors had to be British subjects (see chapter two). Ideas about internationalising BP's management occasionally surfaced, but made little headway. For example, at a BP staff management conference in 1963, Bridgeman was asked by a Swiss manager whether senior management exchanges between companies in the BP Group should be increased in the light of other companies' practices. Bridgeman replied that 'the interchange of staff at the senior level is of doubtful value. As far as other organisations are concerned, you will not find many examples of such interchange outside Shell and Unilever, which are themselves the result of the merger of international companies. Certainly it is not the practice of the major American Oil Companies nor of CFP.'[17] He did not, in short, feel the need to take a lead in developing a more international managerial culture in BP.

Meanwhile, the continuing increase in the numbers of head office staff resulted in the creation of a vast central bureaucracy. In 1950 there had been nine departments in the Company's head office; by the mid-1960s there were more than twenty.[18] As departments continued to set up their own supporting services, there was growing duplication and overlap of tasks. Most staff did not have clear terms of reference or performance targets, with the result that many of them did not feel stretched by their jobs.

Some of these problems came out clearly in a survey of middle managers in five departments, carried out by BP's Staff Department and outside consultants in January 1966. The survey, based on questionnaires and interviews, found that middle managers were generally satisfied with their rates of pay, and extremely satisfied with BP's generous welfare arrangements – sports facilities, social clubs, subsidised lunches, sickness and retirement benefits and other fringe amenities. Relationships between colleagues were found to be congenial. Mirroring the seven 'friendly gentlemen' on the board, 'affairs in BP were said to be conducted in a gentlemanly, pleasant fashion', but with a lack of drive. The typical BP recruit was, it was said, 'the kind of person if you met him at a cocktail party you'd say – awfully nice chap'.

Most of those interviewed considered themselves to be under-used, received little feedback on their performance and felt that promotion was not linked to ability or performance. But they nevertheless stayed with BP, largely because of its non-contributory pension scheme, and because the gentleman's agreement with other oil companies not to poach each other's staff (see chapter two) meant that other oil companies would not consider an applicant from BP until he had resigned. BP managers were reluctant to do that before securing other employment, in case they found themselves unemployed.

One problem, however, stood out as the main source of dissatisfaction. This was that BP's deeply hierarchical structure was over-complex and stifled initiative. It was characterised by 'endless layers of management', slow communications, excessive paperwork, delays in decision-taking and lack of delegation. As the survey reported:

> Underlying most of these [management] problems is that of the strictly hierarchical structure of the organisation – which brings with it difficulties of communication, slowness in response, passing of the buck, and lack of involvement on the part of staff; it is suggested that this structure be re-examined with a view to reducing the number of layers, to delegating greater powers to middle management . . .'[19]

Nothing immediately came of this suggestion. In 1967 BP moved its headquarters from the old Britannic House, designed by Edwin Lutyens in the 1920s, to a new custom-built skyscraper with more than thirty storeys nearby in the City. This new Britannic House (the name was transferred) was both modern and superbly equipped. The external walls of the vast entrance hall were of full-height plate glass set in bronze mullions, the core walls were covered in Arabescato marble, and the doors and panels were veneered in Jacaranda with bronze trim. In the senior dining-room, the panelling was in closely spaced American walnut ribs, and the drum of the domed ceiling was in white plaster, with the curved area covered in gold leaf illuminated by concealed lighting. In the magnificent boardroom, the downlighters were housed in curved plaster pendants to prevent any acoustic flutter between ceiling and table top, which might impair clarity of speech.

But while the new head office was modern, the old hierarchy remained, transplanted to a towering building whose vertical structure in some respects symbolised the multiple layers of BP's management. For example, the chairman's, deputy chairmen's, and managing directors' offices were on the top two floors, together with the boardroom, board dining room, committee room, directors' and senior managers' dining room, and the chairman's dining room.[20] Beneath them, both physically and hierarchically, was the Senior Luncheon Club's dining room, which remained an all-male preserve into which female guests were not permitted, the staff manager commenting that it was unnecessary to promulgate this rule because it was a well-known 'tribal custom'.[21] In this area, at least, a breakthrough was made when, in November 1967, Paula Harris, a senior mathematician in the Computer Department, became the first female member of the Senior Luncheon Club.[22] But for the most part the head office hierarchy remained entrenched in its customs, and growing in number.

One manager who tackled some of these problems in his own area was Christophor Laidlaw in the marketing function. In 1963 the Marketing Department had been renamed the Marketing Development Department and, at the same time, hived off some of its operations to a new Product Sales Department, under Laidlaw. In 1966–7 he reduced the complexity of this arrangement by merging the Product Sales and Marketing Development Departments into a single Marketing Department, which was reorganised, and the number of managers reduced.[23]

Laidlaw also showed enthusiastic approval for the reorganisation of BP's French associate, Société Française des Pétroles BP, which was carried out in 1966 by its president, Jean Chenevier, with the help of the US consulting firm, McKinsey. In a paper on the reorganisation (on which Laidlaw wrote, 'Very good indeed'), Chenevier displayed a much more dynamic, forward-looking approach to organisational problems than could be found in BP's main boardroom. He explained that SFPBP, under pressure from state-subsidised competitors in the French market (see chapter nine), had decentralised its marketing under regional managements, made swingeing reductions in its head office staff, and introduced systems of remuneration, promotion and training linked to performance rather than seniority and length of service. His comments on organisation were so germane that they deserve to be quoted at length. 'Somewhere', he wrote, 'between the creative but sometimes disorderly improvisation involved in starting from scratch and a

48   Paula Harris, Senior Mathematician in the Computer Department, explaining mixed integer circuitry. In 1967, she became the first female employee to be admitted to BP's Senior Luncheon Club

bureaucratic slackness of ageing administration, there is room for a constant effort to adapt and regenerate a living organisation'. Experience, however, had shown how difficult this could be:

> Rather than change the existing structures with the suite of comfortable habits or acquired privileges that invariably follow, one invents reasons, one strengthens walls behind which one has finally built oneself an empire. When new factors appear calling for new requirements, it is so very much easier to add appendices rather than suppress functions or out dated tasks! Little by little one finishes by living in an inextricable forest in which only the initiated know their way about and whose science is all the more appreciated that it is incommunicable . . . amazing, tortuous circuits replace the large open avenues which seem to be obsolete on organisation charts.

A manager's energies could not, he was convinced, 'be harnessed without offering him a challenge that will want to make him [*sic*] surpass himself'. This, he went on, 'presupposes the will to delegate responsibilities to the utmost, to decentralise the levels of decision as much as possible'. Emphasising the need for members of a company to share a 'common philosophy', he concluded that any company, not just SFPBP, 'must increasingly represent a veritable "Cultural Medium" of which we are only just beginning to perceive the potential wealth'.[24]

While SFPBP was reorganising in France, Standard Oil (NJ) and Royal Dutch-Shell were making organisational changes on a larger scale. In 1966 Standard Oil (NJ), which had already set up decentralised regional organisations in the USA, the Far East and Africa, established Esso Europe to co-ordinate the activities of its numerous European affiliates. Previously, these affiliates had reported individually to Standard's head office in New York, but from 1966 they would be accountable to Esso Europe, based in London. Europe was thus fitted into Standard's global system of decentralised regional management, in which the New York head office concentrated on global planning and inter-regional co-ordination, while the regional managements enjoyed substantial local autonomy. By 1970 fourteen nationalities would be represented in the staff of Esso Europe, whose board included the chief executives of Standard's most important European affiliates. Staff exchanges ensured that, as a senior executive in Esso Europe put it, 'we now have a regular <u>flow</u> of European managers, gaining and applying international management experience'.[25]

Royal Dutch-Shell, too, had adopted a decentralised regional structure to replace the old system in which the Group was split into production, refining and marketing functions, and communications were channelled up chains of command to head offices in The Hague and London. In 1966 Royal Dutch-Shell set up Shell Europe on a similar basis to Esso Europe, and each of the managing directors of the parent companies in The Hague and London was given responsibility for a region of the world as well as functional responsibilities. Shell's top executives in London cultivated a 'helicopter view'. 'England', said one, 'is just as much abroad to us as Venezuela'. At the same time, head office staff numbers in The Hague and London were kept under close control.[26]

BP kept track of these moves by its main rivals, but did not immediately follow their example. Towards the end of his chairmanship Bridgeman was thinking about taking BP further towards regionalisation, but decided to leave that to his successor.[27] In 1968, shortly before taking over from Bridgeman, Drake sanctioned the setting up of an internal Manpower Study

Group to investigate and make recommendations on the manpower and organisation of BP's head office.[28] With the help of two outside consultants – John Garrett of Associated Industrial Consultants and Tom Burns, Professor of Sociology at Edinburgh University – the MSG made a thorough-going investigation, interviewing the executive directors and all head office managers and staff above the most junior grades.

When Drake himself was interviewed, he did not seem to feel that there was a need for major changes. 'I feel very strongly', he said, 'that reorgan-isations in radical terms that our competitors have gone through recently have messed them up – and BP is in a much stronger position than ever before'. On the same theme, he remarked: 'What strikes me more and more forcibly is that Shell, with much more regionalisation, has been getting more and more constipated, while BP has been getting livelier'.[29]

The MSG's interim report, issued in November 1969, gave a different impression. It found, overall, 'a management style which has been carried forward from the days when the business was smaller and simpler'. This style had given rise to a 'loose departmental feudalism'. Growth and com-plexity had been accommodated by a head office 'based on large and pow-erful functional departments that remain largely autonomous'. The departments had created 'tall', multi-level hierarchies, in which 'there is a tendency at all levels of management for assistant posts to develop and to the appointment of Assistant General Manager, Assistant Divisional Manager, Assistant Branch Manager and Assistant Section Leader'. Frequently, managers at the assistant levels had ill-defined and unexacting jobs. 'Many of these posts', wrote the MSG, 'could be dispensed with alto-gether'. Associated with this 'phenomenon' was the proliferation of very small units 'of fewer than half-a-dozen staff, complete with manager and assistant manager'.

There were also large overlaps arising from the independent growth of the same functions in different departments. For example, some eight depart-ments had a total of fifteen units or individual posts which had been set up to carry out efficiency or procedural studies. Much the same was true of planning. The Central Planning Department (re-formed in 1964 after the earlier one had been merged into the Supply and Development Division a decade earlier) was mainly concerned with the routine 'development' plan-ning of BP's future business, and insufficient attention was paid, the MSG thought, to more fundamental strategic planning. Outside the Central Planning Department, the Supply and Development Department retained a large planning activity, and there were, in addition, 'development' planning activities in some ten departments. The independence of the departments had also, it was found, led to the evolution of separate departmental staff policies. Overall, 87 per cent of the managers interviewed had no terms of reference and consequently lacked clearly defined accountability for their performance. Relatedly, there was a lack of systematic career planning.

On regionalisation, the authority of the regional co-ordinators in relation to the functional departments was found to be undefined. Supervision of the performance of subsidiaries and associates was shared between the regional co-ordinators and the Marketing, Refineries, Exploration and Finance and Accounts Departments. The division of responsibilities, and the absence of agreed criteria for judging results, was found to create confusion and dissatisfaction both in the London head office and in subsidiaries and associates. A particularly critical deficiency in management control information was the absence of profitability measurement. The Central Planning and Finance and Accounts Departments had been making efforts to fill this gap 'but these have been hampered by important disagreements between the departments on the validity of their respective approaches' (see chapter twelve).

At the top of the head office pyramid, too, the MSG felt that there were structural problems. All six executive directors apart from the chairman/chief executive had functional responsibilities, and all but one had regional responsibilities as well. There was no effective provision for the resolution of inter-departmental and inter-regional matters below the level of the main board. As a result, the directors tended to become excessively involved in resolving anomalies, conflicts and problems presented to them, and there was a danger of them having too little time available for concentration on strategy and policy-making.[30]

To remedy these problems, the MSG recommended a head office reorganisation which, with slight modifications, was approved by Drake and introduced at the beginning of 1971. The reorganisation represented a substantial move towards regionalisation and delegation of authority from the main board. A key feature was the strengthening of the board of BP Trading, which had been created in the 1950s to take over the trading activities of the main parent company without, however, having much real executive authority in its early years. Under the reorganisation of 1971 this was changed. The BP Trading board would in future consist of the seven main board executive directors and, in addition, nine BP Trading directors. These nine would form an Executive Committee to provide an executive and co-ordinating authority between regions and functions.

Of the nine members of the Executive Committee, four would be regional directors, in charge of four regional organisations covering (1) Europe, excluding the UK; (2) the UK and Consolidated areas, in which BP marketed jointly with Shell (see chapters nine and ten); (3) the Western Hemisphere; and (4) the Middle East, Africa (excluding the parts in the Consolidated area), Australasia and the Far East. The existing regional management responsibilities carried out in the Marketing, Finance and Accounts, Refineries and Exploration Departments would, in the main, be transferred to the new regional organisations.

Four other members of the Executive Committee would be responsible for groups of functional departments under the headings of (1) Operations; (2) Finance and planning; (3) Technical; and (4) Administration. The ninth member of the Executive Committee would be the chief executive of BP Chemicals.

The new Executive Committee, immediately below main board level, would help to take some of the weight off the main board, which would be better able to focus on overall strategy and policy. Reporting to the main board would be a newly appointed Policy Planning Adviser, who would concentrate on strategic planning. Development planning would be carried out by a new Central Developmental Planning Department, which would absorb most of the activities of the old Central Planning Department, and planning elements from other departments.

A Staff Development Committee would also be established under one of the deputy chairmen on the main board to strengthen the system of career planning and development. It was expressly stated that 'the Board wishes, wherever practicable, to encourage the movement of staff between associated companies and London Office as a valuable means of developing staff with international experience'. Lastly, an internal Management Consultancy Group was established to undertake continuing studies on organisation, efficiency and modern management practices.[31]

By this reorganisation, BP took a big step towards curbing its feudal functional departments, strengthening its regionalisation and broadening executive authority below the main board. But it had taken BP a long time to reach this stage of organisational development. Looking back over BP's earlier history, the Company had never been a first mover in organisational change. In its early years, from 1909 to the 1920s, it had adopted the nineteenth-century system of managing agencies to run its operations in Iran.[32] On giving that up in the 1920s, the Company introduced centralised functional departments which continued, despite the introduction of regional co-ordination in 1960, to dominate the Company's organisation until the 1970s. Moreover, even after the 1971 reorganisation, BP remained highly centralised. Although there was a greater regional emphasis than before, the regional organisations were not physically divorced from head office, whose headcount continued to grow, rising from about 4,150 at the time of the MSG report in 1969 to more than 4,500 in 1972.[33] The ink was barely dry on the 1971 reorganisation before it became clear that BP would have to move much faster than it had in the past to adapt its organisation to the revolution in the international oil industry wrought by OPEC in the 1970s.

PART III
DIVERSIFICATIONS

14

Alliances in petrochemicals

The rapid growth in oil consumption in the 1950s and 1960s was accompanied by an even greater rise in demand for a burgeoning range of synthetic materials. These appeared in all manner of guises, as rainbow-hued plastics, sheer nylons, wrinkle-proof fabrics, lather-free soaps, gramophone records, toys, floor tiles, carpets, electrical fittings, insulation, packaging, car tyres, rainwater pipes, guttering and countless others. Their ubiquity confirmed the growing market dominance in the industrialised countries of synthetic products over traditional natural materials such as wood, cotton, silk, wool, rubber and leather.

These synthetic materials were not by any means entirely new. Celluloid, a semi-synthetic material, had found applications in, for example, knife handles, buttons and combs in the late nineteenth century. But the effective foundation of synthetic plastics manufacture came in the 1900s with Leo Baekeland's discovery in the USA of a process for making phenol formaldehyde, a thermosetting plastic which could be melted and moulded into products which would keep their shape when cool. Sold under the name of Bakelite by the Bakelite company, phenol formaldehyde found many uses in electrical appliances and other applications.

Other synthetics soon followed. The dark colour of phenol formaldehyde stimulated the search for colourless alternatives and led to the development in the 1920s of translucent urea formaldehydes, which could be coloured by the addition of pigments. Phenol and urea formaldehydes were the chief plastics in Britain before World War II, when many households acquired plastic wireless sets, and the house-building boom of the 1930s brought growing demand for electrical fittings, door knobs and other items which were made of thermosetting plastics. At the same time, the growth of the motor industry expanded the demand for solvents such as acetone, used in paints and lacquers, for ethylene glycol, used as antifreeze, and for rubber tyres. While existing products found expanding markets, new plastics and fibres were also discovered, including nylon, polymethyl methacrylate (perspex) and polyethylene.[1]

All these materials were the products of the organic branch of chemistry, the chemistry of carbon compounds such as sugar, coal and petroleum. The main petroleum raw materials for chemicals were the olefins, ethylene, propylene and butylene. These colourless, odourless and highly reactive gases were obtained from natural gas or as products of refinery cracking plants, originally built to break heavy oil products into lighter liquid products of higher value, primarily gasoline. Cracker gases, rich in olefins, were at first regarded as surplus by-products of liquid oil refining, but in the interwar years it came to be realised that they could be profitably used as feedstock for chemical manufacture.

At that time, the only country to make significant use of petroleum as a raw material for chemicals was the USA, where there were plentiful supplies of natural gas and a large refining industry, in which cracking techniques were widely used to meet the high demand for gasoline from US road users. Seeing an opportunity for the use of by-product cracker gases, some oil companies, most notably Standard Oil (NJ) and Shell, began to manufacture petrochemical products in the USA. So too did US chemical companies such as the Union Carbide and Carbon Corporation and the Dow Chemical Company, who saw petroleum feedstocks as promising new raw materials for their existing chemical products.[2]

In the industrialised countries of Europe, where there were neither significant supplies of natural gas nor a large refining industry before World War II, petrochemicals were virtually unknown. In Germany, coal remained the predominant raw material for organic chemicals produced by the mighty firm of IG Farben, formed in 1925 by a merger of Germany's largest chemical companies, including the three major concerns of BASF, Hoechst and Bayer. More than a match for any other chemical company in the world, IG Farben was held in awe for its pre-eminence in organic chemicals derived from coal.[3]

In Britain, the chemical industry was dominated by Imperial Chemical Industries (ICI), formed in 1926 by the merger of the country's four largest chemical companies, who believed that only by combining could they hope to stand up to IG Farben. In its early years, ICI was preoccupied with the construction of a huge plant at Billingham, on the River Tees in Yorkshire, to manufacture synthetic ammonia for the production of nitrogenous fertilisers in the belief that there would be enough demand from agriculture, principally in the British Empire, to absorb the output. When that hope turned out to be illusory, ICI turned to another project of dubious commercial merit, the production of oil from coal. Absorbed with these projects, which turned out to be unviable, ICI was slow to see the exciting commercial prospects for the new products of organic chemistry that were emerging in the interwar years.

Nevertheless, some interest of a comparatively minor nature was shown. In the late 1920s, ICI negotiated unsuccessfully to acquire British Xylonite Ltd, makers of celluloid. After that failed, other plastics firms were approached and in 1933, following abortive talks with Bakelite Ltd, ICI acquired a controlling interest in the plastics firm, Croydon Mouldrite Ltd. Meanwhile, ICI research staff made important discoveries, finding processes for the manufacture of perspex and polyethylene. Full-scale production of polyethylene was started just before the outbreak of World War II, but the plant capacity was small, reflecting ICI's low commitment to plastics.[4]

In the meantime, another British firm had entered into large-scale production of organic chemicals from a quite different starting point. This was the Distillers Company, which had been formed in 1877 by the merger of six grain distillers to form the largest Scotch whisky concern in the UK. Some of Distillers' founding firms also produced alcohols for industrial use and it was from this base that Distillers entered into the manufacture of organic chemicals. In 1925 Distillers acquired a distillery, located at Hull, which produced industrial alcohol by the fermentation of molasses (the residue left after the extraction of sugar from cane and beet). Soon afterwards Distillers decided to build at Hull a large-scale plant for the manufacture of synthetic organic chemicals, using alcohol from the distillery as the main raw material. The plant, operated by a Distillers subsidiary named British Industrial Solvents, started production in 1929 and became a large supplier of chemicals, most notably acetone and acetic acid. One of the factors that helped to make it competitive was a special fiscal allowance. Unlike whisky and other potable spirits, industrial alcohol was not subject to excise duty. The production of industrial alcohol was nevertheless closely supervised by the excise authorities, to a degree that made continuous working impossible. As compensation for this inconvenience, industrial alcohol producers were granted an 'inconvenience allowance' of 5d (about 2p. in decimal currency) per proof gallon. This allowance, which was not paid in other countries, helped Distillers to establish a profitable organic chemical business based on alcohol made from molasses.

After entering into organic chemicals, Distillers integrated forward into plastics manufacture, acquiring British Resin Products, a small company

49   Early chemical plants at Distillers' Saltend site at Hull in the 1930s

which manufactured resins for the paint trade, in 1937. This was followed in 1939 by the purchase of a 50 per cent interest in BX Plastics, makers of celluloid, from its parent company British Xylonite, which retained the other 50 per cent.

As Distillers expanded its chemical and plastics interests, it inevitably came into contact with other firms engaged in similar activities. The most notable was ICI, which was a major user of solvents and bought large quantities of acetone from Distillers. The ICI–Distillers relationship was not, however, solely a customer-supplier one. There was also the possibility that the two firms might find themselves in competition with each other, a prospect quite contrary to ICI's settled policy of coming to terms with rivals, rather than competing with them. The cornerstone of that policy was the worldwide Patents and Processes Agreement which ICI made in 1929 with the USA's largest chemical company, Du Pont, for the exchange of technical information and the cross-licensing of patents. Distillers, too, came to agreement with ICI to regulate competition. The agreement came into force in 1939 and defined in detail those fields of organic chemicals (including alcohols) which were to be the preserve of Distillers, other fields which were to be left to ICI and a common field which either company was free to enter.[5]

While these events were taking place, the Company had in the late 1920s begun to erect cracking plants at some of its refineries. The first were two crackers using US cracking processes which were brought into operation at the Llandarcy refinery in South Wales in 1927. In 1929, another cracker, using a continuous cracking process developed at the Sunbury research centre, came on stream at the same refinery. It was followed over the next two years by further continuous crackers at Llandarcy and at the Grangemouth refinery in Scotland. At the same time, four much larger US-designed crackers were erected at the Abadan refinery, where they were brought into operation in 1931.[6]

With these new crackers starting up, the Company searched for ways to utilise the cracker gases. Thinking that they might have value as raw materials for chemical manufacture, the Company provided ICI with samples in 1929. In discussions between Dr Albert Dunstan, the Company's chief chemist, and Major A. E. Hodgkin of ICI it was suggested that the two companies might form a joint subsidiary to 'take over the utilisation of this raw material'.[7] Although that did not happen, the Company continued to show interest in the utilisation of cracker gases. In 1930–1, the possibilities of converting them into chemicals were investigated at Sunbury, while similar work was at the same time being done by Distillers at its research centre at Epsom in Surrey.[8]

Sharing a common interest in the possibilities of petrochemical manufacture, the Company and Distillers had talks. The Company's negotiator (identity unknown) was much impressed with Distillers, which was in his view 'the most progressive and enterprising force in the "solvents" market in this country, as well as dominating, completely, British trade in industrial alcohol and its derivatives . . . its activity in the last two years has been almost phenomenal'. Negotiations were taken to the point where draft principles were agreed for the formation of a syndicate between the Company and Distillers to undertake joint research on processes for the production of synthetic materials from cracker gases. 'This alliance', wrote the Company negotiator, 'offers the chance of our doing really <u>big</u> business, if our stuff, so to speak, is the right stuff'. He therefore sought authority to negotiate the final terms of a joint venture with Distillers. However, this suggestion, like the earlier one of a partnership with ICI, was not carried any further at that time.[9]

Yet the idea of the Company forming a partnership with an established chemical company would not go away. In 1932, the Company suggested joint activities in chemicals with ICI, but apparently the Company 'could not offer enough gas to satisfy ICI and their price was high'.[10] The Company made further overtures to ICI later in the 1930s as, too, did Standard Oil (NJ). But they were turned away by ICI, whose policy at the time was not to collaborate with any oil company.[11] In 1938 the Company, Shell and Standard Oil (NJ) all had discussions with Distillers on the large-scale development of oil cracking in Britain, but no definite commitments resulted.[12]

By the time that World War II broke out the Company had thus been actively interested in forming an alliance with a chemical company for about a decade. After the outbreak of war, yet more talks were held with Distillers in 1941–2 regarding possible collaboration in oil cracking,[13] before the Company turned with apparently more serious intent to ICI once again.

NEGOTIATIONS WITH ICI

In February 1943, a top-level meeting, instigated by Sir William Fraser, was held between the Company and ICI at the Dorchester Hotel in London. For the Company, those present in addition to Fraser were Hubert Heath Eves (deputy chairman), James Jameson (a director) and George Coxon (head of refining). ICI was represented by John Nicholson (deputy chairman), accompanied by two other ICI directors, Alexander Quig and William Lutyens, plus Major Kenneth Gordon, who had been closely involved in ICI's work on producing oil from coal. They discussed the possibility of finding a 'common meeting ground' for developments in the field of chemicals. ICI, which was coming round to the view that the future of organic chemicals lay in using petroleum as a raw material, was more receptive to the idea of an alliance with an oil company than previously, and it was agreed there and then at the Dorchester to set up a joint Collaboration Committee under Quig's chairmanship to explore the possibilities.[14]

By May, a 'Draft basis of collaboration' had been drawn up for the formation of a partnership which would ally the Company's abundant supply of raw materials in Iran with ICI's technical know-how in chemical manufacture and its sales outlets for the disposal of chemical products. It was proposed that a 50:50 joint company should be formed to construct a large petrochemical plant alongside the Company's giant refinery at Abadan. The refinery would supply petroleum feedstock for the new plant, whose output of basic chemicals would be transferred to ICI for further processing and sale. The large quantities of liquid light oil fractions which were unavoidably produced in chemical manufacture would be returned to the refinery for reprocessing.

This proposal was not, as the discussions of the Collaboration Committee bear witness, entirely without problems. For example, would Du Pont object to the scheme, for fear that secret information which it gave to ICI under the Patents and Processes Agreement might be passed on to the Company? Might the Company's concession in Iran be infringed if ICI became involved in a joint operation there? And how, with the joint company operating at the uncertain border between the two industries of oil and chemicals, would it be possible to prevent the partners from encroaching on each other's traditional fields of activity? This was a particularly difficult problem, for as Jameson pointed out, 'what is in the chemical field today may be in the petroleum field tomorrow'. However, the forces pulling the two companies together seemed at that time to be more powerful than the forces pulling them apart.[15]

In the summer and autumn of 1943, the scheme of co-operation evolved into modified proposals for a joint company, cumbersomely named Imperial Chemical and Anglo-Iranian Development Company. The idea was that this company, with its own scientific and administrative staff, would promote co-operation between its parents in petrochemicals and in other fields of mutual interest. Subsidiary operating companies would be formed to undertake approved projects, so that, over time, the Imperial Chemical and Anglo-Iranian Development Company would become, in effect, a holding company for joint ICI–Company ventures.[16] This scheme, which had government approval, was set out in 'Notes of proposed arrangement', which were initialled by Fraser and Lord McGowan, the chairman of ICI, in November 1943.[17]

Over the next year, steps were taken to launch the new joint venture, which was given the more concise name of Petroleum Chemical Developments. Kenneth Gordon of ICI was appointed its managing director-designate and recruited a small staff from the two parent companies. In 1944, he and his staff made a tour of ICI and Company operations in the UK before travelling to Iran for further investigations in November.[18] Fraser and McGowan took a close interest in these proceedings, as did Jameson, who wrote in November that there were still 'a few small fences to negotiate' before Petroleum Chemical Developments was registered, but that the 'final stages are now being reached'. 'We can', thought Jameson, 'expect to build up a very large chemical industry based on our Persian operations'.[19]

Meanwhile, the first informal meeting of Petroleum Chemical Developments' directors was held at ICI offices in London. It was anything but a low-level gathering. For ICI, there were present Lord McGowan, Lord Melchett, Lord Ashfield, Sir John Nicholson (newly knighted) and Alexander Quig; for the Company, Sir William Fraser, Hubert Heath Eves, James Jameson and Robert Watson – all of them main board directors – plus Sir Frank Smith, the Company's adviser on scientific research and development. The proceedings were, it would seem, primarily social, for as Jameson recorded, 'We had a very pleasant lunch but no business was discussed.'[20] On the whole, however, the partnership seemed to be moving ahead purposefully.

However, stumbling blocks to Petroleum Chemical Developments were soon encountered. In the winter of 1944–5, while Gordon was in Iran investigating the possibilities of joint petrochemical operations, the Company took legal advice on the concessional aspects of joint activities in Iran. Legal opinion was that while the Company could enter into petrochemical manufacture in Iran on its own, there might be problems if ICI was involved as a partner as this would, in effect, amount to the admission of ICI to work in Iran without a concession. Vladimir Idelson, the Company's legal adviser, thought that 'any action by a subordinate company might affect the rights of the AIOC under the Concession and therefore all actions by all subordinate companies should be under the paramount and complete control of the AIOC'.[21] This finding effectively ruled out joint operations in Iran.[22]

Meanwhile, in 1944 the ICI–Du Pont relationship, defined in the Patents and Processes Agreement, came under scrutiny by the US antitrust authorities, who began legal proceedings in connection with the Agreement.[23] Under pressure, du Pont withdrew the consent it had earlier given to the formation of Petroleum Chemical Developments, which would in effect have extended the coverage of the contentious Patents and Processes Agreement.[24] In that light Fraser and McGowan agreed that antitrust considerations made it 'impossible for ICI to co-operate with AIOC in research', which meant that Petroleum Chemical Developments had to be aborted. However, they agreed that their two companies could still co-operate in specific projects for the manufacture of chemicals.[25]

Such a scheme, codenamed 'Project A', was duly drawn up in autumn 1945. It provided for a jointly owned petrochemical works to be built at Wilton, a new ICI site on the south bank of the River Tees, only a few miles from ICI's existing works at Billingham on the north bank. The joint Wilton works were to include cracking and processing plants for the manufacture of various chemicals including ethylene, ethylene oxide, ethylene glycol and isopropyl alcohol.

From the spring to the autumn of 1946 Project A was extensively discussed by a Joint Policy Panel, which was set up to provide a temporary point of contact between the Company and ICI pending the establishment of a more permanent joint organisation. Although it was intended as a vehicle for co-operation, the Panel quickly became bogged down by differences between the two companies on the scope of their proposed joint venture.

Specifically, ICI insisted that the joint company ought not to produce chemicals which were already being manufactured on a commercial scale by either of the parent companies, or which were made by processes involving information derived from the ICI–Du Pont agreement. These restrictions would, the Company complained, leave the joint venture with an extremely limited field of operation. The Company was particularly aggrieved that two of the chemicals originally included in Project A – ethylene oxide and ethylene glycol – were later withdrawn by ICI on the grounds that ICI already made them, although they did so only on a very small scale and by different processes from those which would be used in Project A.[26]

An outspoken critic of Project A was Douglas Smith, one of the Company's most senior technical men, who had been chief chemist at Abadan and then chief research chemist at Sunbury before he was transferred to New York in 1939.[27] He stayed there for the next six years, gathering information and liaising with US interests on all technical matters connected with refining and research. He thus saw at first hand the methods and processes used in the country which led the world in petrochemicals. In 1945, he returned to the UK to take charge of process development in refining. As secretary of the Joint Policy Panel, he was party to the talks with ICI and by September 1946 he was deeply unhappy with Project A. It had, he wrote, been 'reduced to a shadow of its former self and, in fact, . . . is now mainly a cracking plant to supply ICI with the gases they require . . . it is undoubtedly regarded by ICI simply as a cheap source of ethylene'. ICI, he went on, 'have succeeded in making the scheme most unattractive to us'.[28] In November, Fraser was fully briefed on the situation and wrote to John Rogers, one of ICI's deputy chairmen, expressing his concern that ICI's restrictions would narrow the scope of Project A 'very drastically'.[29]

Shortly after that, another difference emerged between the two companies. By the end of 1946 the Company was questioning its traditional policy of locating new refining plant close to the sources of crude oil production. It was thinking, instead, of erecting additional refining capacity in market locations, including Britain where schemes were afoot for the expansion of its refineries at Llandarcy and Grangemouth and the erection of a new refinery in the Thames area.[30] These plans opened up new possibilities of the Company doing in Britain what other oil companies had for some years been doing in the USA, namely erecting a petrochemical plant adjacent to a refinery so that the supply of raw materials for chemical manufacture would be near at hand. Douglas Smith argued strongly that this would be a more economic arrangement than having a stand-alone petrochemical plant. His observations of the petrochemical industry in the USA had persuaded him that it would be 'most unusual for the production of chemicals to be divorced from a normal petroleum refinery' and that 'an isolated installation for the production of chemicals from petroleum must always be at a disadvantage compared with one alongside a large refinery'. Fraser took up Smith's arguments and conveyed them to Rogers at ICI, commenting that 'any venture on which we should embark should be started on a basis that will always be competitive with others, and I very much fear that the Wilton project does not answer to this description'.[31] In short, the Company believed that the most economical site for Project A was at one of its refineries, not at Wilton.

ICI rejected that argument[32] and after talks between Fraser and Rogers, a meeting of the Joint Policy Panel was called for 30 January 1947 to review the whole question of Project A.[33] The formal language of the minutes cannot conceal the disagreement, bordering on recrimination, between the two sides. The Company representatives complained about the restrictions which ICI wished to place on the scope of Project A and the choice of Wilton as a site. Their ICI counterparts were adamant that they would not agree to a joint venture which was free to manufacture chemicals already made by ICI and that ICI was already committed to erecting a petrochemical plant at Wilton. They said that 'the die was cast and there was no opportunity of altering their plans at this late date'.[34]

This effectively marked the end of Project A as a joint Company–ICI venture. As a matter of form, Fraser and Rogers exchanged letters in February, expressing their regret that the once bright hopes of a petrochemical partnership had come to nothing.[35] ICI duly went ahead on its own with the construction of a cracker and other chemical plants at Wilton. The ICI works there were officially opened in 1949 and the cracker started up in July 1951.[36]

THE PARTNERSHIP WITH DISTILLERS

While the Company's talks with ICI were taking place, Distillers was facing a crisis in its chemical interests. The competitive advantage of fermentation alcohol as a raw material for chemicals was precariously dependent on fiscal advantages, but these were swept away in the Chancellor's 1945 budget. The 'inconvenience allowance' was withdrawn with effect from the beginning of 1946 and at the same time import duty was remitted on oil used for the manufacture of chemicals in the UK.[37] These measures had long been expected, and before their announcement Distillers was already aware that it would have to move from molasses to oil as the raw material for its chemical interests if it were to remain competitive. With the passage of the 1945 budget, that move could no longer be postponed and it became imperative for Distillers to move decisively to secure competitive raw material supplies for its chemical business.[38]

To that end, in 1945 Distillers put forward the idea that it might participate with the Company and ICI in the joint cracking plant at Wilton which they were then considering.[39] With Fraser's agreement, ICI turned down the idea of Distillers' participation.[40] Seeking an alternative, Distillers turned to another oil company, Trinidad Leaseholds, but broke off negotiations because of doubts about the security of raw material supplies from that source.[41] By the autumn of 1946, with the possibilities of a partnership seemingly exhausted, Distillers was considering the construction of its own oil cracking plant at Hull, close to Distillers' existing chemical works. Oil feedstock for the cracker would, it was thought, be purchased on the open market.[42]

It happened at this time, entirely by chance, that C. E. Spearing, who had succeeded Coxon as head of the Company's refining operations, met Graham Hayman, a main board director of Distillers and chairman of its chemical subsidiary, British Industrial Solvents, on a passage home from the USA on board the *Queen Elizabeth* liner. They had informal discussions 'centred around the non-participation of the Distillers Company in the Wilton project'. Hayman had obviously taken Distillers' exclusion to heart, for he was 'of the impression that Distillers had been rather unfairly treated in not being allowed to participate equally with ICI in this scheme'.[43]

After arriving home, Spearing invited Hayman to an informal lunch at which he 'gained the impression that Distillers were determined to secure their own position at all costs, but they would greatly favour an invitation either to participate in the Wilton Project on equal terms with ourselves and ICI or else would consider joint action with AIOC on a similar scale to Wilton'.[44]

This was reported to Fraser, whose interest was evidently aroused, for barely a fortnight later he and Spearing met Hayman for more talks.[45] Fraser was at pains to point out that the earlier exclusion of Distillers from Project A at Wilton should not in any way interfere with friendly relations. He was open in expressing reservations about the Wilton project which, he said, 'suffered from the disadvantages inherent in a scheme which had no oil refinery alongside'. Hayman explained that Distillers had plans to erect a cracking plant at Hull which, he agreed, would suffer the same disadvantage of having no adjacent refinery. Would the Company, he asked, 'consider embarking with them [Distillers] on a scheme to utilise refinery gases from Grangemouth'?[46]

Here, in embryo, lay the possibility of a joint Company–Distillers petrochemical plant alongside the Grangemouth refinery, as an alternative to the Company–ICI scheme for Project A at Wilton.

At the end of this seminal discussion with Hayman, Fraser promised to raise with ICI the possibility of Distillers' participation in the Wilton project or some similar scheme.[47] If he did so, no record of it can be found in the surviving papers and in any event it was by then probably too late to alter the course of the Company's talks with ICI, which were, as seen, on the verge of terminal breakdown.

When the ICI talks were finally called off early in 1947, the Company and Distillers lost no time in making an alliance. After a series of meetings between representatives of the two companies in March, principles of agreement were drawn up and signed in July by Hayman and Neville Gass, one of the Company's managing directors.[48] The agreement was designed to combine the Company's plentiful and cheap supply of raw materials with Distillers' technical know-how and access to chemical markets. This was to be achieved by forming a joint company named British Petroleum Chemicals, in which each parent would have a 50 per cent interest. British Petroleum Chemicals was to erect at Grangemouth cracking and processing plants to produce petrochemicals, mainly ethyl alcohol (ethanol) and isopropyl alcohol (isopropanol) in the first instance. The petroleum feedstock for the cracker was to be supplied by the Company and British Petroleum Chemicals' products were to be sold through Distillers' sales organisation.[49]

This scheme was notably free of the features to which the Company had objected in the earlier proposal for Project A with ICI at Wilton. The plant was to be erected adjacent to a refinery, as the Company favoured; there were no restrictions (such as the ICI–Du Pont agreement had imposed) on the exchange of technical information; Distillers did not demand, as ICI had, that the joint company should stay out of the manufacture of products which would compete with Distillers' established range. 'This understanding', stated the memorandum of agreement, 'is in no way restrictive'.[50] On this most promising basis, British Petroleum Chemicals was incorporated on 14 October 1947.[51]

GROWTH THROUGH JOINT VENTURES

When British Petroleum Chemicals was formed, the petrochemical industry in the UK was in its infancy. Over the next two decades it grew to maturity at an astonishing pace, easily outstripping the general rate of economic growth. Apart from cyclical ups and downs, which often confounded investment planners, petrochemicals enjoyed a virtuous circle characteristic of new high-growth industries. Great increases in demand stimulated the construction of plants of ever larger size and the search for new products and processes. The resultant economies of scale, combined with a high rate of technological innovation, brought down the costs and increased the range of synthetic materials, sold at prices which more and more people could afford. This stimulated further increases in demand, turning the wheel for further rounds of expansion.

The bright prospects of petrochemicals held strong appeal for the Company. There was not only the hope of making profits directly from chemicals, but also the opportunity to gain a new outlet for oil. This would help to relieve the Company's longstanding problem, only temporarily reversed during the Iranian crisis of 1951–4, of having an excess of production capacity over market outlets. By the 1950s that general problem was compounded by the growing imbalance between the products yielded by the Company's refineries and the demands of the market. Since World War II, shortages of coal in Europe had accelerated the long-term trend for heavy oil products, mainly fuel oil, to be used in substitution for coal. As a result the consumption of heavy products had tended to outgrow that of light products such as gasoline, which as a result were produced in quantities surplus to requirements. Looking ahead, the Central Planning Department predicted that this trend would continue, raising the problem for the Company of disposing of surplus light distillates.[52] Petrochemicals, which could be manufactured by cracking the light petroleum distillate, naphtha, therefore offered a particularly desirable outlet.

Though the Company lacked technical and marketing know-how in chemicals, it possessed other attributes needed for success in petrochemicals. One of the most important was size. The economies of scale available from increases in petrochemical plant size could only be realised by massive investments, which, together with the high costs of research, could only be afforded by very large companies. The industry therefore came to be dominated by big firms, which became increasingly multinational as they sought access to foreign markets. Their operations were generally integrated vertically in great chemical complexes which were usually, but not invariably, clustered around oil refineries, the source of their main raw material.

The development of the large petrochemical companies which dominated the industry usually followed one of three paths. One was for oil companies to integrate forward by constructing their own chemical plants; a second was for chemical companies to integrate backwards to secure supplies of raw materials; a third was for oil companies and chemical companies to form

joint ventures, in which as a general rule the oil company supplied the feedstock, the partners co-operated in basic chemical manufacture, and the
chemical company took the chemical products for further processing and
sale. In the UK, these three models could be seen working in practice in the
three firms which emerged as the largest petrochemical concerns. They were
Shell, which integrated forwards into petrochemicals at Stanlow and
Carrington in the Mersey and Manchester districts, and at Shellhaven on the
Thames; ICI, which constructed a great petrochemical complex at Wilton,
and integrated backwards into oil cracking and, later, oil refining; and British
Petroleum Chemicals, in which the Company and Distillers were equal partners.[53] When British Petroleum Chemicals was formed it remained to be seen
how these different types of organisation would fare in one of the most
dynamic sectors of the UK economy.

For the Company and Distillers, the start was slow. Four years elapsed
between their 1947 agreement and the commissioning of British Petroleum
Chemicals' first plants at Grangemouth. Apart from the usual and inevitable time lag between a decision to invest and commissioning, various factors
caused delay. Most notably, there was the need to obtain British government
approval for, among other things, allocations of dollars and steel, both of
which were at the time in extremely short supply.[54] Eventually, in mid-1951,
the first plants came into operation. By cracking petroleum distillate supplied
from the Grangemouth refinery, they produced two olefins, ethylene and
propylene, and converted them into ethanol (by a process licensed from
Shell) and isopropanol.

For the rest of the 1950s, British Petroleum Chemicals, renamed British
Hydrocarbon Chemicals (BHC) in 1956, expanded its capacity at
Grangemouth rapidly by major plant additions (see table 14.1). The first
came in 1953, when a plant came on stream to produce styrene monomer, a
raw material used mainly for polystyrene and synthetic rubber. The plant
was owned by a new associate company, Forth Chemicals, owned two thirds
by British Petroleum Chemicals and one third by the UK subsidiary of
Monsanto, a major US chemical company. The styrene monomer was made
by a Monsanto process, using ethylene supplied by British Petroleum
Chemicals and benzene purchased from other sources. Most of the styrene
was sold under contract to Monsanto for conversion into polystyrene.[55]
Three years after Forth's first styrene plant came on stream a further plant
was added to increase capacity substantially.

Meanwhile, in January 1955 another associate company was added to the
British Petroleum Chemicals group with the formation of Grange Chemicals,
owned two thirds by British Petroleum Chemicals and one third by the

Table 14.1 *British Hydrocarbon Chemicals' plants at Grangemouth, 1951–1960*

Plant	Date on stream	Initial capacity (tons a year)
No. 1 ethylene	1951	30,000
Isopropanol	1951	25,000
No. 1 ethanol	1951	33,600
No. 1 styrene	1953	8,000
Detergent alkylate	1955	12,000
No. 2 styrene	1956	16,000
No. 2 ethylene	1956	30,000
No. 2 ethanol	1956	33,600
Butadiene	1956	8,200
Tetramer	1957	20,000
Cumene	1959	29,000
Polyethylene	1959	10,500
Phenol	1959	13,000
No. 3 ethylene	1960	70,000

Oronite Chemical Company of California, a subsidiary of Socal. Grange Chemicals' plant, sited at Grangemouth, was completed at the end of 1955. It used an Oronite process to make dodecyl benzene (detergent alkylate) via propylene and propylene tetramer. Detergent alkylate was an intermediate for the manufacture of synthetic detergents. Grange's main customers included the soap maker, Thomas Hedley (part of Procter and Gamble), and Unilever.

To meet the needs of these associated companies, and to satisfy other demands, the expansion of British Petroleum Chemicals' cracking capacity became necessary. In late 1956 a second cracker, known as the No. 2 ethylene plant, started up and a No. 2 ethanol plant was also added. Some of the extra ethylene was supplied to Gemec, a subsidiary of Union Carbide, which commissioned a new polyethylene plant at Grangemouth in 1957–8. British Hydrocarbon Chemicals (as British Petroleum Chemicals had been renamed) meanwhile constructed a new plant to extract butadiene from the gases produced by the two crackers. It was commissioned in late 1956, and its output was sold to manufacturers of synthetic rubbers and other polymers. This was followed in March 1957 by a new unit to make propylene tetramer, one of the main feedstocks for the detergent alkylate produced by Grange Chemicals.

50   British Hydrocarbon Chemicals' plant at Grangemouth

In 1959 further new plants were added to the growing Grangemouth petrochemical complex. In November, another propylene-using plant was completed, this time for the manufacture of cumene, partly for use in another new plant which made phenol and acetone by a Distillers process. Also in 1959, British Hydrocarbon Chemicals started up its own 'Rigidex' polyethylene plant at Grangemouth, using a process licensed from Phillips Petroleum of the USA.

With these plants coming on stream, yet more cracking capacity was required. It was provided by the new No. 3 ethylene plant which was commissioned in June 1960. With the capacity to produce some 70,000 tons of ethylene a year, it doubled British Hydrocarbon Chemicals' ethylene capacity and was at the time the largest plant of its kind outside the USA.[56]

As a result of these developments Grangemouth became a major petrochemical complex (see figure 14.1). The expansion of production was dramatic, with output of all products rising approximately tenfold in the first ten years of operation (see figure 14.2). British Hydrocarbon Chemicals' profits also grew rapidly, though they remained very small in the context of the Company as a whole (see tables 14.2, 1.2 and 5.2).

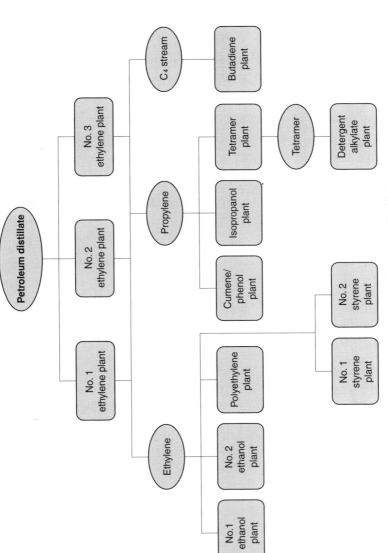

Figure 14.1 British Hydrocarbon Chemicals' plants at Grangemouth, 1960

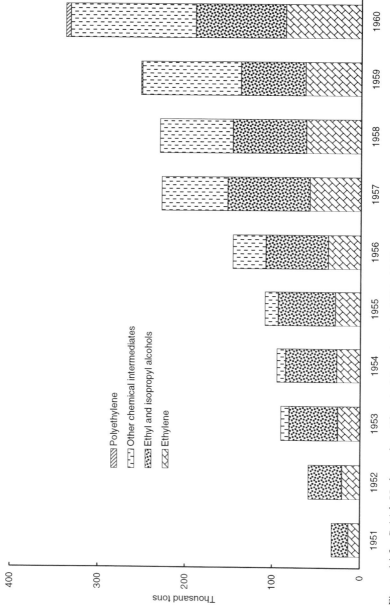

Figure 14.2 British Hydrocarbon Chemicals' production, 1951–1960

Table 14.2 *British Hydrocarbon Chemicals' profitability, 1951–1960*

Year starting in April	Pre-tax profits (losses) (£millions)	Capital employed (£millions)	Return on capital employed
1951	(0·7)	6·8	(10%)
1952	0·2	7·6	3%
1953	1·0	7·0	14%
1954	0·9	6·4	14%
1955	1·2	9·5	13%
1956	1·7	12·2	14%
1957	4·1	13·4	31%
1958	4·2	15·6	27%
1959	3·1	19·4	16%
1960	3·5	23·4	15%

Despite the great expansion of Grangemouth as a petrochemical complex, the unabating growth in demand for synthetics called for still more capacity in basic chemical manufacture. However, BP was against constructing a fourth cracker at Grangemouth because it would not be able to find outlets within economic reach of Grangemouth for the liquid oil by-products. It was therefore decided to establish a second petrochemical complex at Baglan Bay, a few miles from BP's refinery at Llandarcy in South Wales. This decision was announced in October 1960.[57]

The Baglan Bay petrochemical complex was based on a cracker, converting naphtha feedstock from the nearby Llandarcy refinery into gases from which the olefins, ethylene, propylene and butadiene, were separated. Some of the ethylene was transferred to another British Hydrocarbon Chemicals plant on the same site, where it was combined with chlorine (purchased from another source) to make ethylene dichloride, an intermediate used for the manufacture of vinyl chloride monomer, which in turn could be polymerised to polyvinyl chloride (PVC). The remainder of the ethylene was transferred to a new Forth Chemicals plant at Baglan Bay for the manufacture of styrene monomer. Finally, in the initial development of Baglan Bay a butadiene plant was constructed to extract and refine butadiene for sale mainly to manufacturers of synthetic rubber.[58]

The Baglan Bay complex was opened in October 1963. The cracker had the capacity to produce 60,000 tons of ethylene a year, which, combined with the three Grangemouth crackers, gave British Hydrocarbon Chemicals a total ethylene capacity of 190,000 tons a year. This made British Hydrocarbon

Chemicals the largest producer of ethylene in Britain, ahead of ICI, Shell and other firms at that time.[59]

While BP expanded its UK petrochemical operations, it also became interested in overseas ventures, particularly in France and Germany. Before World War II the French chemical industry was concerned mainly with inorganic chemicals and fertilisers and was completely overshadowed by Germany in organic chemicals. However, soon after the war French chemical companies began to show more interest in hydrocarbon chemistry. As early as November 1945, two French firms, Péchiney and Kuhlmann, began examining the possibilities of joining with an oil refiner to manufacture petrochemicals.[60] They approached the Company's French associate, Société Générale des Huiles de Pétrole (later Société Française des Pétroles BP, in which the Company had a 70 per cent share), and negotiations proceeded on the formation of a joint company named Naphtachimie. The essence of the scheme was that Naphtachimie would construct a plant for the manufacture of petrochemicals alongside SGHP's refinery at Lavera, near Marseilles. The refinery would supply the feedstock for the petrochemical plant and the chemical products would be sold by Naphtachimie, with liquid petroleum by-products of the cracking process being sold back to SGHP.[61]

Naphtachimie was duly formed in 1949, owned in equal shares by Péchiney, Kuhlmann and SGHP. The plants decided upon for the initial installations were substantially smaller than the first British Petroleum Chemicals plants at Grangemouth. There was to be a cracker with the capacity to produce 10,000 tons of ethylene a year, units for the production of ethylene oxide (10,000 tons a year), isopropanol (10,000 tons a year) and acetone (5,000 tons a year), and further units for the conversion of ethylene oxide to glycols, ethers and ethanolamines. Construction began in 1951 and the first plants started production in 1953. In 1955 board approval was given to raise Naphtachimie's ethylene capacity to 18,000 tons a year, ethylene oxide capacity to 14,000 tons and ethylene glycol capacity to 20,000 tons. That was followed in 1957 by approval of a further expansion which involved, among other things, the construction of a larger cracker with the capacity to produce 30,000 tons of ethylene a year. Commissioned in 1960, the plant was less than half the size of the new British Hydrocarbon Chemicals cracker which came on stream at Grangemouth in the same year. While these expansions were taking place at Naphtachimie, Kuhlmann withdrew from the joint venture, leaving Péchiney with a majority shareholding and SGHP with 43 per cent of the shares.[62]

In Germany, meanwhile, the victorious Allies were determined to curb Germany's war-making potential. To that end, they agreed that IG Farben, which had made a large contribution to the German war effort, should be broken up. The dissolution was achieved mainly by reconstituting the old firms of BASF, Hoechst and Bayer as independent companies in 1951–2.[63] These three were by far the largest companies in the West German chemical industry and soon began to look to petroleum rather than coal as their main raw material.[64]

At the same time, BP became interested in entering into petrochemicals in West Germany. By the mid-1950s it had the same problem in that country as elsewhere, namely that the balance of demand for oil products was resulting in production of surplus light products. In the search for new markets for these, it was decided that contact should be made with West German chemical firms. However, after offering petroleum feedstock to various chemical companies, BP concluded by mid-1956 that if it wanted to obtain secure outlets in the chemical industry it would have to enter into the production of basic chemicals itself. This could be done either by constructing its own crackers to produce olefins for sale to chemical companies, or by entering into a joint venture with a major chemical firm which would provide a ready outlet for the chemicals produced.[65]

A detailed investigation of these alternatives came firmly to the conclusion that it would be a mistake for BP to make an independent entry into petrochemicals in West Germany. The reasons were several. For one, the major German chemical companies held strong positions and would 'offer formidable competition to a new independent company'.[66] Secondly, the main German companies did not normally favour purchasing more than marginal quantities of basic chemicals from other companies. This effectively meant that independent operations by BP could not be limited to the basic chemicals and would have to extend to the manufacture of finished products such as plastics, which would have to be sold in competition with established firms. This would be difficult, especially as BP had no chemical sales and service organisation, which would have to be built up over time. Thirdly, to avoid competition and patent infringement in existing products, research on new products and processes would be needed. There was no guarantee that this would be successful, especially in the short term. Without ready-made marketing and technical know-how the establishment of an independent operation would thus take time, which BP could not afford, as the West German petrochemical industry was growing extremely fast and it was essential to gain 'a strong position quickly'. The investigation concluded unequivocally that 'collaboration with a German chemical company of standing is the best way of achieving this . . . on the general lines of our association with the Distillers Company in the UK, or with Péchiney in France'.[67]

The number of eligible partners was, however, decidedly limited. BASF could be ruled out, as it had already come to terms with Royal Dutch-Shell for a joint petrochemical venture named Rheinische Olefinwerke GmbH.[68] Hoechst had also been in talks with Royal Dutch-Shell and was reportedly 'extremely picqued' when Shell teamed up with BASF.[69] After that rebuff Hoechst had decided to go into petrochemicals on its own.[70] BP also had discussions with Hüls, another former part of IG Farben, but with no result as Hüls decided to collaborate with Gelsenberg AG and Scholven AG, two companies operating refineries and already having ownership links with Hüls.[71] This left only Bayer, to whom BP made overtures with little hope of a positive outcome as it was known that Bayer was having talks with Standard Oil (NJ) regarding possible collaboration in petrochemicals.[72] However, in November 1956 these talks broke down and Bayer quickly turned to BP as a possible partner.[73]

By spring 1957 representatives of the two firms had worked out a scheme of collaboration, subject to approval by the boards of the parent companies. It provided for the formation of a new German company, owned 50 per cent by Bayer and 50 per cent by BP, to construct petrochemical plants at Dormagen, near Cologne in the Ruhr. BP would have the exclusive right to supply the petroleum feedstocks and would be responsible for disposing of the liquid petroleum by-products of the new plants. The chemicals to be manufactured would generally be large-tonnage basic chemicals, using processes which for the most part had been developed by Bayer. About half of the chemicals produced would be purchased by Bayer for further processing in its own works, and the remainder would be marketed by Bayer as selling agent for the joint company.[74] This scheme was very similar to BP's existing joint ventures in chemicals in the UK and France.

After the outline of the German scheme had been drawn up, John Pattinson, one of BP's managing directors, visited Bayer in early May and was highly impressed by 'the great size of the organisation . . . and its vast ramifications in the chemical manufacturing business'.[75] He promised Bayer that he would discuss the whole project with his colleagues soon, writing enthusiastically that 'we will proceed on the lines which we discussed, I hope without any delay'.[76] Maurice Bridgeman, another BP managing director, shared his enthusiasm, noting 'I don't think we should miss the opportunity . . . Let us work for a case to the Board on the 23rd provided no snags arise'.[77]

A case was duly prepared and approved by the board on 23 May, signalling that detailed negotiations were to proceed.[78] In September, formal agreement was reached for the formation of the joint company, named Erdölchemie GmbH, and funds were approved for the construction of plants to crack petroleum feedstock and manufacture a range of petrochemicals including ethylene oxide, propylene oxide, ethanol, isopropanol and butraldehydes.[79] BP's shareholding in Erdölchemie was to be held through its German subsidiary, BP Benzin und Petroleum AG. The initial two cracking plants were to have the capacities to produce 15,000 and 30,000 tons of ethylene a year respectively. More detailed agreements followed in March 1958 and over the period 1958–60 the first Erdölchemie plants came into operation. At the same time, BP constructed a new refinery at Dinslaken in the Ruhr, from where feedstock was transported by pipeline to Erdölchemie's works at Dormagen.[80]

While BP was expanding in petrochemicals overseas through Naphtachimie and Erdölchemie, Distillers was also developing its own chemical activities outside British Hydrocarbon Chemicals. After acquiring British Resin Products and 50 per cent of BX Plastics before World War II, Distillers had also purchased a 50 per cent share in the plastics business of F. A. Hughes Ltd in 1941.[81] In 1947 Distillers acquired the remainder of Hughes' share capital and merged Hughes with British Resin Products, moving their combined manufacturing activities to Barry in South Wales.[82]

Meanwhile, in 1945 Distillers and the US chemical company, B. F. Goodrich, jointly formed British Geon Ltd to produce PVC. This company was 55 per cent owned by Distillers and 45 per cent by Goodrich, the leading US producer of PVC. Goodrich contributed its know-how in PVC manufacture and the brand name, 'Geon', while Distillers put in its technical knowledge of a PVC paste which it had developed in World War II for uses such as the coating of textiles to make artificial leather cloth. British Geon's first plant, located at Barry, started up in 1948. The new company was immediately successful and quickly captured some 40 per cent of the UK market, the only other British PVC producer at that time being ICI, which started production in 1942.[83]

The success of the British Geon partnership to produce PVC encouraged Distillers to make similar arrangements for polystyrene manufacture. Distillers had been interested in polystyrene since before the war, when a small pilot plant for polystyrene manufacture had been erected at Tonbridge in Kent. In the early years of the war the Tonbridge plant kept up production, but it was later shut down as large quantities of polystyrene became available for import from the USA. Distillers nevertheless retained its polystyrene trademark, 'Distrene' (short for Distillers' polystyrene), and restarted production at Barry after the war ended. The Barry plant was, however, small in scale, high in costs and its product was not of the highest quality. When it was put out of action by an explosion in 1953, Distillers decided to form a partnership with the US company, Dow Chemicals, in polystyrene.[84]

Dow was a pioneer in polystyrene, having been the first company to start commercial production of styrene monomer and polystyrene in about 1930. Its polystyrene brand name, 'Styron', had been widely advertised throughout the world and had achieved an enviable reputation. Indeed, it was the benchmark against which the quality of other polystyrenes was measured. Dow had in the past shown no desire to invest directly overseas, but underwent a change of heart because its export business from the USA was threatened by the postwar shortage of dollars, which made it difficult for foreign customers to pay for US goods. In 1954 Dow therefore joined with Distillers in forming Distrene Ltd, in which Dow had a 45 per cent share and Distillers a 55 per cent share (changed to 50 per cent each in 1963). A new polystyrene plant, which took styrene monomer supplies from Forth Chemicals, was constructed at Barry and came on stream in 1955. It rapidly became profitable, with Styron 475, a high-impact grade, gaining a leading position in the UK market.[85]

The expansion of Distillers' plastics interests did not end with Distrene. As seen, Distillers had acquired 50 per cent of BX Plastics, a subsidiary of British Xylonite, before the war. In 1961, hearing rumours that Courtaulds was contemplating buying the rest of British Xylonite, Distillers moved first and bought the shares that it did not already own so that it had 100 per cent ownership.[86] The next year Distillers reached agreement with Union Carbide to form a 50:50 joint company in which British Xylonite was merged with Union Carbide's majority shareholding in Bakelite Ltd and with the polyethylene operations at Grangemouth owned by the Union Carbide subsidiary, Gemec. At the same time an offer was made and accepted for the publicly held minority shareholding in Bakelite Ltd so that the whole of this company was absorbed into the Distillers-Union Carbide joint venture, which was named Bakelite Xylonite Ltd from 1 January 1963. It was one of the largest plastics companies in Europe, with interests in the manufacture of, among other things, polyethylene, polystyrene, PVC, phenolics and nitro-cellulose.[87]

The outcome of these developments was that Distillers held extensive chemical and plastics interests outside its partnership with BP in British Hydrocarbon Chemicals. These included Distillers' wholly owned chemical operations, centred at Hull and organised in British Industrial Solvents; the wholly owned plastics operations of British Resin Products; and plastics partnerships with three major US companies in British Geon, Distrene and

51 Man using a dip to check the level after filling a British Industrial Solvents' road tanker at Hull in 1950

Bakelite Xylonite. While these were the most important of Distillers' chemical and plastics interests, there were also others of less significance, both overseas and in Britain, which are unmentioned here to avoid excessive complication.[88] An overall view of the main chemical and plastics interests of BP and Distillers is shown in figure 14.3.

Through this complicated pattern of partnerships Distillers built up a group of chemical and plastics interests which were linked not only financially, but also operationally in the physical movement of products between plants, from the cracking of petroleum feedstocks to the manufacture of finished plastic goods.[89] To give added cohesion to this vertically integrated operation, in 1963 Distillers reorganised its chemical and plastics interests by merging them in a new Chemicals and Plastics Group under the chairmanship of George Ashford, who was also a main board director of Distillers and a member of the internally powerful Management Committee.[90]

Yet there were almost constant doubts in Distillers about the competitiveness of its partnership system, compared with the more unitary organisations

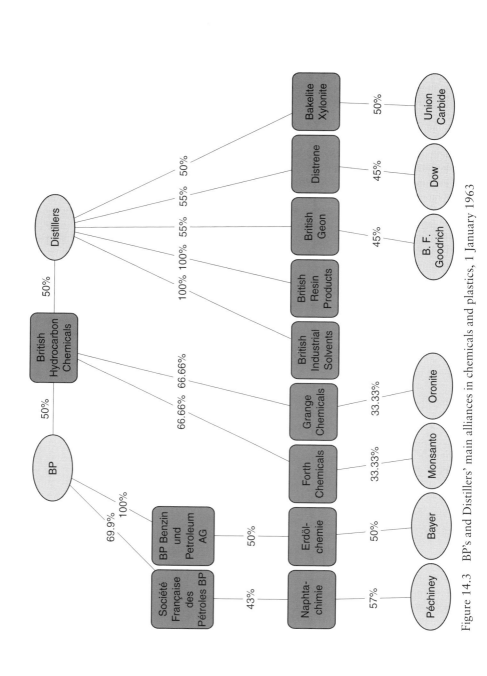

Figure 14.3   BP's and Distillers' main alliances in chemicals and plastics, 1 January 1963

52 Men refuelling Distillers' road tankers at Hull chemical site in the 1950s

of ICI and Shell in the UK. Partnerships had, it was acknowledged, some advantages. Notably, they allowed Distillers to enter into areas where it had no independent position and to gain access to brand names and technical know-how across a wide range of chemicals and plastics developed by world leaders in specific sectors. This enabled Distillers to build a substantial stake in the industry in a shorter time and at lower expenditure than would otherwise have been possible. The partnership system also economised on managerial resources, in the sense that a large central organisation was not required. In Ashford's words, it 'provided a framework which has allowed a whisky company to manage successfully an interest in an unrelated field'.[91]

However, the system of partnerships also had shortcomings which, as time went by, tended to be seen as outweighing the initial advantages. A major problem was that as the number of partnerships grew there were increased overlaps, which gave rise to conflicts of interest, distrust and fears

Treat it rough!

It's made with GEON PVC

Survival of the fittest is the rule for bargain-hunters in pursuit of Neohide bucketbags and handbags. Neohide is tough Geon PVC on the outside, soft suede-like material on the inside. Made into a handbag it provides the dual texture demanded by the discerning shopper... a lustrous finish and a luxurious touch.

In the rough and tumble of everyday wear, handbags made with Geon PVC, the waterproof long-wearing plastics material, outlast those made with conventional materials.

Geon PVC is widely used in the production of leather-goods. For further information about Geon PVC please write for descriptive booklet No.   free on request.

"Neohide" by Storey Brothers and Co. Ltd.; bucketbag by Peter Black (Keighley) Ltd.

'Geon' is a regd. trade mark

BRITISH   GEON   LIMITED

53   'Treat it rough!' Advertisement for Geon PVC in 1957

54 Petula Clark, the film and TV star, at the wheel of her car in the 1960s. The upholstery, interior trim and hood were all in Geon PVC leathercloth

of competition between Distillers' different partners. For example, in the 1950s British Xylonite (before it became part of Bakelite Xylonite) was unhappy that Distillers was developing a competing interest in British Resin Products.[92] When Distillers acquired 100 per cent ownership of British Xylonite in 1961, Dow complained that this would give Distillers an incentive to give British Xylonite priority in its polystyrene plans.[93] Later, when Distillers joined with Union Carbide to form Bakelite Xylonite and thereby acquired a stake in Union Carbide's polyethylene plant at Grangemouth, BP complained that this was a potentially competing interest with British Hydrocarbon Chemicals' 'Rigidex' polyethylene.[94] BP's displeasure was increased by the fact that Distillers did not inform it about the plastics merger with Union Carbide until 'literally a couple of hours before the press release'.[95]

The effects of overlapping went deeper than mere occasional bickering. It was thought to have 'an extremely bad psychological effect on staff at all levels who often come to regard other companies within the Group not as

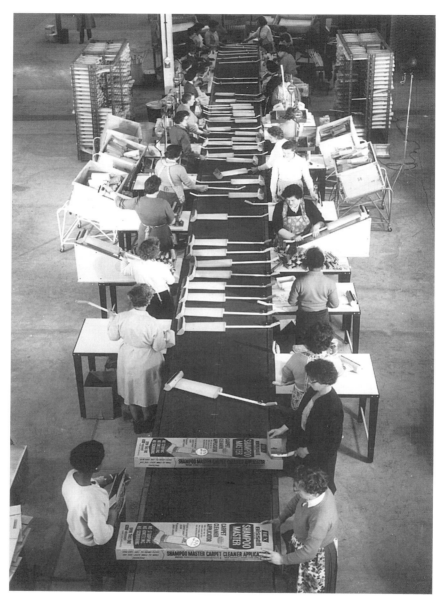

55 Assembly line for Bex-Bissell carpet cleaners in the plastics fabrication works of the British Xylonite Company at Highams Park, London, in 1961

collaborators but as dangerous competitors'.[96] It also had the effects of duplicating staff, preventing the maximum exploitation of economies of scale, hindering the rationalisation of parallel operations and confusing customers. The competitive power of the group as a whole was reduced by the lack of free exchange of market information and technical know-how. Some of the advantages of vertical integration were lost on account of sometimes difficult negotiations on transfer prices between different parts of the group. In general, negotiations between partners were liable to lead 'to messy compromises on location, process and scale'.[97]

As the Distillers Chemicals and Plastics Group grew it also became increasingly difficult to manage. In the words of T. F. Anthony Board, one of Distillers' directors, 'the multiplicity and varied nature of these partnerships present a problem in control and co-ordination which is quite exceptional'.[98] There were complaints about the amount of management time that was taken up with negotiations on transfer prices and conflicts of interest, diverting attention from the primary aim of keeping ahead of outside competitors.[99] The management problem was nicely summed up by John Hunter, who was made a director of Distillers' Chemicals and Plastics Group when it was formed in 1963, with responsibility for liaison with Distillers' main chemical and plastics associates in the UK.[100] He later wrote, 'any company or group of companies needs a coherent and consistent overall policy and a clear definition of objectives which should be made known to all those responsible for operations. It is particularly important that this should be done in the case of jointly owned companies so that the operating management is not confused by the possibly differing objectives of the owning partners. Wholly owned companies are easier to manage than joint companies.'[101]

The most important and largest of Distillers' chemical and plastics partnerships was, without doubt, its alliance with BP in British Hydrocarbon Chemicals. As was seen earlier in this chapter, it expanded very rapidly in the 1950s. Yet, despite its great growth, British Hydrocarbon Chemicals remained comparatively undeveloped as an organisation. It had been brought into existence largely as a by-product of its parents' core businesses, which lay in oil (BP) and potable spirits (Distillers). In Ashford's words, written in 1960, 'BP is fundamentally an oil company . . . BP's top management still thinks of petroleum chemicals mainly as a way of selling petroleum'. On the other hand, he went on, Distillers 'is primarily a Scotch whisky company. Its second love is gin and its third chemicals and plastics'.[102]

As an intermediary organisation, acquiring its feedstock from BP and selling its chemicals output through Distillers, British Hydrocarbon Chemicals was, as Anthony Board described it, a 'truncated corpse', with no selling or research departments or commercial management of its own.[103] Its lack of a selling organisation was seen as a drawback, for as Ashford commented, 'when production and sales do not come under one management it is almost impossible to achieve maximum co-ordination'.[104] Similarly, with regard to research, Ashford thought that there was 'bound to be some disadvantage in lack of exchange between BHC and DCL Research Department unless they come under one management'.[105] In short, British Hydrocarbon Chemicals had only one function: production. It was not by any stretch of the imagination a fully fledged commercial organisation. Ashford, for one, doubted whether such a company, which bore 'all the marks of compromise', could 'develop the kind of dynamism which is essential in an increasingly competitive industry'.[106]

Doubts about British Hydrocarbon Chemicals' competitive strength were fuelled in the 1950s by persistent concern about the rapid progress being made by other firms, particularly ICI and Shell, with their wholly owned vertically integrated operations. ICI developed its Wilton petrochemical complex on a very large scale and was easily the largest manufacturer of petrochemicals in Britain. Shell also achieved rapid growth and was vying with Distillers for second place in the UK chemical industry.[107] Faced with this competition, Distillers kept coming back to the question: would it not be better to develop a more unitary vertically integrated organisation in petrochemicals to compete with the likes of ICI and Shell?

THE LURE OF INTEGRATION

That question first came up in July 1953, only two years after British Petroleum Chemicals' first petrochemical plants started up at Grangemouth. It was raised by Board, who asked Harold Snow, one of the Company's managing directors, whether the Company might be interested in taking a 50 per cent share in Distillers' chemical activities outside British Petroleum Chemicals.[108] Board's reason for raising the subject was, he explained, that he was concerned about Distillers' competitive position compared with ICI and Shell. In particular, those two companies, being vertically integrated under unitary managements, could set their internal transfer prices (at which chemicals were exchanged between plants) so as to minimise tax on their overall operations. However, the prices at which chemicals were transferred from British Petroleum Chemicals to Distillers were market based, which could result in higher taxes being paid than would be necessary if they operated as a single integrated whole. Snow was unmoved by Board's argument and told him that it was the Company's policy 'to confine our "petroleum chemical" activities to the production of intermediates and not to become concerned with the finished chemicals field'. Moreover, Snow added, the Company already had sufficient capital commitments for some time ahead 'not to undertake new ones in the non-essential category'.[109] In other words, the Company saw its core business as oil and did not wish to venture further into chemicals than the bulk production of basic intermediates.

This attitude was tested again in autumn 1955, when Sir Graham Hayman (knighted in 1954) suggested to Pattinson that the partners in British Petroleum Chemicals would benefit from a 'fuller integration' of the two companies' chemical interests. The exact form of Hayman's fuller integration was not defined, but his idea was broadly speaking to enlarge British Petroleum Chemicals to include not only the production of basic chemicals, but also Distillers' chemical interests further downstream. His reasoning was that 'in the growing field of petroleum chemicals, a fully integrated business . . . was generally likely to yield better results than the present arrangement. The Shell's organisation in this respect is a case in point.' Pattinson could see the force of this argument. He noted that 'in favour of the proposal is an undoubted advantage for an organisation to be able to take petroleum products and process them right through to a stage where they appear in the very numerous forms (hundreds, perhaps thousands) of chemicals of which petroleum is the only or main constituent. It can be seen that the flexibility of manufacture and marketing could be great. The selling price of these chemicals is high, frequently very high, compared to petroleum and although the unit manufacturing costs are also high compared to oil refining, a good profit can be made.' He promised Hayman that he would give the matter consideration.[110]

Pattinson duly discussed it with BP's top directors, Fraser (chairman), Jackson (deputy chairman) and Gass. 'They all felt', Pattinson told Distillers, 'on present information that BP should not go so far into the production of chemicals for distribution in the form of large numbers of fine chemicals in comparatively small quantity . . . BP was accustomed to bulk business, and whereas they felt they knew the BPC [British Petroleum Chemicals] present business, they would be completely out of their depth in trying to exercise any control over a business such as BIS [British Industrial Solvents]'. For those reasons, BP would not be prepared to consider further 'such far-reaching proposals as those broadly outlined by Sir Graham Hayman'. Pattinson also reiterated the point made by Snow in 1953, that BP's capital was fully committed to the development of its core oil business.[111]

The subject of integration would not, however, go away. Four years on, in February 1959 Distillers raised it again, suggesting a scheme that was more ambitious than earlier ideas. The proposal this time was to form a new joint company in which would be merged all of Distillers' chemical and plastics interests, British Hydrocarbon Chemicals, and BP's interests in Naphtachimie and Erdölchemie. In short, a worldwide amalgamation of all BP's and Distillers' chemical and plastics operations into an autonomous vertically integrated organisation with its own production, selling and research functions.[112]

Three factors appeared to lie behind this far-reaching scheme. One was that Distillers was feeling the need to develop internationally, particularly in continental Europe. Secondly, Distillers wanted to expand its partnership with BP because it could not alone find the capital to keep up with the fast pace of development in the petrochemical industry. Thirdly, there was, Pattinson noted, 'the need to make the fullest use of the advantages of a completely integrated operation' which would encompass sales and research as well as production. 'In short', wrote Pattinson, 'the only way to meet competition from others in the petroleum chemical field who are fully integrated, is to be fully integrated oneself. Shell Chemical Company is the obvious challenger and DCL are beginning to feel very disturbed by Shell's increasing activities. ICI are, by putting up their own oil cracking plants, also beginning to be a serious menace.'[113] This theme was a constant refrain throughout the negotiations on Distillers' proposals.[114]

These were discussed by BP and Distillers at the highest level, with Gass, who was by this time BP's chairman, declaring that he was 'deeply interested' in the scheme.[115] However, despite lengthy talks and much ingenuity in devising variations on the basic scheme, no way could be found around the main stumbling block, which was that BP would not agree to put its interests in Naphtachimie and Erdölchemie into the merger. Also, BP insisted on retaining its freedom to enter into other petrochemical ventures overseas without Distillers as its partner. There was nothing deliberately obstructive in this stance: BP simply felt that it could not tie itself to Distillers overseas because situations could arise where it would be necessary, for political or other reasons connected with its core oil business, to make alliances with other partners. As Snow and Pattinson (both deputy chairmen of BP by this time) explained to Distillers: 'Primarily, BP's policy was based on petroleum' and they could not promise even to consult Distillers on overseas petrochemical opportunities 'because BP would have to make decisions in relation to their world-wide petroleum policy'.[116] As Distillers was not prepared to go ahead with a merger that would effectively limit its petrochemical interests to the UK, the scheme was abandoned. There were no hard feelings on either side and both companies agreed at their last meeting on the scheme in August

1960 that their association in British Hydrocarbon Chemicals was 'most satisfactory to both sides and everything should be done to encourage its development'.[117] On that cordial note, discussions on expanding British Hydrocarbon Chemicals into a more integrated organisation receded. It would not, however, be long before ideas about integration came to the fore again.

15

Integration in petrochemicals

By the 1960s pronounced cyclical swings had become an unnerving feature of the fast-growing petrochemical industry. The basic reasons for their existence were not hard to find. Economies of scale encouraged, indeed required, firms to build very large plants if they were to be competitive. Because of their high capital costs, these plants had to be run at high throughputs if they were to be profitable. The combination of large plants and the need to operate at high throughputs tended to create a glut of supply over demand when new plants came on stream. This in turn put pressure on prices and profits, prompting a downturn in investment until rising demand put pressure on supply, prompting new investment and a repetition of the cycle.

The timing of the chemical cycle was coupled to wider macro-economic fluctuations in national and international economies. In the UK, the economic cycle peaked in 1951, dipped in 1952, peaked again in 1955, dipped again in 1956, peaked in 1960 and fell again in 1961–2.[1] The deflation of the early 1960s coincided with worldwide chemical overcapacity and price cutting. British Hydrocarbon Chemicals (BHC) could not escape the consequences and its profits in the year starting in April 1961 were lower than they had been since 1956.[2] However, BHC shared in the subsequent recovery, achieving record levels of profits and production in the mid-1960s (see table 15.1 and figure 15.1).

With the opening of the new Baglan Bay chemical complex in October 1963, BHC became the largest ethylene producer in the UK, but not for long. Encouraged by the economic upturn, chemical companies indulged in a capital spending spree, leap-frogging one another with new plants of ever larger size.[3] In 1963–4 Shell and ICI announced projects to construct new crackers on a scale not seen before in Europe. Shell's plant was to have the capacity to produce 150,000 tons of ethylene a year, while ICI's was to be

Table 15.1 *British Hydrocarbon Chemicals' profitability, 1961–1965*

Year starting in March/April	Pre-tax profits (£millions)	Capital employed (£millions)	Return on capital employed
1961	2·8	28·2	10%
1962	3·7	33·5	11%
1963	3·6	33·8	11%
1964	4·5	35·5	13%
1965	5·7	38·7	15%

still larger. When these plants came on stream Shell's total ethylene capacity would be 250,000 tons a year and ICI's would be 300,000 tons. These easily topped BHC's capacity of about 190,000 tons.[4]

In July 1965, before either of the new Shell or ICI crackers came on stream, BHC announced an even larger project to construct a fourth cracker at Grangemouth with an ethylene capacity of 250,000 tons a year.[5] The capacity of this plant alone was nearly twice the combined capacity of BHC's existing three Grangemouth crackers. Not to be outdone, ICI a few months later announced that it was going to build a giant new cracker at Wilton with an ethylene capacity of 450,000 tons a year.[6] This was fifteen times the size of BHC's first two crackers at Grangemouth, which had started up in 1951.

These massive increases in the scale of plants went hand in hand with the continuing trend towards vertical integration. ICI, for example, integrated backwards to secure its raw material supplies, announcing in 1964 that it was going into partnership with the US oil company, Phillips Petroleum, to build a large new oil refinery on the River Tees to supply feedstock for ICI's petrochemical operations.[7] The rationale behind ICI expanding from chemicals into oil could also be applied to BP expanding from oil into chemicals. In the words of Alastair Down, one of BP's directors, 'if ICI can profitably enter on oil operations with no sources of and no markets for oil of its own, simply because of the advantage of combining oil refining and chemical manufacture, it is unlikely that we, an integrated oil company, cannot do the same'.[8]

A step in that direction was taken in 1963, when BP, Distillers and ICI formed a new joint company, Border Chemicals, in which they held equal shares. Border was to construct a new plant at Grangemouth to use a Distillers process to make acrylonitrile, the raw material of acrylic fibre, used in clothing and other textile goods. The main raw materials for acrylonitrile

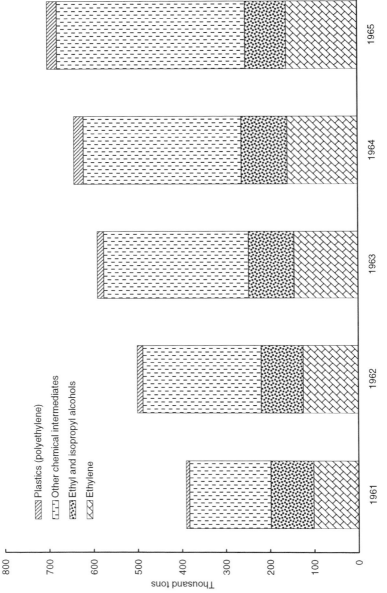

Figure 15.1 British Hydrocarbon Chemicals' production, 1961–1965

were propylene and ammonia, which were supplied by BHC and ICI respectively. Border's plant came on stream in 1965.[9]

In that year BP made a more adventurous move when it acquired, on its own, the UK plastics interests of the US oil company, Socony Mobil, and placed them in a new wholly owned subsidiary, BP Plastics Ltd. The acquired assets consisted mainly of a new polystyrene plant and plastics fabrication works at Stroud in Gloucestershire. A wide range of plastic articles was produced there by moulding and extruding polystyrene, PVC, polyethylene and cellulose acetate. The output included food containers, packaging, radio cases, refrigerator linings, motor car parts, curtain rails, light fittings, tool handles and box liners. In addition, casein blanks and sheets were produced for button manufacture. A smaller works at Wokingham in Berkshire was also included in the purchase from Socony Mobil. It turned acrylic sheet (purchased mainly from ICI) into a wide variety of items, mostly for light fittings.

The purchase of Socony Mobil's plastics interests appeared to fit with BP's policy of moving downstream in petrochemicals. As an internal report stated,

> When the BP Group branched out into petroleum chemicals it was right and natural that it should start from the provision of a suitable feedstock and turn this into the basic chemical intermediates. To a large extent it has been unable itself to progress beyond this point and has had to rely on its petroleum chemical associates (Distillers, Monsanto, Bayer, Péchiney) to take the basic intermediates and process them into marketable products. The reason for this is clear – they had the fabricating plants, the sales organisations, the brand names and the market outlets. To start from scratch and build up our own integration forward though not impossible has been judged too big a task in view of BP's other expansion commitments, but the lack of such a position has been keenly felt. Now we have an opportunity at one step to acquire a large modern plant, an experienced and well-trained staff . . . and a sales organisation with established outlets and internationally known brand names.[10]

This strategy of forward integration represented a departure from the policy of the early and mid-1950s, when BP had restricted itself to participation in the manufacture of basic chemicals and had been wary of going downstream towards the final consumer. By the mid-1960s, however, BP was following a course in petrochemicals which had obvious parallels with the development of its mainstream oil business. In both oil and chemicals the Company started out as an upstream producer, far removed from the end customer. In both, it looked largely to partnerships for downstream operations. In oil, the main downstream partner was Shell, with whom the

Company had joined forces in the Consolidated Petroleum Company in 1928, and in Shell-Mex and BP in 1932. In chemicals, the main partnerships were with Distillers in BHC, Bayer in Erdölchemie and Péchiney in Naphtachimie, all formed after World War II. In both oil and chemicals, however, BP sought to establish a more independent downstream position in the 1960s.

While the trends towards ever larger plants and vertical integration continued, so too did the concentration of the chemical industry in the hands of very large companies. This happened partly by the organic growth of firms like ICI and partly by mergers. In France, Péchiney, whose main business was in aluminium, and Saint Gobain, whose core business was glass, combined their chemical interests in a new 50:50 partnership, Péchiney-Saint Gobain in 1962.[11] Péchiney, however, retained separately its majority share in Naphtachimie, in which BP's French subsidiary, Société Française des Pétroles BP, held a 43 per cent stake (see chapter fourteen). In 1965–6 there were more large chemical mergers, of Ugine and Kuhlmann in France[12] and of Edison and Montecatini in Italy. These last two formed Montedison, the third largest chemical company in Europe (after ICI and Bayer) and the sixth largest in the world.[13]

In the UK, BP and Distillers had considered consolidating their chemical interests more than once in the 1950s. However, nothing had come of it and in August 1960 the idea of an international merger of BP's and Distillers' chemical interests had been shelved (see chapter fourteen). This left unresolved the problems of the partnership system, in which the parts were bolted together under different managements instead of being cast in a single, integrated mould.

Meanwhile, technological forces were pulling chemicals closer to oil than to drinks. Distillers had originally gone into making chemicals as a natural extension of its core drinks business. The two activities, based on industrial and potable alcohol respectively, shared the same basic technology. However, as petroleum rose to become the predominant raw material for organic chemicals in the UK, the use of industrial alcohol for this purpose declined, and Distillers' chemical and drinks businesses grew apart. This posed a dilemma for Distillers. Should it try to continue in both industries, and, if it did, could it afford the huge investments in ever larger plants which were required to be a major player in chemicals? Or should it concentrate on its core business and divest its chemical interests?

BP was in an opposite position. Chemicals were becoming more closely related to the oil industry. The remark that Jameson had made to ICI in 1943 – 'what is in the chemical field today may be in the petroleum field tomorrow' – had come true.[14] Moreover, for tax reasons which will be explained later, BP was actively seeking to invest in profit-making activities in the UK. Given these factors, it was inevitable that BP and Distillers would before long reopen discussions on their relationship in chemicals and plastics. The tone of the discussions, their timing and direction, were also, however, influenced by personalities on both sides.

THE BP–DISTILLERS MERGER

At the end of March 1963 Sir Graham Hayman retired after a remarkable career in which he had risen from army sergeant to become not only chairman of Distillers, but also chairman of the Association of British Chemical Manufacturers (1950–3), president of the Federation of British Industries (1955–7) and chairman of both BTR Industries Ltd and British Plaster Board (Holdings) Ltd. Virtually all his career at Distillers had been spent in the chemical and plastics side of the business, rather than in whisky and gin, and he had masterminded Distillers' entry into petrochemicals.[15]

His successor, Ronald Cumming, was quite different. His family had been associated with the Scotch whisky industry for four generations and his career with Distillers had been spent entirely on the drinks side.[16] The change at the top seemed, therefore, to point in an obvious direction. Cumming, having less interest than Hayman in chemicals and plastics, would surely be more likely to divest Distillers' petrochemical interests and concentrate on drinks.

That, however, was not the way that Distillers' policy was formed. Just two months after Cumming took over as chairman, Distillers' internally powerful Management Committee held a special meeting to review the company's future in chemicals and plastics. Cumming was present, and so too was George Ashford, the head of Distillers' Chemicals and Plastics Group. The fundamental issue discussed was what Distillers' future policy should be, now that its traditional production technology based on fermenting molasses was effectively obsolete. With that development 'it seemed inevitable', the minutes of the meeting recorded, 'that DCL's influence in the chemical world would wane. The chain of events was started by the formation of BHC and did not seem capable of reversal.'[17]

Ashford doubted whether this fatalistic vision had to be accepted. In 1959–60 he had supported the scheme (described in chapter fourteen) for Distillers and BP to merge all their chemical interests into an enlarged BHC. Now, however, he 'conceded that his previous view had been wrong' and declared that he was against Distillers giving up its independence in chemicals by putting its wholly owned chemical interests into an enlarged BHC. This, he believed, would leave Distillers in the position of a mere investor, a 'sleeping partner' in BHC, with no power to influence events. He suggested that instead Distillers should retain a nucleus of managerial, research, planning and commercial expertise in chemicals. This, he argued, could be 'valuable in sustaining the growth of . . . [Distillers'] chemical interests and maintaining DCL's status'. The Management Committee concurred and came to the conclusion that Distillers should retain its chemical and plastics interests.[18] The change in chairman from Hayman to Cumming did not, therefore have the result that might have been expected. There was no reduced commitment to retaining Distillers' chemical and plastics interests: if anything, the opposite.

BP, on the other hand, was growing keener on enlarging BHC by merging it with Distillers' wholly owned chemical interests (a 'roll-up', as it was called at the time). The lead in that direction came from Alastair Down, who had masterminded the establishment of BP's vertically integrated oil operations in Canada, where he had been posted in 1954 (see chapter four). In 1962 he was promoted to BP's board and returned to London to take up his new duties as a managing director. He was made responsible for petrochemicals, to which he brought the same unswerving sense of strategic purpose that had characterised his spell in Canada. In both cases, vertical integration was his aim. As mentioned, the purchase of Socony Mobil's UK plastics interests was a move in that direction. However, the possibility of merging Distillers' chemical interests into an enlarged BHC was a larger and more sought-after objective. It would take BP 'down the line' towards the market in chemicals and it would reduce the overlaps between BP and Distillers in chemicals and plastics.

Down was also keen to take fuller advantage of the system of double taxation relief, which allowed BP to deduct the foreign tax that it paid to the governments of oil-producing countries from its UK tax liability. As BP did not have a high enough UK tax bill to offset all its foreign tax, it had unused tax relief. This meant that income from petrochemicals (or any other activity in the UK) had the enormous attraction of being effectively tax-free. For Down, a chartered accountant, this was a powerful incentive to expand BP's UK chemical operations.

Down put the idea of a 'roll-up' to Ashford in mid-1963, but Ashford, adhering to Distillers' new policy, was against it.[19] By March 1964 Shell and ICI had announced their projects for huge new crackers and Down was worried that BP, tied to its partnership with Distillers in BHC, would be left behind.[20] However, the idea of a 'roll-up' still foundered, as Distillers stuck to the line that it wished to retain an independent capability in chemicals.[21]

The matter went to a higher level at the end of July, when Sir Maurice Bridgeman and John Pattinson had dinner with Cumming and Alexander McDonald (Distillers' deputy chairman) at Cumming's London flat. Bridgeman said that BP 'were somewhat disappointed by the progress of BHC, as they wanted to develop much faster in the petrochemical field than seemed to be envisaged . . . They saw on the one hand the Shell Company making progress over a wide field at a rapid rate and on the other ICI breaking into the oil refining business while still dominating the chemical field. They felt that BP, through its association with DCL, was now being hampered. They . . . were now very willing to consider the acquisition of a half interest in DCL's own chemical business.' Pattinson went even further, stating that 'they really wanted a petrochemical complex at every one of their refineries'. Cumming and McDonald responded with the now familiar argument that if Distillers' wholly owned chemical interests were merged into BHC, Distillers would become merely an investor, effectively a sleeping one, in the enlarged partnership with BP. In any case, they argued, Distillers did not want to reduce its investment in chemicals and was not, therefore, interested in receiving money in return for a share in its chemical business. If BP wanted a merger, it would have to contribute assets rather than cash to balance the assets put in by Distillers. At that, Pattinson expansively suggested that BP might take over Fisons and put it into the merger as its balancing investment.[22]

After that meeting some discussions were held 'through a third party' with Fisons, codenamed 'Z Company' in the records.[23] At the same time BP, disappointed by the talks with Distillers, considered going into petrochemicals on its own in the UK. Within BP, Pattinson stressed again that the growth of Shell and ICI in UK petrochemicals made it essential for BP's chemical interests 'to be considerably expanded, otherwise our position would grow progressively weaker'. There were, he went on, 'doubts whether BP could rely on the BHC route to provide the expansion they needed in their own interests unless some fundamental change in approach were to take place'. BP, he said, might have to set up its own large petrochemical complex and he asked for estimates to be prepared of the scale of plants that would be required if BP 'were to make a major entry by themselves into the UK market'.[24]

News of BP's thinking reached John Hunter, who worked under Ashford at Distillers.[25] He reported it to Ashford, with the suggestion that Distillers should be more open to compromise with BP. In particular, Hunter thought that Distillers should agree to merge its wholly owned chemical interests (located at Hull) into BHC without demanding a simultaneous balancing investment from BP. In a note appropriately entitled 'Partnership problems'

he explicitly warned Ashford that if BP were frustrated in its desire to enlarge BHC, it would turn elsewhere for new opportunities in petrochemicals.[26]

The thought of BP turning from partner into competitor must surely have worried Distillers and may well account for the adoption of a more compromising line by Distillers' top management. In any event, Hunter's suggestion was taken further when Cumming, McDonald and Ashford had dinner with Bridgeman, Pattinson and Down at Bridgeman's London flat on 26 November. The Distillers representatives said that 'the right thing seemed to be for BP to take a half share in Hull' without necessarily contributing balancing assets. They insisted, however, on some other conditions. The most important were that Distillers should manage the enlarged BHC partnership; that BHC's Rigidex polyethylene operations should be transferred to Bakelite Xylonite, the plastics company in which Distillers and Union Carbide were partners (see chapter fourteen); that the enlarged BHC should have opportunities to expand overseas and that BP would not compete with BHC by setting up its own chemical operations in Europe.[27] The BP representatives generally welcomed the suggested compromise and Bridgeman 'clearly appreciated that DCL were making a concession. Just at the moment he was not too clear what BP had to offer as a quo for DCL's quid.' It was agreed to form a joint working party to study the matter further.[28]

The working party duly met and in January 1965 Ashford reported bullishly to the Distillers' Management Committee that 'the conviction was growing on both sides that the operations at Hull and BHC should be merged'.[29] 'We are agreed', wrote Down a few days later, 'that the incorporation of your Hull activities within an enlarged BHC is very necessary as it will allow the whole BHC/DCL chemical activities (as distinct from plastics) to be the undivided responsibility of one management'.[30]

Yet, despite the large area of agreement, there were problems. Most notably, BP would not agree to part with Rigidex polyethylene or make commitments to BHC in overseas markets, as Distillers demanded.[31] By the end of March, Ashford told Down that he was feeling very discouraged about the whole scheme for an enlarged BHC because of the failure to reach agreement on these points, for which he blamed BP.[32]

The talks, already cooling, were soon frozen by financial chills. Down was worried about BP's financial position, and especially about the Labour government's 1965 budget proposals to introduce a new Corporation Tax, and to end the double taxation relief which BP was accustomed to claiming on income tax withheld from dividends (see chapter twelve). Uncertain about the financial outlook, Down was unwilling to commit BP to buying a half share in Distillers' chemical interests.[33] Ashford angrily reported to the

Distillers' Management Committee that this was a matter of 'not a little irritation'.[34] Down, however, confirmed in October that 'BP could not at this stage foresee their future financial position and accordingly the enlarged BHC proposition had been shelved *sine die*'.[35]

It remained so while BP arranged to strengthen its finances by making a £60 million rights issue of new shares, announced at the end of January 1966.[36] At Distillers, meanwhile, a working party under Ashford was making another review of the future of Distillers' chemical and plastics interests. All sorts of ideas were considered, including a 'horizontal' merger with other medium-sized chemical companies such as Albright and Wilson, Fisons and Laporte; a 'vertical' merger reaching forward into fibres and textiles by selling out to Courtaulds; a 'vertical' merger reaching backwards into oil refining by selling out to BP; or, perhaps, a merger with BHC, which in turn might merge with Albright and Wilson, Fisons and Laporte to form a 'second ICI'. Narrowing down this plethora of ideas, Ashford favoured a sale to BP. Distillers' chemical and plastics interests were, he thought, 'likely to expand and thrive more as part of an international oil company, where the two sides could be mutually supporting, than as part of a whisky company, however nice'.[37]

At BP, Down was enthusiastic about buying all Distillers' chemical and plastics interests. The rights issue had fortified BP's finances, BP would for several years be allowed to claim transitional relief from the extra tax burden imposed by Labour's 1965 budget (see chapter twelve), and with its overseas tax payments far exceeding its Corporation Tax liabilities, BP would be able to claim double taxation relief against new sources of profits in the UK. Such profits would therefore be tax-free and the chemical industry was, Down thought, 'one of the few activities in which it would make sense to expand in the UK without going too far away from our main oil business'.[38]

With both Ashford and Down in favour of BP taking over Distillers' chemical and plastics interests, the prospects of striking a deal looked good. Down knew that within Distillers there was 'some divergence of opinion between McDonald and Ashford', but he thought that the split might be just 'one of method' rather than a fundamental matter.[39] He did not seem unduly concerned by it and after a top-level meeting at which Sir Ronald Cumming (knighted in 1964) and Bridgeman were present, negotiations got under way as winter turned to spring in 1966. The Distillers' negotiator was McDonald and his opposite number at BP was Down.

They were both tough negotiators and after some early skirmishes it became apparent by the end of the summer that they were miles apart in their valuations of Distillers' chemical and plastics interests. At a meeting in

September, Down told McDonald that BP's valuation was about £55 million. McDonald found this 'quite unacceptable' and countered with £85 million. Down dismissed this as 'far too high' and told McDonald that 'it would be useless for us to continue talking' on that basis. Down afterwards noted, 'I should say that our discussions took place in an entirely friendly atmosphere.'[40] They did not long remain so.

At the end of September Down put forward a revised offer of £64 million, split between £20 million in cash and £44 million in new BP shares.[41] He suggested that McDonald and he should meet as soon as possible to take the matter further.[42] McDonald, however, replied uncompromisingly that he had discussed the offer with Cumming and Distillers' Management Committee and 'in view of our unanimous reaction to the terms proposed, there really would be little point in our meeting to discuss the subject'.[43]

Down, taken aback by the finality of McDonald's response, turned to Bridgeman and suggested that the matter should be 'taken off the McDonald/Down level' and that Bridgeman should speak to Cumming. 'I am still not clear', wrote Down, 'whether there is a gap that can be bridged or not'.[44] He also wrote to McDonald expressing his surprise that 'you appear . . . to consider that our discussions are at an end, treating a draft proposal as a final offer'.[45]

As Down had suggested, the matter was taken up at a higher level on 11 October, when a meeting was held at Cumming's flat. Cumming and McDonald represented Distillers, opposite Bridgeman and Down for BP. Cumming thought that the Distillers board would accept £80 million, and Bridgeman thought that the BP board would go up to £75 million.[46] This was not a huge gap. It was agreed that McDonald and Down should meet again a few days later, after their respective boards had considered the matter.[47]

Down duly met McDonald on the 18th and said that BP would be prepared to pay £80 million (£25 million in cash and £55 million in shares), although this figure was 'very high'. McDonald accepted this as a fair figure for Distillers' assets at 31 March 1966, but said that another £8 million should be added to cover Distillers' net investment between that date and 31 December 1966. This would make a total price of £88 million. Down was simply 'astonished'. The discussions at Cumming's flat had led him to believe that an offer of £80 million would be acceptable and he found it 'impossible to understand' how McDonald could go back on that. He said that BP's final offer was £80 million, while McDonald said that Distillers could not accept less than £88 million.[48] They were at complete loggerheads.

As Bridgeman was abroad in Australia, Down consulted the other BP managing directors who were in London – Eric Drake, Bryan Dummett,

Billy Fraser and David Steel. They agreed not to go above £80 million and that the most senior of them, Drake, should see Cumming.[49] Meanwhile, McDonald had his own consultations in Distillers and telephoned Down to confirm that Cumming and the Management Committee would not take less than £88 million.[50]

With the negotiations seeming to be on the verge of terminal breakdown, Down telephoned Bridgeman in Australia to explain the situation. Bridgeman 'considered that McDonald's behaviour seemed inexplicable unless he really did not want to sell'.[51] Down shared Bridgeman's suspicions, doubting whether McDonald was 'really ready to sell, although it is apparent that the other Directors are'.[52] Was McDonald so against the whole idea of selling Distillers' chemical and plastics interests to BP that he was seeking a breakdown in negotiations? If so, it was all the more important that Drake should see Cumming.

He did so on 25 October, finding Cumming 'in a guarded but friendly mood'. Drake told him that 'there was a strong impression that Mr McDonald was a somewhat unwilling seller'. Cumming 'too felt rather unhappy about the way things had turned out'. He said that 'the trouble was that the matter had been wrongly put to his Board'. Both he and Drake thought that there should be broader participation in the negotiations, which should not be left to McDonald and Down on their own. That agreed, they parted 'on excellent terms'. Drake recorded that 'the general tone of our meeting was so friendly and Sir Ronald was evidently anxious to get the business back on the right lines that I am hopeful that all may not yet be lost'.[53]

Drake's timely intervention was crucial. From that point on, one-to-one confrontations between McDonald and Down were avoided by ensuring that their respective seniors were present at key meetings. This more constructive approach resulted in an agreement in principle, reached on 3 January 1967.[54] The price to be paid by BP was £25 million in cash and 19 million ordinary shares with a value of £61 million. The total consideration was therefore equivalent to £86 million. This was only just short of McDonald's demand and although it was not admitted in BP, it could not be denied that he had succeeded in raising the price by his brinksmanship. His handling of the negotiations apparently did him no harm in Distillers, indeed arguably the opposite, for he succeeded Cumming as chairman later in the year.

In return for the payments of cash and shares to Distillers, BP was to take over all Distillers' chemical and plastics interests apart from some minor carbon dioxide operations. The assets to be acquired included Distillers' wholly owned chemical interests, located mainly at Hull; its wholly owned plastics interests in British Resin Products, located at Barry; its half share in BHC; its holdings in joint companies, most notably British Geon, Distrene and Bakelite Xylonite and its relatively small overseas interests in Australia, France, India and South Africa. With regard to the joint companies, the consent of Distillers' partners was generally required for the transfer of Distillers' holdings to BP. If consents were refused, appropriate reductions would be made in the price that BP paid to Distillers.

In addition to these assets, BP would also acquire Distillers' chemical sales and research organisations, plus the management and service departments concerned with chemicals and plastics. Key managers, most notably Ashford and Hunter, would join BP. In short, BP would at long last possess all the elements of an integrated chemical and plastics company.[55]

In terms of size, the acquisition of Distillers' chemical and plastics interests made BP the second biggest British chemical company, after ICI.[56] The gap between the two was, however, immense. ICI was one of the three largest chemical companies in the world, with sales of some $2,349 million in 1967, but BP ranked only sixtieth in the world, with estimated chemical sales of $286 million.[57] Of the oil companies, Shell and Standard Oil (NJ) had easily the largest international chemical interests. BP came seventh (see figure 15.2).

Size apart, the takeover of Distillers' chemical and plastics interests presented BP with a considerable challenge. As Down told the BP board in January 1967, 'we have an immense task ahead in getting unscrambled this extremely complex petrochemical bag of tricks, but I am confident that if we tackle the problem in an orderly fashion it may take some years but we should have a very strong and powerful position in petrochemicals in this country and in fact in Europe and elsewhere'.[58]

RATIONALISATION

The agreement in principle of 3 January was followed by three months of intensive activity, preparing for the actual merger.[59] That took place on 31 March, when Distillers' wholly owned chemical and plastics interests, plus some of its partly owned ones, were transferred to BP. At the same time, Distillers' chemical staff became BP employees. The transferred interests were put in BP Chemicals Ltd, a wholly owned subsidiary of BP, which also held BP Plastics and BP's shares in BHC, Naphtachimie and Erdölchemie.[60] Ashford, who joined BP's main board in July, was also appointed managing director of BP Chemicals from 1 August.[61] The merger had worked out well for him.

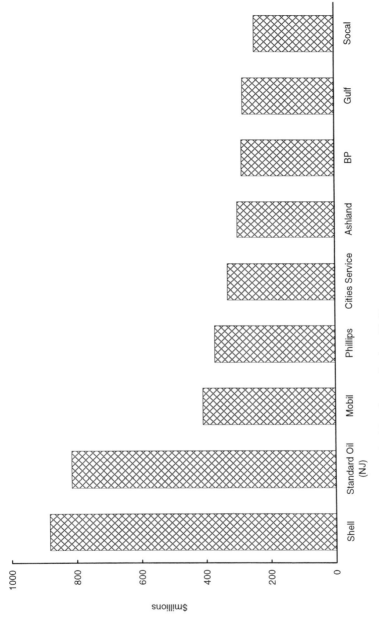

Figure 15.2   Oil companies ranked by chemical sales, 1967

However, not everything went so smoothly. In January, Down and Ashford had flown to the USA for talks with Distillers' partners in its three main plastics partnerships. These were B. F. Goodrich in British Geon, Dow in Distrene and Union Carbide in Bakelite Xylonite.[62] After further negotiations, Goodrich agreed to sell its share in British Geon to BP. This went through in September 1967, when British Geon became a wholly owned subsidiary of BP Chemicals.[63] Dow, however, refused to consent to the transfer of Distillers' share in Distrene to BP. Distrene therefore remained in joint Distillers/Dow ownership until 1968, when Dow bought Distillers' share.[64] Similarly, Union Carbide would not agree to the transfer of Distillers' share in Bakelite Xylonite to BP. As a result, Bakelite Xylonite remained in joint Distillers/Union Carbide ownership until 1973, when Union Carbide bought Distillers' share.[65]

In the meantime, BP Chemicals went through a difficult time as petrochemicals, the great growth industry of the previous twenty-odd years, began to experience the problems of being a mature, but still highly cyclical, industry. By the autumn of 1966 the spurt in chemical industry investment which had started in 1963 was past its peak. The government's macro-economic deflationary measures were squeezing demand with the usual result that chemical companies deferred or scaled down their investment plans.[66] The downturn lasted through 1967, when sterling was devalued from $2·80 to $2·40, and into 1968.[67]

In the summer that year, BP's new 250,000-ton ethylene plant at Grangemouth, originally announced by BHC in 1965, was commissioned. It made two of the three old plants obsolete and they were closed down, the third being retained as a stand-by unit.[68] The new plant was expected to remain competitive for most of the next decade, but the same could not be said of BP's other UK cracker at Baglan Bay. Commissioned in 1963, it had the capacity to produce only 60,000 tons of ethylene a year, which by 1968 was too small to be economic.[69] Indeed, it had probably been uneconomically small from the outset.[70] This, Hunter noted in September 1968, was an 'economic weakness which must be eliminated'.[71]

The chemical cycle was by then recovering from its latest trough and new investment projects were being hurried forward, not only by BP. At the end of September, ICI announced plans to spend £100 million to double its plastics output by 1980.[72] BP quickly followed, announcing in October a major new project at Baglan Bay. At the heart of the project was a huge new cracker with the capacity to produce 340,000 tons of ethylene, 190,000 tons of propylene and 60,000 tons of butadiene a year. The main outlet for the ethylene would be the manufacture of PVC, for which new PVC, vinyl chloride

56   Night view of BP Chemicals' Baglan Bay site, South Wales, in 1969

monomer, chlorine and electric power plants were to be constructed. In addition, a new isopropanol unit would provide an outlet for propylene. The estimated capital costs of the project were just short of £60 million. After claiming investment grants from the government, BP would have to put up £37½ million. That, however, was only the first stage, to be followed by more new plants to produce ethanol, vinyl acetate and styrene (through Forth Chemicals).[73] This was by far the largest chemical project that BP had undertaken and would, Hunter warned, 'tax, at all stages, our manpower and experience'.[74]

So it turned out. Beset by innumerable labour disputes and high cost inflation, the project fell far behind schedule and went well over budget. By April 1972 the estimated capital costs had risen by over 40 per cent and construction of the plants was on average 10 months behind schedule.[75] Then, in 1973 the ethylene plant was severely damaged by fire during commissioning and following temporary repairs only operated at reduced output until a new labour dispute brought virtually all operations at Baglan Bay to a halt in December. The plant therefore had to be recommissioned in January 1974, an event that finally marked the completion of the project.[76]

Table 15.2 *BP Chemicals' UK profitability, 1967–1974*

	Pre-tax profits/ (losses) (£millions)	Capital employed (£millions)	Return on capital employed
1967	5·1	81·4	6%
1968	2·5	86·7	3%
1969	3·2	93·7	3%
1970	(0·8)	99·6	(1%)
1971	(8·7)	145·4	(6%)
1972	(7·4)	187·0	(4%)
1973	1·0	192·5	1%
1974	38·2	185·6	21%

While BP was struggling with problems at Baglan Bay, its UK chemicals profits were plunging (see table 15.2). By the early 1970s the whole chemical industry was suffering from overcapacity and depressed profits. It had, however, to be admitted that BP Chemicals was doing worse than its main competitors.[77] Traditional cost-cutting remedies were applied.[78] The number employed in chemicals was reduced from 12,500 to 10,500 between the beginning of 1971 and mid-1973.[79] Capital expenditure continued on projects to which BP was already committed, but was drastically cut for new projects.[80] There was a shedding of peripheral and unprofitable operations in speciality chemicals at Carshalton, Surrey, and in plastics at Stroud, Gloucestershire (the ex-Socony Mobil works).[81] Meanwhile, chemical activities were reorganised as BP Chemicals International Ltd, with Denis Bean as its managing director (Ashford remained on the main BP board).[82] These measures, combined with a general economic upturn, had the desired effect and BP's UK chemical operations returned to modest profit in 1973 before making a spectacular gain in 1974 (see table 15.2).

The chemical industry cycle also affected BP's main overseas chemical partnerships in France and Germany. In France, Société Française des Pétroles BP (SFPBP), retained its 42·8 per cent share in Naphtachimie, but there were some quite complicated changes in its partners in the 1960s. The result was that by the end of 1971 Péchiney's shareholding had passed to another French chemical company, Rhône-Progil.[83] Meanwhile, Naphtachimie's fortunes fluctuated. After substantial losses in the early years, a dividend was paid for the first time in 1963 and progressively increased in succeeding years to 1970. However, there was a sharp fall in profits in 1971–2 and the dividend, having been reduced in 1971, was cut to zero in 1972.[84] In the meantime, work proceeded on a major project, announced in 1969, for the construction at Lavera of new plants, including a cracker with an ethylene capacity of 400,000 tons a year. The new plants came on stream in 1972, in time for the recovery of 1973–4, when Naphtachimie's fortunes picked up again.[85]

In Germany, BP's 50:50 partnership with Bayer, Erdölchemie, had also been growing fast. The first two crackers were replaced by a third, commissioned in 1963, and a fourth, commissioned in 1970, with ethylene capacities of 80,000 and 360,000 tons respectively.[86] The growth in profits received a setback in 1970–2, caused by the industry-wide recession, but resumed with the general upturn of 1973–4.[87]

Meanwhile, BP Chemicals and its main UK competitors discussed the possibilities of co-operating to reduce the cyclical volatility of the industry. In March 1968 ICI, BP and Shell agreed to form a joint study group to examine the cracker expansion plans of the three companies with a view to recommending the co-ordinated phasing of new capacity on the broad assumption that their market shares would remain unchanged.[88] The study group's report, completed in October, estimated that the UK's demand for ethylene would be about 1½ million tons in 1975. To be economic, new ethylene plants would have to be very large, with capacities of 400,000–500,000 tons a year. One new plant would therefore meet about a third of the whole of the UK's forecast ethylene demand. If more than one company constructed new plants to come on stream at about the same time, there would inevitably be surplus capacity and consequently a poor return on the investment. There would be less surplus capacity if companies constructed smaller plants, but they would then be uncompetitive with larger new ethylene plants built elsewhere in Europe. These problems could, it was suggested, be resolved by co-operative 'rationalisation' of capacity. In particular, if the major UK chemical companies agreed to take it in turns to construct new plants of competitive size, the commissioning of new capacity could be phased to meet total UK demand without creating surplus capacity.[89] Attracted by this idea, ICI, BP and Shell drew up a draft agreement to consult on the construction of new ethylene plant. They laid it before the Board of Trade, who responded favourably.[90]

Ideas about co-operation took a quantum leap in November 1971, when Bean and Hunter visited ICI's Wilton works at the invitation of John Harvey-Jones, head of ICI's Petrochemicals Division. Harvey-Jones told them that he was in favour of close co-operation and was 'clearly inclined' to amalgamate the whole of ICI's Petrochemicals Division with BP Chemicals in a new joint company. This was no off-the-cuff suggestion. Detailed papers and charts were produced, showing the 'fit' between ICI and BP. Bean could see that Harvey-Jones had 'made a fairly deep study of a merger' and that his approach was a 'very serious one', although it had not yet been put to ICI's board of directors.[91]

After an interlude of some months, Harvey-Jones got in touch again. His ideas had been put to ICI's executive directors, who, he said, were prepared to consider closer co-operation between ICI and BP, 'including even a major roll up of the two organisations'. Representatives of the two companies agreed to take things further by exchanging information under a secrecy agreement, using the code name 'Seagull'.[92] At the same time, it was agreed that the tripartite rationalisation discussions between ICI, BP and Shell should go ahead separately. 'In any case', noted Hunter, 'it looked unlikely that these discussions would bear fruit, largely because Shell do not seem prepared to practice what they preach'.[93]

Under the Seagull code name, it was agreed to study a 'Grand Design', defined as a merger of parts of ICI and BP into a 50:50 joint company, with BP supplying the feedstock and ICI having a 'special position' in purchasing the products.[94] These ideas harked back to BP's earlier partnership with Distillers in BHC. If they were carried out, it would mean, in the words of an internal BP report, that 'BP would have reverted to their former strategy of having three joint petrochemical interests in Europe but with an even larger investment, bringing back in an even more acute form the problems of managing them'. It would also mean a BP withdrawal from downstream chemicals.[95]

Whether for these reasons, or others, the Seagull project foundered. In January 1973 Ashford had a 'quiet word' with Maurice Hodgson, an ICI director, about the talks that were taking place between BP and ICI's Petrochemicals Division. Hodgson told him that Harvey-Jones 'was always showing great initiative. The Board had no objection to his having discussions on a broad front but nothing had been seriously considered by the ICI Directors.'[96] The 'Grand Design' did not materialise and nor did tripartite rationalisation between ICI, BP and Shell.

Instead, talk turned to the idea of building a new joint ICI/BP cracker. BP would have liked the cracker to be at Grangemouth, but ICI wanted it to be at Wilton.[97] The pattern of the Company's talks with ICI in the 1940s seemed to be repeating itself. Then, the big idea of a joint petrochemical company, Petroleum Chemical Developments, had given way to the more limited 'Project A' for joint petrochemical plants at Wilton, which was in turn abandoned (see chapter fourteen). However, in the 1970s things turned out differently. In 1974 the two companies announced that they were going to build a new joint cracker at Wilton with an ethylene capacity of 500,000 tons a year. An ethylene pipeline would connect the new plant to BP's petrochemical complex at Grangemouth. The two companies would have equal shares in use of the new plant.[98] The cracker came on stream in 1979, turning into reality the idea of joint BP/ICI basic chemical manufacture which had been discussed and discarded more than thirty years before.

Meanwhile, in the mid-1970s BP's main petrochemical interests outside the UK remained alliance based. In Germany, Erdölchemie, though owned equally by BP and Bayer, was effectively an 'integral part' of the Bayer group. In fact, BP had 'to a large extent taken on the role of supplier of relatively low cost capital and base chemicals to Bayer'.[99] In France, Naphtachimie, in which BP had a minority shareholding, was 'increasingly . . . becoming part of the Rhône-Poulenc Group of companies'.[100] Since these major partnerships in Europe were not managed by BP Chemicals, neither control nor strong co-ordination was possible. Internationally, therefore, BP's chemical interests consisted essentially of three separate national packages in, but not sufficiently co-ordinated to take full advantage of, the European Common Market which Britain had joined in 1973.

There remained, however, good reasons for BP to be in chemicals. In the 1970s, as OPEC countries took over most of BP's low-cost crude oil reserves, BP felt that it could no longer rely on its traditional competitive advantage of low-cost crude oil production. It was therefore seeking new opportunities to add as much value as possible to the raw material, crude oil. The manufacture of petrochemicals was one way of achieving this goal, and therefore constituted a core element in BP's business.

=== 16 ===
Computing

In the 1950s and 1960s BP's business spread rapidly. Exploration resulted in new discoveries, increasing the variety of BP's crude oils, each unique in its chemical composition. New refineries were built in a number of market locations, in contrast to the earlier concentration of refining capacity at Abadan. Technical advances in, for example, catalytic cracking, catalytic reforming and desulphurisation increased the range of products that could be produced. BP entered new geographical markets with varying patterns of demand. With this expansion, BP's activities extended across a vast and diverse network of oil fields, refineries, markets and transportation systems. At the London centre of this burgeoning multinational, the Supply and Development Division (a Department from 1958) was responsible for co-ordinating the flow of oil throughout BP's far-flung operations.

The nature of the supply problem is illustrated in figure 16.1 which shows a hypothetical small integrated company with two sources of crude, two refineries and three markets. Crude oils A and B could be supplied to either refinery in any proportion. Each refinery could manufacture products for any of the three markets, which in turn could take products from either of the refineries. In this integrated system, a change in one component would affect others, requiring a recalculation of the entire pattern of operations. For simplicity, the possibilities of selling crude oil and of purchase or exchange of crude and products from third parties are omitted.

BP's actual operations were far more complicated than this simplified illustration. As the number of its oil fields, refineries and markets grew (see table 16.1), the task of determining which combination of crude oils, refineries and markets represented the optimum, or 'best', pattern of operations became increasingly difficult. The matter was not one of mere physical logistics, it

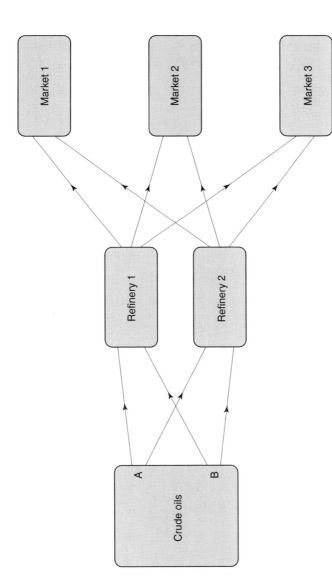

Figure 16.1   Simplified illustration of supply programming in a small integrated oil company

Table 16.1 *Growth of BP's oil operations, 1958–1968*

	Number of oil fields	Number of wholly owned refineries	Number of partly owned refineries	Sales of crude and products (million barrels)
1958	14	8	7	450
1960	18	10	8	550
1962	27	10	10	640
1964	28	13	17	790
1966	41	14	20	890
1968	55	16	21	1070

was, more importantly, one of economics. How, in short, could BP programme its operations to the best economic advantage?

BP's supply planning was based on Medium and Short Term Estimates, drawn up by Supply and Development (S&D). The Medium Term Estimates were produced quarterly and covered the period from three months to two years ahead. They were based on the premise that BP's medium-term marketing requirements could be predicted with reasonable accuracy. The forecast demand for crude oil and products was therefore regarded as a fixed commitment, to be met by drawing up a supply programme covering, for example, the allocation of crude oils to refineries, product yields, purchases and exchanges, processing arrangements with other refiners, shipping, bulk sales and stock levels.[1]

The Short Term Estimates were produced monthly and covered a period up to four months ahead.[2] They were foreshortened in day-to-day operations in which deviations from forecast inevitably occurred. For example, ships might not arrive on schedule, processing plants might fail, or the weather might suddenly change and alter the demand for products. To some extent, unforeseen demands could be met by drawing on stocks, but when sizeable variations from plan occurred there was no alternative but to deal with them by improvised changes to the supply programme.

The Medium and Short Term Estimates were subject to two important constraints. In the first place, they were limited to the utilisation of existing capital equipment, as there was not time to construct significant new plant and facilities within the medium- and short-term planning horizons. Secondly, as sales were taken as given, the Estimates did not offer a plan which would maximise profits by selecting the most remunerative pattern of trade. The economic objective of supply planning was rather to meet a given pattern of trade at minimum cost.[3]

In the mid-1950s, BP's supply planning was carried out largely on the basis of the judgement and accumulated experience of S&D personnel. They were no doubt numerate and competent in simple arithmetic, but few S&D staff had much understanding or formal training in mathematics. They may have applied homespun economics to supply problems, but they were not in general familiar with economic concepts such as marginal costing. Certainly, there was no rigorous technique by which they could assess whether their decisions represented an optimum solution to the problem in hand. In essence, S&D operated by rule of thumb, intuition and experience.[4]

Although these methods would later seem archaic, S&D had dealt effectively with the supply disruptions during the Iranian nationalisation crisis in 1951–4 and the Suez Canal crisis in 1956–7. The solutions may have been of the 'seat of the pants' variety, but they had worked. Lifted by success, the S&D cadre had attained elite status in BP and was generally unaware of shortcomings in its methods.[5]

However, as BP's operations proliferated, the chances of reaching an optimum plan by the old rough and ready methods were reduced. The number of variables was becoming too great for solution by trial and error. Although no ready-made alternative lay to hand, exciting new possibilities were beginning to open up in the new technologies of operational research and electronic computing.

The term operational research (OR) had been coined in World War II to describe mathematical and statistical techniques that were applied to problems such as how to make the best use of radar coverage and how to achieve the optimum balance between firepower and speed in fighter aircraft.[6] OR techniques fell broadly into two categories. One, based on simulation, was aimed at producing a range of solutions to alternative futures and was therefore suitable for dealing with problems of uncertainty. The other was deterministic, seeking the optimal solution to a single future.[7] It was, of course, only effective in situations where the future could be accurately predicted.

Of particular interest as far as BP was concerned was the deterministic technique of linear programming (LP), which could be used to produce optimal solutions to problems involving many variables. It required that the objective, say cost minimisation, could be expressed in a linear relationship with other variables and that limits, such as plant capacities or raw material availabilities, were specified.[8] The linear relationships between variables within the limits were set out in an array of equations, forming a mathematical model from which it was possible to calculate the optimum solution, i.e.

that combination of variables which minimised costs.[9] Reports on the early use of linear programming for optimising refining operations and blending aviation spirits in the USA were published in 1952–3.[10]

The early publicists and practitioners of OR stressed that optimisation was inseparable from integration. Optimisation, they pointed out, was not an additive procedure, so that the results of optimising sub-units of operation and adding them together would not produce the same results as optimising the whole unit as one.[11] In the words of the chief OR protagonists in BP:

> a large complex of refineries and markets linked together . . . is a more efficient economic unit than a number of independent refineries and markets fulfilling their tasks separately. Economists are familiar with the concept of the economy of scale but in what we are discussing more than the benefits of size are involved.
>
> The added benefit consists in the mathematical possibility of sorting out dissimilar requirements in different areas of either production or supply and balancing them out between the different manufacturing centres in such a way as to utilise the total resources in the fullest way and according to the most suitable and economic pattern.
>
> This adds another dimension to the increase in efficiency normally attributable to scale, namely that of integration.[12]

However, large-scale LP applications involved an enormous number of calculations. To be of practical value, they had to be made more quickly than was possible using the standard office equipment of mechanical desk calculators, which could not perform sequences of calculations automatically.[13]

Since the 1940s various research groups, mainly in the USA and Britain, had been working on the newer and more powerful technology of automatic electronic computers.[14] These offered great advantages over mechanical calculators. Their electronic circuitry was much faster in operation and they could perform sequences of calculations (programmes) automatically.[15] After the design and prototype phase, automatic electronic computers such as Remington Rand's UNIVAC 1 and the IBM 701 began to become commercially available from the early 1950s.[16] Being based on thermionic valve technology, they consumed a great deal of electrical power, generated a lot of heat and were physically gigantic, but by the standards of their day they could perform calculations at prodigious speeds.

BP was one of the first British firms to purchase an electronic computer, placing an order for a model known as the DEUCE (Digital Electronic Universal Calculating Engine) in 1956. The DEUCE was one of the first electronic computers to become commercially available in Britain. It was manufactured by the English Electric Company as the production version of the Pilot ACE computer which had been built to the design of Dr Alan Turing at the National Physical Laboratory, Teddington, in the late 1940s. The first DEUCE had been delivered in 1955.[17]

Within BP, the proposal to purchase the DEUCE had emanated from the Refineries and Technical Department and it was envisaged that the computer would be used for purposes such as the formulation of motor gasolines. There was no intention, it seems, of using it for operational research.[18] That, however, was before R. J. (Bob) Deam arrived in London and made a great mark as a pioneer of OR.

THE RISE OF OPERATIONAL RESEARCH

Deam was a man of extremes who championed OR with enormous energy and conviction, mixed with outspoken intolerance and disrespect for managerial authority. His considerable powers of creative originality and innovation went hand in hand with a strong adversarial streak which was fuelled by his sense of being an outsider pitted against the Company's 'Establishment'. He was extremely hard to manage, especially as it could be virtually impossible for general managers with little understanding of OR to separate the real worth of Deam's ideas from the bombast that sometimes surrounded them. In any case, few dared to challenge such an overbearing personality. It was, therefore, a test of both judgement and nerve to decide whether he should be treated as some sort of elemental mathematical genius and given the untrammelled freedom he demanded, or whether he should be brought to heel and forced to perform more routine tasks.

The torrent of expletives which Deam poured on the managerial hierarchy at BP's head office was in part no more than harmless 'pommy-bashing', for Deam was Australian. After studying chemistry and mathematics at Adelaide University, he had worked in the chemicals industry, in academia, and as a process engineer at Caltex's Bahrain refinery before joining BP in 1955 to work as a production estimator at the Kwinana refinery.[19] The priority at the time was to maximise the yield of gas oil, which was in short supply. Deam therefore applied LP techniques to devise a production programme to increase gas oil output. Lacking a computer, he used manual calculations and a good deal of trial and error, by which means he succeeded in turning BP's shortage of gas oil in Australia into a surplus available for export.[20]

His work impressed Maurice Banks, at that time deputy general manager of Refineries and Technical Department, who was thinking, by late summer in 1956, of moving Deam from Kwinana to BP's head office in the UK. With that in mind, arrangements were made for Deam to visit England to give a paper at BP's annual technical conference which was to be held at Sunbury in September 1956.[21] At the conference, Deam presented a simplified mathematical model of a refinery and explained his ideas on the use of LP to optimise operations.[22] He made quite an impact and no time was lost in offering him a job in Refineries and Technical Department at head office.[23] Deam accepted and shortly afterwards returned to Australia to dispose of his accommodation there and collect his belongings before returning to England with his wife and children in February 1957.[24] With a touch of sentimentality he settled in to his new address, 'Kwinana', 2 Sackville Road, Cheam.[25]

At head office, Deam was put in charge of a Computer Group in the Research and Development Division of Refineries and Technical Department.[26] He was full of ideas on the application of OR techniques and was itching to get his hands on the DEUCE computer which, though ordered, was not yet installed. 'Using the speed of the electronic computer', he hoped that 'we shall be able to select the optimum feasible programme under given conditions and hence overcome the inadequacies of the older methods'.[27]

To that end, work began on the construction of a mathematical model of a single refinery, Llandarcy.[28] This required the collection of the necessary data and the compilation of a set of equations to specify the operation of the refinery as completely as possible. By careful formulation, only linear equations were found to be required to set out the relationships of quantity and quality, the operations of processing and blending, and restrictions such as throughput limits.[29] Within ten days of the DEUCE computer being commissioned in late July 1957 the mathematical model of Llandarcy refinery was substantially complete and by the end of September it had been test run on the computer.[30] That month, Deam's manager, Dr Alan Rawlings, reported that Deam had made an 'excellent beginning', but 'is rather impatient and tends to be too outspoken. Full of ideas.'[31]

In 1958 the first successful application of a computerised refinery model was made at Llandarcy.[32] In May that year, Deam reported that so far as he could tell, the Texas Company, Shell and Standard Oil (NJ) were lagging behind BP in their efforts to use LP for refinery programming. 'It would appear', he wrote, 'that the ability to use linear programming techniques has eluded them inasmuch as they have not mastered the art of describing nonlinear relationships in linear form by the selection of activities as we have done. There seems little doubt that we are well ahead in our philosophy.'[33]

Between 1958 and 1960 further models were constructed and computer routines became available not only for Llandarcy, but also for the Grangemouth, Kent, Aden and Kwinana refineries. By the end of 1960 LP was an established technique for programming short- and medium-term operations to yield the desired products at minimum cost at these five refineries. Only variable costs were taken into account because in the short and medium terms fixed costs were unalterable and therefore irrelevant to programming decisions.[34]

Importantly, the new LP techniques not only minimised costs for a given pattern of product output, they also showed that there was considerably more scope for adjusting the yield of products than had previously been exploited. In the late 1950s, BP continued to be generally short of gas oil and had to resort to substantial outside purchases to meet demand. As purchases from other companies were more costly than production, BP's refiners were instructed to produce as much gas oil as they could, while also meeting demand for other products. The application of LP revealed that there was significant potential for doing this. The amount of additional gas oil offered by the computer solution did not represent the maximum physical quantity that could be produced at any price. The computer generated a much more vital calculation: the maximum quantity that it was economic to produce. This was determined by allowing output to increase until the marginal cost of extra production was equal to the procurement cost of outside purchases.[35]

So far, LP had been applied to the five refineries as separate, single units. Although substantial benefits had been gained, it was realised that there was much further to go, particularly in the direction of integrated refinery/supply programming.[36] This was the term used to describe the programming of an integrated operation covering the flow of oil between a number of oil fields, refineries and markets. Deam must have been aware that by venturing into these areas he would be moving outside the territory of Refineries and Technical Department and into the domain of the powerful Supply and Development Department. This, however, did not deter him. He thought that S&D was 'probably the most fruitful department for Operations Research in the Group' since it was responsible for deciding crude oil allocations and refining programmes 'alarmingly enough from quite lowly staff levels without adequate technical training'. Deam was convinced that by demonstrating the superiority of LP techniques over existing methods he would be 'allowed ready access to solving the Group integration problem, now currently handled by Supply Division'.[37]

Deam's desire to push ahead with integrated refinery/supply programming was typical of his urge to expand the frontiers of OR at every opportunity.

As soon as he had solved one OR problem he liked to move on, paying little attention to the duller tasks of consolidation and routine operation. As a result, his ideas tended always to be one or more steps ahead not only of the comprehension of most of BP's management, but also of the available computing power and human resources.

The earliest electronic computers had much less power than later models. The DEUCE was no exception and could not handle more than a single refinery solution at a time. The solution of the much larger mathematical models involved in integrated refinery/supply programming was well beyond its capabilities. To remedy this situation a more powerful Mercury computer was ordered from the British electronics firm, Ferranti, in 1959.[38]

BP was also finding that people with suitable training and experience for OR work were in short supply. The essential requirement, as put by a member of Accounts Department, was for people who 'through long training and practice, automatically and almost unconsciously translate the activities in the world about them into curves, parameters, tables and equations'.[39] Such people were a rarity not only in the commercial departments of BP, but also in Britain at large. As a manager in BP's Refineries and Technical Department put it: 'there is a critical shortage of qualified people to pursue each and every phase of this [OR] work. This is the Number One Problem – difficult to solve because of the nationwide demand for just such people.'[40]

To this was added British management's widespread ignorance of mathematics and the resultant tendency of British managers 'to go in awe of the high order of mathematics involved in operational research studies'.[41] BP was no exception, but was quick to seek a remedy by training management and staff in OR. The courses ranged from short two-day introductory courses for senior management to intensive ten-week courses for lower level managers and staff who were thought to be suitable for OR work.[42] The ten-week course held at the London School of Economics in 1959 was the first of its kind to be held at a university in the UK.[43] It was reserved solely for BP managers from various departments such as S&D, Refineries and Technical, Markets, Accounts and Office Management. By these means, knowledge of OR began to spread into BP's main departments and, conversely, experience of departmental operations began to be brought to bear on OR projects. Interaction was increased by the secondment of some of the course members to work on OR projects in the Computer Group after the course was over.[44]

The secondment of departmental staff to the Computer Group was essentially an ad hoc measure which left unresolved the larger question of how

OR was to be accommodated in BP's organisational structure. In 1958–9 this was the subject of sundry papers, reports and discussions, none of them conclusive.[45] The fundamental problem was that OR techniques were being applied to a wide range of BP's activities which were carried out by various departments. OR therefore cut across the functional departmental structure in which there tended to be strict lines of demarcation between the managerial hierarchies of the departments. In April 1960, a new OR Branch was established in the recently formed Research and Technical Development Department.[46] The technical manager of the new Branch was Deam, reporting to W. J. (Bill) Newby.[47] However, while some of the OR staff were retained in the OR Branch, others were seconded to OR groups which were embedded in the main operating departments. The implant of OR staff in the user departments was designed to foster the build-up of OR groups with technical know-how in the departments. At the same time, it was decided that the provision of a computer service should be split from OR and handed over to Office Management Department.

At the time OR Branch was formed, Deam was close to achieving the first successful computer solution of an integrated refinery/supply problem. The DEUCE was still the only computer available for OR work in BP, as the Mercury had yet to be commissioned. However, the problem of computer capacity was overcome by hiring time on a powerful IBM 7090 computer from the US computing firm, the Corporation of Economic and Industrial Research (C-E-I-R), in Washington.[48] Arrangements were made for members of OR Branch to visit Washington late in 1960 to attempt a simultaneous solution of an integrated five refinery matrix which covered the operations of BP's three UK refineries (Llandarcy, Grangemouth and Kent) and the refineries at Aden and Kwinana.[49] Despite problems with the IBM computer programme, the results were encouraging and further trial runs followed in the first half of 1961, still using the IBM computer in Washington, but with an improved programme.[50]

The success of these trials was enthusiastically received in BP. Stuart McColl, the assistant general manager in S&D with responsibility for short- and medium-term programming of refinery operations, wrote that 'we were very pleased to hear of the promising outcome of the tests conducted on a large computer in Washington . . . and we would most certainly want to see everything done to bring this highly desirable project to fruition and to establish these improved methods as the regular means for preparing the Medium Term Estimates and as the basis for the Short Term Programme'.[51]

In July, the first fully operational run of the Five Refinery model was carried out using computer time on a new IBM 7090 in London, hired from

C-E-I-R (UK) Limited, the British subsidiary of C-E-I-R in the USA.[52] The results were jointly examined by the OR Branch, S&D and Production Control Branch (in the Refineries Department). As was to be expected in a first run, some errors and inconsistencies were found, but they were of a relatively minor nature. The general verdict was that 'as a whole, the operation had been very successful'.[53] The next month solutions of the Five Refinery model were used for programming refinery and supply operations.[54] Thereafter, their use became an established routine.[55] According to an internal report on OR, most oil companies regarded 'the planning of overall operations in this way as offering a very substantial prize', but 'our lead in this field appears to be quite substantial'.[56]

The integrated Five Refinery model was indeed a quantum leap from the earlier single refinery models. Solutions of the new model provided an optimal plan for supplies of crude oils to the five refineries, the individual refinery programmes and the allocation of products to marketing areas. For these purposes, the optimal plan was defined as that which met a given pattern of demand at minimum variable cost. In the computation of the model this was achieved by requiring the refineries to engage in marginal cost competition for the available crude oils and product markets. The point of optimisation was that at which marginal costs were the same at all refineries and no product was manufactured at a marginal cost greater than the cost of procuring it from outside.[57]

The selection of Llandarcy, Grangemouth, Kent, Aden and Kwinana as the five refineries was no accident. They accounted for about 45 per cent of BP's refining capacity, they were wholly owned, flexible in operation and their seaboard locations gave them ready access to a wide range of markets.[58] These characteristics made them particularly suitable for the new integrated model, which was concerned with balancing BP's overall operations after allowing for supplies from other refineries.[59] For this reason, the Five Refinery model later came to be known as the Balancing Refineries model.[60]

The economic gains achieved by use of the new integrated model could never be precisely quantified, but were held to be substantial. The largest single benefit was undoubtedly the discovery that the scope for altering the product yield from BP's refineries was significantly greater than had previously been thought.[61] In the words of Newby and Deam, 'results show that the ability of a single refinery to alter the production of any particular product can amount to about 5% on crude, whereas with five or more refineries this flexibility is increased to about 10%. This is achieved solely by optimising the use of existing facilities and does not require the investment of additional capital. This represents a most important economy in the operations of a company'.[62]

By exploiting its new-found flexibility, BP was better able to adjust its output to meet the pattern of demand, thereby reducing outside purchases of products in short supply. Additional economies were had from, for example, reductions in freight costs by the elimination of tanker back-hauls, the increased use of low-cost crudes and reductions in product stocks now that there was greater scope for matching output to swings in demand.[63]

Although the economies to be had from the use of the Five Refinery model were impressive, there was considerable scope for making further savings by extending LP to cover other parts of BP's refinery/supply operations. The logic, broadly speaking, was that the more operations were included, the larger the pay-off.[64] However, as had earlier been the case with the DEUCE computer, BP's progress in the OR field was again straining the limits of the available computing power. Even C-E-I-R's powerful IBM 7090 computer lacked the capacity to handle significantly larger matrices than the Five Refinery model. Indeed, the decision to limit the model to the five refineries mentioned had been influenced by the capacity of that computer.[65] As for BP's Mercury computer, it was less powerful than the IBM 7090 and by the time it was commissioned in 1961 its capacity had already been overtaken by the demands of OR work.

BP therefore decided to increase its computing capacity by an arrangement with the University of London. Specifically, in autumn 1961 BP decided to make a financial contribution to the University's purchase of a new Ferranti Atlas computer. In return, BP was to have an annual allocation of 1,250 hours on the Atlas for five years and 500 hours per annum for a further two years.[66] The University was only the third organisation to order an Atlas, claimed to be the most powerful computer in the world.[67]

Pending the installation and commissioning of the Atlas, BP continued to hire time from C-E-I-R (UK) on an IBM 7090 and, from 1964, an IBM 7094.[68] At the same time, BP improved its integrated refinery/supply programming techniques and broadened their application. In 1962, the Five Refinery model was expanded to include the shared refineries at Abadan and in Kuwait, followed by the Company's new wholly owned refineries at Gothenburg (Sweden) and Westernport (Australia) in the mid-1960s.[69] Thus the Five Refinery model came to include nine BP refineries (wholly and partly owned) as well as the Milazzo refinery in Italy where BP had processing rights.

Meanwhile, a second major refinery/supply model was introduced, known initially as the Central European Supply Programme (CESP) and later as the European Refineries Supply Programme (ERSP).[70] The CESP was centred on operations in Central Europe, where the traditional pattern of supply (shipping products from large refineries in the Middle East) had been changed by

57   BP's Mercury computer in 1963

the construction of new inland refineries which were located close to the main areas of consumption and fed with crude oil through pipelines from deep-water ports.[71] The project to construct an integrated refinery/supply model for this area originated in a study of internal distribution in Switzerland which was carried out in about 1960. It was evident from the study that it was impracticable to consider one part of the Central European marketing area in isolation from others since variations in the Swiss pattern of supply involved corresponding changes in supply patterns in neighbouring countries.[72] It was therefore decided to broaden the study by setting up an inter-departmental team to devise methods for optimising the supply and distribution of products in Germany, Switzerland and the Benelux countries.[73] The result was the new CESP model, which included BP's main refining interests in these countries, together with partial representation of its operations in France and Italy.[74] After successful trials in 1962, the model went into routine use the following year.[75] It was based on the same principles as the Five Refinery model in that it enforced marginal cost competition between alternative refinery/supply possibilities to arrive at optimum solutions to the problem of meeting given market requirements at minimum variable cost.[76] Like the Five Refinery model, the CESP was extended over the next few years to include new refineries, namely those at Rotterdam and Vohburg. It was also expanded to include full representation of BP's French operations by incorporating the integrated refinery/supply model for France which was originally separately developed and came into operation in 1964.[77] As a result of its expansion, the CESP came to include eleven BP refineries (wholly and partly owned) as well as others where BP had processing rights.[78]

It was a logical extension of the principles behind integrated refinery/supply programming that the Five Refinery and CESP models should be combined into a single matrix covering most of BP's operations. This was an explicit objective, which it was hoped to achieve when the new Atlas computer was commissioned.[79] Meanwhile, the two models were run in a cycle which permitted correcting factors to be injected into one, based on the latest solution from the other. By this method the two systems were brought into approximate balance with surpluses/deficits in one system being met by imports/exports in the other. This, however, was a very imperfect and clumsy procedure which, as Deam acknowledged in 1965, 'is time-consuming, liable to failure, takes a great deal of computing and highly skilled staff time, and yields a solution which is not optimal'.[80]

Notwithstanding these shortcomings, there was no question that the application of LP techniques had come a very long way since the first computer optimisation of the Llandarcy refinery in 1958. By 1964, the Five Refinery and CESP models encompassed the massive integrated network which is illustrated in figure 16.2. By 1966, after further expansion, crude oil throughputs at BP refineries which were included in the two models amounted to nearly 90 per cent of the crude oil throughputs of all BP's wholly owned and shared refineries and for a similar proportion of BP's sales of refined products.[81]

Meanwhile, the rapid expansion of BP's OR activities continued to outrun both its computing capacity and its ability to recruit and retain suitably skilled staff. Deficiencies in these areas were made good largely by purchasing computer time and services from C-E-I-R (UK). This company had been

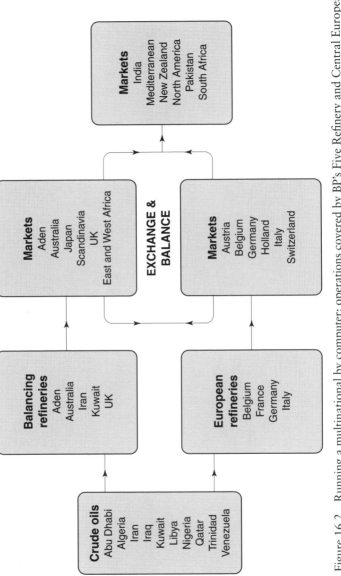

Figure 16.2   Running a multinational by computer: operations covered by BP's Five Refinery and Central European linear programming models, c. 1964

incorporated in 1960 as a consultancy firm offering a range of computer services from the hire of computer time to computer programming, statistical analysis, data processing and other activities based on mathematical and computer techniques.[82] From its beginning, C-E-I-R (UK) had close personal and business connections with BP. The personal contacts were largely those of Professor Maurice Kendall, who was an OR consultant to BP in the late 1950s when he held the chair of statistics at the London School of Economics.[83] On the formation of C-E-I-R (UK) he gave up his chair to join C-E-I-R, initially as technical director and, from 1961, as managing director.[84] The same year, his personal consultancy with BP was terminated to avoid conflicts of interest arising from C-E-I-R's activities as consultants to other oil companies.[85] However, the commercial relationship between BP and C-E-I-R (UK) remained close and BP was a large user of C-E-I-R's services, both in the hire of computer time and in other activities such as analysis and programming. In 1963, about 40 per cent of C-E-I-R (UK)'s revenue came from its business with BP.[86]

While BP was important to C-E-I-R as a customer, so too was C-E-I-R important to BP as a supplier of computer time and services, in which it had considerable expertise, having built up a large specialist staff.[87] BP had of course also built up its own OR organisation, which was given additional status in 1962 when the OR Branch became the OR Division (under Deam who continued to report to Newby).[88] At the same time, a new Operational Research Directing Group (ORDG) was set up as an inter-departmental body consisting of senior managers to direct OR activities to the overall advantage of BP.[89] But despite BP's efforts to build up its OR strength, it continued to experience difficulty in recruiting and retaining suitable staff for OR work. As Newby commented at the beginning of 1964, 'because of the abnormal demand by industry and commerce for people with computer and OR experience, it is becoming increasingly more difficult [sic] to recruit suitable staff and there is now no guarantee of retaining them after training. This position is likely to get worse and it seems possible that specialists will tend to concentrate in consultant organisations where the rewards and professional interests appear greater.'[90] Given this prognosis, BP would not have wished to lose C-E-I-R's services.

However, BP was becoming concerned about C-E-I-R's future. It was expected – wrongly, as it turned out – that the Atlas computer would be operating by the middle of the year and that BP's hire of computer time from C-E-I-R would then cease.[91] This, Newby believed, would leave C-E-I-R in an uncertain financial position.[92] Rather than jeopardise C-E-I-R's future and risk the loss of its expertise, BP acquired a 60 per cent interest in C-E-I-R

(UK) from its US parent in March 1964.[93] The acquisition was consolidated by purchases of further blocks of C-E-I-R shares, as a result of which C-E-I-R became a wholly owned subsidiary in 1966.[94] Although owned by BP, C-E-I-R was not integrated into the departmental structure of head office, but retained its separate existence as a computer firm which did business with other parties as well as BP. One of the outcomes of BP's growing activities in OR and computers was, therefore, a significant diversification into the computer services industry. In 1968, the name C-E-I-R was changed to Scientific Control Systems, Scicon for short.[95]

The 1964 acquisition of a majority shareholding in C-E-I-R, the participation in the University of London's purchase of an Atlas computer, and the development of the huge LP models added up to a very substantial commitment by BP to computer applications. At the same time, BP's conviction that it was leading the field in OR was confirmed by outside recognition. For example, it was reported in *The Statist* in July 1964 that BP's LP models were 'perhaps the most advanced industrial application of linear programming in this country'.[96] Three months later a report in *The Times* praised Newby for his presentation to a conference on 'Computable models in decision making'. According to the report, 'the star turn of the day was undoubtedly Mr W. J. Newby of British Petroleum. The reason for that was simple. He presented a method of production and planning and distribution control that was actually working and achieving majestic savings. Many of the problems that members of the audience were trying to tackle or thinking of tackling had already been approached and resolved by this firm.'[97]

It was, however, on Deam that most of the glory was showered. In 1963 he was unofficially approached to take a new chair in operational research which Standard Oil (NJ)'s UK subsidiary was proposing to set up at the University of Cambridge. According to a colleague, Deam was 'not prepared to consider it' because he wished 'to continue his constructive research in the stimulating atmosphere which BP gives him'.[98] Two years later a draft BP citation supporting Deam's nomination for the Britannica Australia Awards stated that:

> In our opinion Mr. Deam has perhaps done more than/as much as any other individual in the development of methods of applying these modern techniques [of OR] to the complexities of modern industry and commerce and that as the knowledge of these techniques becomes more widespread the substantial economic benefits inherent in their application will result in Mr Deam becoming recognised as one of the great innovators.[99]

However, alongside the glamour and prestige of innovation there were also risks, as was amply demonstrated in the commissioning of the Atlas computer. The Atlas resulted from a conscious, though ultimately unsuccessful, attempt to build a British super-computer capable of competing with the new US super-computers which were being designed and developed in the late 1950s and 1960s.[100] When BP agreed to participate with the University of London in the purchase of the Atlas in 1961, the new computer was scheduled to be commissioned by January 1964.[101] In fact, it was not delivered until May 1964 by which time it was, according to a later history of this episode, 'effectively obsolete', having been upstaged by its US rivals, in particular the Control Data Corporation (CDC) 6600 computer.[102] Although the Atlas still had to be commissioned, it was officially opened by the Queen Mother in June 1964.[103] Over the next six months, the commissioning of the computer was plagued by technical problems which became acutely embarrassing and inconvenient to the various parties involved, namely the University of London, BP and International Computers and Tabulators (ICT), who had acquired Ferranti's computer division in 1963. By the middle of December, the Atlas had become, in the words of an article in *The Observer*, 'the focus of a first class row' at the University, whose academic staff were angry about the delays in providing them with the computing facilities they had been led to expect.[104] However, as the Principal of the University pointed out in his Report for 1964/5, 'we have been the victims of the risks inherent in the ordering of what is virtually a prototype of a new machine'.[105] By 1965 the hardware of the computer had been commissioned, which is to say that it was operating as a piece of physical machinery.[106] However, before BP could make operational use of the computer it had to commission the software, i.e. the programmes without which the computer was incapable of running BP's LP models. This proved to be something of a nightmare which absorbed most of OR Division's time in 1965 and early 1966.[107]

The problems with commissioning the Atlas did much to sour relations between the OR Division and the main operating departments, especially S&D which had to deal with the practical problems of using the unwieldy Five Refinery and CESP models. Colin Williamson, who had recently joined the OR Division as assistant manager, thought that the experience with the Atlas had 'naturally frightened a lot of people off methods of super-integration' and reported that 'the gulf is widening between the "practical" and the "theoretical" O-R teams'.[108] The situation was aggravated by Deam, who claimed in March 1965 that the Atlas was showing fewer faults than the IBM 7094 and noted that a problem which would have taken about thirteen hours on the DEUCE and three hours on the Mercury was solved in little more than a minute on the Atlas. 'The Atlas', he wrote, 'is indeed a most satisfactory machine'.[109]

Others were more cautious. It was not until a year later that Newby reported that the first trial Five Refinery case had been solved on the Atlas and that 'after all the setbacks it now seems we are out of the wood'.[110] Even this assessment seems to have been premature, for it was a further two months before Newby informed Derek Mitchell, the chairman of the ORDG, that the Atlas was at last beginning to perform satisfactorily.[111] This infuriated McColl, who was by then the general manager of S&D. On reading Newby's remarks, he wrote to Mitchell early in June 1966:

> I feel that a quite unjustified gloss continues to be put on the current perfor-
> mance and promise of Atlas. I could not, for example, agree that 'it is begin-
> ning to work satisfactorily'; to my mind, it is beginning to work, but far from
> satisfactorily . . .
>   I am constrained to express these views for the reason that in this
> Department our staff have been strained for a period of many months by the
> shortcomings of the computer facilities available. Not only does this mean that
> we have had to take a number of inadequately informed decisions, but also
> such strains can only be borne for a limited period of time and I attach the
> greatest importance to steps to put the operations on a satisfactory basis as
> quickly as possible and to minimise the difficulties in the interim.[112]

The performance of the Atlas was still, therefore, a vexed issue in June 1966, which was the first full month of its operational use.[113] Even thereafter, S&D continued to be unhappy with its reliability and performance.[114]

McColl was not exaggerating the strain on S&D. Concerned that his department was being swamped with the massive volume of data that was required for the Five Refinery and CESP models, he had requested C-E-I-R to examine the problem.[115] A joint C-E-I-R/BP study team was set up and reported in June 1966. They found that S&D received vast quantities of data from other parts of BP (figure 16.3). The data covered all aspects of BP's operations including crude oil supplies, the availability and disposal of prod-ucts, sales estimates, shipping details and so forth. They were presented to S&D in all manner of shapes, sizes and layouts and sometimes by telephone, so that the system depended 'on the memory and personal jottings of indi-viduals for a detailed record of transactions'. S&D was reported to be 'creak-ing under the workload'. While the Department's use of LP techniques was 'at a significantly high level of development' it was 'supported by an anti-quated data collection and analysis system'. In fact, the study team found that 'input procedures are so laborious that by the time a program is run the input provided is out of date'. S&D staff were found to achieve objectives 'by sheer determination rather than by a smooth operation' and were 'men-tally harrassed' by broad responsibilities and excessive workloads. To relieve

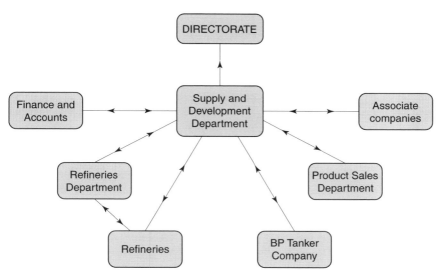

Figure 16.3   Data flows between BP departments, 1966

the situation, the study team recommended the introduction of a computerised data processing system. That was put in hand and later came into operation as the 'Moving Data File'.[116] However, the move towards automated data processing dealt with only one side of the problem. The other, more fundamental solution to S&D's problems was to simplify the clumsy operation of the Five Refinery and CESP models, to which no solution was yet in sight.

Newby, meanwhile, was trying to bring order to BP's OR and computing activities. Since the creation of the OR Branch in 1960, OR and computing had been under separate managements, with OR being located in the Research and Technical Development Department (as the OR Division from 1962) and computing being a part of the Office Management Department until 1965 when a new Computer Division was formed in the Engineering Department.[117]

Deam was in charge of the OR Division, but he had no interest whatsoever in lines of authority, systems of reporting or formal allocations of responsibilities. By 1966, Newby felt that the Division had become 'too large to be run as a happy family' and he was aware of 'weaknesses in the present OR organisation' which could be 'traced to lack of effective management'.[118] At his instigation a new Computer Department was formed in August 1966, bringing the previously separate OR and Computer Divisions together under a single management. The head of the new Department was Newby, with Bob Deam and Colin Williamson under him as assistant managers in charge of the OR Division and Computer Division respectively.[119]

Newby hoped that the establishment of an improved structure for the control of OR and computing would help to improve relations with the main user departments.[120] However, it was impossible to contain Deam whose 'outstanding ability to innovate' was, as Newby noted with understatement in 1967, combined with 'difficulty in being confined in any framework'.[121] The fact was that having made the breakthrough in integrated refinery/supply programming, Deam was by now throwing his energies into the next challenge of applying OR methods to BP's long-term planning. This was no answer to S&D's prayers for rationalisation of the refinery/supply models.

After the Atlas had finally come into operation, S&D was keen to take advantage of the new computer's capacity by merging the Five Refinery and CESP systems into a single model.[122] The operation of the two models, bad enough in normal conditions, was found to be totally unworkable when BP's supply operations were disrupted by the Arab–Israeli Six-Day War in June 1967. During the emergency, which was aggravated by the loss of Nigerian oil supplies after the commencement of the Nigerian civil war in July, BP's supply operations had to be speedily rearranged. The unwieldy and highly detailed Five Refinery and CESP models were, in McColl's words, 'totally impracticable in the tempo of the crisis' and resort had to be made to hand calculations on the outbreak of the emergency.[123] At the same time, efforts were made to develop a simplified model, integrating the most significant parts of the Five Refinery and CESP systems. This appears to have been mainly the joint work of S&D and Refineries Departments and McColl was none too pleased with what he described as 'the lack of help from Computer Department in integrating the CESP and Balancing Models'.[124]

Nevertheless, rapid progress was made in developing the new model, which was called the Simplified Linear Integrated Model (SLIM).[125] It was first used in July 1967 and was formed by selecting sections of the Five Refinery and CESP models and integrating them into a compact and simplified model.[126] Although it excluded some of the supply areas included in the more elaborate models from which it was derived, the advantages of being able to formulate a single problem for speedy computation were found to outweigh by far the disadvantages of loss of detail. As a result, SLIM continued to be used, with modifications, after the crisis was over.[127]

However, the new model had shortcomings. It had been hurriedly put together during the crisis when there was no time for extensive remodelling.

Consequently, part of BP's operations, especially those in Germany and the UK, were represented in too much detail while other parts, most notably the refineries at Abadan and Aden, were inadequately represented. Other areas, namely France, Italy, Australia and Canada, were excluded altogether. The geographical area covered therefore needed expanding, and extensive re-modelling was necessary to arrive at a better balance between areas.[128]

Steps to increase BP's computing power had already been taken in July 1967 when a new UNIVAC 1108 computer, manufactured by Sperry Rand, was ordered by C-E-I-R. This computer was to be shared by BP and C-E-I-R and was considerably more powerful than the Atlas.[129] It was yet to be commissioned when, in February 1968, the ORDG decided to set up a sub-committee to exploit the power of the new computer for short- and medium-term supply planning.[130] The sub-committee, which included representatives of S&D, Refineries and Computer Departments, decided to design a new Group Resource Allocation Model (GRAM) to replace SLIM. In designing the new model, S&D had no desire to return to the mass of detail which had been included in the Five Refinery and CESP models. It was more important for S&D to have a model small enough for quick computing than to have a completely comprehensive model which was slower in use. As had been demonstrated in the 1967 Middle East crisis, changes of situation during the period of data input were a larger risk than omission of detail from the model.[131] The need was, therefore, to design a single integrated model which embraced the total supply problem (or as much of it as was critical), but which was also compact enough to be fast in operation.

It was also, however, recognised that detail which was unnecessary in the Group model was significant in local supply areas.[132] Rather than include local detail in the head office model, it was decided to develop local area models (LAMs) which had already been introduced by some of BP's main associates such as those in France, Germany, Australia and Canada.[133] The concept of GRAM was, therefore, that it would be a condensed head office model, covering BP's overall operations, while associates operated more detailed LAMs.[134] The new head office model was duly constructed on these lines and tested on the UNIVAC 1108 computer, which was commissioned in July 1968.[135] After testing, GRAM replaced SLIM for the computation of the Medium Term Estimates in September 1968 and for the computation of the Short Term Estimates in March 1969.[136]

It had thus taken over a decade of continuous straining at the very limits of computer applications to advance from the introduction of single refinery optimisation at Llandarcy to the single integrated GRAM model for refinery/supply programming, which unlocked the economies of integration yet was sufficiently flexible to avoid the diseconomies of the excessively cumbersome and detailed Five Refinery and CESP models.

THE DEMISE OF OPERATIONAL RESEARCH

For some time before GRAM came into operation, Deam's main interest had lain not in the improvement of the refinery/supply models, but in extending the use of OR for long-term planning. As has been seen, short- and medium-term planning was limited to the utilisation of existing capital equipment because the planning horizon was too close to allow for new capital projects. Long-term planning, on the other hand, looked several years ahead and was concerned with investment in new plant and facilities for BP's longer-term development.

In the early to mid-1960s considerable OR work was done on extending the short- and medium-term refinery/supply models so that they could be used for the Long Term Estimates, which were compiled annually to define the new plant that was required to meet forecast demand five years ahead.[137] For that purpose it was necessary to include capital expenditure as well as variable operating costs in the computer model. This was difficult to achieve with the available computing power and it was not until 1966 that computer procedures came into use for long-term supply planning.[138] The integration of most of BP's operations in a single model was achieved when the Group Estimating Model (GEM) came into use for long-term supply planning in 1967.[139]

In Deam's view, the main shortcoming of this system was that it failed fully to integrate marketing with refining and supply.[140] He therefore came up with a new concept, named Integrated Marketing and Refining (IMR), which he claimed was a 'revolutionary method'.[141] It went beyond BP's previous planning models not only by integrating marketing with refining and supply, but also by seeking a different objective. Whereas the earlier models sought to meet the forecast demand at minimum cost, IMR's objective was to select the most profitable pattern of trade. In short, it sought profit maximisation instead of cost minimisation. Simplistically, the method was to offer an IMR model the total sum of capital which was available, without any explicit ration between refining and marketing, in order to arrive at a detailed integrated investment plan which maximised profits.[142] However, like the earlier models, IMR was a deterministic technique which used LP to calculate the optimum solution for the future that was forecast.[143] Faith in the technique therefore depended on the certainty with which BP could look ahead. The less certain the outlook, the less confidence could be placed in deterministic planning.

The first IMR model covered BP's Canadian operations and was formulated in 1963.[144] After further research work, the development of IMR gathered considerable momentum during the second half of the 1960s when it created quite a stir in BP.[145] While managers in various departments struggled to understand and evaluate the technique, Deam was characteristically impatient to force his ideas through. In 1966 he argued that 'the only chance of success is to run the [IMR] project in the first instance under a dictator', to which he added that 'given the appropriate dictatorial leadership, departments must display very considerable sympathy for the project'.[146] The Canadian IMR model became a pilot exercise for a larger scheme under which a Group Macro-IMR model was to be linked to more detailed Micro-IMRs for local associate companies.[147] By the end of the 1960s, a Micro-IMR model had come into use for BP Canada and further Micro-IMRs were being developed for other associates. It had also been decided that the Macro-IMR model would be used in 1970 in parallel with existing procedures for the preparation of the 1971–6 Group long-term plan.[148]

In the meantime, Deam was also engaged in other OR projects which brought more and more of BP's activities into his ever expanding OR empire. One project was Joint Venture Analysis (JVA), which sought to apply OR methods to BP's joint operations, such as Shell-Mex and BP. Another was a financial planning model, known as FIRM, an extension of IMR, which was aimed at maximising BP's discounted cash flows.[149] By the late 1960s there seemed to be no end to the expansion of Deam's LP models, which had successively integrated refining and supply; marketing (through IMR); and finance (through FIRM). Increasingly, it looked as though Deam would not stop short of controlling BP's entire operations by LP.

Deam was so convinced by his own ideas that he was blind to the thought that he might have overreached. However, in the late 1960s and early 1970s a crescendo of criticism was mounting against him. In extending his LP models Deam had alienated a growing number of executives in the departments concerned. Apart from objections to his dictatorial style, there was unease about his seemingly insatiable demands for huge volumes of data to feed his LP models. Moreover, executives in the operating departments felt that the models were too rigid. Fundamentally, they preferred the exercise of managerial flexibility and judgement to obeisance to a deterministically planned economy.[150]

Although opposition to Deam was widely diffused in BP, there was one person who emerged as his most formidable critic: Peter Walters. He had joined BP in 1954 and risen with great speed through the ranks of S&D in London, with two spells in BP's New York office in 1960–2 and 1965–7.[151]

He had a calm, quietly confident manner and was not intimidated by Deam's volcanic personality. Nor was Walters overawed by the mystique surrounding Deam's mathematical abilities, which had tended to inhibit others from challenging him.

When Walters returned from the USA in May 1967, it was intended that he would go into the Central Planning Department, but in the emergency of the Middle East crisis he was appointed instead as assistant general manager in S&D in joint charge of operations. That experience convinced him that the LP models were incapable of speedy, flexible response to rapidly changing external conditions. He was also concerned about the lack of management information on the profitability of individual parts of BP's integrated business. How, he wondered, could he decide on the allocation of scarce supplies to marketing associates when he had no idea what margins they were making? He was, moreover, convinced that the centralised control of integrated operations diminished the responsibility and motivation of operational executives.[152]

Walters' belief that Deam and his OR ideas were taking BP in the wrong direction was strengthened by an idea which Deam put forward in 1968 for the 'beneficiation' of crude oil. The idea was to construct a huge, but simple, refinery with a throughput capacity of one million barrels per day at an entrepôt site where crude oil could be split into distillate and residue. The distillate was to be used to enrich or 'beneficiate' crude oil shipments to market-located refineries, enabling them to produce the extra quantities of light and middle distillates which were in demand, especially in European markets. The residual fuel oil produced at the entrepôt refinery could, it was suggested, be sold in areas where fuel oil demand was high and, in addition, 'it may be found possible to utilise the fuel oil in the neighbourhood of the entrepôt by constructing a large power station and/or steel works'.[153] In a paper co-authored by Deam it was suggested that 'this operation has every chance of being highly successful and we would be surprised if it does not show a large pay-off'.[154]

The scheme was regarded with complete scepticism by Walters, who believed it was ludicrously counter-economic for BP to haul crude oil to the entrepôt refinery for part-processing and then to haul the 'beneficiated' crude to market-located refineries for further processing.[155] As for the idea of using fuel oil produced at the entrepôt in an adjacent power station, McColl drily pointed out that 'this adjacent power station is for the time being hypothetical'.[156] The proposal apparently got no further, but it helped to convince Walters that the implementation of Deam's ideas, or at least some of them, had to be stopped.

Deam's position was soon further undermined by the findings of the Management Study Group (MSG), which was set up in 1968 to examine head office and make recommendations for its reorganisation (see chapter thirteen).[157] The MSG members who reported on the Computer Department were extremely critical of Deam and his OR Division. Their report made repeated references to the appalling state of relations between the OR Division and OR user departments. It stated, for example, that 'attitudes on both sides have frequently deteriorated to a point where co-operation has been minimal' and that Deam had 'developed over-bearing attitudes towards customer departments, which have created unco-operative responses with these departmental managements'. In interviews with departmental managers and staff, the MSG found that there were allegations of Computer Department 'over-selling projects and attempting to force them prematurely in to the implementation stage'. Furthermore, Deam had 'attempted to take operational and executive responsibilities for activities which more rationally lie with other departments'. To some degree, these problems were attributed to weak control by the Operational Research Directing Group (ORDG), which the MSG thought needed strengthening.

While the MSG acknowledged Deam's pioneering work on large LP models, they were also concerned that 'considerable dangers have developed', partly because the large LP models 'seem to be generating even larger ones' and partly because under Deam's influence the OR Division had concentrated almost exclusively on the deterministic technique of LP, largely ignoring other OR methods such as those concerned with simulation. The MSG recommended that the OR Division should be encouraged to broaden its activities and to place greater emphasis on non-LP techniques.[158]

Newby, hurt by the MSG's report on his Department, complained that it failed to show understanding of the nature of OR work.[159] However, he was unable to turn the tide of criticism and in May 1971, upset and disillusioned by the direction of events, he decided to retire from BP.[160] The announcement of his retirement was immediately followed by a reorganisation of the Computer Department, in effect a break-up. Computer services and systems remained in the Computer Department, headed by Colin Williamson. However, the OR Division was transferred to the recently formed Central Developmental Planning Department (CDPD, the successor to the Central Planning Department), where it was put under Robert Horton. He was a rising star who at the age of thirty-two had just completed a year as a Sloan Fellow at the Massachusetts Institute of Technology after holding a series of positions in oil supply, marketing, finance and planning during the 1960s.[161] As for Deam, he was removed from OR Division and made research adviser to Geoffrey Searle, the BP Trading director responsible for finance and planning.[162]

At the same time, or thereabouts, the MSG's recommendation that the ORDG should be strengthened was put into effect. The chairman of the reconstituted ORDG was, significantly, Peter Walters, who had succeeded McColl as general manager of S&D before being further promoted to the board of BP Trading early in 1971.[163] The new ORDG over which he presided was a formidable body, whose membership included five other BP Trading directors and Robert Belgrave, the policy planning adviser to the board. Deam, who had been a member of the old ORDG, was excluded from the new body.[164] Although the composition of the new ORDG was not totally one-sided (Belgrave was closely associated with IMR as its management 'spokesperson' from the mid-1960s), its formation was ominous for Deam.

It took some time fully to define Deam's new terms of reference, but the basic conditions were clear from the outset. He was to be concerned essentially with ideas and to have no responsibility for the day-to-day direction or management of the OR function, which, as seen, was under Horton's wing in CDPD. Deam's proposals for new research and development work were to be submitted to the new ORDG, which was to have sole and final authority for approval or disapproval and for the sanctioning of the necessary resources.[165]

Isolated and outpowered, Deam remained characteristically unsubdued. According to John Boxshall, the general manager of CDPD, Deam's attitude to the new ORDG was initially a 'point blank refusal to cast his pearls before swine'.[166] Both Boxshall and Horton had 'arduous sessions' with Deam in which they attempted to move him from an 'openly mutinous' to a more co-operative attitude.[167] Deam could not, however, be restrained from expressing his views and in the middle of July 1971 he wrote to Searle setting out what he saw as the 'essential fundamentals' of OR in the Company. According to Deam:

> Science in general has found that the major advancements have been made where problems were assumed with great confidence not to exist. The same has been true of the history of operational research in BP. Sixteen years ago no-one was interested in the idea of integrating Refinery/Supply programmes; indeed, it was known that such integration was completely unnecessary and that there was no problem. It has been found that the closer an individual was to an operation the more likely he was to consider that no basic problem existed . . .
>
>   It is a fact that all projects begun in the OR Division were not only internally generated, but at the time of their generation were against what was generally accepted in the Company as 'common sense' . . .

It is, of course, possible to run research on the basis of an authoritarian administrative control unit, but the quality of work produced under such circumstances is bound to suffer, and in particular the big problems will be overlooked, since the work being done is likely to be concerned with the small problems recognised by operations people.[168]

This thinly veiled attack on the ORDG (the 'authoritarian administrative control unit') and on the alleged myopia of its members fell on deaf ears.

The new ORDG had already decided at its first meeting in June that Horton should review the future organisation and scope of OR in BP.[169] His report, issued at the beginning of September, recommended that OR user departments should have a greater say in OR projects and should be the sponsors of OR work. He also felt that BP's OR was biased towards deterministic techniques and neglected simulation methods. He therefore recommended that OR research should be divided between two units, one concerned with optimisation, the other with simulation. The purpose of this suggestion was 'to redress the balance between the two broad areas of OR'.[170] He also recommended the establishment of a third unit to provide consultancy services to OR users in the company. Horton's recommendations were approved and the OR Division was duly split into three branches dealing, respectively, with optimisation projects, simulation projects and OR user services.[171] The ORDG also agreed that as a general principle OR projects should not be carried on for longer than six months without the support of a sponsoring department or departments.[172]

Meanwhile, during the second half of 1971 existing OR projects were reappraised, resulting in greater emphasis on simulation techniques and the simplification or curtailment of Deam's grand plans for Group optimisation through IMR and FIRM.[173] The position was summed up by Newby who wrote to Searle early in November 1971 after returning from what was presumably a retirement trip around various overseas associates of the company. According to Newby, 'IMR and the use of LP for planning purposes and most of Deam's ideas appear to have been fully discredited by London. Deam now has very few friends in the field and his personal position is probably becoming untenable.'[174]

As Deam watched his old empire being dismantled, he continued to complain to Searle that he was being sidelined and had 'no direct contact with the ORDG. This means that I have no right to discuss progress in this field.'[175] When Deam did attend the ORDG's next meeting to give a presentation on the FIRM project, his ideas were called into question and failed to win acceptance. There is more than a hint in Horton's handwritten notes on

Discussion

D Gerwick – agrees with major principles but has some doubts on obj fn.
– agrees with Time – phasing and eventually with Finance into IMR.
has doubts on optimum – is it achievable? Practicability :· of circumstances.

Is it worth the money (what are the costs)

Work shd continue but FAA shd be more involved

M. Rennell .. define objectives of company – can it be done ... "Maximise value of coy to shareholder" (W Greaves)

D Greaves : doubts about validity of answers because of restrictions in model.

O Grassud – net cash flow not divs.

R. Da $NPV = D + E = \dfrac{NOPAT}{K} + I\left[\dfrac{E^* - K}{K}\right] \theta$

58   Extract from Robert Horton's notes on BP's Operational Research Directing Group meeting on 9 November 1971

the meeting (see illustration 58) that the members of the ORDG and Deam were not even speaking the same language.[176]

The reorganisation of OR Division under Horton was a temporary measure pending a further investigation of OR by the Management Consultancy Group (MCG), which had been set up in 1970–1 as an internal unit for organisational studies.[177] The MCG recommended what was, in effect, a final purge of Deam and his old OR Division. They suggested that the post of research adviser should be abolished and that most of the OR Division should be merged with CDPD's Planning Systems Division, which already used mathematical techniques for planning and forecasting.[178] All that would remain of the OR Division was a small OR Services Unit with the passive role of providing support, liaison and consultancy services to OR users. The sovereignty of user departments and associates over OR was affirmed by the suggestion that where departments and/or associates had a need for OR applications, these should be performed by OR units embedded in the departments concerned. As if to complete the exorcism of Deam, it was further suggested that 'in order to lessen misunderstanding and suspicion it is desirable to avoid the title "OR" altogether'.[179] Consistent with these recommendations, the MCG also proposed that the Operational Research Policy Directing Group (as the ORDG had been renamed) should be incorporated into a new Corporate Systems Directing Group which would concern itself not specifically with OR, but with planning and information systems.

The MCG's recommendations were endorsed by the ORPDG in August 1972 and were implemented in September, except for the formation of the Corporate Systems Directing Group which was delayed for three months.[180] Under the new arrangements, Deam was put into dignified exile as a BP-sponsored professor at Queen Mary College, University of London, where he was made director of a new Energy Research Unit working on a world energy model.[181]

Deam's departure from BP was but one of the signs of the end of an era. For nearly twenty years BP had operated in a relatively stable environment. There had been no major concessionary crisis since the upset in Iran in 1951–4 and there had been no major recession in the world economy. On both the supply and demand sides BP had enjoyed quite a smooth ride, with the occasional bump, but no large crash. However, by the early 1970s, with BP's concessionary position looking insecure, the future was highly uncertain. This new environment was no place for deterministic planning, or indeed for integration. But although Deam and his LP techniques had fallen out of favour, BP retained the wholly owned Scicon through which BP had made a substantial diversification into computing.

⟵ 17 ⟶
Nutrition

One of the concerns of the postwar world was that, as the human population grew, mankind would be faced with a rising shortage of proteins. Natural protein supplements, mainly fishmeal and soyabean, would not, it was thought by the UN and others, be enough to fill the 'protein gap'. Synthetically produced substitutes would have to be found to prevent the occurrence of a classic Malthusian check on world population growth.

The idea that a solution to this problem might come from an oil company would have seemed far-fetched and was certainly not in the minds of the oilmen at BP. By tradition and convention, oil and food did not go together, indeed were best kept apart. That seemed obvious, until new possibilities were opened up by research which initially was aimed at improving oil products, not producing food.

GENESIS

In 1957, BP's 70 per cent owned French associate, Société Française des Pétroles BP (SFPBP), began microbiological research to try to solve the problems of removing sulphur from some crude oils. For that purpose, a microbiology section was set up at SFPBP's Lavera refinery near Marseilles in September 1957. Professor Jacques Senez, who ran a government laboratory in Marseilles, affiliated to the Centre National de la Recherche Scientifique, was recruited as a consultant soon afterwards.[1] He had already worked on the metabolism of hydrocarbons by bacteria. Results of the research at Lavera and at Senez's laboratory were, however, disappointing, causing the director of refining at SFPBP to request a reduction of expenditure on the microbiology project.

Senez was keen to continue the collaboration and suggested an alternative line of research. He had found that certain micro-organisms could feed on hydrocarbons, consuming only the wax (n-paraffins) that was present. This interested SFPBP because the high wax content of gas oils from some crude oils caused high pour points, which were a problem. Research was therefore continued into the possibilities of using micro-organisms to dewax gas oils. The results confirmed that this was scientifically possible, but that the process was uneconomic unless a use could be found for the protein-rich biomass which was created as a by-product.

The genesis of the idea that the process could be used for the production of protein remains somewhat shrouded, and opinions differ on who should take the credit for the original concept. The view that came to prevail was that the person with the strongest claim was Alfred Champagnat, director of research at SFPBP. He had apparently already thought of producing proteins from oil before a lunch at the Hotel Terminus in Marseilles in November 1959, which by all accounts was a crucial event in the birth of BP's proteins project. The lunch, at which Senez and Champagnat were present, was organised as a '*repas d'adieux*' to bring to an end the collaborative work on microbiological dewaxing of gas oils. Accounts of who said what at the lunch vary, but one way or another the idea was put forward that the dewaxing process could be used to produce proteins not as a useless by-product, but as food. The outcome was a new initiative to produce proteins from oil and so to help solve '*la faim du monde*'.[2]

RESEARCH AND DEVELOPMENT

Research proceeded on a process in which gas oil was fed into a fermenter, where yeast micro-organisms converted the waxy n-paraffins in the gas oil into protein. In a later harvesting stage, the dewaxed gas oil was separated from the protein. The products of the process were thus dewaxed gas oil and protein concentrate.[3]

The first patents covering the process were taken out in August 1960 by Champagnat. His role in the early stages of the protein project was similar to Bob Deam's in operational research (see chapter sixteen). Both men possessed the priceless asset of originality, were fired by enormous enthusiasm for their work, and saw almost limitless scope for the application of their ideas. At the same time, they tended to make exaggerated claims for their achievements and were conspicuously bad at fitting into a corporate hierarchy. It was these corporate nonconformists, individualists rather than team-workers, who provided the early creative drive for BP's pioneering work in the separate fields of operational research and proteins. They were both

59　Alfred Champagnat, pioneer of BP's proteins-from-oil process, at the Lavera refinery in France

helped by Maurice Banks, the most technically minded (and qualified)[4] of BP's executive directors at the time. He recognised the value of Deam's ideas about operational research, and was instrumental in transferring Deam from Australia to the centre of BP's operations in London. As for Banks' role in proteins, Champagnat later recalled that 'for me he was "*le bon dieu*" – if I had any hierarchical problems I went to see Banks – it was of vital importance to my morale'.[5] For his part, Banks in 1960 regarded the proteins project as 'a good example of Champagnat's original approach . . . the work is a real invention and one never knows where an invention of this type might lead – I suppose that sentence sums up our opinion of Champagnat pretty well. You never know what he might produce.'[6]

By autumn 1962, laboratory work at Lavera had confirmed the potential of the protein produced from gas oil as a food supplement for animals and, possibly, humans. Further development required a pilot plant to provide larger product samples, to help in assessing engineering factors, and to yield information on the economics of the process. It was therefore decided to build a pilot plant at Lavera, with the capacity to treat about two tons of gas oil a day.[7]

Shortly afterwards, Champagnat delivered a paper on the protein process at BP's annual technical conference. His presentation in broken, emotional English, accompanied by slides of victims of malnutrition, was a memorable occasion. He received a standing ovation, and his vision of BP helping to solve the problem of malnutrition in the world inspired great enthusiasm. David Llewellyn, assistant manager in the Operational Research Division, spoke in reply:

> The results so far achieved at Lavera are, I believe, only the beginning of what will become a major branch of technology. Mon. Champagnat has shown us the way. It seems that this new development could lead to the creation of a whole new industry, in a similar manner to the Petroleum Chemical Industry and its development in the 1930's. Those companies which were first in the Petroleum Chemical Industry to see the long range implications have reaped larger benefits than those who entered later. In this new field we believe that we are first and are at present alone. We cannot expect to remain alone but we must remain in front.

Pat Docksey, the general manager of BP's Research and Technical Development Department, closed the session, saying, 'a major break-through had been made, which would enable science to make a new contribution to this very important world problem'.[8] Shortly after the conference, Llewellyn was appointed co-ordinator of the protein project. He later recalled that 'the project from that point was unstoppable, more or less unlimited amounts of money, blank cheques etc'.[9]

The project may have been unstoppable, but it could not be allowed to run out of control. This was an awkward management problem, which at times caused tension between SFPBP in France and BP in London. The French, having discovered the protein process, were keen to push ahead quickly with its development and wished to preserve a special position in its exploitation. Indeed, the subject of proteins was a very emotional one for the French and they were quick to resent any proposal or decision which seemed to them to fail to acknowledge their contribution to the project. In London, on the other hand, BP was concerned to ensure that the process was exploited for the benefit not only of SFPBP, but of the whole of BP.

The initial research on the process had been done at Lavera, but in May 1963 BP's board approved the establishment of a new Research and

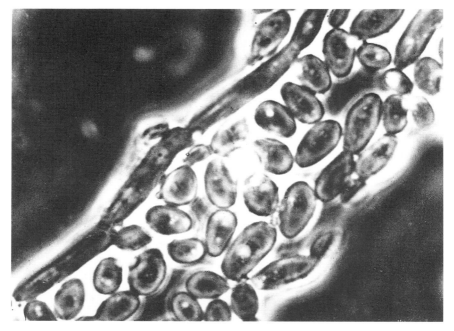

60   Making proteins from oil: the breeding of yeast on gas oil

Development Group to work exclusively on proteins at the Grangemouth refinery in the UK. Funds were therefore sanctioned for a microbiological laboratory and a small pilot proteins plant at Grangemouth. One of the purposes of this move was to ensure that knowledge of the proteins process was available in the UK 'and to the BP Group as a whole'.[10]

The Grangemouth team worked on an alternative to the Lavera gas oil process, which had been found to have drawbacks. The main one was that after the yeast metabolised the waxy n-paraffins in the raw gas oil feedstock, there remained a gas oil residue which was difficult to remove in the harvesting stage. At Grangemouth, waxy n-paraffins which had already been separated from gas oil were used as feedstock and were almost completely metabolised by the yeast in fully sterile conditions, which made the harvesting stage easier. The added cost of using n-paraffin feedstock was balanced by savings in the simpler harvesting procedure.

Meanwhile, the pilot plant at Lavera commenced operations in March 1963, but not without problems. A consultant, Professor S. R. Elsden of the

Department of Microbiology at Sheffield University, visited Lavera and reported that in his opinion those working on the project had insufficient knowledge of fundamental microbiology. He thought that 'they have attempted to run before they can walk'.[11] In London, it was agreed that the proteins work in France needed to be better organised and that steps should be taken to 'integrate it satisfactorily with the work of the rest of the Group'.[12] To that end, a new section was set up in Research and Technical Development Department to co-ordinate the proteins project. The new section would, among other things, arrange for feeding tests on BP proteins and report the results to a newly formed Protein Research and Development Steering Committee.[13]

The laboratory chosen for the feeding trials was in the Netherlands, the Centraal Instituut voor Voedingsonderzoek (CIVO) of the Technische Nederland Organisatje (TNO). It was well equipped, and had an excellent international reputation. Under the new testing procedures, samples from Lavera and Grangemouth were sent to Llewellyn at Sunbury, from where they were despatched to CIVO. The results were sent back to Llewellyn, who decoded and forwarded them to the research teams. This procedure took control of testing away from the French, in particular Champagnat, whose reports on early feeding trials were thought to be excessively optimistic.[14]

There was also some divergence between SFPBP in France and BP in London on the aims of the project. On the French side, much of the emotional commitment to the project came from the idea that they were working to help solve the problems of starvation and malnutrition in the world. They therefore assigned a high priority to developing BP protein as a food for humans as well as animals. This idealistic vision also had some influence in London, where Fyfe Gillies, assistant general manager in the Research and Technical Development Department, thought that the BP process held out 'the only short term possibility of eliminating malnutrition from the world'. He assumed that this was 'of over-riding importance'. Profits, it seems, were a secondary consideration. As Gillies put it, 'although the goal to be wished for is the credit of removing starvation from the world, and perhaps a Nobel prize, one can contemplate with some confidence that BP will not be out of pocket at the end of the day'.[15] Generally, however, BP's London management was cautious about the idea of using BP proteins for human consumption, at least for some years. The short-term contribution of BP proteins to human nutrition would come indirectly, through its use in feeding animals, which in turn would be eaten by humans.[16] Bridgeman himself 'laid down the ruling that we should go the animal feed route'.[17]

Another problem in controlling the proteins project was that Champagnat was not temperamentally suited to lead the patient work of development. Although he was partly responsible for the deficiencies in the organisation of the project in France, he resented what he regarded as interference from London. He complained that BP in London was putting brakes on research at Lavera by slowness of decisions and concentration on details. Llewellyn, he wrote, 'is becoming more and more difficult to bear'. The Research and Technical Development Department was, he argued, making technical errors, basing decisions on theory rather than the practical experience gained at Lavera. There was, he thought, no need for the Grangemouth team to attempt to work under sterile conditions. He complained that it was impossible to get access to the TNO (CIVO), that there was a danger of their over-technical reports being interpreted in London by people with inadequate knowledge of nutrition, that London intervened too much in monitoring external relations, and that censorship of what could be published in articles and at conferences was too severe.[18]

Jean Chenevier, the president of SFPBP, understood the problem with Champagnat, describing him as 'remarkably enthusiastic and excellent for research work, but by the same token not always easy to manage . . . he was always very difficult to control − his temperament was suited to his working by himself, pursuing his ideas, and not to obeying orders . . . he had about 10 new ideas per day − an extraordinarily inventive person − very original − it is true, though, that he was not made for life in a large organization, having little notion of discipline, co-ordination'.[19] The idea had already been mooted in mid-1963 of taking Champagnat away from the protein project at Lavera and putting him in charge of a new laboratory owned by BP rather than SFPBP, but probably situated in France.[20] This idea was carried out in 1964, when BP set up a new microbiological research laboratory at Epernon, near Paris, to undertake microbiological research with an emphasis on fundamental rather than applied work. Champagnat was put in charge on the basis, laid down by Banks, that he 'must have freedom to develop his ideas in any interesting direction'. However, Banks went on, 'this enterprise is bigger than any single individual and . . . it must be set up firmly for the interest of the Group as a whole'. Champagnat was therefore to report to the general management of the Research and Technical Development Department in London, who would be 'the ultimate authority for the coordination of the operation'. Banks also insisted that the new company which was set up to own the Epernon laboratory should be 90 per cent owned by BP and only 10 per cent by SFPBP,

61 Sir Maurice Bridgeman, BP's chairman and chief executive (right), and Maurice Banks, one of BP's two deputy chairmen (left), examining apparatus used in microbiological research at Epernon, France, in the mid-1960s

despite requests from Chenevier for a higher French shareholding.[21] The establishment of the Epernon laboratory was thus intended to keep microbiological research firmly under London's control, while removing Champagnat from direct involvement in the development of the Lavera process for the bulk manufacture of proteins from oil. Instead, he would undertake the more free-ranging, fundamental research to which he was better suited.

DEMONSTRATION UNITS

By 1965, BP had made a substantial commitment to proteins research and development. Three microbiological laboratories had been set up at Lavera,

Grangemouth and Epernon respectively, and there was also a biological section at the Sunbury research centre. Pilot plants had been installed at Lavera and Grangemouth. Two processes were under development: the gas oil process at Lavera and the n-paraffin process at Grangemouth. Product testing was being carried out in the Netherlands.

BP could not, however, expect to maintain its lead in proteins without challenge. SFPBP's discovery of 'petroproteins' encouraged other companies to search for ways of making proteins from non-conventional sources. BP, wanting to stay ahead of the competition, was keen to commercialise its proteins discovery as quickly as possible, despite continuing problems with the processes it had under development.

In mid-1965, serious reservations about the viability of BP's proteins project were being expressed by some of those closely involved in it. In July, Docksey doubted whether it would be economically viable to produce proteins by the BP processes.[22] Dr Frederick King, scientific adviser in the Research and Technical Development Department, was of like mind about the economics, and also doubted whether farm animals, apart from broiler chicks, would 'thrive for their normal lifespan on diets fortified with yeast'.[23] In October, he wrote that, after reviewing the latest information from Lavera, Grangemouth and the TNO, 'one cannot but feel serious anxiety for the future of the Protein Project'.[24] This statement was echoed by Gillies in November.[25]

Less than three months later, however, Gillies, spurred by reports that other organisations were planning to manufacture proteins from oil, suggested that BP should announce plans for a commercial plant within six months. 'Realistically', he wrote, 'we must now accept that time is not with us, if we hope to obtain the credit for demonstrating the commercial practicability of producing food from oil'.[26] A few months later, an announcement by Standard Oil (NJ) and Nestlé that they were engaged in a joint research programme on petroproteins added to the pressure.[27] Gillies thought that 'when two companies of this standing make an announcement it is obviously serious business'.[28]

By late February 1967 Bridgeman had agreed in principle, though not confirmed, that BP should construct at least one proteins plant with the capacity to produce 16,000 tons of protein a year.[29] The Central Planning Department was asked to examine the matter, and in October issued a study on the commercialisation of BP proteins. It concluded that:

> There is little doubt . . . that sooner or later the world will see the development of a protein manufacturing industry. Technologically speaking, BP appears not only to be well in the lead, but also to be the only organisation in the world

with a viable process within its grasp; however, it cannot be assumed that this will be maintained . . .

. . . It would be a pity to find that BP had been denied the privilege of building the world's first commercial protein plant and so provide the embryo of a new industry which must contribute substantially to the wellbeing of the human race.

There is enough positive evidence to justify BP taking a decision of principle today to enter this field of activity. It is accordingly recommended that this should be done as such a decision of principle would protect the Group's position of leadership.[30]

Other oil companies, apart from Standard Oil (NJ), were by this time known to be actively seeking ways of making proteins. They included Shell, Gulf, Sun Oil and Mobil. BP also knew that in the communist bloc the Russians, Czechs, East Germans and Chinese were working on processes to make petroproteins.[31] BP therefore hurried ahead and in November 1967 announced that it intended to build a £2 million gas oil protein plant at Lavera, with construction expected to start in 1968 and finish in 1970.[32] At the same time, news came out that Standard Oil (NJ) and Nestlé had successfully tested a pilot plant and were going to build a larger plant in the USA.[33]

Although BP had issued a statement of intent to build a plant at Lavera, financial approval for the plant had yet to be granted by the BP board. Comparisons between Lavera's gas oil process and the n-paraffin process developed at Grangemouth indicated that there was not much difference in the economics of the two processes. This meant that there was as good a case for commercially proving the Grangemouth process as the Lavera process. The first to get approval, by a board decision in mid-1968, was the n-paraffin process, which was to be used in a new plant to be erected at Grangemouth. The plant would be designed to produce 4,000 tons of protein a year from a single fermenter, and would hopefully demonstrate the feasibility of constructing larger plants with multiple fermenters.[34] It was for that reason called the Demonstration Unit. Approval for the gas oil plant at Lavera followed at the end of 1968. This plant was to have the capacity to produce 16,000 tons of protein a year. Although the projected economic return was low (a discounted cash flow return of 4 per cent), it was hoped that the plant would demonstrate the economic viability of the process.[35]

The spur of competition that had pushed BP to announce its plans to build the Lavera plant also hastened its first licensing agreement. For several years, BP had been in intermittent contact with the Kyowa Hakko Kogyo Company of Japan concerning possible collaboration in proteins. Kyowa

Hakko was seen as a desirable partner in the Japanese market, which was thought to have great promise. The national diet was low in protein for an industrial country, Japan had a large refining industry (for feedstock), and was well placed to supply the region of Asia, which had the largest protein deficiency in the world. Kyowa Hakko was a leading fermentation company, indeed its main fermentation plant was said to be the largest in the East, and possibly in the world. Despite these advantages, BP held back from granting a petroproteins licence in 1965 because it wished, first, to construct a commercial proteins plant of its own.[36]

Talks with Kyowa Hakko were resumed in 1967,[37] and were hurried on by the announcement that the Kanegafuchi Chemical Company, one of Japan's larger chemical companies, intended to manufacture proteins from oil. Gillies noted this 'with considerable concern'.[38] After negotiations, BP and Kyowa Hakko signed a licence agreement on 1 November 1968. It provided for Kyowa Hakko to make a commercial demonstration of BP's n-paraffin proteins process. Kyowa Hakko's demonstration unit was not, however, to start up until BP's own demonstration unit had been commissioned.[39] Gillies wrote anxiously, 'the advanced knowledge of the process which Kyowa Hakko have acquired on their own must warn us that others are very close on our heels. It is therefore absolutely vital to use all possible speed to mount the Grangemouth demonstration if we are to secure the greatest technological prize and prestige which BP has ever had within its grasp.'[40]

LARGE-SCALE COMMERCIALISATION

After the decisions to build the demonstration units at Grangemouth and Lavera, a review began of the future strategy and organisation of the proteins project. The prevailing mood was one of great optimism. Gillies thought that the 'economic manufacture of protein from oil offers BP the possibility of a major diversification and the chance, over say the next twenty years, of becoming one of the great food companies of the world'.[41] The French, in a long paper on the future of petroleum microbiology, took an exalted tone:

> Is this not a signal that a third age in the history of man is about to begin: surviving initially by hunting and gathering . . . living today by husbandry and agriculture . . . now we see that having mastered the chief factors of biology thanks to industrial techniques, mankind is making sure of biosynthetic food production entirely under his own control, thus preparing the era of future nutritional practice.

The paper went on in underlined text, 'the unacceptable reality of hunger in the world places on the shoulders of the inventors of such a discovery the moral obligation of developing it to the best of their ability within the limits in which this development can be profitable'.[42]

At a high-level meeting in October 1969, George Ashford, a BP director, emphasised that proteins were considered to be 'one of the major lines of development for the BP Group and stressed the Management's and his own enthusiasm in the matter'. In this expansive mood, ambitious targets were set for BP's future protein operations in Western Europe: an investment of £105 million to construct several plants with the combined capacity to produce 700,000 tons of protein in 1974. As for organisation, 'it was time', noted Ashford, 'to organise the protein project as a business'.[43] That was done by the formation of a new subsidiary, BP Proteins, which was announced on 1 December 1969, with the formal incorporation following in March 1970.[44]

The targets agreed in October 1969 were soon reduced as various obstacles were encountered. The main purpose of the demonstration units at Grangemouth and Lavera was to enable BP to acquire the necessary data for the design and operation of large-scale (100,000 tons a year) plants.[45] It was assumed, when the targets for future production capacity were agreed in October 1969, that the Grangemouth demonstration unit would be commissioned in July 1970, and the Lavera unit in April 1971. In practice, commissioning was delayed by mechanical problems, with the result that the Grangemouth and Lavera units were not commissioned until 1971 and 1972 respectively. Even then, they did not operate at full capacity, mainly because of continuing technical problems. As these units came on stream, commercial sales of Toprina (the brand name for BP protein) began in the UK in late 1971 and in France in early 1972. Different grades were sold as milk replacer in calf feeds and as replacements for fishmeal and soyabean in pig and poultry feeds.[46]

Experience with the demonstration units showed that the earlier cost estimates for large-scale plants were much too low.[47] While cost estimates rose on the one hand, the availability of capital fell on the other. BP had been under growing financial pressure for several years (see chapter twelve), and had been able to fund its growing capital expenditure only by increasing its debts. In the early 1970s, its top investment priorities were to develop its oil discoveries in the North Sea and Alaska. Capital constraints therefore had to be applied to other parts of BP's business. In June 1971, the board of BP Trading, which ran BP's operations under the strategic oversight of the main board, agreed that BP could not afford capital expenditure on new developments in either chemicals or proteins. Only a 'really worthwhile project' could lead to a reconsideration of this policy, especially if it could be financed in a way which would not reduce the funds available for investment on higher priorities.[48]

These financial constraints were very important in BP's decision to build its first large-scale proteins plant in Italy in partnership with ANIC, a subsidiary of the Italian state-owned company, ENI. In June 1971, BP and ANIC reached agreement to form a 50:50 joint company, Italproteine, which was to construct a plant at Sarroch in Sardinia, with the capacity to produce 100,000 tons of protein a year, using BP's n-paraffin process.

Before proceeding with this project, a study group was to report on its viability.[49] BP wished to hold back from committing to the project until the Grangemouth demonstration unit was operating satisfactorily, the availability of Italian government grants and loans was confirmed and the Italian government had approved Toprina for use in animal feeds.[50] As had happened before with proteins, however, the appearance of a competitor hurried things along. In this instance, it was the Italian chemical company, Liquichimica, which in late 1971 announced plans to build a 100,000 tons a year protein plant in Calabria, southern Italy, using Kanegafuchi's n-paraffin process under licence. ANIC was worried about the appearance of a rival, mainly because Liquichimica would compete with Italproteine for financial 'incentives' from the Italian government.[51] To satisfy ANIC, BP agreed to a press release in Italy in June 1972, announcing that contractors had presented bids for the construction of Italproteine's plant and that an order would be placed within the next few months.[52]

By November, a case for the Italproteine plant had been drawn up for consideration by BP's board. Ashford's notes for his verbal presentation in support of the case stated clearly that in view of BP's capital constraints 'we cannot afford to give much if any weight to non-economic factors such as the starving millions, BP's public image and pressure from our possibly over-enthusiastic French colleagues. We may even have to ignore the fact that this is BP's only major successful research project for a number of years.' However, after 'sober assessment' of the economics, and in spite of BP's 'present financial stringencies', Ashford argued that it would be wrong to abandon the protein project. If BP held back Liquichimica would get into the market first and BP would be at a disadvantage if and when it followed suit.[53]

The proposal which was laid before the board did, indeed, look appealing. The estimated capital costs of the plant were £20 million, of which three quarters would be provided by Italian government grants and subsidised loans. BP's equity contribution would be a mere £2 million. The board approved it on 9 November.[54] Less than a week later, the Italian decree giving provisional approval for the production and sale in Italy of proteins produced from n-paraffins was published.[55] The way at last seemed clear for the construction of a large-scale proteins plant, and contracts for its construction were placed with the contractor, Foster Wheeler, in February 1973.[56]

Disturbingly, however, the estimated costs of the plant were soon revised upwards. In July, when they were raised to £26 million, BP agreed to continue with the project, pending a full review at the end of the year.[57] By then, the capital cost had risen again, to £29 million.[58]

That, however, was nothing compared with the impact of the revolution that was taking place in the international oil industry at the time. In the last three months of 1973, the OPEC countries had dramatically taken control of crude oil prices and almost quadrupled them from about $3 to nearly $12 a barrel. Rocketing feedstock costs posed a stiff, perhaps insuperable, challenge to the economic viability of producing proteins from oil.

In this context, BP Trading's Executive Committee met in mid-December to consider the Italproteine plant. According to the minutes, 'we now had little hope of much profit in this project . . . Attention was drawn to the continuing technical risks (with the Grangemouth plant not having worked properly since May); to the steeply rising cost of oil and the chance that protein prices (fishmeal and soyabean) might not rise so fast. Further that oil-based protein might prove not to be an economic source of food.'[59] On seeing these minutes, BP's chairman and chief executive, Eric Drake, doubted 'the whole future of the present protein process'.[60]

In January 1974, the board considered the review of the Italproteine project. Oil, it was pointed out, was not the only commodity that had gone up in price in 1973. There were, at the same time, dramatic increases in protein commodity prices, especially fishmeal and soyabean, which was expected to continue rising in price. More decisively, perhaps, withdrawal from the Italproteine project would have involved cancellation penalties that would have exceeded BP's share contribution, even at the revised cost estimates. Briefed on these lines, the board agreed to continue with the Italproteine project. Construction of the plant in Sardinia began early in 1974.[61]

Although the Italproteine project was going ahead, the sharp rise in oil prices punctured the idea of building other large-scale protein plants in Europe. BP looked instead to the possibilities of protein joint ventures and licensing agreements with oil-producing countries, which had cheap feedstocks and the desire to industrialise through oil-related technology. It was hoped that the use of BP's proteins technology in partnership with producing countries would not only be viable in its own right, but would also help

BP to gain a 'preferred position' in the oil-rich states which possessed the world's largest and cheapest oil reserves.[62] Talks on protein ventures were held with Iran, Saudi Arabia, Venezuela, Indonesia and the USSR, but nothing resulted.[63]

In Japan, meanwhile, BP's association with Kyowa Hakko in Japan had proved disappointing. Two other Japanese companies, Kanegafuchi and Dainippon Ink and Chemicals, were much more energetic in seeking official approval of their petroprotein processes than Kyowa Hakko, who seemed to lack enthusiasm for the project.[64] In early 1973, strong protests from consumer groups were sparked off by reports that the Japanese Ministry of Health was about to grant approval to Kanegafuchi and Dainippon for the use of their protein products as animal feeds. So strong was the public outcry that these two companies abandoned their plans for protein plants in Japan. There was some speculation that Japanese interest in petroproteins would revive, but it did not, and BP's agreement with Kyowa Hakko came to nothing.[65]

DEMISE OF BP PROTEINS

Failing developments elsewhere, BP's only extant project for large-scale protein manufacture was Italproteine's plant at Sarroch in Sardinia. This soon ran into deep problems with product approval and environmental concerns. On both these issues, BP became convinced that its problems were political, rather than scientific.

The main institutional protagonists were, on the one side, BP, ANIC and ENI, who were meant to be partners but did not always behave like it; and on the other side, the Italian Ministry of Health. This Ministry was advised by the Consiglio Superiore della Sanita (CSS), in effect a higher health council. Below the CSS, and reporting to it, was the Istituto Superiore della Sanita (ISS).

At an individual level, Cyril Shacklady, BP Proteins' nutrition manager, was the BP man most directly involved with trying to gain product approval for Toprina. He was a scientist, highly respected within BP for his competence and integrity, but he had little sense of diplomacy and was temperamentally quite unsuited to dealing with Italian bureaucracy. Shacklady expected the meticulous scientific evidence which he presented to demonstrate the safety of Toprina to be accepted, and he came to regard the questioning of the evidence as an attack on his personal integrity. A former colleague later described him as 'very much a loner, difficult to persuade him of other views. Approached the Italians by himself at his insistence. Wore bowler hat, monocle and umbrella . . . and shouted at the Italians. Constantly referred to the World War II and the Italians' part in it.'[66] Another ex-colleague recalled that he refused to learn Italian, 'never knew the names of relevant people in Rome, did not adopt any Italian interests, or express appreciation for their culture etc. Quite the opposite attitude was taken.'[67]

On the side of the Italian authorities, the main protagonist was Professor L. Bellani, Director General of the Veterinary Service in the Ministry of Health. He proved very difficult to deal with, even for those more diplomatic than Shacklady. Bellani was responsible for, among other things, setting up tests on Toprina to be carried out in Italy. He was an active socialist, and may have been influenced by the political opposition to petroproteins which came from a group in the socialist party. According to an ENI employee, Bellani was 'our sworn enemy, and would try every possible means to obstruct us'.[68]

After the issue in November 1972 of the Italian decree provisionally approving Toprina, the ISS conducted its own tests on the product, and in February 1974 a second decree was issued, enlarging on the earlier one.[69] However, a month before the second decree came out, an article by Professor Biocca, a medical parasitologist, was published in a national paper, *Paese Sera*. The article implied that there might be serious health risks for people consuming Toprina-derived products.[70] Further articles began to appear in the press, largely directed against Liquichimica's protein project, but also criticising Italproteine. The thrust of these articles was that proteins from oil were unsafe, perhaps carcinogenic, and that Italians were being made guinea pigs for them because corrupt Italian politicians and bureaucrats were ready to permit production and sale of a product that other countries would not allow. For example, in April 1974 the left-wing newspaper, *Il Manifesto*, carried the headline, 'A story of multinationals, millions and ministers. Liquigas [Liquichimica] will produce bioproteins that fatten cattle and kill the people who eat them'.[71]

Much of the press comment was stimulated by an environmental pressure group, Gruppo Ambiente, led by an ambitious socialist magistrate, Gianfranco Amendola. Liquichimica was the prime target because of its association with the Calabrian socialist, Mancini, who was under attack by rivals within the socialist party. A working group formed by the Health and Agriculture committees of the party called for a ban on the commercialisation of petroproteins, the withdrawal of grants already made, and strict testing to be carried out by the ISS.[72] Other interests thought to be behind the anti-proteins campaign included the importers of soya, who were seeking to protect their share of the animal feed market, and communists holding the monopoly of Italian meat imports from Eastern Europe.[73]

By April 1975 Liquichimica's plant was nearing completion, but because of the attacks in the press, Liquichimica announced that it was reconsidering plans to build a bigger plant in Italy, and that it was being offered foreign incentives to build plants in other countries. This was perceived by the left-wing press as attempted blackmail, and caused a renewed outcry.[74] By the end of June, the Liquichimica plant had been completed, but Liquichimica had still not received permission to produce or sell its protein.

With such controversy raging, the ISS was reluctant to issue a report favourable to Toprina.[75] The Ministry of Health therefore called a meeting of the higher council (CSS) to consider the ISS test results. The council met in June and recommended that further tests should be carried out, and that in the meantime the approval decrees already granted should be suspended.[76] Shacklady was incandescent with rage. He wrote to an Italian professor in Rome,

> I know of no other European country where the antics of mini Mussolinis and bureaucratic Buonapartes could, in the space of two days destroy a position which it had taken over four years to attain.
>
> You know and I know that this whole business is so enmeshed with politics, with self advancement and with the protection of those who feel their interests may be threatened by our successful incursion into this field that logic, science and integrity are not going to be the deciding factors in determining the outcome . . .
>
> This latest performance of the CSS would have been more appropriately conducted in La Scala than in the Ministry of Health . . .
>
> As it is unlikely that the University of Rome will confer a Doctorate honoris causa upon me I am now contemplating presenting for their consideration a thesis, the title of which will be, 'The discharge of responsibility in Italian State Institutes; a study in invertebrate physiology'.[77]

The only point on which the ISS reported that it had doubts about Toprina was its finding of 71 parts per million (ppm) of n-paraffins in the fat of pigs fed on the product. BP countered that the n-paraffins used for the manufacture of Toprina met US specifications for food-grade mineral oil, which was permitted at levels of up to 2,000 ppm in the preparation of bread and confectionery in the UK and USA, and up to 950 ppm in frozen meat in the USA. Moreover, analyses of food products bought at random in shops in Rome revealed n-paraffin levels of up to 80 ppm in grissini (bread sticks) and up to 1,400 ppm in rice. The argument that there were no scientific objections to Toprina was, BP pointed out, supported by the French government and by

Professor Nevin Scrimshaw, who was chairman of the United Nations Protein Advisory Group, a director of the World Health Council, and head of the Nutrition Department at the Massachusetts Institute of Technology.[78]

The problems with product approval strained the relations between ANIC and BP, the partners in Italproteine, and there was considerable bickering about who was responsible for the unhappy state of affairs. One of the worst bouts of recrimination occurred in September 1975, when Camillo d'Amelio, president of Italproteine and vice-president of ANIC, made some off the record comments which were highly critical of BP. His main accusations were that BP had presented documents on Toprina containing false information, failed to disclose the presence of n-paraffins in Toprina and given inadequate support to ANIC in the effort to gain product approval.[79] Shacklady was furious. He wrote to Hector Watts, the managing director of BP Proteins, that (with his underlining), the first of d'Amelio's accusations was,

> a direct attack upon my personal and professional integrity and this I refuse to allow to pass unanswered.
>
> I insist that d'Amelio be made to substantiate, to my satisfaction, his justification for this remark or to withdraw it as publicly as he made it. I intend to pursue this particular point to the end, no matter how unpleasant the consequences. You may care to take this up with d'Amelio or I shall do so myself, if you prefer it, but taken up it must be; there is a point beyond which I am not prepared to concede anything and this is it.[80]

Watts insisted on a formal discussion on d'Amelio's remarks at the next Italproteine board meeting, after which there was an ostensible (if temporary) clearing of the air.[81]

In February 1976, the product approval position took a further turn for the worse, when the Italian government formally suspended the approval decrees of November 1972 and February 1974.[82] By springtime 1977, the Sarroch plant had been completed, but it could not be commissioned, because of the suspension decree.[83] Pressure from ENI helped to secure authorisation in October to start the plant, but conditions were attached: production was to be for experimental purposes only, and was limited to running the plant at the rate of 40,000 tons of feedstock a year; and the sale of Toprina from the plant was forbidden.[84]

When the plant was started in December, there were soon protests about it on environmental grounds. Problems were encountered with, among other things, the effluent system, which became overloaded. Modifications were required to rectify this problem, and to reduce emissions of yeast into the atmosphere around the plant. The operation of the plant was therefore stopped by mid-February 1977.[85] In Sardinia, a petition to the local authorities was drawn up and attracted thousands of signatories. It claimed that the production of proteins at Sarroch would constitute a grave danger to those living near the plant, or working at it.[86] The Minister for Industry came under attack for issuing the production permit to Italproteine and announced that the permit was suspended.

With the plant lying idle, the BP board approved further capital expenditure of £4·3 million on Italproteine, partly for modifications to meet environmental concerns.[87] At Sarroch, however, the local mayor withheld permission for the extension of the effluent system, announcing that he would not expose the local population to emissions from the plant until they were declared innocuous.[88] The ISS requested expensive modifications to incinerate waste gases from the plant to prevent the possible emission of live yeasts into the atmosphere. This would, BP thought, 'impose a heavy and fatal burden on the plant'.[89] In September, BP and ANIC, at the end of their tethers, agreed to take steps to liquidate Italproteine unless all the approvals necessary for the operation of the plant had been granted by the end of January 1978.[90] A BP Proteins strategy review in November went further and recommended that if the necessary approvals were not obtained in Italy in time for the Sarroch plant to be commissioned in mid-1978, then all BP's protein activities, including those at Grangemouth, Lavera and Sunbury as well as in Italy, should be closed down. More generally, the review commented that on the basis of normal economic criteria and prevailing oil prices, the large-scale and widespread adoption of petroprotein technology, which had been expected before the 1973 oil crisis, was unlikely to take place.[91]

The chances of obtaining approval for the product and modifications to the plant were poor. Although there was some concern about job losses if the plant closed, the mayor of Sarroch still refused permission for the extension of the effluent system. He was standing for re-election in February 1978, and the issue was central to his campaign.[92] The CSS was due to discuss Toprina on 16 November, but the meeting was largely taken up with discussion of Liquipron, the Liquichimica product. Conflicting reports were presented by two committees of experts, which led to the setting up of yet another committee to consider the problem.[93] This new committee was meant to reach a final decision by the end of January, but it failed to do so, and was due to meet again on 14 February. Before that date arrived, the BP board met on 2 February and decided to liquidate Italproteine.[94] This decision was confirmed at an Italproteine board meeting on the 28th.[95]

Formal procedures were begun in April, and liquidators officially appointed in May. It took several months for ANIC/ENI and BP to resolve their differences. At last, in November, it was agreed that BP would transfer its shareholding in Italproteine to ANIC and be released from all obligations arising from Italproteine, in return for a payment which, together with a loan reimbursement, brought the cost of withdrawal to £8 million.[96] The withdrawal from Italproteine was followed by the winding up of BP Proteins at the end of 1978.[97]

By that time, not many of BP's competitors in petroproteins were left. Shell had given up its methane process. Liquichimica had announced that it was going to liquidate its proteins subsidiary. ICI, who had started research into protein manufacture in 1968, and had subsequently announced plans to construct a full-scale plant, were described in 1978 as the sole survivor in petroprotein technology in the West.[98] By 1984, there appeared to be no commercial petroprotein plants in operation in the non-communist world.[99]

BIRTH OF BP NUTRITION

With a twist of strategy, BP was still, however, in the animal feed business. In 1973–4, expecting that it would soon be a large-scale producer of proteins, BP had decided that it should extend its proteins activities downstream. This decision stemmed from the conviction that BP should not limit itself to being a large-scale producer of an intermediate product, sold in bulk to feed compounders who would incorporate it into animal feeds. BP wished to share in the added value of the more finished product. The easiest way to do this was in alliance with established compounders, who already had knowledge of the market, established sales teams and management experience in an industry about which BP knew very little.[100] This was the classic BP strategy: to start in the upstream sector, and then to move downstream in association with established marketers. BP had followed this route in its mainstream oil business and in petrochemicals. Now it was the turn of proteins.

BP's first choice was to make a single major investment which would enable it to achieve its downstream objectives at a stroke. The only suitable candidate was Spillers, which had a large animal-feed business and various agri-business ventures in the Middle East, which fitted with BP's desire to promote BP proteins in oil-producing countries. In talks in May 1974, it became clear, however, that Spillers were only interested in joint agri-business ventures with BP in developing countries, and that they did not favour a wider association which would include their UK animal-feed business. As an alternative to an alliance, BP considered making a takeover bid for Spillers. A full takeover would, however, have cost about £50 million and would have included Spillers' large flour milling and bakery business, which BP did not want. Selling it would be difficult, because the Monopolies Commission would almost certainly not agree to it falling into the hands of either of the two other large UK flour and baking concerns, Rank Hovis and Associated British Foods. As there was no single alternative to Spillers, BP looked instead for a combination of smaller concerns.[101]

BP did not feel that it could afford to dally. Shell, which was working on the manufacture of proteins from methane, was reported to have made an offer for the Colborn Group, a UK feed company.[102] In addition, BP thought that ICI would soon try to enter the downstream sector.[103] BP therefore moved quickly to make two downstream acquisitions. One was a two-thirds shareholding in Trouw and Company NV, a Dutch firm which was a leading European producer of milk replacer, fish feeds and other specialities. The second was the 100 per cent purchase of Cooper Nutrition Products, a UK subsidiary of the Wellcome Foundation, with a 30 per cent share of the UK market for mineral and vitamin animal feed supplements and a 70 per cent share of the UK fish feed market. The combined purchase price for Trouw and Coopers was more than £8 million, which was thought to be on the high side. However, they had both also been approached by Shell, and BP felt that it could not afford to miss the opportunity of buying them.[104] In addition to BP's interests in Trouw and Coopers, SFPBP held a 50 per cent shareholding in Jouy, which specialised in fish feed, and Deutsche BP (formerly BP Benzin und Petroleum) owned HAKRA,[105] a company specialising in pig feeds.[106] From the beginning of 1977, these animal feed companies became parts of BP Nutrition, which had overall responsibility for BP's proteins and animal feed interests.[107]

With the abandonment of the idea of making proteins from oil, BP's downstream nutrition business would, however, have to find a new strategic purpose. BP Nutrition, conceived as an extension of BP's core oil business, had become a diversification unrelated to oil.

PART IV
OPEC TAKES CONTROL

=== 18 ===
'An avalanche of escalating demands'

By 1970 the political foundations of the oil majors' dominance of the international oil industry were weakening. British and American power, which had for so long underpinned the positions of the Western oil companies in less developed countries, was no longer a guarantee of security. After years of comparative economic decline and nearly a quarter-century of decolonisation, Britain was reduced to a secondary world role, clinging to its 'special relationship' with the more powerful USA.

While Britain declined, the USA stood out as the pre-eminent power in the West in the postwar years. The champion of capitalism, the USA was home to a growing number of US multinational companies, which expanded vigorously in the 1950s and 1960s, when they dominated the world's flows of foreign direct investment. It was in this period that the term 'multinational' was invented, and that the role of multinationals in the world economy became a matter of controversy.

To their supporters, the multinationals were engines of international economic growth, transferring capital, technology and skills to countries that lacked them. By promoting interdependence in the world economy, they enriched, it was claimed, both themselves and the host countries in which they operated, contributing to global prosperity for universal benefit. This view won sufficient acceptance in North America and Western Europe for multinationals to be able to operate with comparative freedom and security in most major economies in these areas.

Elsewhere, however, multinationals encountered greater obstacles and opposition. Japan was generally resistant to foreign control over Japanese industry and was not very receptive to multinationals. The communist world comprising the USSR, its satellite states and China, was closed to multinationals altogether. In the Third World, multinationals were prominent, especially in petroleum and other natural resource industries, but there was strong opposition to them. Many Third World countries were new nations, created in the process of decolonisation, and determined to assert their national sovereignty. Other less developed countries, though not former colonies themselves, associated themselves with the anti-imperialist cause.

In many cases, the less developed countries depended on a single commodity for most of their foreign exchange and government revenues. When that commodity, the commanding height of the national economy, was controlled by multinationals, the host countries tended to see themselves in a dependent rather than an interdependent situation. In their view, multinationals represented a form of economic colonisation, operating in enclaves, under expatriate management and appropriating the raw materials of less developed countries for the benefit of the more prosperous economies where the multinationals were usually domiciled. In the Third World, therefore, multinationals were widely seen, not as desirable engines of international economic growth, but as undesirable symbols of a new imperialism, emanating mainly from the USA, and superseding the old system of European colonial rule.[1]

By the late 1960s protests against American neocolonialism and economic imperialism were gaining ground even in the USA, where young Americans, disturbed by their country's involvement in the Vietnam War, increasingly questioned the USA's role in the world.[2] The costs of the war in Vietnam also contributed to the weakening of the USA's economic leadership, which was eroded by growing budget deficits, rising imports causing balance of payments deficits, and the growth of foreign competition from the resurgent European and Japanese economies, which were catching up with the USA in their technologies and labour productivity.[3] Its power apparently waning, the USA adopted a more limited world role, which President Richard Nixon announced in 1969 in what became known as the Nixon Doctrine. The USA would continue, Nixon declared, to provide a nuclear umbrella for its allies, but it would not fight their conventional wars for them. Instead, it would provide economic and military support for friendly regional powers to act as regional policemen.[4]

OIL, 'THE ACHILLES HEEL OF THE WEST'

The limits of the USA's power to shoulder the problems of its allies were also showing in the world of oil. The USA had become a net oil importer in 1948, but for the next two decades it had retained enough spare capacity to raise its output substantially in emergencies. This enabled it to help out other oil-consuming countries during the oil shortages caused by the Suez crisis in 1956 and the Arab–Israeli Six-Day War in 1967.

The reserve capacity of the USA was not, however, inexhaustible. Fast economic growth during the prolonged postwar boom in Western Europe and Japan used massive amounts of energy. Oil was much the cheapest fuel and in a dramatic transformation of the energy market, oil imports, principally from the OPEC countries, displaced locally produced coal as the dominant source of energy. By 1970 OPEC oil accounted for more than half of Western Europe's total energy consumption and nearly two thirds of Japan's.[5]

The US oil industry could not, meanwhile, keep up with even domestic US oil consumption. Americans continued to be the world's most profligate oil consumers, buying gas-guzzling cars, and looking upon cheap oil as virtually a birthright, part of the fabric of American society. 'Gasoline', said Senator Frank Church, 'is something like steak – people have got to have it'.[6] So much of it, in fact, that by 1970 the US oil industry had no spare production capacity left. It was working flat out to provide 77 per cent of the nation's oil consumption,[7] and could no longer provide emergency cover for the needs of Western Europe and Japan. Henry Kissinger, who was at that time Nixon's national security adviser, would later write that oil had become 'the Achilles heel of the West'.[8]

While the industrialised countries were becoming increasingly dependent on oil imports, the major oil companies were losing their grip on the international oil industry. Between 1950 and 1970, in the face of growing competition from many new private and state-owned entrants to the industry,[9] the seven majors' share of international crude oil production fell from 90 to 70 per cent.[10] At the same time, the bargaining power of the oil-exporting countries was fortified by the formation and consolidation of OPEC, which strengthened their capacity for joint action. By 1970 the number of countries belonging to OPEC had risen to ten.[11] Their oil exports accounted for more than 80 per cent of the world's international oil trade.[12] With the virtual disappearance of spare production capacity in other countries, the OPEC members were beginning to be more powerful than the majors.

This change in the balance of power went largely unnoticed at the time. The great size and global reach of the largest multinationals fuelled the idea, in the West as well as the Third World, that they were as, or more, powerful than nation states. One Western commentator wrote that 'future students of the twentieth century will find the history of a firm like General Motors a great deal more important than the history of a nation like Switzerland'.[13] Another that 'Standard Oil of New Jersey is much more powerful vis-à-vis Indonesia . . . than ever the British Empire was against the German'.[14] Still others claimed that the interests of multinationals determined the foreign policies of their parent governments.[15] In the next few years these ideas would lose much of their credibility as the tensions between nationalism and multinational enterprise came to a dramatic climax which shattered the concessionary system in the international oil industry and demonstrated the power of nation states to wrest control of their resources from the multinationals.

THE LIBYAN BREAKTHROUGH

The spearhead of the oil-exporting countries' campaign against the oil companies was Libya. After Qaddafi seized power from the ageing pro-Western King Idris on 1 September 1969, the political landscape of the international oil industry was transformed. Before then, the dominant voices among the oil-exporting nations were those of the pro-Western, conservative governments of Saudi Arabia, Iran, Kuwait and Libya, whose interest in increasing their oil revenues was balanced by their dependence on Western powers for protection against external, and perhaps internal, threats.[16]

Qaddafi most emphatically did not belong in that camp. He was an avowed radical who quickly established his anti-Zionist, anti-imperialist and anti-Western credentials. Literally overnight, Libya became one of the revolutionary republics, alongside Iraq and Algeria, the former French colony which had won its independence in 1962 after a bitter eight-year struggle against France.[17] Although Qaddafi's regime quickly announced that it would respect Libya's existing agreements with the oil companies, it was obvious that oil would be a political weapon in Qaddafi's armoury against the West, and especially against the USA. As the Libyan Oil Minister said, 'oil must be used against our enemies; it does not make sense to sell oil to America which supplies Phantom planes to Israel, which are aimed to kill Arabs'.[18]

Qaddafi's regime was in a uniquely strong position from which to challenge the Western powers and their oil companies. From a standing start in 1961, Libyan oil exports had grown at such a pace that by 1969 Libya was the largest Arab oil producer, ahead even of Saudi Arabia and Kuwait in the volume of its output.[19] This phenomenal growth reflected the strong demand for Libyan oil, which was high in quality, low in polluting and corrosive sulphur, and cheap to transport to the nearby markets of Western Europe. Its transport cost advantages over Persian Gulf crudes were increased by the closure of the Suez Canal, which had been out of operation since the Arab-Israeli Six-Day War in 1967. While Persian Gulf oil had to be shipped all the way round Africa to Europe, Libyan oil needed only to be moved across the Mediterranean. Production leapt ahead in 1968–9, by which time Libya was supplying about a quarter of Western Europe's oil requirements.

It was not only the strong demand for Libyan oil that put Qaddafi in a powerful position. The Idris regime had deliberately encouraged competition in the Libyan oil industry by granting concessions to many companies, independents as well as majors. The most successful independent in Libya was Occidental, led by the legendary entrepreneur, Dr Armand Hammer.[20] Occidental burst on the Libyan scene in 1966, securing concessions by bids wrapped in the Libyan national colours of red, green and black and including an offer to invest 5 per cent of the profits from the sale of crude oil from the concessions in an agricultural project at the Kufra Oasis, where the King's father was buried. Occidental was fantastically successful in its Libyan exploration drilling and quickly made major discoveries which it hastily developed. By late 1969 Occidental was the largest oil producer in Libya, just ahead of Standard Oil (NJ), and had acquired downstream operations in Europe to provide outlets for its Libyan oil. The other large producers of Libyan oil were the Oasis group, consisting of Shell and three US independents, Continental, Marathon and Amerada-Hess; Amoseas, a partnership of Texaco and Socal; Mobil in partnership with the German firm, Gelsenberg; and BP in partnership with the US independent, Bunker Hunt. These and other, smaller producers made up a fragmented industry structure in which the majors accounted for less than half of Libya's production. Qaddafi did not, therefore, have to deal with a dominant concession holder like the IPC in Iraq, the KOC in Kuwait, the Consortium in Iran, or Aramco in Saudi Arabia. The Libyan government was dealing with competing firms, which did not necessarily share the same interests.[21]

After securing the US withdrawal from its Libyan Wheelus Air Force Base, the Libyan government told the oil companies in January 1970 that it wanted a rise in posted prices.[22] Qaddafi, typically, was not in a mood for compromise and declared to the twenty-one oil companies operating in Libya that 'the Libyan people who have lived without oil for five thousand years can live without it again for a few years in order to attain their legitimate rights'.[23] It was an echo of Musaddiq, who more than fifteen years earlier had said that 'it is better to be independent and produce only one ton of oil a year than to produce 32 million tons and be a slave to Britain'.[24] 'You have never understood', he had told the Americans, 'that this is basically a political issue'.[25] But unlike Musaddiq, Qaddafi really was in a position to forgo current oil revenues to achieve his ends. Libya's oil income had risen to levels which exceeded the short-term development needs of the desert state, with its small population. Indeed, by 1970 Libya's per capita income was twice as high as Saudi Arabia's and nearly four times as high as Iran's.[26] The traditional dependency relationship had been largely reversed: in the short term, Libya could live with reduced oil exports more easily than Western Europe could live with reduced imports.

After meetings with BP and other companies in February 1970, the Libyans picked the two largest producers of Libyan oil, Standard Oil (NJ) and Occidental, as their principal targets.[27] Pressure on these two companies continued throughout March, with both of them rejecting the Libyan price demands.[28] In April, the Libyan pressure was transferred wholly to Occidental, which was in a much weaker position than Standard, as it was wholly dependent on Libyan oil to feed its European downstream outlets. Keeping up the political rhetoric, Qaddafi made speeches on the theme that the oil companies were part of 'an Imperialist-Zionist plot to enslave Libya' and warned that US interests in the Middle East were 'balanced on a razor's edge' because of US support for Israel.[29]

In May, Libya's hand was further strengthened when the Trans-Arabian Pipeline (Tapline), which normally delivered 500,000 barrels per day of Saudi crude to the Sidon terminal on the Mediterranean, was ruptured by a bulldozer working on a telephone cable in Syrian territory. Syria, one of the radical Arab states, refused to allow the pipeline to be repaired until January 1971. In the meantime, the loss of short-haul Saudi crude put further strain on the world's tanker fleet, already fully stretched by the closure of the Suez Canal.[30] Freight rates rose rapidly and Libya's short-haul crude became even more in demand.

In Libya, Occidental was threatened with expropriation, nationalisation and shutdown.[31] In May, it was ordered to reduce its production of some 800,000 barrels per day to 680,000 barrels.[32] Meanwhile, the three radical oil-exporting states of Libya, Algeria and Iraq got together to form a common front against the oil companies.[33] The pressure continued to rise. In June the Libyans ordered more production cuts from Occidental and this time also Amoseas; Algeria nationalised the assets of Shell, Phillips and other smaller companies which refused to accept higher posted prices; and early in July Libya nationalised the local marketing companies of Shell, Standard Oil (NJ) and ENI.[34] Occidental, though, remained the main pressure point. Feeling the effects of the production cuts, Armand Hammer turned to Standard Oil (NJ) for relief. In a crucial test of oil company unity, he asked Standard to provide replacement oil at cost. But Standard was no friend of its upstart competitor and turned him down.[35]

Disunited, the oil companies soon came under still more pressure. In July Algeria unilaterally announced a rise in the posted prices for French companies from $2·08 to $2·85 a barrel, while in Libya, more production cuts were ordered between July and September from Occidental, Oasis, Mobil/Gelsenberg and Standard Oil (NJ).[36] Occidental, with its pre-May production rate virtually halved, could not stand the strain. Early in September it cracked and agreed to a rise in the posted price of Libyan crude from $2·23 to $2·53 a barrel, retroactive to the commencement of Occidental's exports from Libya, plus annual escalation of 2 cents a barrel for five years starting from the beginning of 1971. To cover the retroactive element, and to replace Occidental's payment for the Kufra project which was taken over by the Libyan government, the rate of Libyan income tax paid by Occidental was raised from 50 to 58 per cent.[37]

Having broken Occidental's resistance, the Libyan government quickly turned the screws on other oil companies. On 12 September the members of the Oasis group (Continental, Marathon, Amerada-Hess and Shell) were told that they also were expected to accept a retroactive price rise.[38] The three Oasis independents agreed to the Libyan terms on 21 September. Shell, however, held back. It held only a one-sixth share in Oasis and, unlike the independents, it had extensive interests in other oil-producing countries which it had to consider. It knew that concessions granted to Libya would also be demanded by other oil-producing nations. When the Iranian Prime Minister, Abbas Hoveyda, heard that the Oasis independents had accepted Libya's demands, but that Shell had not, he warned the Consortium that 'if the major companies agree to higher posted prices and higher taxes in Libya they will have to do so here'.[39] David Barran, chairman of Shell Transport and Trading, was particularly concerned about the knock-on effects of accepting the principle of retroactivity, which he thought might undermine the 'whole nexus of relationships between producing Government, oil company and consumer'.[40] Shell, therefore, rejected the Libyan demand for retroactivity and suggested arbitration. The Libyan answer was that 'the Revolutionary Government will never consent to arbitration. The concessions themselves are illegal.' A meeting between Shell and the Libyans lasted only ten minutes before Shell was summarily told: 'There is no need to enter into further discussion. Shell's production will stop from tomorrow.' On that same day, 22 September, Texaco, Socal, Bunker Hunt, ARCO, Grace Petroleum and Gelsenberg were given until 27 September to accede to Libya's terms on pain of the same shutdown that was applied to Shell.[41]

With the deadline looming, the heads of the oil majors agreed to meet in New York on 25 September to discuss the situation.[42] The day before, on the

24th, BP's board met and conducted its own review of the position. BP, so far, had got off lightly. It was the only one of the large oil producers in Libya that had not been ordered to cut back production and, in company with Standard Oil (NJ) and Mobil, it had not been given a deadline to accede to Libya's demands. A deadline was, however, expected at any time and BP, like Shell, realised that if Libya got its way, other oil-producing countries would quickly make similar demands. Sir Eric Drake, BP's chairman and chief executive, favoured strong resistance. He 'hoped there would be a firm rejection of the [Libyan] Government's unilateral demands, despite the threat to stop supplies'. He accepted that the companies might have to increase the posted prices of Libyan and other short-haul crudes to reflect their freight advantages, but they should not, he thought, go further than that by granting Libya's demands for retroactivity and an increase in the tax rate.[43]

After the board meeting, Drake flew to New York, hoping to obtain the support of the other majors and of the UK and US governments for a policy of rejecting Libya's unilateral demands. He was to be disappointed. In New York, he and Barran saw the British Foreign Secretary, Sir Alec Douglas-Home, who was there on United Nations business. Douglas-Home agreed that every effort should be made to stand firm in support of the principle that posted prices and tax rates could be amended only by mutual agreement, not by unilateral fiat. He recognised the need to strengthen the resolve of European governments to face up to the possibility of a fuel shortage and undertook to speak to them.[44] Later, however, he reported back that other governments had no enthusiasm for action which would result in a reduction in Europe's oil supplies.[45] The British companies in Libya were told to be guided by 'their commercial judgement'.[46]

In Washington, Drake and the heads of the other oil majors met officials of the State Department. Drake told them that he favoured making a stand against Libya's demands. A short-term oil shortage in Europe would, he said, 'be insignificant compared with the long term costs which would result from submission to Libyan demands'.[47] Barran held the same view.[48] The State Department did not, however, feel able to support them. There was, State Department officials said, no economic or political leverage which the US government could apply in Libya to help the oil companies. Libya had substantial financial reserves and was in a strong position to hold out for a long time against the majors. There was 'little that the US Government could do'.[49]

Left to their own devices, the majors discussed the situation among themselves. Drake stressed that if one major yielded in Libya it might be difficult to prevent a reaction on all companies in other producing countries. He urged the majors to stand together. Shell agreed, and there was also some support from Gulf and Mobil. But Standard Oil (NJ) was too concerned about US antitrust laws to commit itself to a united stand, while Texaco and Socal, which were facing a Libyan deadline which was due to expire in two days' time, were in no mood for a fight with Libya. Socal's chairman, Otto Miller, had earlier even 'seemed to be inviting the State Department to bless a capitulation in Libya'. The majors ended their discussion without any agreement on what action to take.[50]

Divided and bereft of backing from their parent governments, the majors were in no position successfully to resist Libya's demands. By the end of September, Texaco and Socal, plus various independents, had given in. Standard Oil (NJ), Mobil and BP raised their Libyan posted prices voluntarily, but resisted on retroactivity and higher taxes until 8 October, when they also gave in. Shell followed suit a few days later. The terms, in all cases, were an initial posted price rise of 30 cents a barrel, with 2 cents a barrel annual escalation for five years. The higher price was made retroactive, with most companies (including all the majors) agreeing to pay off the arrears by increased rates of income tax, which varied slightly, but on average were raised to about 55 per cent.[51]

Libya's triumph was a decisive moment in the relations between the majors and the OPEC countries. In the 1960s the majors had made some concessions on royalty expensing, but they had generally managed to preserve the stability of the concession system, without increases in posted prices or tax rates. Now, though, Libya had broken through on prices and taxes, and it was inconceivable that other producing countries would not try to follow suit. The majors would have to face, in Barran's phrase, 'an avalanche of escalating demands' which they would try to stem in the vain hope that they could restore some stability to the industry and avoid the liquidation of the concession system.[52]

THE TEHRAN AGREEMENT

The Libyan settlement had an immediate impact in other oil-producing states. The Shah was furious that he had been upstaged by Qaddafi and was adamant that his prestige would not allow him to accept a lower tax rate than Libya.[53] The Consortium agreed to raise the Iranian tax rate to 55 per cent and the same terms were quickly offered to other oil-producing states in the Persian Gulf. The posted prices of some Gulf crudes were also raised. In Venezuela early in December, legislation was introduced raising the tax rate even higher, to 60 per cent, and providing for the unilateral setting of prices.[54]

These piecemeal measures were quickly followed by concerted OPEC action. On 9–12 December 1970, OPEC met in Caracas and called for a general increase in posted prices, the elimination of the remaining royalty expensing allowances, and acceptance of 55 per cent as the minimum tax rate in all member countries. To achieve these goals, the Persian Gulf members were to start negotiations with the oil companies in mid-January 1971. If they failed to achieve a settlement by 3 February, OPEC would enforce its demands by 'concerted and simultaneous action' by all members.[55] Although the OPEC resolutions were not officially published until later, BP could see that new demands were coming. As David Steel noted on 11 December, 'the first shots in the second round are now being fired'.[56]

Once again, the initiative was taken by Libya, where on 2 January 1971 the Deputy Prime Minister, Major Abdesselam Jalloud, summoned representatives of the oil companies and demanded a further rise in posted prices and a 5 per cent increase in the Libyan tax rate. These demands were accompanied by the now-familiar threats of shutdown or nationalisation, and by the political message that the Libyan Revolutionary Command Council was determined to coerce the oil companies to put pressure on the US government to change its foreign policy in the Middle East.[57] A week later, Hunt and Occidental were given until 16 January to accept Libya's demands.[58]

The oil companies were extremely worried. In the autumn of 1970, divided and lacking political support, they had been picked off one by one by the Libyan government. Now, Libya was making new demands and repeating its previously successful tactic of singling out the most vulnerable independents for pressure. It was, thought the oil companies, as if the OPEC countries exporting in the Mediterranean were playing leap-frog with the OPEC countries exporting in the Persian Gulf, taking turns to vault over each other in their demands.[59]

To counter this, the oil companies tried to form a united front with the backing of their parent governments. Already, on 5 January, Drake had warned the British Prime Minister, Edward Heath, that the oil companies could not resist OPEC's demands without concerted support from the governments of the main oil-consuming nations.[60] A week later, Drake and David Steel were in New York, where a common front was quickly put together. On 15 January the seven majors and eight of the independents signed the Libyan Producers' Agreement, pledging that none of them would make an agreement with Libya without the assent of the others, and agreeing that if any of them was ordered by Libya to cut production the others would help to provide replacement oil at cost price. It was hoped that this agreement, known as the Libyan 'safety net' would prevent the oil companies being split by the Libyan tactic of picking on the most vulnerable among them.[61]

The day after setting up the safety net, the oil companies issued a 'Message to OPEC'. It proposed an 'all-embracing' negotiation between the companies and all the OPEC members, with the objective of reaching an agreement on posted prices and tax rates which would last for five years.[62] The Message was initially signed by thirteen oil companies, but another eleven companies later joined them. They included majors and independents from the USA, BP from Britain, Royal Dutch-Shell from the Netherlands and Britain, CFP from France, the Arabian Oil Company from Japan, Belgium's Petrofina, Spain's Hispanoil and Germany's Gelsenberg. Indeed, the only internationally significant oil companies that did not sign were the state-owned ENI of Italy and Elf/ERAP of France.[63]

When the Message to OPEC was delivered to OPEC members on 16 January, the oil companies believed that consumer-country governments were behind them. The British Foreign Office sent instructions to British diplomats in the OPEC countries to support the companies.[64] In the USA, both the Libyan Producers' Agreement and the Message to OPEC were approved by the antitrust authorities. Immediately after the publication of the Message, Under-Secretary of State John Irwin went on a special mission to Iran, Saudi Arabia and Kuwait with Nixon's authorisation. The oil companies presumed that he was going to express the USA's support for the collective negotiating strategy proposed in the Message.[65] A powerful bloc of oil companies and oil-consuming countries seemed to be lined up against OPEC.

It proved, however, to be a fragile solidarity. In Libya, Jalloud called the Message a 'poisoned letter' and told Hunt and Occidental to dissociate themselves from the other companies and start separate negotiations. 'Libya', he said, 'will defeat the consuming countries and also the old companies'.[66] Meanwhile, Irwin arrived in Tehran to be told by the Shah that the companies were making a 'most monumental error' in seeking to negotiate with OPEC as a single bloc because the OPEC moderates like Iran would not be able to curb the demands of the radicals like Libya. The result, he said, would be a settlement which reflected the highest common denominator of OPEC demands. If, however, the companies would agree to negotiate separately with the Persian Gulf states, excessive demands would not be made, and a five-year agreement could be reached with firm assurances that there would be no leap-frogging. Irwin was won over. He recommended to the State Department that the companies should be urged to proceed with negotiations with the Persian Gulf states, and to drop the idea of the global negotiation. Douglas MacArthur II, the US Ambassador to Tehran, agreed with

Irwin. So did the British Ambassador, Sir Denis Wright. The US Secretary of State, William Rogers, accepted Irwin's recommendation. After returning to Washington, Irwin reported to Nixon that in all three countries he had visited – Iran, Saudi Arabia and Kuwait – he had stressed that the US government would not become involved in the negotiations with the oil companies.[67] Without US government support the idea of global collective bargaining between the companies and OPEC was dead. Henry Kissinger would later record, 'our hands-off policy ordained the result: the companies yielded'.[68]

When the companies learned that Irwin had accepted the Shah's arguments they suggested a compromise whereby 'separate but connected' negotiations would be held with two groups of OPEC countries: one group would include oil exporters in the Persian Gulf, while the other would include oil exporters in the Mediterranean. Although the negotiations would be split between the two groups, the companies still intended to put forward a single comprehensive proposal to the two OPEC groups, and any settlement would have, they said, to cover both groups. To implement this approach, the companies planned to present their proposals simultaneously in Iran and Libya by splitting their negotiating team into halves. The Iranian half would be led by BP director Lord Strathalmond, son of the chairman who had led BP through the Iranian nationalisation crisis in 1951–4. The Libyan half would be led by George Piercy, a director of Standard Oil (NJ). After the proposals were presented, neither half would enter into negotiations without authorisation from the companies as a whole. To co-ordinate their negotiations, the companies had already set up a London Policy Group, which met at BP's London headquarters, and a parallel group in New York.[69]

The companies' new plan quickly came unstuck. When Piercy presented the companies' proposals to the Libyans in Tripoli on 28 January, he was rebuffed. The Libyan Oil Minister simply refused to receive the proposals and indicated that Piercy could send them to him. Piercy did that, but a day or two later the proposals were sent back to Standard Oil (NJ) without comment. Faced with the Libyan refusal to negotiate, Piercy and his half of the companies' team flew back to London 'undramatically'.[70] The companies' hopes of reaching a simultaneous settlement with all the members of OPEC now lay in tatters. The Libyans were going to wait for the outcome of the negotiations with the Persian Gulf oil exporters before making their own move. In other words, they were going to leap-frog.

In Tehran, meanwhile, on 28 January Strathalmond's half of the oil companies' negotiating team presented the companies' offer to the OPEC countries in the Persian Gulf. OPEC's negotiators were easily a match for the oil companies in education and expertise. They were Iranian Finance Minister, Jamshid Amouzegar, a multilingual cosmopolitan diplomat educated at Cornell and the University of Washington; Iraqi Oil Minister, Saadoun Hamadi, a respected technocrat with a PhD in agricultural economics from the University of Wisconsin; and Saudi Oil Minister, Zaki Yamani, a superb negotiator educated in part at New York University and Harvard Law School.[71]

It quickly became apparent that the gap between the two sides was very wide indeed. The companies were offering a general posted price rise of 15 cents a barrel, indexed against inflation, compared with 54 cents a barrel demanded by the Persian Gulf oil exporters.[72] There was very little time, a mere five days, in which to bridge this gap before the 3 February deadline which OPEC had set at its Caracas conference in December 1970. Moreover, it was not only the financial terms that had to be settled. Another thorny issue was the assurances which the companies were seeking against future leap-frogging and production shutdowns. These issues became so vexatious that before long the very use of the word assurances made the Iranians 'absolutely furious'.[73] Strathalmond cabled from Tehran that he doubted whether a settlement could be reached by the deadline, when in his view OPEC would probably decide on a worldwide one-week shutdown of production.[74] In London, Sir Eric Drake told the British Minister for Trade and Industry that the companies would rather face a shutdown than submit to the 'current exorbitant demands'.[75] In Tehran, the negotiations remained deadlocked: the companies came up with an improved offer which the OPEC negotiators turned down; they in turn put forward a revised offer, which the companies turned down.[76]

On 2 February, the day before the deadline, Drake and David Barran called on Sir Alec Douglas-Home and said that in the absence of the assurances which they sought against leap-frogging and shutdowns 'it was not worth signing an agreement'. At Drake's suggestion, on the morning of the deadline, 3 February, Edward Heath sent a message to the Shah asking for more time for negotiation. Similar representations were made by the USA, France, Italy, the Netherlands, Japan and West Germany.[77] Indeed, on 3 February in Tehran virtually every foreign ambassador from the industrialised world was pressing for more time for negotiations.[78]

Thanks to these diplomatic interventions, the deadline was extended to 15 February, by which time the companies had been persuaded to change their stance. On 5 February they still generally agreed, with the possible exception of the French, that they would rather leave OPEC to take unilateral action than accept its terms. But over the next few days British and US

62   Oil company negotiators leaving the Iranian Ministry of Finance in serious mood during the negotiations between OPEC and the oil companies in Tehran in 1971. From right to left: W. P. Tavoulareas (Mobil); John E. Kircher (Continental); Lord Strathalmond (BP)

government officials made it clear that they favoured a policy of accommodation rather than confrontation with OPEC. Most of the governments of oil-consuming countries were, they pointed out, more concerned about getting oil supplies than about the price. They definitely did not want a shutdown.[79] The companies were won round and decided 'in the light of reaction from the consuming countries' that it was better to make an agreement, even at great financial cost, than to face unilateral action by OPEC.[80]

The outcome was that on 14 February 1971 the oil companies signed the Tehran Agreement with the Persian Gulf members of OPEC.[81] The five-year Agreement gave the Gulf states an immediate posted price rise of 35 cents a barrel, with annual increases of 5 cents a barrel and 2½ per cent for inflation. Prices of heavy crudes would be further increased under a new scale of gravity differentials and the royalty expensing allowances were finally eliminated.

63   OPEC negotiators looking cheerful during the negotiations between OPEC and the oil companies in Tehran in February 1971. From right to left: Saadoun Hamadi (Iraq); Jamshid Amouzegar (Iran); Zaki Yamani (Saudi Arabia)

The overall effect of these various elements was that from 1 June 1971 the price of an average Gulf light crude went up by nearly 50 cents a barrel. These terms were, the Foreign Office noted, only 'marginally better' for the companies than the OPEC countries' original demands for 54 cents a barrel.[82] The companies did, however, at least gain some assurances of stability insofar as the 55 per cent tax rate and the new posted prices were to last for five years, during which there would, the Persian Gulf states promised, be no leapfrogging or embargoes in the Persian Gulf.[83]

But despite the assurances, the Tehran Agreement could not honestly be presented as a victory, or even a draw, for the oil companies in their contest with OPEC. When the companies were drafting a press release on the Agreement, one of their American executives commented, 'there is not much we can say – we cannot say there was a fair, reasonable settlement after fair, reasonable discussions – we came to agreement after having all hell bludgeoned out of us'.[84]

64   Lord Strathalmond (right), for the oil companies, and Jamshid Amouzegar (left), for OPEC, shake hands after signing the Tehran Agreement in February 1971

LIBYA AGAIN

The ordeal was not yet over, for the oil companies still had to come to terms with the OPEC countries with exports in the Mediterranean. They were Libya, Algeria, Iraq and Saudi Arabia. The last two were members of both the Persian Gulf group who signed the Tehran Agreement and the Mediterranean group because they exported oil from terminals both in the Gulf and, via their trans-desert pipelines, in the Mediterranean. Three of the Mediterranean exporting states – Libya, Algeria and Iraq – belonged in the radical, anti-Western Arab camp, and the oil companies must have known that they were in for a hard time negotiating with these states.

The oil ministers of the four OPEC Mediterranean exporters met on 23 February and delegated Libya to negotiate on their behalf. As usual, there was a deadline and a threat, in this instance that if no settlement was reached by 10 March the oil ministers would meet again to decide what action to take, including the stoppage of oil exports. It was also agreed that Libya would negotiate with the companies individually, and not as a group.[85] This, of course, was the tactic which had worked so well for the Libyans in the autumn of 1970.

For the oil companies, the prospect of another round with the Libyan negotiators, mainly Jalloud and Oil Minister, Ezzedin Mabrouk, was unsettling. John Sutcliffe, a senior BP executive, was one of the company negotiators and knew what to expect after his experience negotiating with the Libyans in autumn 1970. Mabrouk was, he thought, 'something of a bully', and as for Jalloud, 'if you bore him or touch one of his many exposed nerves he becomes very excitable, his arms thrash around and he resorts to threats'.[86] The Libyans' methods of unsettling their opponents by sudden changes of temper, dramatic gestures, threats and ultimatums made it virtually impossible to conduct what the oil companies regarded as reasonable discussions. But the differences between the two sides went much further than matters of negotiating style. At heart, the oil companies wanted stability in new arrangements which would be no more than a modification of the status quo. The Libyans wanted radical change.

On 24 February the negotiations started with a bang. That day, Algeria nationalised 51 per cent of French oil interests in Algeria, and Jalloud presented Libya's demands for a $1·20 rise in the posted price, from $2·55 to $3·75 a barrel.[87] The companies thought that this was 'ridiculously high' and came back a few days later with a counter-proposal for a five-year agreement and a posted price of just under $3, escalating with inflation over the five-year term. The Libyan reaction was immediate and typically dramatic: 'If that is all you have to say you can leave', said Jalloud after less than ten minutes of discussion. 'I want you to know', added Mabrouk, 'that you are playing with fire'.[88] After a few more days, the companies raised their offer to $3·15 and offered to bring forward the first escalation payment of 12 cents a barrel so that the posted price would rise immediately to $3·27.[89] Jalloud was 'excited and upset' by this offer, which the Libyans described as 'ridiculous and deceitful'.[90] The companies wanted a five-year agreement in the hope that this would provide some stability, but that did not interest Jalloud. He told Sutcliffe that he was prepared to reach agreement only 'for the past and the present and not the future'. Sutcliffe replied that it was necessary to have a stable basis for future investment, to which Jalloud countered, '[we don't] want you to invest, we want you to get out of the country . . . we don't want to be bound by a long-term agreement . . . we do not care whether Europe gets our oil or not'. He said that he would give guarantees, but must still have the right to make changes according to circumstances. Sutcliffe argued that 'this was no guarantee'.[91]

On 8 March, with just two days to go before the deadline, Sutcliffe and Piercy briefed the US, British, French, Spanish, Dutch and West German embassies in Libya. They explained that the oil companies and Libya were far from reaching agreement, that the deadline was nearing, and that although the Libyans had not specified what action they would take when it arrived, they had spoken of unilaterally raising posted prices and of imposing an oil embargo jointly with the other Mediterranean exporters. 'The Ambassadors had no suggestions to make.'[92]

In the evening of 10 March, with just hours to go before the deadline at midnight, Texaco and Socal representatives delivered proposals, including the same price as before of $3·15 plus 12 cents accelerated escalation, to the Libyan Oil Ministry. They were told to go away while the Libyans thought about it. Half an hour later they were recalled, but when they entered the room Jalloud threw the papers at them and 'told us to get out, we understood him to mean we should get out of Libya'. They left, only to be recalled again later that evening, when Jalloud said that if no deal was agreed by 11.30 pm, Libya 'would finish off the companies one by one starting with Socal and Texaco'.[93]

The deadline was extended, but the atmosphere remained tense. The companies raised their offer to $3·19 plus 12 cents as accelerated escalation, to which Jalloud said, 'you can leave and not just the room', adding, 'we could take action without warning'.[94] The Libyans said that they would accept $3·32, compared with the companies' offer of $3·19. Only 13 cents a barrel separated them, but Piercy said that the companies 'could give no more', and Jalloud replied that he would have to take 'irrevocable action'.[95]

On 15 March the oil ministers of OPEC's four Mediterranean exporters met again in Tripoli. The Libyans told Yamani that they were going to nationalise Standard Oil (NJ)'s Libyan operations as an example, which they thought would bring the other companies to heel. The Algerians thought that the Libyans would 'sleep soundly at night after they take this step'. Hamadi and Yamani both supported the Libyan position. The four ministers agreed on minimum price demands and warned that they would shut down production if the demands were not met.[96] Hamadi and Yamani told Piercy that the companies were being completely unreasonable. Hamadi painted a 'black picture' of what would happen if the companies did not agree to Libya's demands. Without being specific, he mentioned legislation, nationalisation and embargo. Piercy thought it was obvious that Hamadi and Yamani had 'been given the job of working him over'.[97]

Under this pressure, the companies agreed to raise their offer to $3·30 plus 12 cents accelerated escalation, making a total of $3·42 a barrel.[98] Still more negotiations followed until, by the end of March, the gap between the two sides had nearly been closed. On 31 March Lord Strathalmond, resting in the Caribbean on medical advice, cabled David Steel, asking, 'No news. What gives?' To which Steel replied, 'We do – a further two cents should make a deal.'[99]

An extra 2 cents a barrel was, indeed, enough to clinch agreement. On 2 April the oil companies operating in Libya signed agreements which raised the posted price of 40° Libyan crude by virtually 90 cents a barrel, from $2·55 to $3·447 ($3·32 plus 12·7 cents accelerated escalation). In other respects, the new Libyan agreements (a separate one for each Libyan operator) followed the same lines as the Tehran Agreement. They were five-year agreements, starting with effect from 20 March 1971 and running to the end of 1975. They provided for annual posted price escalation of 5 cents a barrel plus 2½ per cent of the basic posted price. The royalty expensing allowance, which had been suspended since 1967, was permanently eliminated. The extra tax which the companies had agreed to pay to cover the retroactive element of the autumn 1970 price rise was changed into a supplemental payment of about 9 cents a barrel. The basic tax rate was raised from 50 to 55 per cent, except for Occidental, which was liable for an extra 5 per cent to pay for the Kufra agriculture project.[100] These terms became the benchmark for the Mediterranean oil exports of Iraqi, Saudi Arabian and Algerian crudes, whose prices were duly raised to levels equating to the new Libyan price after adjusting for quality and freight differentials.[101]

At the signing ceremony for the Libyan agreements on 2 April 1971, Jalloud assured the assembled company representatives that all points in dispute had been resolved and, referring to leap-frogging, that there would be no further demands from Libya.[102] The companies certainly hoped so. They had achieved some success in their efforts to restore stability to the concession system by obtaining agreements which ran until the end of 1975. But at the same time, they had given a lot of ground on posted prices and taxes. The posted price of 40° Libyan crude had gone up from $2·23 before September 1970 to virtually $3·45 in March 1971. In the Persian Gulf, postings had risen by less, with the key 34° Saudi crude going up from $1·80 to $2·18 a barrel over the same period.[103] In addition, the longstanding 50:50 sharing of profits between companies and host governments had been replaced by a new standard 55:45 split in favour of the hosts by the raising of the tax rate.

In the summer of 1971 the renowned oil consultant, Walter Levy, published an article on the impact of these changes. He calculated that under the new five-year agreements the revenues of the oil-exporting states would rise from $7 billion in 1970 to $18½ billion in 1975; that Europe's bill for oil imports would rise by $5½ billion in the same period; and Japan's by $1½ billion. Levy thought that the terms of world trade had been radically altered, that there had been a decisive shift in the balance of power between oil-exporting and oil-consuming nations, and that the winds of change in the international oil industry had reached 'hurricane proportions'.[104] He could not have known that compared with what was coming, they were a light breeze.

=== 19 ===
The end of an era

PARTICIPATION AND NATIONALISATION

When they signed the Tehran and Tripoli agreements in February and April 1971, the oil companies were hopeful that the agreements would help to stabilise the international oil industry. Posted prices and tax rates up to the end of 1975 had been agreed, the OPEC countries had given assurances against leap-frogging and embargoes, and they had confirmed that the existing concessions, as amended, would continue to be valid.[1]

Stability, however, proved to be ephemeral. In July 1971 the OPEC countries threw down a new challenge to the international oil companies when they resolved to implement the objective, which they had adopted three years earlier, of gaining participation in concession-holding companies, such as Aramco in Saudi Arabia and the Kuwait Oil Company in Kuwait. In September, OPEC decided to open negotiations with the oil companies on participation, another word for partial nationalisation.[2]

For years, national governments in oil-producing countries had sought larger shares in the ownership and control of their national oil industries. In many cases, they had set up state-owned oil companies which co-existed with private foreign companies. In others, producing countries had taken the more extreme step of nationalising the oil-producing operations of private foreign companies, as happened in the Soviet Union (1918), Bolivia (1937 and 1969), Mexico (1938), Iran (1951), Burma (1962), Egypt (1962), Argentina (1963), Indonesia (1963) and Peru (1968). These nationalisations, few and far between, put some parts of the world outside the reach of the private companies, but left most of the world's oil production under the ownership and control of private oil multinationals, who opposed participation by the OPEC countries.[3]

467

The oil companies initially objected to OPEC's calls for negotiations on participation on the grounds that they broke the assurances of stability which had been given in the Tehran and Tripoli agreements.[4] The OPEC countries disagreed, as did some people in the West, but in any case the legalistic arguments did not matter much. What counted was bargaining power, and there was no doubt that OPEC held the stronger hand. That was the unequivocal message of a 100-page State Department report, which argued that the prospects of finding substantial new oil reserves outside the Middle East were low, that the West could not escape from becoming almost completely dependent on OPEC oil, and that the companies' strength in dealing with producing governments had largely gone. 'The high trumps', wrote the State Department, 'are all in the hands of the producing countries and will be for the next twenty years'. There was very little that the US government could do to help the companies because it had little leverage in the oil-producing nations, with the exception of the US client state of Iran. Most of the others received no US aid, while the world needed their oil. If OPEC members decided to enforce their demands for participation, resistance would be ill-advised, because it might provoke OPEC countries to go straight for full nationalisation. On the basis of this analysis, the State Department concluded that the companies should follow a policy of accommodation by immediately offering the OPEC countries a new relationship to go into effect in 1976, after the Tehran and Tripoli agreements expired.[5]

The radical Arab states were, however, forcing the pace. In Algeria, President Boumedienne had already nationalised 51 per cent of the French-owned concessions in February 1971.[6] Later in the year, before negotiations on participation had started, Libya took unilateral action against BP in reaction to events in the Persian Gulf that lay wholly outside BP's control. In 1971 Britain went ahead with its plans, announced in 1968, to withdraw British forces from the Persian Gulf, removing British protection from the nine little sheikhdoms of Bahrain, Qatar and the seven Trucial states. As these miniature kingdoms were thought to be too small to survive independently, Britain encouraged them to form a federation, the United Arab Emirates. Bahrain and Qatar opted out and decided to go their own ways, while six of the Trucial states (followed soon afterwards by the seventh) decided to join in the UAE. Britain's longstanding protective treaties with the Trucial states were to be terminated on 1 December 1971, and the UAE was to be formally inaugurated the next day.

The Shah of Iran, determined that Iran should fill the power vacuum that would be left by Britain's withdrawal, took a close interest in this situation. In the negotiations over Britain's withdrawal from the Persian Gulf, he had agreed to give up his claim to Bahrain, but insisted that Iran should take possession of three tiny islands – Abu Musa and the Greater and Lesser Tunbs – which lay near the entrance to the Persian Gulf, a strategic waterway which the Shah wished to control. Ignoring the Arab Trucial states of Sharjah and Ras al-Khaimah, who claimed Abu Musa and the Tunbs as their respective possessions, the Shah's forces seized the islands with just a day to go before the formal end of British protection. Britain's lack of resistance to Iran's action aroused the wrath of the Arab world. Iraq broke off diplomatic relations with Britain, and on 7 December Libya nationalised BP's Libyan operations in retaliation against 'the plot mechanised by Britain with the puppet government of Iran, against the Arab nation'.[7]

Diplomatic protests were made, arbitration was sought, and compensation was claimed, but nothing could undo the act of nationalisation. A subsidiary of Libya's state-owned oil company took over the operation of BP's concession and succeeded in selling nationalised oil to the Soviet Union, Eastern Europe, and in increasing quantities to Western purchasers after BP lost a legal claim to ownership of a nationalised cargo in Sicily. The eventual compensation settlement, reached in November 1974, was for far less than BP sought.[8]

Meanwhile, in January 1972 negotiations on participation started between the oil companies and OPEC negotiators, led by Yamani, who demanded participation starting at 20 per cent and rising to 51 per cent over a period of years. Described by a BP executive as 'very ambitious both for himself and for Saudi Arabia's position in the Gulf and the Arab world', Yamani toyed with the apprehensive company representatives in the first negotiating session.[9] But although the company representatives clearly felt unsettled, they nevertheless resisted the principle of participation until King Faisal personally intervened and told Aramco that it must speedily accept Saudi Arabia's demand for participation of 20 per cent, as a minimum. The royal intervention, followed by a warning from Yamani that OPEC was about to take unilateral action, persuaded the companies to change their stance. In March 1972 they accepted the principle of 20 per cent participation.[10]

Acceptance of the principle was not, however, the same as agreement on the terms. For four more months, negotiations between Yamani and the companies continued on the level and timing of participation, on the compensation that would be payable, and on the terms on which the companies would buy back some of the nationalised oil for refining and marketing in their downstream facilities.[11]

While Yamani and the companies continued to talk about participation, Iraq and Iran went other ways. In Iraq, the IPC and the government had been locked in dispute for years, mainly over Law 80 of 1961, whereby Iraq had expropriated 99½ per cent of the IPC's concession area without compensation. In the autumn of 1971, relations were even worse than usual. Only about six months had passed since the OPEC oil exporters in the Mediterranean had won big rises in posted prices, reflecting the freight cost advantages of short-haul Mediterranean oil in a tight tanker market. Freight rates were, however, extremely volatile, and in the autumn of 1971 an easing of the tanker market made long-haul oil exports from the Persian Gulf more competitive against oil shipped from Mediterranean ports. The companies began to lift more oil from the Gulf, and less from the Mediterranean, where the bulk of Iraq's oil was exported. This made the Iraqis very angry.

At the same time, Iraq's attitude towards the IPC hardened with the emergence of the uncompromising Saddam Hussein as the strong man in the Iraqi government. He announced that there would be 'decisive' negotiations with the oil companies, and that no 'leniency' would be shown. In fact, the talks had got nowhere by the spring of 1972, when the IPC offered to increase its Mediterranean exports if Iraq would agree to reduce its tax 'take' in the Mediterranean. This outraged the Iraqis, who saw the offer as an attempt to undo the price and tax rises which OPEC had achieved in 1971. In mid-May, they issued an ultimatum, giving the IPC two weeks in which to increase exports and to make a 'positive offer' to settle other matters in dispute. Preparing for a showdown, the Iraqi government adopted an economic austerity programme, put on a vigorous campaign to arouse popular support, and took two leading communists into the Cabinet as a 'necessary political requirement for the confrontation' with the IPC.

When the deadline arrived on 31 May the companies made a new proposal, which included compensation for the oil rights expropriated under Law 80, a vexatious issue. The Iraqis would not hear of it, and rejected the terms out of hand. The next day, 1 June 1972, Iraq nationalised the IPC. The Iraqis set up their own company to operate the IPC's oil fields, and found customers for their oil in the Eastern bloc and through a contract with CFP. In February 1973 they agreed to a settlement of all outstanding claims with the IPC. Through a sister company, the Basra Petroleum Company, the IPC members retained concessionary rights and oil-producing interests in southern Iraq, from where oil was exported via the Persian Gulf.[12] But they had lost their main oil fields in Iraq, most notably the famous Kirkuk field, which was one of the biggest oil fields in the world.

In Iran, meanwhile, the Shah upheld his image as the regional champion of the West. He complained that he was politically isolated in the Middle East, where he regarded Iraq as openly hostile and the rest of his Muslim neighbours as internally unstable and susceptible to radical subversion. He dissociated himself from OPEC on participation which, he argued, was irrelevant for Iran because its oil industry had been nationalised since 1951. Instead of participation, he sought to negotiate a 'lasting partnership' with the Consortium. At the same time, he continued to turn to the USA for support, which Nixon was more than willing to provide to his loyal ally, the most reliable bastion against communism in the Middle East. Nixon and Kissinger (national security adviser to the President) visited the Shah in Tehran in May 1972 and agreed to permit him to purchase virtually any type of conventional weapons from the US arsenal, including F-14 and AWACS aircraft, Phoenix and Maverick missiles, Spruance-class destroyers and a $500 million electronic surveillance system. Although their discussions barely touched on oil, Nixon was fully briefed on the state of negotiations between the Consortium and the Shah, who announced on 24 June 1972 that they had reached an agreement in principle. It provided, in essence, for the extension of the 1954 Consortium agreement, with revisions, up to 1994. Pleased by this outcome, Nixon personally congratulated the Shah on his moderation.[13]

The other OPEC countries were not so pleased. Indeed, King Faisal of Saudi Arabia was positively angry. The participation talks between Yamani and the companies had got nowhere, and Yamani was accusing the companies of stalling.[14] Faisal had staked his authority on participation and suspected that the companies were trying to use Iran as a wedge to split OPEC and weaken the movement for participation. He issued a statement reaffirming his determination to 'take all measures' to achieve his demands. The companies took this as an ultimatum: the King, they believed, was prepared to nationalise if no progress were made on participation.[15] Faisal also adopted the Shah's tactic of appealing to the USA to put pressure on the companies to come to terms. He not only wrote to Nixon, but also sent one of his sons, Prince Saud, the Deputy Oil Minister, to Washington to tell the State Department that Saudi Arabia would order a cut in production unless the companies agreed to participation.[16]

With added urgency, the companies and Yamani continued their negotiations, which resulted in a draft agreement on participation in October, and final agreement in December 1972.[17] It provided for an initial 25 per cent participation on 1 January 1973, rising by stages to 51 per cent on 1 January 1982. The companies would receive compensation (though not nearly as much as they sought) and would buy back most of the producing countries' share of crude production on a reducing scale for four years.[18] The agreement, known as the General Agreement on Participation, was signed by

Saudi Arabia, Abu Dhabi and Qatar. The Kuwaiti Oil Minister approved it, but there was strong opposition in the Kuwaiti National Assembly, which refused to ratify it.[19]

Influenced by their radical neighbours in Iraq, who had nationalised the IPC in June, the Kuwaiti opposition wanted something more than gradual participation. It was in the hope of winning them over that the companies had agreed to raise the initial level of participation to 25 per cent, instead of the 20 per cent which had been Yamani's opening demand.[20] But as happened so often in the fractured and fractious Middle East, the attempt to placate one interest failed, while provoking another. In this case, the Kuwaiti opposition was not won over, and the Shah was aroused.

When the Shah had negotiated his 'lasting partnership' with the Consortium, he had made it clear that the terms must be at least as favourable to Iran as the 20 per cent participation which the companies were at that time prepared to concede to Yamani.[21] But when the draft agreement on participation between Yamani and the companies came out in October 1972, the Shah learned that Yamani had obtained participation starting at 25 per cent and rising to 51 per cent. Not only would participation now start 5 per cent higher than the 20 per cent benchmark, but the provision for 51 per cent suggested that control of operations would be transferred from the companies to the host governments. In Iran, on the other hand, although the oil industry was in national ownership, the companies would retain managerial control of operations under the 'lasting partnership'. This, said an Iranian representative, was 'a definite blow to the Shah's prestige', which he could not tolerate.[22] To add insult to injury, Yamani announced on the radio that his agreement was four times better than the Shah's.[23] The Shah was 'burned up' when he heard it.[24]

Determined to do better than Yamani, he demanded a new agreement which would make the National Iranian Oil Company not only the owner, but also the operator, taking control from the Consortium. The Consortium, he suggested, could purchase oil from the NIOC on a long-term contract, at a price which would give financial parity with Yamani's participation agreement. It would be a new seller/buyer relationship. The Shah was insistent that since the companies had 'given 51 per cent to the Arabs, a partnership was quite unacceptable'.[25] He must have control.

True to form, the Shah tried to influence the US government to take his side. In November 1972 he sent an envoy, Riza Fallah, to Washington to put his case to high US government officials; and in January 1973 the Iranian Minister of Economy, Hushang Ansari, raised the matter with the Secretary of State, William Rogers, and the Vice-President, Spiro Agnew.[26]

The companies, worried that acceptance of the Shah's demands would escalate the rivalry between Iran and Saudi Arabia, also lobbied the US government to help them resist the Shah's demands.[27] The State Department weighed up US national interests, balancing the desirability of protecting US foreign oil investments against the USA's broader political, economic and security relations with Iran, 'a pivotal country in the Middle East'.[28] Secretary of State Rogers concluded that the Shah's proposals for a new relationship with the Consortium did not seriously endanger US national interests, and advised Nixon accordingly.[29] The USA took up an even-handed approach of encouraging negotiations and trying to bring the two sides together. Both Nixon and the British Foreign Secretary, Sir Alec Douglas-Home, wrote to the Shah urging moderation, while the State Department encouraged the companies to be flexible in talking about possible changes in their relationship with Iran.[30]

The two sides reached a new twenty-year agreement, which was finalised in July 1973. With retroactive effect from 21 March 1973 the NIOC would take over the Consortium's operations, produce the crude oil, refine and market a portion of it at home and abroad, and sell the rest to the Consortium at a price calculated to achieve financial parity with the General Agreement on Participation negotiated by Yamani. The Consortium would set up a new service company (OSCO, the Oil Service Company of Iran) to carry out operations for the NIOC under a five-year contract. The companies would thus become service contractors and crude purchasers, while Iran would be, for the first time, the owner and operator with the final say on capital expenditure, development capacity, oil recovery methods and the volume of oil exports.[31] With these terms agreed, the Acting Secretary in the State Department reported to Nixon that 'the Shah's overriding political objectives have been met. He can now proclaim that Iran is again second to none in protecting its sovereign interests and attaining its national aspirations'.[32]

Meanwhile, other countries were asserting their sovereignty over their national oil industries. In Nigeria, legislation in 1968 and 1969 had provided for greater state control and 'Nigerianisation' of oil operations.[33] After emerging victorious from the Biafran Civil War, Colonel Gowon's military government blamed 'the expatriates, the ex-imperialists and the oil companies' for the country's many problems. A BP executive visiting Nigeria found that companies with colonial or imperialist connections were deeply distrusted, the administration inherited from British colonial rule was breaking down, and 'under the relentless pressure to Nigerianise, the few capable and experienced expatriates who have been underpinning the existing system, are being squeezed out'.[34]

In May 1971, Nigeria and the oil companies operating in the country signed a five-year agreement modelled on the Libyan settlement reached in Tripoli in April. It provided for an increase in the posted price of Nigerian 34° crude from $2·42 to $3·21 a barrel with effect from 20 March 1971. In July that year, Nigeria joined OPEC. In March 1973, it reached agreement with the companies, providing for participation to start at 35 per cent, rising to 51 per cent in 1982.[35]

In Libya, Qaddafi followed up the nationalisation of BP with further measures of a similar nature. In June 1973 he celebrated the third anniversary of the expulsion of the 'US imperialist forces' from the Wheelus Air Force Base by nationalising Bunker Hunt's Libyan operations. He announced to cheering crowds that this step had been taken to deal the USA 'a big hard blow . . . on its cold insolent face'.[36] In August, Libyan demands for 51 per cent nationalisation were accepted by Occidental, Continental, Marathon and Amerada-Hess. On 1 September, the fourth anniversary of Qaddafi's seizure of power, the Libyan government issued a general decree nationalising 51 per cent of the remaining majors and independents in Libya. Gelsenberg and Grace promptly accepted, but the others stood out.

THE YOM KIPPUR WAR

Since coming to power, Qaddafi had made no secret of his wish to pressure the more moderate Arab leaders to use the oil weapon against the pro-Israeli USA. King Faisal, the leading Arab moderate, who controlled more oil reserves and production capacity than any other Arab leader, was not easily converted to the idea. But by the spring of 1973 he was feeling the pressure from the radicals, and from Anwar Sadat, successor to Nasser as Egypt's leader. Early in May, Faisal told Aramco executives that he could not stand alone much longer against the tide of Arab opinion, which was running against America. He asked Aramco to help lobby the US government for a change in US policy.[37]

On 23 May, Aramco executives met Faisal again. Since their last meeting the pressure on him had increased. The radical states of Libya, Algeria and Iraq, and the increasingly radical state of Kuwait, had stopped the flow of oil to the West for one hour as a symbolic protest against the USA.[38] Faisal himself had been to Cairo where, Yamani said, he had had a 'bad time' with Sadat, who was putting pressure on Faisal for support against Israel. Speaking to the Aramco representatives, Faisal repeated his worries that Saudi Arabia was becoming increasingly isolated in the Arab world. He said

65   King Faisal of Saudi Arabia (right) with President Sadat of Egypt during Sadat's visit to Arabia in 1972

that he would not let that go on, and warned the Aramco representatives: 'time is running out . . . you may lose everything'.[39]

At Faisal's bidding, the US majors who were parents of Aramco reported his comments to the State and Defense Departments and the White House, but US officials responded with complacency; they did not believe that any drastic action was imminent.[40] The Aramco parents also appealed, at Faisal's request, direct to the US public. Mobil advertised in the *New York Times*; Socal sent a letter to its 300,000 shareholders; Howard Page, formerly a director of Standard Oil (NJ) (renamed Exxon in 1972), made a public speech – all with the message that the USA should try to improve its strained relations with Arab countries.[41]

In August, Sadat told Faisal that he was considering going to war with Israel and asked for support, which Faisal agreed to give, pledging that he would use the oil weapon.[42] Faisal tried to convince the USA of the seriousness of his intentions in interviews with *Newsweek*, the *Washington Post* and the *Christian Science Monitor*, warning that he would feel obliged to curtail oil supplies to America unless the US government reduced its support for Israel.[43]

But none of this made any difference to US foreign policy. Since the 1960s the US oil majors had been warning the US government that by supporting Israel it was running the risk of radicalising the moderate Arab states (see chapter seven). The warnings had been ignored, the oil companies' fears had come to pass, and the Middle East was on the brink of war.

At the same time, concerns about energy supplies were growing. The industrialised countries were in an economic boom which raised the demand for energy. But the coal and nuclear industries were producing less than had been expected, the commercialisation of Alaskan oil was obstructed by native land claims and conservation interests, the USA's crude production was declining and its oil imports, already 35 per cent of consumption, were rising without restriction after Nixon announced in April that he was ending the mandatory import quotas which had been in place since 1959. Meanwhile, books like *The Limits To Growth*, written for the Club of Rome's project on the predicament of mankind, popularised the idea that the world was running out of vital raw materials; and geologists like BP's Harry Warman were making gloomy forecasts of the amount of oil that remained to be discovered (see chapter eight).[44] This situation gave rise to a mixture of fatalism and alarmism. The idea set in that the tight energy market was the symptom of an irreversible trend towards the exhaustion of the world's raw materials, rather than the temporary result of a major economic boom which could be corrected. The title of a celebrated article by James Akins, Director of the Office of Fuels and Energy in the State Department said it all: 'The oil crisis: this time the wolf is here'.[45]

Even the oil companies suggested that people should use less oil. Mobil, for example, announced in June 1973 that it was discontinuing gasoline advertising and redirecting its efforts to persuading people to conserve energy. 'Smart drivers make gasoline last', ran one advertisement; 'She plans to save a gallon a week', ran another. 'We as a nation', announced Mobil's chairman, 'must adopt long-term approaches to conserve energy, because the energy shortages will be with us for some time'.[46] There were some dissenting voices, but the feeling of scarcity was becoming increasingly prevalent.[47]

As rising demand put pressure on supply, market prices, which for years had been at a discount to posted prices, went up. Producing countries selling oil which they obtained from nationalised concessions and from participation, found that it fetched prices equal to, in some cases even above, the postings set by the 1971 Tehran and Tripoli agreements. These postings, which were meant to endure, with inflationary adjustments, to the end of 1975 had already been raised twice to take account of devaluations of the US dollar in 1971 and 1973. By the summer of 1973, BP was worried that OPEC would demand another rise in postings and also a rise in the buy-back prices of participation crude to reflect the new market conditions. Calls for the revision of existing agreements had been coming from the radicals in Algeria, Libya and Iraq for several months. In September, Yamani joined them, declaring that the Tehran Agreement was 'dead or dying' and needed to be changed. OPEC agreed, and called on the companies to meet representatives of the Persian Gulf oil exporters, headed by Yamani, for negotiations in Vienna on 8 October.[48]

The arrival of the negotiators in Vienna was, however, overshadowed by a much greater drama: on 6 October, the Jewish holy day of Yom Kippur, Egypt and Syria launched concerted surprise attacks on Israel, starting the fourth Arab–Israeli war in twenty-five years. As Israel fell back before the advancing Arab armies, the company negotiators in Vienna found that they too were on the defensive, offering a mere 15 per cent increase in posted prices, whereas OPEC was asking for 100 per cent. The companies knew that they were out of their depth. An increase of that order would have a huge impact on the industrialised countries, for which the companies could not take responsibility. They therefore asked for a two-week recess in which to consult the governments of oil-consuming countries. Amid confusion, no reply was formally given by the OPEC negotiators beyond the parting remark of Yamani on 12 October that the companies would hear their decision about the application for a recess on the radio. What the companies actually heard on the radio just four days later, 16 October, was an announcement that OPEC had decided to raise posted prices by $2 a barrel. This 70 per cent rise took the price of Arabian Light 'marker' crude from $3·011 to $5·119 a barrel.[49] This was a landmark, not only because of the amount of the increase – itself quite momentous – but also in the way that it was decided unilaterally by OPEC, without reference to the companies. Their grip on price-setting had long been weakening; now they had lost hold of it altogether.

Meanwhile, events were moving fast on other fronts. On the day that the company and OPEC negotiators broke off their talks in Vienna, the heads of Exxon, Mobil, Texaco and Socal sent a message to Nixon warning that King Faisal intended to impose a cut in production because of the USA's support for Israel in the war so far. A much bigger cut by both Saudi Arabia and Kuwait would follow if the USA increased its support for Israel; and a snowballing effect would in all probability result in 'a major petroleum supply crisis'. It was not, the oilmen warned, only US commercial interests that were at stake; 'the whole position of the United States in the Middle East is on the way', they wrote, 'to being seriously impaired'.[50] The oilmen received no reply from Nixon, who was so mired in the Watergate scandal that he had handed virtually complete control of foreign policy to his newly appointed Secretary of State, Henry Kissinger.[51]

While the warnings of the oilmen were brushed aside, the USA launched a massive airlift, bigger even than the Berlin airlift of 1948–9, of military supplies to support an Israeli counter-offensive. In Syria, Israeli forces crossed the prewar lines and stopped their advance only 20 miles from Damascus. In the Sinai, the Egyptian offensive was put into reverse after a major tank battle. The USA's support for Israel was, however, very much a solo effort; it was not a shared action by the Western countries, or by NATO. Indeed, most European countries dissociated themselves from it. Pro-Arab France would not take part. Nor would Britain, whose Prime Minister, Edward Heath, rejected Nixon's request for US landing rights at a British base in Cyprus. Spain refused to allow the Americans to use a chain of military bases just built and paid for by the USA. Greece and Turkey denied the USA access to military bases. West Germany at first co-operated with the USA, but later stopped the Americans from supplying Israel via West Germany. Of the European countries, only the Netherlands came out in support of Israel, while Portugal allowed US aircraft to refuel en route to Israel.[52]

America's support for Israel precipitated the use of the Arab 'oil weapon', as the oilmen had warned. On 17 October, the day after the announcement of the 70 per cent rise in posted prices, the Arab oil states decided to reduce their oil exports by 5 per cent from September levels and to make further cuts of 5 per cent each month until Israel had withdrawn from the territories occupied in 1967 and the 'rights of Palestinians had been restored'. The cutbacks were soon increased: on the 18th, Saudi Arabia announced that it was doubling the initial cutback to 10 per cent; on the 19th, Nixon asked Congress to approve $2·2 billion in emergency aid for Israel to pay for military equipment; on the 20th, Saudi Arabia retaliated by declaring a complete embargo on oil exports to the USA. Other Arab countries followed Saudi Arabia's example, and they also extended the embargo to the Netherlands and Portugal, accomplices of the USA.

While the USA, the Netherlands and Portugal were completely embargoed, the treatment of other oil-consuming nations depended on how 'friendly' they were to the Arabs. 'Friendly' nations, which included Britain and France, were to receive deliveries at the same rate as in the first nine months of 1973; others would have their supplies reduced, but not completely cut off. The only Arab oil exporter which dissented from these measures was Iraq, which nationalised the American and Dutch interests in the

Basra Petroleum Company, and joined in the outright embargo of the Arabs' enemies, but did not impose other restrictions on supplies.[53]

In the meantime, with Israeli forces encircling the Egyptian Third Army, a ceasefire was negotiated and came into effect on 22 October, only to break down almost immediately. The Israelis completed the encirclement of Egypt's Third Army and cut off its supplies. The tension escalated when, in response to a Soviet threat to intervene to save the Egyptians, the USA put its worldwide forces on the highest level of peacetime alert on the night of 24/25 October, without consulting its European allies. The superpowers were on the brink of a showdown. Soviet airborne troops were on alert, and so were East German forces. The number of Soviet ships in the Mediterranean was at an all-time high, and a Soviet flotilla was heading for Alexandria. On the other side, US airborne troops were also ready to move, and US aircraft carriers were ordered to proceed rapidly to new positions in the Mediterranean. Three days of frenetic negotiations defused the situation, a new Arab–Israeli ceasefire came into effect, the Egyptian Third Army was resupplied, and Egyptian–Israeli peace talks began.[54]

THE OIL CRISIS

Although the MAD threat of Mutually Assured Destruction was over, the oil crisis was not. The cutbacks imposed by the Arabs reduced the availability of Arab oil from 20·8 million barrels per day before the crisis to a low point of 15·8 million bpd in December 1973. Production increases elsewhere offset some of the loss, but the net cut in supplies was still 4·4 million bpd, about 9 per cent of the non-communist world's pre-crisis supplies. That was enough to throw the oil-consuming countries into panic and disarray as they strove to secure supplies and to curry favour with the Arabs. When the oil companies sent a letter to the producing countries protesting, albeit mildly, about the unilateral price rise, the consuming countries were informed about the letter and privately supported it. But they made no protest themselves because, David Steel told the BP board, 'they were frightened of doing anything to upset the Arabs'.[55]

In the scramble for oil the consuming countries risked falling out between themselves. On 6 November, the foreign ministers of the nine countries in the European Economic Community issued a direct challenge to US policy in the Middle East by making a pro-Arab statement in the hope that this would prevent a deterioration of the EEC's oil supply position. It worked: on 18 November the Arab countries cancelled the further cuts in supplies to the EEC scheduled for December, though the embargo on the Netherlands stayed in place. Japan quickly followed the EEC's example and received the same reward. This was the first time since World War II that Japan had openly broken with the USA in its foreign policy.[56]

Within the EEC, autarkic policies overrode the Treaty of Rome's provisions for the free flow of commodities within the Common Market. Belgium, the Netherlands, Italy, Spain and Britain, all of which were large exporters of refined oil products, imposed export controls to try to cushion their domestic economies against the oil shortage. The Netherlands, embargoed by the Arabs, requested other EEC countries to pool supplies and share them, but France and Britain, anxious to preserve their favoured status with the Arabs, obstructed such a move. The Dutch, though, were not without bargaining power of their own. They warned that if the EEC failed to respect its own rules, they might curb their exports of natural gas to other EEC countries. As the Netherlands supplied 40 per cent of France's natural gas consumption, including much of the gas used in the Parisian region, this was no trivial threat. A quiet rapprochement seems to have taken place, for on 20 November the Dutch expressed satisfaction with an unspecified 'common position' which had been reached to ease the embargo.[57]

While trying to secure supplies, the consuming countries also took steps to reduce oil consumption. In the USA, on 7 November Nixon announced proposals to curb US oil use by measures such as car pooling, lower speed limits on the roads, reduced air conditioning and heating in homes and at work and a halt to the switching of utilities from coal to oil. He called for a new national undertaking, Project Independence, for the USA to become self-sufficient in energy by 1980. He compared it to the Apollo project which had put men on the moon. In Europe, governments introduced bans on Sunday driving, lower speed limits, restrictions on heating and lighting and other measures. In Britain, the oil crisis was compounded by a labour dispute in the coal industry. The National Union of Miners, the most powerful of Britain's militant trade unions, started an overtime ban in November, and later went on complete strike. In December, the Chancellor of the Exchequer, Anthony Barber, unveiled a crisis budget to cope with what he described as 'the gravest situation by far since the end of the war'. Offices and factories were allowed only five days' energy consumption in the last fortnight of the year, followed by a three-day working week in the New Year.[58]

The oil companies, meanwhile, worked overtime to allocate oil supplies on the principle of 'equalising the misery' of the oil-consuming nations, as they had done in the Suez crisis in 1956 and the Arab-Israeli Six-Day War in 1967. Apportioning the cutbacks evenly between the consuming countries, while at the same time observing the selective Arab embargoes, was a highly complex task. The companies' integrated supply systems were designed for optimal efficiency in normal patterns of trade. Refineries were designed to convert specific crude oils into products in proportions that matched the demands of particular markets. They were not general purpose plants that could convert any crude oil into any yield of products. There was some scope for adjustment to meet variations in conditions, but the companies' integrated operations were not infinitely flexible. In the event of a major supply disruption, they could not easily be rearranged.

The oil majors had coped with these problems before, but not in such difficult circumstances. In the 1956 Suez Canal crisis, once the Anglo-American schism was repaired, the companies had worked in close co-operation with the OEEC (later OECD) Oil Committee, which set up the OEEC Petroleum Emergency Group (OPEG) to allocate oil to OEEC members on the basis of fair shares for all (see chapter three). In 1967 the International Industry Advisory Board (IIAB) had been set up as, in effect, a successor to OPEG, although it played only a limited role in the comparatively slight 1967 oil crisis.[59] In 1973, however, the selective targeting of the Arab embargoes and restrictions helped to split the consuming nations, which failed to activate a co-ordinated allocation system. Instead, it was everyone for himself in pursuit of oil. This meant, in practice, governments putting pressure on the oil companies to give their national needs priority over others, even in the same economic community of nations.

The chief perpetrators of the 'me first' approach were Britain and France. On 21 October, only four days after the Arabs announced their first oil export cuts, BP's chairman, Sir Eric Drake, and his Shell counterpart, Frank McFadzean, met Prime Minister, Edward Heath, and other senior British ministers for dinner at Chequers, the Prime Minister's country residence. As Drake's notes at the time and Heath's later memoirs make clear, it was not a cordial party. Heath, despite being an ardent public advocate of European co-operation, kept, as Drake noted, 'reverting to the problem of why British companies should not look after Britain and forget about everybody else'. Drake and McFadzean 'patiently tried to get him off this', but without success. Referring to Winston Churchill's role in championing the British government's purchase of a majority shareholding in the Company in 1914, Heath remarked that 'Churchill had bought a share in BP to ensure that the country's demands were met'. Drake could not deny that this was true, but argued that 'the circumstances were a little different', and showed no sign of yielding to Heath's demands.[60]

Drake was convinced that a politically agreed system of international allocation was needed, but to BP's disappointment a meeting of the OECD Oil

66   BP's chairman and chief executive, Sir Eric Drake (right), with British Prime
Minister, Edward Heath (left), at BP's head office, Britannic House, London, in
June 1973

Committee on 25/26 October failed to activate the IIAB because of opposition from countries on the Arabs' 'friendly' list, who were afraid that they
might lose their favoured position if they joined in an allocation scheme
which would help the embargoed nations.[61]

On 29 October, Drake met the British Defence Secretary, Lord Carrington,
who repeated the government's view that Britain, being on the Arabs'
'friendly' list, should not suffer any reduction in oil supplies. At this and
further meetings with ministers and government officials, Drake repeated

that BP was obliged under its international contracts to treat all customers fairly and equally in the event of an oil shortage. This commitment could only be overridden by a public government directive ordering BP to keep up its supplies to Britain, which would enable BP to claim *force majeure* for the non-fulfilment of its overseas contracts. But even this would not solve the problem, for as Drake explained, any country insisting on the full maintenance of normal supplies would only be worsening the situation for others, who would probably retaliate with their own measures of discrimination, resulting in the 'total fragmentation of European oil policy' and, quite possibly, the nationalisation of BP's overseas operations. What was needed, Drake repeatedly insisted, was an OECD allocation scheme based on fair shares for all.[62] Other oil majors took the same line.[63]

Britain was not the only oil-consuming country that put pressure on the companies. The French government, traditionally *dirigiste* in oil policy, was generally perceived to be the leading practitioner of *sauve qui peut* in the oil crisis. Refusing to accept that France should lose any oil at all due to the Arab cutbacks, the French government instructed the state-owned companies, Elf/ERAP and CFP, to give priority to France, and also warned other oil companies, including BP's French associate, that their refining licences would be revoked if they cut supplies.[64]

Whether French government pressure had any effect is difficult to say, but in Britain it is clear that although Drake undoubtedly resisted Heath's requests, he did privately bend to the government's demands. In November, he met a senior minister and officials, who said that they could 'conceal some disparities' in the official returns and wondered if BP and Shell might be prevailed upon 'to bend their supplies slightly in the UK's favour'. Drake replied that 'he was always willing to do anything . . . to assist the UK within the strictures of our international responsibilities, but in the present circumstances, any such help could only be small. He strongly advocated immediate allocation of UK oil supplies and an OECD sharing arrangement as the only practicable course for the UK to follow.' The minister suggested that a government official might be seconded to BP 'to help us programme our supplies in a less fair manner to other countries', an offer which BP did not accept. It was, however, apparent at the end of the meeting that the minister was 'still hoping that we might make some marginal and undetectable additional contribution towards the solving of the UK problem'.[65]

BP evidently obliged. In December, a secret BP note, addressed to Drake, referred to 'the private maintenance by BP of its normal UK programme to end year which you instructed us to achieve without publicity'. Drake forwarded the note to a minister with the request that 'the bit about BP's under-the-counter supplies to Britain' should not be known to anyone except the Prime Minister because BP would be in 'awful trouble' if 'our successful under-the-counter measures become known'.[66] BP's role in keeping Britain supplied with oil was not publicly revealed at the time, nor indeed twenty-five years later, when Heath wrote in his memoirs that he 'was deeply shamed by the obstinate and unyielding reluctance of these magnates [Drake and McFadzean] to take any action whatever to help our own country in its time of danger'.[67]

Although Drake yielded to government pressure, BP still abhorred the scramble for supplies that was taking place. This resulted not only in intense political pressure from oil-consuming countries, but also in bilateral state-to-state oil deals which reduced the role of the companies. The French, characteristically, were in the forefront of this movement. They signed a series of bilateral deals with OPEC countries, including a huge twenty-year contract for 5 billion barrels of Saudi crude in return for French fighter planes and other heavy weapons. Britain also made bilateral deals, including one with Iran for oil in exchange for industrial goods. Japan was extremely active in offering financial and technical assistance for industrial projects in numerous countries as a trade-off for guaranteed access to oil.[68]

With buyers flocking to them, the oil-exporting states had no trouble in selling participation crude at fantastic prices. In November, an auction of Nigerian crude fetched bids of up to $16·85 a barrel, and the total volume covered by the bids exceeded the whole of Nigeria's annual production. In December, auctioned Iranian crude fetched $17·04 a barrel, more than three times the posted price which had been announced in October.[69] The excess of market prices over posted prices had the same effect as it had done earlier in the year: it encouraged the OPEC countries to raise their postings. On 22–3 December OPEC oil ministers met in Tehran. Saudi Arabia, fearing that a very large increase would plunge the world into a depression so deep that even OPEC would be affected, argued in favour of moderation; Iran, seeking as always to maximise its revenues to finance the Shah's grand ambitions, argued for a steeper increase. Iran prevailed, and OPEC decided to raise posted prices again, lifting the posting of Arabian Light marker crude to $11·65 a barrel from 1 January 1974.[70] The posting of this crude had now risen from $1·80 in 1970, to $2·18 in 1971, to $3·011 by the autumn of 1973, to $5·12 in October 1973 and to $11·65 in January 1974. The cost of production? About 12 cents a barrel.[71]

OPEC's decision to raise postings in December 1973 was, to use Kissinger's phrase, 'one of the pivotal events in the history of this century'.[72] Oil was at the heart of world commerce, and the OPEC countries which now controlled it were regarded as the new masters of the world economy. The increase in the oil price dealt a colossal blow to the economies of the industrialised oil-consuming nations. The USA's gross national product fell by 6 per cent between 1973 and 1975, while unemployment doubled. In 1974, Japan's GNP fell for the first time since World War II. The less developed oil-consuming countries suffered even more, as the high price of oil made their attempts to escape from poverty seem more hopeless than ever.

In the oil-exporting countries, on the other hand, the rise in oil prices generated dramatic increases in income, which led to overheated economic development, characterised by great spending binges on all sorts of goods and services from basic infrastructure items to sophisticated weapons and extravagant private luxuries.

There was nothing that the oil-consuming countries could do in the short term to reverse the price rise. Nixon sent a message to the Shah, expressing his great concern about the destabilising impact of higher oil prices and the 'catastrophic problems' they could pose to the international monetary system. But the Shah would make no concessions. His advice to the industrial nations was that they 'will have to realise that the era of their terrific progress and even more terrific income and wealth based on cheap oil is finished. They will have to find new sources of energy. Eventually they will have to tighten their belts; eventually all those children of well-to-do families who have plenty to eat at every meal, who have their cars, and who act almost as terrorists and throw bombs here and there, they will have to rethink all these aspects of the advanced industrial world. And they will have to work harder . . . Your young boys and young girls who receive so much money from their fathers will also have to think that they must earn their living somehow.'[73]

Belatedly, the main oil-consuming nations met in Washington for an energy conference in February 1974, and agreed to take steps which would lead to the creation later in the year of the International Energy Agency, which the USA, Japan and the main European oil-consuming countries, apart from France, agreed to join. By the time the IEA was set up the Arab embargoes had been lifted, but the new agency would help to establish a more harmonised response to future crises.[74]

The oil companies, meanwhile, were losing nearly all their remaining concessions in OPEC countries. Kuwait, which had not ratified the General Agreement on Participation, took 60 per cent participation in the Kuwait Oil Company from the beginning of 1974, and 100 per cent from March 1975. Qatar followed suit, while Abu Dhabi stopped at 60 per cent participation. Saudi Arabia took a 60 per cent share in Aramco in June 1974 and settled the terms of full nationalisation in 1976. Iraq decided to nationalise the remaining private interests, including BP's 23¾ per cent, in the Basra Petroleum Company in December 1975. In Africa, Nigeria raised its participation to 55 per cent in 1974; and Libya took either 51 per cent or 100 per cent, depending on the circumstances, of those companies which had stood out against the nationalisation decree of September 1973.[75]

These nationalisations were but a few examples of the uprising against multinationals that was taking place across the Third World, where the radical Arab oil states had become role models for less developed countries seeking to assert national sovereignty over their natural resources. Between 1970 and 1976 at least eighteen countries nationalised oil-producing operations,[76] while many more nationalisations took place in other natural resource industries, particularly in minerals such as copper and bauxite. In 1975, when the number of nationalisations peaked, twenty-eight less developed countries carried out eighty-three acts of expropriation (see figure 19.1). Never before, nor since, had multinationals been expropriated on this scale.[77]

While oil concessions fell like ninepins in the OPEC countries, the traditional flows of foreign investment were partly reversed, to the consternation of the new target countries. In the USA, concern about investments from OPEC countries led to the creation of the Committee on Foreign Investment in the United States to monitor inward investments. One of the cases it reviewed was a plan by the Iranian government to acquire a 9 per cent stake in Occidental, which did not come to fruition because Iran and Occidental failed to reach agreement on the terms.[78] In West Germany, there were fears that the OPEC countries would use their new-found wealth to dominate some sectors of German industry. Iran's acquisition of a 25 per cent stake in Krupp aroused concern, but it was the acquisition of 14 per cent of Daimler-Benz by the government of Kuwait that really touched off a German reaction. German firms organised a holding company to acquire Daimler shares, and the government established an informal notification system whereby banks and major firms would report on impending sales of companies or large blocks of shares to foreigners, particularly OPEC governments and their agencies.[79] In Switzerland, the government, worried about Swiss firms falling under OPEC control, made it clear that if voluntary methods failed to preserve the Swiss character of local companies, it would consider taking other measures.[80] In Italy, meanwhile, the Libyan government acquired 14½ per cent of Fiat in return for a capital injection of $400 million.

BP itself was not immune to acquisitive moves from OPEC countries. In 1975, Iran made several attempts to buy BP shares from the Bank of England, which had acquired Burmah Oil's large BP shareholding in the

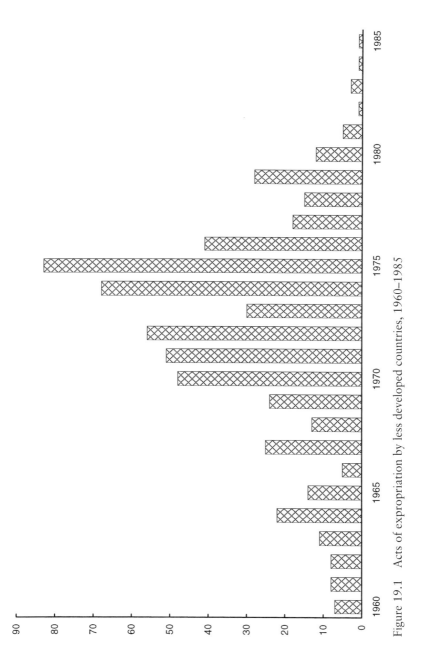

Figure 19.1 Acts of expropriation by less developed countries, 1960–1985

course of rescuing Burmah from bankruptcy in 1975. Iran's approaches came to nothing because the Bank made it clear 'that there was little prospect of an early disposal of the shares'.[81]

It was not only in the OPEC countries that the oil companies were unpopular. In their home countries it was widely assumed before the crisis that the oil majors were 'flag' companies, which would act on behalf of their home countries in emergencies. There was widespread disillusionment when the public found itself having to forego some normal comforts and to pay high oil prices, while the oil companies were making record profits, owing to stock-holding gains and higher margins on their remaining share of crude production. There were even allegations, popularised in a series of articles in the *Washington Post*, that the companies had conspired with the oil-exporting states to raise prices. But it was not only the companies' high profits that fuelled suspicions that they had changed sides and were unpatriotically acting with the OPEC countries against their home countries. Had they not, after all, undertaken a propaganda and lobbying campaign to try to persuade the USA to change its Middle East policy to a more pro-Arab stance? And had they not carried out the embargoes on the instructions of the Arab states?[82] When Americans were asked in a Gallup poll in January 1974: 'Who or what do you think is responsible for the energy crisis?', the most common response was to blame the oil companies, not the Arab nations.[83]

The companies came under attack from politicians as well. In the USA, there was a flurry of Congressional investigations into the oil industry, the best known being Senator Frank Church's Senate Subcommittee on Multinational Corporations. Reviving interest in antitrust allegations against the oil majors, which had lain virtually dormant for twenty years, the Subcommittee grilled US oil executives with hostile questions about their operations. In a mood reminiscent of that which had led to the dissolution of Rockefeller's Standard group in 1911, Congress debated, but did not pass, proposals for breaking up the integrated companies.[84]

In Europe, also, the oil majors were beginning to look like an endangered species. The West German Federal Cartel Office launched proceedings against several companies, including Deutsche BP, on charges that they had held up the prices of products. After much-publicised hearings, the proceedings were dropped in the summer of 1974. In France, a commission of enquiry criticised the companies for various forms of 'anti-social' behaviour. In Italy, hostility towards the companies was strengthened when, in February 1974, the offices of Unione Petrolifera, the industry trade association, were raided and documents were found which suggested that the Unione was acting as a channel for payments to a wide range of political parties and individuals (see chapter nine), and that it was compiling false information on oil supplies with the intention of obtaining price increases. In Belgium, there was a confrontation between the industry and the government when practically the whole industry went on an import strike in March 1974 to protest against price controls. The government allowed prices to be increased, but to allay left-wing concerns it also created a special commission with sweeping powers to investigate the oil industry. In Japan, disagreement over prices led the oil companies discreetly to threaten to cut back supplies, and an anti-trust case culminated in May 1974 with the indictment of twelve leading companies, not including BP.[85]

BP was, however, deeply involved in political issues in the UK, where a Labour government came to power in February 1974. The left-wing Tony Benn was soon appointed Secretary of State for Energy, responsible for extending state control over the British North Sea through the newly created state oil company, the British National Oil Corporation. Drake and Heath may not have seen eye to eye, but at least they shared a belief in private enterprise. Benn did not, and made no secret of his dislike for Drake, whom he dubbed 'the most Tory of Tories'.[86] The negotiations for state participation in the North Sea held many problems for BP, which, Benn argued, ought to do what the government wanted because of the large BP shareholdings held by the state and by the publicly owned Bank of England.[87]

Amid these pressures from all sides, it was difficult to see what role the oil majors would play in the future. They were no longer regarded as national champions; and with the growth of bilateral state-to-state oil deals, their role as buffers between the oil-producing and oil-consuming nations had been greatly reduced. They could no longer control the international flow of oil, now that OPEC had taken control of the upstream sector and partially de-coupled it from the companies' downstream operations. The prospects of making major new oil discoveries in accessible areas were generally held to be poor. Old-style concessionary relationships were obsolete and the OPEC countries, not the oil companies, were setting prices which, virtually everybody agreed, would go on rising as the world's oil supply was depleted. Never before had the oil majors faced such fundamental challenges.

Retrospect and conclusion

In the first three quarters of the twentieth century there were four main phases in BP's development. In the first phase, stretching from the Company's incorporation in 1909 until 1950, the Company concentrated on Iran (see figure 20.1). From its start in that country, the Company exploited the competitive advantage it gained from being a first mover in discovering and producing Middle East oil, and became one of the world's seven oil majors. They were the dominant players in the international oil industry and belonged in the premier league of the world's largest industrial firms. Such was the Company's concentration on Iran that in 1950 the Company's giant Iranian oil fields and the vast Abadan refinery – the largest refinery in the world – accounted for about three quarters of the Company's crude production and refining throughputs.

Meanwhile, although the Company had ventured into marketing, primarily in Western Europe, its marketing capabilities remained comparatively undeveloped. This reflected its late entry into markets where the world's two largest oil companies, Standard Oil (NJ) and Shell, were already entrenched with powerful incumbency advantages. Rather than compete head-on with such formidable rivals, the Company entered into market-sharing agreements with them in, for example, the Achnacarry and Burmah-Shell Agreements of 1928 and the Far Eastern Agreement of 1948. These agreements, restrictive in nature, limited the Company's scope to develop its own marketing capabilities. The Company also accepted junior positions to Shell in their two main joint marketing ventures. One of these covered Britain through Shell-Mex and BP, in which Shell was the more powerful partner. The other covered the much larger territory, including eastern and southern Africa, in which the jointly owned (but Shell-managed) Consolidated Petroleum Company marketed Shell and BP products. This marketing strategy, though unadventurous,

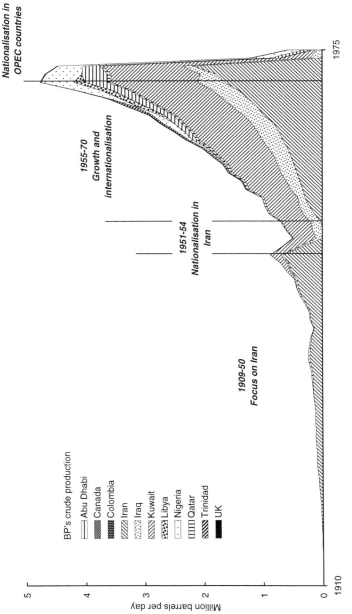

Figure 20.1   Four phases in BP's history, 1909–1975

was realistic. In oil, as in other industries, joint ventures provided a means for late comers to gain access to capabilities that they lacked, and could be an excellent way for late starters to enter markets. The same was true in petrochemicals, where the Company, a late comer, made its entry through a partnership with Distillers.

The second phase in the Company's development, lasting only from 1951 to 1954, was a period of crisis, brought on by the nationalisation of the Company's assets in Iran, and the stoppage of the Company's operations in that country. Though short in duration, this phase witnessed a fundamental restructuring of the Company, which took urgent action to overcome the immediate shrinkage of its operations, and decided to pursue an energetic policy of diversifying its sources of crude oil and its refining operations to reduce its dependence on Iran and the Middle East as a whole. The Company emerged from the crisis with a reduced position in Iran, and a new determination to develop a more international spread of interests.

This inaugurated the third phase in BP's development, the period of growth and internationalisation from 1955 to 1970. During these years, BP grew very quickly. Outside its traditional oil business, BP expanded in chemicals, primarily by taking over Distillers' chemical and plastics interests, and diversified into nutrition and computing. In its mainstream oil business, BP grew very rapidly, achieving a double-digit average growth rate of 10 per cent a year in the volumes of its crude production, refining throughputs, and sales.[1] BP was not, however, alone in achieving a high growth rate, and in 1970 its position in relation to the other majors in the league of the world's biggest industrial firms was slightly lower than in 1956 (see table 20.1). BP shared in the expansion of the international oil industry, but did not achieve a distinctive growth performance.

At the same time, BP extended its geographical reach. The towering move was its large-scale entry into the USA, which fundamentally shifted BP's geographical balance, and gave it a substantial presence in both hemispheres of the globe. But BP was also extending its reach in other directions. Between 1955 and 1970 the number of countries, excluding the USA, in which BP held oil-producing interests rose from 6 to 10,[2] and the number in which it held refining interests rose from 12 to 27.[3] In marketing, BP entered into negotiations which would end its joint marketing with Shell in Britain and in the Consolidated area, leaving BP with its own marketing organisations in Britain and parts of Africa. Meanwhile, BP reached out from Western Europe and Australasian markets into Canada, the USA and, in a comparatively small way, the Far East. By these moves, BP became a far more international

Table 20.1 *The oil majors' rankings among the world's largest industrial firms in 1970, compared with 1956*

	1956		1970	
	Gross sales (including excise taxes) ($billions)	Ranking (by sales) among world's largest industrial firms	Net sales (excluding excise taxes) ($billions)	Ranking (by sales) among world's largest industrial firms
Standard Oil (NJ)	7·13	2	16·55	2
Royal Dutch-Shell	6·50	3	10·80	4
Mobil	2·75	8	7·26	7
Gulf	2·34	14	5·40	13
Texaco	2·05	16	6·35	11
British Petroleum	2·02	17	4·06	19
Socal	1·45	23	4·19	17

business than it had been in the days when it was focused on Iran, though it remained well behind the two 'super-majors', Standard Oil (NJ) and Shell, in its degree of globalisation.

Opportunities for emulating Standard Oil (NJ) and Shell were limited not only by BP's resources and capabilities, but also by an external environment which was not, on the whole, favourable to the spread of international business in oil. Before BP's incorporation in 1909, Rockefeller's Standard and Royal Dutch-Shell had already established themselves as the first movers in the globalisation of the oil industry. They did so in an open world economy, around which capital, labour, management and technology could be moved with little, if any, hindrance. It would be many years before such conditions returned.

After the disruption of World War I, hopes of returning to prewar conditions were shattered by the autarkic policies of economic nationalism that were widely adopted in the 1930s. World War II brought further disruptions and devastation. While it raged, the USA and Britain laid plans for a postwar liberal international economy, which came into being, in modified form, after the end of the war. It was, however, a far cry from the open world economy that had existed before World War I. With the onset of the Cold War, an 'iron curtain' fell between the communist and the non-communist worlds. The Soviet Union, its satellites in Eastern Europe, and communist China became closed areas for international business.

In the developed non-communist world, Japan was resistant to foreign multinationals, but in Western Europe and America there were generally fewer barriers to international business, which spread very rapidly, mainly from the USA, in the 1950s and 1960s. Most of the American multinationals spreading into Western Europe were, however, in manufacturing industry. Oil had become such a vital commodity that even in the comparatively open economies of the developed world, governments adopted nationalistic oil policies.

At the same time, the liquidation of the old European empires brought into being a host of newly decolonised nations, mostly less developed countries, and the world became divided between rich and poor nations. Many of the less developed countries adopted nationalistic economic policies, seeking to assert their independence from the old imperial powers, and the multinational companies that were associated with them. Radical nationalism became a potent force, not least in the Middle East, where BP's oil concessions were the main source of its crude oil.[4]

In this environment, the tensions between internationalisation and nationalism posed fundamental problems for BP. With its emphatically British identity, and the British government as its main shareholder, BP stood out more than most companies as a symbol of British imperialism, and a target for nationalists. Aware of its high exposure to political risks, particularly in the Middle East, BP adopted a dual strategy. On the one hand, BP tried to retain its self-acknowledged 'very strong interest in the status quo' by adopting a 'fabian policy of delay and immobility' in trying to hold on to its Middle East oil concessions;[5] while on the other hand it used the exploration capabilities it had earlier learned as a first mover in Iran to explore for oil in politically more secure areas.

In seeking to uphold the status quo, BP tried, with considerable success in the 1960s, to persuade the moderates in OPEC to restrain the radicals. This was achieved by convincing the Shah of Iran, seen as the weak link in OPEC, that he had more to gain by co-operating with the oil companies than by turning against them. To keep Iran on the side of the West, the Consortium had to deliver very large increases in Iran's oil exports, on which the Shah depended to finance his economic and military ambitions. With other oil-exporting countries also pushing for increases in production, the companies found themselves on a tightrope as they endeavoured to keep a balance between the demands of the oil-producing states.

Pushed into maximising crude oil production to placate the oil-exporting countries, BP needed every market it could find. But in the downstream sector, as in the upstream, BP's opportunities were limited by economic nationalism.

Less developed oil-importing countries such as India, released from the imperial yoke, established state oil companies which took a large share of the market from the majors. More important, insofar as they consumed much more oil, were the oil-importing countries of the developed world, which adopted various measures of economic nationalism in oil, ranging from import quotas in the USA, comprehensive state controls in France, state backing for ENI in Italy, to milder forms of intervention in West Germany.

In markets that remained open, BP invested very heavily in downstream assets, but so did other oil companies. With growing competition and an excess of supply over demand, oil prices were depressed. Although the volume of BP's sales grew fast, BP's profits stayed flat in real terms. As the funds generated by BP's operations were insufficient to finance its heavy capital expenditures, BP's finances came under considerable strain. To sustain the growth in production without concomitant downstream investment, BP sold a growing proportion of its production as unrefined crude oil. Even so, BP's borrowing increased, and its debt ratio rose inexorably. The policy of maximising production to satisfy the oil-exporting countries where BP held concessions was proving very costly.

Far fewer resources were dedicated to the other side of BP's dual strategy, the search for new sources of oil in politically secure areas. The results, literally and metaphorically, were striking. The discovery of vast petroleum reserves in BP's acreage at Prudhoe Bay in Alaska in 1969 and the opening up of the North Sea as a major new petroleum province gave BP substantial oil and gas reserves in comparatively safe areas at a most opportune time.

In the first half of the 1970s the mounting tensions between the international interests of the oil majors and the national interests of nation states reached a climax. In this fourth phase in BP's history, the OPEC countries took control of crude oil prices and production and nationalised their oil industries. In the space of a few months in 1973–4, crude oil prices were raised dramatically from, in round terms, $3 to $12 a barrel. While the Arab states denied or reduced oil supplies to countries that were friendly to Israel, the oil companies, taking an international view, tried to equalise the impact of the reduced supplies on the oil-consuming nations. But governments in oil-consuming countries put their national interests first as each scrambled for supplies. International economic co-operation was at a low ebb and the international oil companies found themselves under intense pressure from all sides, operating in a more unfavourable political environment than ever before.

The magnitude of these events, and the speed with which they occurred, took the world by surprise and posed formidable problems of adjustment.

At BP, the challenges were fundamental. Throughout its history, BP had developed on the basis that its concessions gave it access to virtually unlimited reserves of Middle East crude oil, which could be produced more cheaply than any other crude in the world. By 1970 sales of unrefined crude oil accounted for about 45 per cent of BP's sales tonnage,[6] and half of BP's profits.[7] The rest of BP's production was transferred to BP's downstream operations for refining, marketing and sale. In most of BP's main markets, these operations had been developed as a means of providing guaranteed outlets for BP's production, rather than as a means of adding to the profits that could be made from selling crude oil.[8] As stand-alone operations, divorced from production, they were not viable. Indeed in Western Europe, BP's main refining and marketing area, they were making heavy losses, as BP's internal downstream profitability studies revealed beyond doubt in the early 1970s.[9] As BP's position as a crude oil producer fell away owing to participation or nationalisation by OPEC countries, the justification for holding downstream assets was no longer to dispose of crude, but to make profits.

Such a turnaround would be an extremely difficult task, especially in the economic environment that prevailed in the mid-1970s. At the beginning of the decade, in a world conditioned by the continuous economic growth of the 1950s and 1960s, BP had committed itself to large expenditures on new shipping and refining capacity, based on forecasts that Western Europe's demand for oil would rise by about 7 per cent a year for the next five years. That, however, was not how it turned out. The dramatic increase in oil prices in 1973 helped to cast the industrialised world into a deep economic recession, the first since World War II. Western Europe's oil consumption, whose earlier growth rate was already slowing in the early 1970s, actually fell in 1974 and 1975. BP, in common with other oil companies, was caught out, and found itself taking delivery of new VLCCs which it had ordered earlier, but now had no use for. It was left with a very large surplus of shipping capacity.[10] In refining the story was the same. The great expansion in capacity since 1970 left BP, and other European refiners, with a great overhang of spare capacity when the rising demand for oil products went into reverse in the mid-1970s (see figures 20.2 and 20.3).

Upstream, by contrast, BP's Alaskan and North Sea oil reserves were the jewels in BP's assets. With the rise in crude oil prices, their value rocketed and their prospective profitability was transformed. BP's top priorities were, therefore, to bring these reserves on stream as quickly as possible to make profits, while restructuring downstream operations to stem losses.

Looking beyond that, however, BP was a company in search of a strategy. For years, BP had done much detailed forward planning, but little funda-

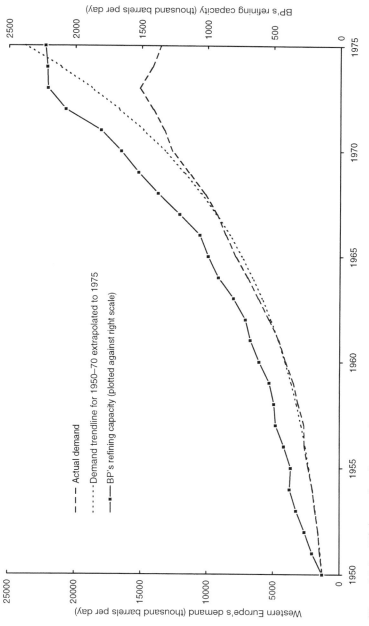

Figure 20.2　Oil demand and BP's refining capacity in Western Europe, 1950–1975

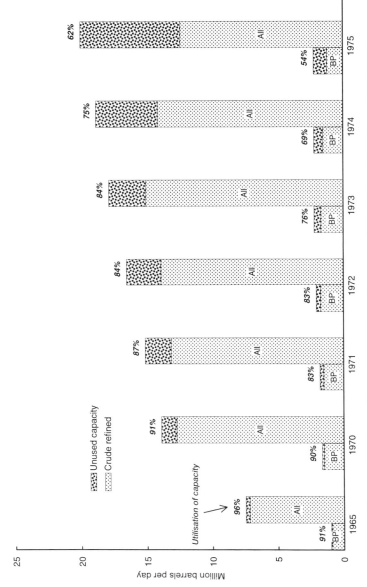

Figure 20.3   The growth of unused capacity in Western European refining: BP and all refiners, 1965–1975

mental strategic reappraisal. Now, the name of BP's Planning Committee was changed to the Strategic Planning Committee, and there was much discussion about strategy.[11] Many questions were raised about what BP should aim to be: What was its future as an oil company? Should it be an energy corporation, going into coal, synthetic fuels, or even nuclear power? Or should it build on its existing diversifications, enter new industries, and become more of a conglomerate? Beyond question, a seismic upheaval had transformed the relationships between nation states and international oil companies. The OPEC countries were in control of the commanding heights of the world oil industry, and the once-dominant oil majors would have to find new roles in the new order of global oil.

Notes to the text

INTRODUCTION

1 The leading study of the rise and growth of corporate capitalism is Alfred D. Chandler, *Scale and Scope: The Dynamics of Industrial Capitalism* (Cambridge, Mass., 1990). For a study of the response in the USA to the rise of big business, see Louis Galambos, *The Public Image of Big Business in America, 1880–1940* (Baltimore, Md., 1975).

2 For general studies, see Anthony Sampson, *The Seven Sisters: The Great Oil Companies and the World They Made* (London, 1975); Daniel Yergin, *The Prize: The Epic Quest for Oil, Money, and Power* (New York, 1991), chs. 1–19.

3 Edith T. Penrose, *The Large International Firm in Developing Countries: The International Petroleum Industry* (London, 1968), p. 78; Neil H. Jacoby, *Multinational Oil: A Study in Industrial Dynamics* (New York, 1974), p. 211.

4 For detailed histories, see R. W. Ferrier, *The History of The British Petroleum Company: Volume I, The Developing Years, 1901–1932* (Cambridge, 1982); J. H. Bamberg, *The History of The British Petroleum Company: Volume II, The Anglo-Iranian Years, 1928–1954* (Cambridge, 1994).

5 For a corporate history of Rockefeller's Standard Oil, see Ralph W. and Muriel E. Hidy, *Pioneering in Big Business: A History of Standard Oil Company (New Jersey), 1882–1911* (New York, 1955). On Rockefeller himself, Ron Chernow, *Titan: The Life of John D. Rockefeller* (New York, 1998).

6 For brief histories of these and other international oil companies, see Penrose, *The Large International Firm*, ch. 4.

7 On the emergence of Royal Dutch-Shell, see Stephen Howarth, *A Century in Oil: The 'Shell' Transport and Trading Company, 1897–1997* (London, 1997), chs. 1–4.

8 Ferrier, *British Petroleum, Volume I*, chs. 1–5.

9 See, for example, P. H. Frankel, *Essentials of Petroleum* (London, 1940); John M. Blair, *The Control of Oil* (New York, 1976), esp. pp. 241–6; M. A. Adelman, *The World Petroleum Market* (Baltimore, Md., 1972), ch. 3.

10 Penrose, *The Large International Firm*, pp. 96 and 130.

11 T. A. B. Corley, *A History of the Burmah Oil Company: Volume II, 1924–1966* (London, 1988), p. 25; Bamberg, *British Petroleum, Volume II*, p. 107; BP 9249, Strictly confidential memorandum, 19 December 1949; BP 106234, Board minutes, 20 December 1949; BP 113811C, Agreement between AIOC and Shell, 31 January 1951; BP 59545, Burmah-Shell Agreements, 30 March 1955.

12 George Philip, *Oil and Politics in Latin America* (Cambridge, 1982), chs. 1–3, 10 and 17; Bamberg, *British Petroleum, Volume II*, pp. 143–6.

13 On the Iranian concession, see Ferrier, *British Petroleum, Volume I*, chs. 1, 9 and 13; Bamberg, *British Petroleum, Volume II*, chs. 2, 9 and 15. On Iraq, Edith and E. F. Penrose, *Iraq: International Relations and National Development* (London, 1978), chs. 1, 3 and 6; Helmut Mejcher, *Imperial Quest for Oil: Iraq, 1910–1928* (London, 1976); Marian Kent, *Oil and Empire: British Policy and Mesopotamian Oil, 1900–1920* (London, 1976). On the Kuwait concession, A. H. T. Chisholm, *The First Kuwait Oil Concession Agreement: A Record of the Negotiations, 1911–1934* (London, 1975).

14 For a detailed study, see Irvine H. Anderson, *Aramco, the United States and Saudi Arabia: A Study of the Dynamics of Foreign Oil Policy, 1933–1950* (Princeton, N.J., 1981).

15 Peter R. Odell, *Oil and World Power*, 7th edn (Harmondsworth, 1983), pp. 12–13 and ch. 3.

16 Angus Maddison, *The World Economy in the Twentieth Century* (Paris, 1989), ch. 6; A. G. Kenwood and A. L. Lougheed, *The Growth of the International Economy, 1820–1990*, 3rd edn (London, 1992), chs. 17–21; James Foreman-Peck, *A History of the World Economy: International Economic Relations since 1850*, 2nd edn (Hemel Hempstead, 1995), ch. 13.

17 John G. Clark, *The Political Economy of World Energy: A Twentieth Century Perspective* (Hemel Hempstead, 1990), p. 193.

18 Joel Darmstadter and Hans H. Landberg, 'The economic background' in Raymond Vernon (ed.), *The Oil Crisis* (New York, 1976), pp. 19–20.

19 Clark, *World Energy*, p. 101.

20 R. H. K. Vietor, *Energy Policy in America since 1945: A Study of Business–Government Relations* (Cambridge, 1984), chs. 5–6.

21 Geoffrey Jones, *The State and the Emergence of the British Oil Industry* (London, 1981).

22 Leslie E. Grayson, *National Oil Companies* (Chichester, 1981), chs. 2–4; US Federal Energy Administration (FEA), *The Relationship of Oil Companies and Foreign Governments* (Washington D.C., 1975), pp. 47–51. For a different view of French oil policy, which challenges the orthodox 'strong state' interpretation, see Gregory P. Nowell, *Mercantile States and the World Oil Cartel, 1900–1939* (New York, 1994).

23 Grayson, *National Oil Companies*, chs. 5–6; FEA, *Oil Companies and Foreign Governments*, pp. 57–62 and 85–96.

24 See especially, Jacoby, *Multinational Oil*.

25 For general studies of decolonisation, see, for example, R. F. Holland, *European*

Decolonization, 1918–1981: An Introductory Survey (London, 1985); John Darwin, *Britain and Decolonisation: The Retreat from Empire in the Postwar World* (London, 1988).

26 Darwin, *Britain and Decolonisation*, p. 13.

27 On the changing conditions in which foreign enterprise operated, see, for example, Charles Lipson, *Standing Guard: Protecting Foreign Capital in the Nineteenth and Twentieth Centuries* (Berkeley, Calif., 1985).

28 For an excellent study of Britain's position in the Middle East, see Elizabeth Monroe, *Britain's Moment in the Middle East, 1914–1971*, 2nd edn (London, 1981).

1   'THE STRUCTURE AND SINEWS OF THE COMPANY'

1 *Parl. Debates*, House of Commons, vol. 55, col. 1477, 17 July 1913, Churchill.

2 BP 102759, Management report for board meeting, January 1951.

3 Stephen Howarth, *A Century in Oil: The 'Shell' Transport and Trading Company, 1897–1997* (London, 1997), pp. 121–2.

4 BP 64040, Staff manual, *c.* 1949.

5 BP 90865, Memorandum and articles of association embodying amendments up to 4 October 1979.

6 BP 64040, Staff manual, *c.* 1949.

7 Political and Economic Planning, *Graduate Employment: A Sample Survey* (London, 1956), p. 142.

8 BP 64040, Staff manual, *c.* 1949. The housing scheme also applied to the Glasgow staff of Scottish Oils, a Company subsidiary.

9 BP 78468, Mayhew, AIOC inter-managerial conference, April 1951.

10 BP 64040, Staff manual, *c.* 1949.

11 J. H. Bamberg, *The History of The British Petroleum Company: Volume II, The Anglo-Iranian Years, 1928–1954* (Cambridge, 1994), p. 347. See BP 67593 and 106299 for data on the total numbers employed in all countries.

12 Geoffrey Jones, *The Evolution of International Business: An Introduction* (London, 1996), pp. 233, 236 and 256–8; Julian Bharier, *Economic Development in Iran, 1900–1970* (London, 1971), p. 160; Fereidun Fesharaki, *Development of the Iranian Oil Industry: International and Domestic Aspects* (New York, 1976), pp. 142–50; Mira Wilkins, *The Maturing of Multinational Enterprise: American Business Abroad from 1914 to 1970* (Cambridge, Mass., 1974), pp. 122–7.

13 BP 62017, Staff manual (overseas), *c.* 1949.

14 BP 62039, Overseas staff, *c.* 1949.

15 BP staff file on Drake.

16 BP 70346, Biographical details of Basil Jackson, 28 July 1950.

17 BP 72322, Pattinson to Elkington and Jameson, 28 March 1946.

18 *Ibid.*, Organisation charts, 7 January and February 1948; *ibid.*, Mayhew to

Fraser enclosing staff table, 1948; *ibid.*, Mayhew to Jameson enclosing organigram of the Production Department, 23 March 1948.

19 Emphases by Jackson. The three managing directors to whom he referred were James Jameson, Neville Gass and Edward Elkington. The deputy director was John Pattinson.

20 BP 96478, Jackson's untitled note, 30 December 1949.

21 On the Company's production and profits at this time see Bamberg, *British Petroleum, Volume II*, pp. 273–6 and 352.

22 Bamberg, *British Petroleum, Volume II*, chs. 16–17.

23 BP 70943, Coxon, Programme Committee, 10 July 1951.

24 BP 96455, Interim report of the Future Programme Committee, 25 July 1951. On the long-term crude oil contracts with Standard Oil (NJ) and Socony, see Bamberg, *British Petroleum, Volume II*, pp. 303–7.

25 Bamberg, *British Petroleum, Volume II*, p. 298.

26 AIOC, *Annual Report and Accounts*, 1951.

27 BP 96455, Interim report of the Future Programme Committee, 25 July 1951.

28 Bamberg, *British Petroleum, Volume II*, pp. 242–6.

29 BP 96455, Hubbard note to Fraser, 25 July 1951.

30 BP 10811, Monthly returns of crude oil availability.

31 On the negotiations with Standard Oil (NJ) and Socony see, generally, BP 51701, 77969, 71339, 66824 and 68169. On the specific matter of the quantities to be delivered in 1952 see BP 77969, Smith to Standard Oil (NJ) and Socony, 28 January 1952.

32 Comment Gulliver to Bamberg, October 1987.

33 BP 10811, Monthly returns of crude oil availability.

34 BP 96455, Notes on action to be taken resulting from the discussion of the Future Programme Committee's interim report in the chairman's room on 31 July 1951; *ibid.*, Hubbard, Aden refinery, 23 November 1951; *ibid.*, Second interim report of the Future Programme Committee, 7 December 1951; *ibid.*, Coxon, Notes on meeting held in chairman's room on 19 December 1951; *ibid.*, Coxon, Future refineries programme, 20 February 1952; *ibid.*, Note on Qatar as a possible refining centre, 6 March 1952; BP 70943, Butler to Coxon, 18 September 1951; *ibid.*, Aden refinery project, Economic assessment, 7 November 1951; *ibid.*, Steering Committee, 5th meeting on 8 November 1951; BP 66259, Snow to Smith, 19 February 1952; *ibid.*, Snow to Coxon, 3 March 1952.

35 BP, *Annual Reports and Accounts*, 1954–5.

36 BP 71379, Distribution information memorandum no. 28, 14 May 1952; BP 96455, Coxon note on long-term policy, 8 January 1953.

37 AIOC, *Annual Reports and Accounts*, 1951–2.

38 Details of the processing arrangements can be found in BP 65613, 43321, 15637, 95579–95582.

39 BP 12417, Papers entitled 'Current supply position' and 'BP Aviation Service –

marketing', Proceedings of AIOC marketing management conference, May 1954; BP 61677, Papers entitled 'Current Activities' and '1955', Aviation marketing meeting, September 1954; BP 66260, Unsigned note on aviation gasoline.

40 BP 95579–80, Final accounts papers, 1951–2; BP 71376, Notes for Drake prior to his return to New York, 23 July 1953.

41 On the success of this move, see BP 66239, Anderson to Jackson, 17 September 1954.

42 BP 71379, Distribution information memorandum no. 28, 14 May 1952; BP 96455, Coxon note on long term policy, 8 January 1953; BP 69507, unsigned note, Analysis of freight cost of marginal tonnage, 31 January 1952; *ibid.*, Cooper note to Fraser, Supply/shipping programme, 7 February 1952; BP 106235, Minutes of board meeting on 28 February 1952 (Snow, 'shipping factor'); BP 12417, Paper entitled 'Current supply position', Proceedings of AIOC marketing management conference, May 1954; AIOC, *Annual Reports and Accounts*, 1952–3.

43 BP 102771, 102796 and 102810, Management reports for board meetings, February 1952, January 1954 and January 1955; Bamberg, *British Petroleum, Volume II*, p. 293. In Iraq the Company's sales became almost negligible following its agreement with the Iraqi government that from the beginning of 1952 the government would take over sales of main products, leaving the Company with a small volume of sales in special products such as aviation spirit and lubricating oil. See BP 71379, Distribution information memorandum no. 28, 14 May 1952.

44 BP 102799, Management report for board meeting, April 1954.

45 BP 106235, Minutes of board meeting on 29 November 1951.

46 BP 30580, Shelbourne to Morris and Snow, 25 March 1952.

47 BP 102810, Management report for board meeting, January 1955.

48 BP 4313, Unsigned note on British staff employed in Iran, n.d.; BP 72323, Unsigned aide-memoire on policy governing review of the Company's staff requirements, 23 October 1951.

49 BP 72323, Pattinson to Jameson, 11 October 1951.

50 BP 4313, Baxter, Notes on the reactions of overseas staff to the present situation, 5 November 1951.

51 BP 7142, Elkington to Baxter, 2 October 1951.

52 Minutes of the meetings of the Staff Redeployment Committee and the Resettlement Panel are in BP 7142, 3226, 64910, 64912 and 4304. See also BP 72323, Staff Redeployment Committee, Report no. 1, 12 October 1951; *ibid.*, Elkington to heads of departments, 22 October 1951; *ibid.*, Unsigned aide-memoire on policy governing review of the Company's staff requirements, 23 October 1951.

53 BP 4313, Particulars of staff reductions following on evacuation from Persia, No. 15; BP 4352, Foreign staff absorbed into AIOC organisation as at 28 January 1953.

54 Details can be found in BP 4313.
55 BP 41348, Report on AIOC Resettlement Panel, October 1951–July 1953.
56 BP 70346, Biographical details of Basil Jackson, 28 July 1950.
57 BP 92785, Jackson, Memorandum on Refineries Department, 13 March 1951.
58 BP 72322, Pattinson, Memorandum on Production Department, 3 May 1951.
59 BP 72323, Pattinson, Production Department, 11 October 1951.
60 BP 1967, Table of departmental staff numbers, 21 May 1952. There is a variance between the number of staff employed in the Production Department given here, and the authorised staff numbers given in chapter two from another source. The variance is not material to the arguments presented.
61 For organisation charts of the department see BP 22993, Whitehead report, 1950. The Supply Division was at that time known as the Petroleum Supply Division.
62 BP 66259, Unsigned note on the Distribution Department organisation, 20 March 1952.
63 BP 66874, Smith to Snow, Petroleum Supply Division, 2 December 1952.
64 BP 69507, Snow to Fraser, 7 July 1953.
65 BP 68947, Hubbard, Central Planning Department, 12 October 1953. On the genesis of the Central Planning Department and its early work see also Bamberg, *British Petroleum, Volume II*, pp. 226, 229, 278 and 288–9.
66 BP 59571, Pattinson telegram to H. W. Coxon, 1 September 1953; BP 68947, Hubbard, Central Planning Department, 12 October 1953; BP 66261, Snow to Berkin, 26 January 1954.
67 BP 69507, Snow to Fraser, 7 July 1953.
68 Despite the loss of some of its functions, mainly the compilation of statistics, the Trade Division retained a separate existence, dealing with prices and special agreements.
69 BP 68947, Staff memorandum 54/2, 22 January 1954; *ibid.*, Snow and Gass, Note on departmental organisation, 22 January 1954.
70 IOP 60.04/4/1, Drake's biographical details, 16 September 1975; BP 68947, Staff Memorandum 54/2, 22 January 1954; BP 66236, Unsigned letter to Drake, 28 December 1953. On Drake as general manager in Iran see Bamberg, *British Petroleum, Volume II*, pp. 420, 425, 428–30, 432–5.
71 BP 66261, Snow to Berkin, 26 January 1954.
72 BP staff file on Drake.
73 Alfred D. Chandler, *Strategy and Structure: Chapters in the History of the American Industrial Enterprise* (Cambridge, Mass., 1962), pp. 168 and 189–93.
74 On the shift from tonnage royalties to profit sharing see, for example, G. W. Stocking, *Middle East Oil: A Study in Political and Economic Controversy* (London, 1971), ch. 6.
75 Bamberg, *British Petroleum, Volume II*, pp. 339–41, 344 and 508.
76 See generally documents in BP 34883. Also, AIOC/BP, *Annual Reports and Accounts*, 1952–5.

77 AIOC, *Annual Reports and Accounts*, 1952–3.

78 See the letter sent to the Company in 1914 by Sir John Bradbury, Joint Permanent Secretary of the Treasury, reprinted in appendix 6.1 in R. W. Ferrier, *The History of The British Petroleum Company: Volume I, The Developing Years, 1901–1932* (Cambridge, 1982), pp. 645–6.

79 Bamberg, *British Petroleum, Volume II*, pp. 326 and 387. Also, PRO T 236/4755, Unsigned paper on 'Relationship between HM Government and the British Petroleum Company', n.d.

80 AIOC, *Annual Reports and Accounts*, 1947–51; BP 95593–95595 and 95578–95579.

81 *Parl. Debates*, House of Commons, vol. 491, col. 674, 26 July 1951, Gaitskell.

82 *Parl. Debates*, House of Commons, vol. 493, col. 204, 7 November 1951, Butler.

83 PRO T 236/4755, Unsigned paper on 'Relationship between HM Government and the British Petroleum Company', n.d.; AIOC, *Annual Report and Accounts*, 1952; BP 95580, Final accounts schedules.

84 PRO T 236/4742, Playfair to Armstrong, 7 April 1954.

85 *Parl. Debates*, House of Commons, vol. 526, col. 213, 6 April 1954, Butler.

86 PRO T 236/4742, Playfair to Armstrong, 7 April 1954; *ibid.*, Playfair to Bridges, 22 April 1954; *ibid.*, Bridges note, 23 April 1954.

87 AIOC, *Annual Report and Accounts*, 1953; BP 95581, Final accounts schedules.

88 Bamberg, *British Petroleum, Volume II*, p. 488.

89 Stevens papers 1/33, Stevens to his parents, 26 August 1954.

90 *Ibid.*, 28 September 1954.

91 AIOC and BP, *Annual Reports and Accounts*, 1950–4.

92 BP, *Annual Report and Accounts*, 1954.

93 BP 18808, AIOC notice of extraordinary general meeting, 23 November 1954; *ibid.*, Fraser's statement reprinted from *The Times*, 17 December 1954.

94 PRO T 236/4742, Armstrong to Playfair, 7 April 1954; *ibid.*, Bridges to Brittain, Rowan and Playfair, 14 April 1954.

95 *Ibid.*, Playfair to Gilbert, 7 October 1954.

96 BP 18808, Fraser's statement reprinted from *The Times*, 17 December 1954.

97 *Ibid.*, Extract from *Daily Express*, 17 December 1954.

98 BP 106236, Minutes of board meeting on 16 December 1954; BP, *Annual Report and Accounts*, 1954.

99 BP 18808, Extracts from *Evening News* and *The Star*, 20 December 1954.

100 *New York Times*, 2 January 1951 and 17 December 1954; *Financial Times*, 1 January 1951 and 17 December 1954.

101 On criticism of Fraser, see Bamberg, *British Petroleum, Volume II*, pp. 213, 280–1, 326–8, 443, 451–2, 460–3, 470–1, 488–9, 499, 511 and 520–1.

102 BP 96455, Interim report of the Future Programme Committee, 25 July 1951.

103 *Ibid.*, Coxon, Widening of our sources of crude oil, May 1953.

104 BP 59341, Central Planning Department memorandum to Fraser, December

1953 [emphasis in original]; BP 44391, Coxon memorandum on exploration to Pattinson and Bridgeman, 17 June 1953.
105 BP 70359, Jackson to Morgan, 20 October 1954.

2 MANAGEMENT AND CULTURE

1 BP staff file on Gass.
2 J. H. Bamberg, *The History of The British Petroleum Company: Volume II, The Anglo-Iranian Years, 1928–1954* (Cambridge, 1994), ch. 15.
3 Stevens papers 1/33, Stevens to his parents, 13 April 1954.
4 Harmer had shipping and banking directorships and was thought to be 'very good indeed' on the financial side. He had worked for the Treasury during World War II and was described as 'probably the most successful temporary civil servant in the Treasury' (BP 106600, Unsigned note, n.d.). Munro had earlier been managing director of the banking firm of Helbert, Wagg and Co. From 1941 to 1946 he acted as financial adviser to the UK High Commissioner in Canada. He was subsequently the Treasury representative in the USA.
5 P. W. Brooks, 'Ronald Morce Weeks', in D. J. Jeremy (ed.), *Dictionary of Business Biography* (London, 1986), vol. V, pp. 715–18.
6 R. W. Ferrier, 'Sir Maurice Richard Bridgeman' in D. J. Jeremy (ed.), *Dictionary of Business Biography* (London, 1984), vol. I, pp. 440–3; Bridgeman obituary, *BP Shield*, July 1980.
7 PRO T 236/4755, Lee minute, 25 March 1960.
8 See generally Stevens' correspondence in Stevens papers 1/33–8, 1954–8.
9 The full text of the Bradbury letter is printed as Appendix 6.1 in R. W. Ferrier, *The History of The British Petroleum Company: Volume I, The Developing Years, 1901–1932* (Cambridge, 1982), pp. 645–6.
10 PRO T 236/4754, Anderson to Armstrong, 16 January 1957.
11 Ferrier, *British Petroleum, Volume I*, pp. 312–14 and 336.
12 PRO T 236/4755, Wilson to AIOC, 23 May 1941 quoted in full in unsigned paper on the 'Relationship between H.M. Government and the British Petroleum Company Limited', n.d.
13 *Ibid.*
14 *Parl. Debates*, House of Commons, vol. 537, col. 1867, 1 March 1955, Butler; *ibid.*, vol. 545, cols. 1996–7, 10 November 1955, Boyle; *ibid.*, vol. 546, col. 2111, 29 November 1955, Butler; *ibid.*, vol. 568, col. 31, Written answers, 2 April 1957, Thorneycroft.
15 PRO T 236/4754, Unsigned minute, 7 January 1954; *ibid.*, Playfair note, 2 June 1954; *ibid.*, Bridges to Playfair, 4 June 1954.
16 At the time of writing the Treasury file, reference PRO T 273/361, on the chairmanship remains closed. Applications for access to it for the purposes of this history have been turned down.

17 BP 116677, Press release, 12 January 1956. Details were reported in the press the next day. See, for example, *The Times*, 13 January 1956.
18 PRO T 236/4754, Playfair note, 2 June 1954.
19 *Ibid.*, Makins record of conversation with Weeks and Harmer on chairmanship of BP, 11 January 1957.
20 BP 106236, Minutes of BP board meeting on 20 December 1956; PRO T 236/4754, Jackson to BP secretary, 9 January 1957.
21 PRO T 236/4754, Makins record of conversation with Weeks and Harmer on chairmanship of BP, 11 January 1957.
22 *Ibid.*; *ibid.*, Makins to Chancellor, 19 January 1957.
23 BP 106236, Minutes of BP board meeting on 10 January 1957.
24 PRO T 236/4754, Makins record of conversation with Gass on chairmanship of BP, 11 January 1957.
25 *Ibid.*, Makins record of conversation with Cobbold on chairmanship of BP, 11 January 1957; *ibid.*, Makins record of conversation with Leggett and Waterfield on chairmanship of BP, 15 January 1957; *ibid.*, Makins record of conversation with Harper on chairmanship of BP, 16 January 1957; *ibid.*, Makins to Thorneycroft, 19 January 1957.
26 PRO FO 371/127200, Thorneycroft to Macmillan, 22 January 1957.
27 *Ibid.*, Makins record of conversation with Gass on chairmanship of BP, 23 January 1957; BP 106236, Minutes of BP board meeting on 24 January 1957.
28 BP 106236, Minutes of BP board meeting on 25 July 1957.
29 PRO T 236/4754, Makins record of conversation with Gass, 12 July 1957.
30 'An unexpected career', *BP Shield*, October 1972.
31 PRO T 236/4747, Makins record of conversation with Harmer, 13 March 1959.
32 PRO T 236/4755, Lee record of conversation with Harmer, 9 February 1960.
33 *Ibid.*, Lee minute, 25 March 1960.
34 *Ibid.*, Heathcoat Amory to Macmillan enclosing Lee's minute, 30 March 1960.
35 BP 106237, Minutes of BP board meeting on 7 April 1960.
36 PRO T 236/4755, Lee minute, 25 March 1960.
37 BP Archive shelf file on Distribution Department organisation.
38 BP 92785, Jackson, Memorandum on Refineries Department, 13 March 1951; BP 72322, Jackson, Memorandum on Refineries Department, 26 March 1952.
39 BP 69698, Falcon, Head office organisation of geological work, 6 January 1953.
40 Bamberg interview with Harmer, 18 January 1990.
41 BP 59571, Cox to Bridgeman on organisation of D'Arcy/Production Department work, 9 November 1953.
42 *Ibid.*
43 *Ibid.*, Review of head office strengths, Production Department, March 1952.
44 *Ibid.*, Dowling to Jackson, 13 January 1953; *ibid.*, Cox to Bridgeman on organisation of D'Arcy/Production Department work, 9 November 1953.

45 *Ibid.*, Dowling to Jackson, 13 January 1953; *ibid.*, Dowling note on Production Department – organisation, 28 August 1953; *ibid.*, Dowling to Pattinson, 28 August 1953; *ibid.*, Dowling, Production Department organisation and establishment head office, 15 December 1953; *ibid.*, Pattinson to Jackson, 29 December 1953; *ibid.*, Pattinson to Fraser, Production Department, 5 February 1954; *ibid.*, Pattinson to Fraser, 17 November 1954.

46 *Ibid.*, Pattinson to Fraser, 17 November 1954.

47 *Ibid.*, Cox to Bridgeman on organisation of D'Arcy/Production Department work, 9 November 1953; BP 69698, Cox to Bridgeman on Production Department/D'Arcy organisation, 29 December 1954.

48 BP 59571, Jackson, Organisation of D'Arcy and Production Department, 11 November 1953.

49 BP 92785, Mayhew memorandum, 19 April 1956.

50 BP 69698, Dowling, BP Exploration Department circular no. 2, 21 June 1956.

51 BP staff file on Fraser.

52 BP 92781, Mayhew memorandum on head office organisation and appointments, 1 June 1956; BP 74769, Pattinson note on head office organisation, 9 August 1957.

53 BP Archive shelf file on Supply and Development Department organisation.

54 BP Archive shelf file on Refineries and Technical Department organisation.

55 Alfred D. Chandler, *Strategy and Structure: Chapters in the History of the American Industrial Enterprise* (Cambridge, Mass., 1962)

56 *Ibid.*, ch. 4. and pp. 352–62.

57 Derek F. Channon, *The Strategy and Structure of British Enterprise* (London, 1973), pp. 115–16; Graham Turner, *Business in Britain*, revised edn (London, 1971), pp. 109–23.

58 The seven regions were North Europe, South Europe, North America, Australasia and the Far East, West Africa, the Middle East, and finally the areas covered by the joint marketing arrangements with Royal Dutch-Shell, namely the UK, the Consolidated area (covering eastern and southern Africa, parts of the Middle East and Ceylon) and the Burmah-Shell area (covering India and Pakistan).

59 BP 62359, Davies, A preliminary plan for the establishment of a system of regional co-ordination, 15 July 1960.

60 BP 75016, Memorandum no. 1, Regional co-ordination, 7 November 1960.

61 BP 62359, Davies, A preliminary plan for the establishment of a system of regional co-ordination, 15 July 1960.

62 BP 62560, Belgrave, Co-ordination, 16 June 1961.

63 *Ibid.*, Steel to Drake, Regional co-ordination, 11 September 1961.

64 *Ibid.*, Regional co-ordinators to Bridgeman, Regional co-ordination, 16 November 1961.

65 *Ibid.*, Minutes of meeting of the chairman and managing directors with the regional co-ordinators on 20 November 1961.

66 *Ibid.*, Minutes of chairman's meeting with regional co-ordinators on 31 July 1962.

67 On the widespread adoption of the multi-divisional structure in postwar Britain, see Channon, *Strategy and Structure*.

68 BP, *Annual Report and Accounts*, 1955.

69 On the disturbances, see Bamberg, *British Petroleum, Volume II*, pp. 76–80 and 377–9.

70 BP 72322, Cooper to Smith, Sunbury expansion, 4 October 1954; *ibid.*, Smith to Fraser, Research and Development Division, 6 October 1954; BP 22450, Gass to Fraser and Jackson, Engagement policy for Distribution Department, 1 April 1955; BP 41393, Elkington to Fraser, The manning of the BP refineries, 14 April 1955; *ibid.*, Mayhew to Fraser, 3 May 1955; BP 69698, Cox to Pattinson, Production Department establishment, 22 April 1955.

71 BP 66936, Organisation Committee – training, 24 June 1949 meeting.

72 Federation of British Industries archives, MSS 200/F/4/65/1, Report of the conference on industry and the universities, November 1949. The FBI was the predecessor of the Confederation of British Industries (CBI), the main representative organisation of British business.

73 BP 22450, Mayhew to Mylles, 9 February 1953; BP 12608, Training Division annual report, 1953.

74 BP 72322, Smith to Fraser, Research and Development Division, 6 October 1954.

75 Cambridge University Appointments Board archives, APTB 1/60, Men's board annual report, 1954.

76 BP, *Annual Report and Accounts*, 1955.

77 BP 12608, Training Division annual report, 1953.

78 BP, *Annual Report and Accounts*, 1955.

79 BP 59704, George Courtauld to Jackson, 8 September 1955; *ibid.*, Mayhew to Ashton, 16 September 1955; *ibid.*, Pattinson to Jackson, 16 September 1955; BP 59702, Progress report up to 27 September 1956.

80 BP 57253, Female staff recruitment, 1955; BP 57354, Female staff recruitment, 1956.

81 BP 15868, Female staff section memorandum to Green, 6 March 1957.

82 BP 106299, Head office staff strengths.

83 BP 7141, Mullaly to the chairman, 28 June 1961.

84 BP 65850, Forsyth to Mullaly, Schools, 11 September 1962.

85 *Ibid.*, Degrees by universities at 31 December 1962.

86 BP 59559, Pattinson, Address for probationer course, 17 October 1955.

87 BP 7141, De Quidt, International staff, 18 July 1961.

88 BP 6417, File for Taylor's visit, January 1956.

89 BP 49862, Papers and booklets issued for Pattinson and Banks' visit to Aden, 1957.

90 Frank Rickwood, *The Kutubu Discovery: Papua New Guinea, its People, the Country, and the Exploration and Discovery of Oil* (Victoria, Australia, 1992), pp. 126–30.

91 BP 49670, Matthews, Note on visit to Nigeria, 23–29 March 1963.

92 BP 36867, Form for appraisal of overseas staff, May 1957.

93 Mira Wilkins, *The Maturing of Multinational Enterprise: American Business Abroad from 1914 to 1970* (Cambridge, Mass., 1974), pp. 399–400.

3 THE SUEZ CRISIS

1 PRO PREM 11/1177, Eden to Eisenhower, 1 May 1956.

2 *Parl. Debates*, House of Commons, vol. 233, col. 2047, 23 December 1929, Eden.

3 D. C. Watt, 'Britain and the Suez Canal' (Report for Royal Institute of International Affairs, London, 1956).

4 Details of Europe's pattern of energy demand and supply in the mid-1950s can be found in publications by the Organisation for European Economic Co-operation (OEEC), notably *Europe's Growing Needs of Energy: How Can They Be Met?* (Paris, 1956); *Oil: The Outlook for Europe* (Paris, 1956); *Europe's Need for Oil: Implications and Lessons of the Suez Crisis* (Paris, 1958). For data on oil's share in energy consumption, see Joel Darmstadter, *Energy in the World Economy: A Statistical Review of Trends in Output, Trade, and Consumption since 1925* (Baltimore, Md., 1971), p. 91.

5 BP 41390, Mitchell to Gillespie, 10 April 1956; BP 102837, Management report for board meeting on 24 January 1957.

6 BP's liftings of Iraq crude in 1955 taken from BP 102837, Management report for board meeting on 24 January 1957.

7 OEEC, *Europe's Need for Oil*, p. 12.

8 PRO PREM 11/1473, ME (O) (56) 17, 10 April 1956.

9 BP 42541, Unsigned note of meeting held at the State Department on 15 June 1956.

10 PRO POWE 33/2164, OSAC comments on Ministry of Fuel and Power memorandum, 8 May 1956.

11 H. Lubbell, 'Middle East crises and world petroleum movements', *Middle Eastern Affairs* 9 (1958), 339–40 and *Middle East Oil Crises and Western Europe's Energy Supplies* (Baltimore, Md., 1963), p. 5; M. S. Venkataramani, 'Oil and US foreign policy during the Suez crisis, 1956–1957', *International Studies* 2 (1960–1), 113; David S. Painter, *Oil and the American Century: The Political Economy of US Foreign Oil Policy, 1941–1954* (Baltimore, Md., 1986), pp. 179–81; Bennett H. Wall, *Growth in a Changing Environment: A History of Standard Oil Company (New Jersey), Exxon Corporation, 1950–1975* (New York, 1988), pp. 459–61.

12 PRO PREM 11/1177, Eden to Eisenhower, 27 July 1956. Cited in

Venkataramani, 'Oil and US foreign policy', 120 and A. Eden, *Full Circle* (London, 1960), p. 427.

13 Venkataramani, 'Oil and US foreign policy', 121–2; US House of Representatives, 85th Congress, 1st Session, *Petroleum Survey: 1957 Outlook, Oil Lift to Europe, Price Increases* (Washington D.C., 1957), p. 15; BP 58675, CASO 230, 7 August 1956; BP 20545, Plan of action under voluntary agreement relating to foreign petroleum supply, 10 August 1956.

14 BP 42366, Minutes of meeting no. 1 of MEEC Statistical Sub-Committee, 6 September 1956; BP 53332, Fraser to Gass, 13 September 1956.

15 In the weeks following Nasser's nationalisation of the Suez Canal OSAC was the main channel for liaison between the MEEC and European oil companies. Representatives of OSAC attended MEEC meetings as observers and it was through OSAC that the European companies submitted information on their operations to the MEEC (BP 36083, Extract from *Financial Times*, 3 September 1956; BP 62906, Loudon to Jackson, 24 September 1956; BP 42541, Stacey, Suez Canal, 5 September 1956).

16 BP 62906, Loudon to de Metz, 20 August 1956.

17 BP 42541, Minutes of meeting held at St Helen's Court on 18 September 1956. OSAC remained in existence, but without Trinidad Leaseholds which resigned with effect from 14 September 1956 (BP 59138, FPSC/OSAC telegram, 18 September 1956).

18 BP 20549, Jackson to Maud, 4 October 1956; BP 53332, Maud to Jackson, 15 October 1956.

19 BP 62906, Loudon to Jackson, 24 September 1956; BP 59138, Jackson to Wilkinson, 26 September 1956; *ibid.*, Wilkinson to Jackson, 26 September 1956; *ibid.*, Wilkinson to Seaton, 27 September 1956; BP 53332, Minutes of inaugural meeting of OELAC Executive Committee, 3 October 1956; *ibid.*, Minutes of 2nd meeting of OELAC Executive Committee, 15 October 1956.

20 See documents in BP 20548, 20549, 20550, 42541, 42543, 53332, 59138, 62906.

21 PRO POWE 33/2197, Note [unsigned] on Suez Canal oil supplies, International aspect, undated [but must be August 1956]; *ibid.*, Washington Embassy to Foreign Office, 10 August 1956; *ibid.*, Vaughan, Summary notes on 2nd US-UK discussion on Middle East oil on 15 August 1956; PRO CAB 134/1216, EC (56) 15th meeting, 14 August 1956; PRO PREM 11/2014, EC (56) 15th conclusions, minute 5, 14 August 1956.

22 BP 20548, Meeting at Ministry of Fuel and Power to discuss Suez Canal/OEEC, 18 September 1956; BP 59138, FPSC/OSAC telegram, 19 September 1956; BP 62906, Report of meeting of the OEEC Oil Committee on 20 September 1956.

23 OEEC, *Europe's Need for Oil*, pp. 27–8.

24 For a detailed account, see Keith Kyle, *Suez* (London, 1991), chs. 7–18.

25 PRO PREM 11/1177, Eden to Eisenhower, 6 September 1956. Cited in Eden, *Full Circle*, p. 464 and R. Rhodes James, *Anthony Eden* (London, 1987),

p. 506. See also H. Macmillan, *Riding the Storm, 1956–1959* (London, 1971), p. 132; J. Selwyn Lloyd, *Suez 1956: A Personal Account* (London, 1978), p. 95.

26 Eden, *Full Circle*, p. 426.

27 PRO PREM 11/1135, EC (56) 35, August 1956.

28 *Ibid.*, Macmillan to Eden, 26 August 1956. Macmillan reiterated his concern about oil supplies to the Egypt Committee of senior ministers on 27 August and to the Cabinet on the 28th. See PRO CAB 134/1216, EC (56) 22nd meeting, 27 August 1956; PRO CAB 128/30, CM (56) 62nd conclusions, 28 August 1956.

29 PRO CAB 128/30, CM (56) 64th conclusions, 11 September 1956; Kyle, *Suez*, p. 243.

30 Of the numerous published accounts of these events, see, for example, Kyle, *Suez*, chs. 17–25.

31 OEEC, *Europe's Need for Oil*, p. 28; BP 59138, Unsigned note on Suez Canal blockage, 13 November 1956.

32 BP 45680, Note on Syria, 5 November 1956; *ibid.*, Note dictated over the telephone by Gibson's secretary, 5 November 1956; *ibid.*, Reuter message, 5 November 1956; BP 59138, Bridgeman to Fraser, 15 November 1956; *ibid.*, Lawson, Syria, 15 November 1956.

33 BP 45680, Reuter message, 6 November 1956; *ibid.*, Reuter message, 7 November 1956; BP 41356, Stockwell to deputy chairman, 8 November 1956; BP 36081, Extract from *Financial Times*, 10 November 1956; 'Chronology', *Middle East Journal* 11 (1957), 87; 'Chronology', *Middle Eastern Affairs* 8 (1957), 36.

34 PRO PREM 11/2014, Washington Embassy to Foreign Office, 31 October 1956; PRO POWE 33/2198, Ministry of Fuel and Power to Washington Embassy, 31 October 1956.

35 Diane B. Kunz, *The Economic Diplomacy of the Suez Crisis* (Chapel Hill, N.C., 1991), p. 124.

36 PRO PREM 11/2014, Washington Embassy to Foreign Office, 31 October 1956; *ibid.*, Foreign Office to Washington Embassy, 31 October 1956; *ibid.*, Washington Embassy to Foreign Office, 2 and 6 November 1956.

37 Selwyn Lloyd, *Suez*, p. 206.

38 Rhodes James, *Anthony Eden*, pp. 566–7.

39 Macmillan, *Riding the Storm*, p. 164; Eden, *Full Circle*, p. 557.

40 PRO CAB 128/30, CM (56) 80th conclusions, 6 November 1956.

41 For an account of the events of 6 November see Rhodes James, *Anthony Eden*, pp. 573–5; Kyle, *Suez*, ch. 25.

42 Rhodes James, *Anthony Eden*, pp. 576–85; Macmillan, *Riding the Storm*, pp. 167–9; Eden, *Full Circle*, pp. 563–73.

43 BP 102835, Management report for board meeting on 22 November 1956.

44 *Ibid.* Also, BP 45680, Admiralty to British shipping, 30 October 1956.

45 BP 53332, Drake to New York Office, 31 October 1956; BP 23613, Bridgen to Clark, 1 November 1956; *ibid.*, Clark to Bridgen, 2 November 1956; *ibid.*,

Unsigned note, Re-routing of tankers – no. 2, undated [but clearly early November 1956]; *ibid.*, Clark, Note on re-routing of tankers – no. 3, 3 November 1956. Two ships on consecutive voyage charter to BP, the SS *Brigitte* (Panamanian flag) and the MV *Hektoria* (Norwegian flag) were detained in the Canal until 8 January 1957 when they sailed from Port Said for Italian ports, to be drydocked before resuming service. See BP 102835 and 102837, Management reports for board meetings on 22 November 1956 and 24 January 1957. The fleet size of 330 vessels included owned and chartered ships, in addition to which a fleet of ten tankers was owned by the Lowland Tanker Company, in which BP held a 50 per cent interest. For confirmation of the approximate number of vessels in BP's owned and chartered fleet see BP 102840, Management report for board meeting on 25 April 1957.

46 BP 53332. Drake to New York Office, 31 October 1956; BP 23613, Bridgen to Clark, 1 November 1956; *ibid.*, Clark to Bridgen, 2 November 1956; *ibid.*, Unsigned note, Re-routing of tankers – no. 2, undated [but clearly early November 1956]; *ibid.*, Clark, Note on re-routing of tankers – no. 3, 3 November 1956; BP 20550, Minutes of 4th meeting of OELAC Executive Committee, 2 November 1956; BP 42367, Bridgen to Clark, 8 November 1956; *ibid.*, Hill to Smith, 8 November 1956; *ibid.*, Hill to Bean, 12 November 1956; *ibid.*, Stratton to McColl, 12 November 1956; *ibid.*, Clark to Hill, 13 November 1956; BP 102835, Management report for board meeting on 22 November 1956.

47 BP 28475, Unsigned note, 21 November 1956.

48 BP 4844, Unsigned note on the supply of oil to Europe since the disruption in the normal flow of oil from the Middle East, 16 January 1957.

49 BP 28475, Unsigned note, 21 November 1956.

50 BP 4844, Unsigned note on the supply of oil to Europe since the disruption in the normal flow of oil from the Middle East, 16 January 1957. In the case of BP, the first Cape-routed vessels to reach Britain were the chartered *Athina Livanos*, which docked at the Isle of Grain (Kent refinery) on 28 November, and the BP-owned *British Realm*, which docked at Finnart (Grangemouth refinery) on 30 November. See BP 102836, Management report for board meeting on 20 December 1956.

51 PRO POWE 33/2198, Political aspects of cuts in Middle East oil production, Draft paper [unsigned] for Middle East (Official) Committee, 23 November 1956. Emphasis in original.

52 BP 102847, 102849, 102850 and 102851, Management reports for board meetings on 19 December 1957, 27 February, 27 March and 24 April 1958.

53 PRO POWE 33/2198, Political aspects of cuts in Middle East oil production, Draft paper [unsigned] for Middle East (Official) Committee, 23 November 1956. Emphasis in original.

54 BP 58722, Supply and Development Division to New York Office, 2 November 1956.

55 *Ibid.*, New York Office to Distribution Department, 2 November 1956.

56 BP 53332, Fraser to Gass, 3 November 1956.

57 PRO T 234/79, ES (EM) (56) 2 – Rampton, Oil supplies for Europe and the UK, 16 November 1956; *ibid.*, Rampton, Oil, 21 November 1956; *ibid.*, Rampton, Dollar oil, 3 December 1956.

58 BP 59138, Fraser to Drake, 12 November 1956.

59 BP 102837, Management report for board meeting on 24 January 1957.

60 BP 58722, Unsigned note on western crudes, 11 December, 1956; BP 24573, Bell to SFPBP, 26 November 1956; *ibid.*, Bell to BP Benzin und Petroleum, 23 November 1956.

61 Documentation on the Suez crisis is full of references to fuel oils being the products most urgently required in Europe. See, for example, BP 20548, Unsigned note on meeting held at the MOFP, 31 July 1956; BP 59745, Minutes of 20th refineries/supply general managers' meeting on 29 November 1956; BP 36081, Extract from *Financial Times*, 8 December 1956; BP 4844, Notes on meeting between OPEG and the OEEC Oil Committee on 17 December 1956; *ibid.*, Armstrong, Report on meeting of OEEC Oil Committee/OPEG on 3 January 1957; BP 59158, Toronto to New York, 4 January 1957.

62 BP 36081, Extract from *The Times*, 9 January 1957; BP 59158, Department of the Interior, Information service, for release on 27 January 1957; BP 36083, *Petroleum Press Service*, January 1957.

63 BP 102837, Management report for board meeting on 24 January 1957.

64 BP 102836, Management report for board meeting on 20 December 1956.

65 BP 30580, Monthly bunker statements.

66 BP 8945, Watts to western marketing associates, 2 November 1956.

67 BP 42366, Davies to western marketing associates, 6 November 1956.

68 BP 58675, Snow to general managers of European marketing associates, 13 November 1956.

69 *Ibid.*, Davies to general managers of European marketing associates, 20 November 1956. On the increase in freight rates see OEEC, *Europe's Need for Oil*, pp. 24–6.

70 BP 58675, Davies to general managers of European marketing associates, 20 November 1956.

71 BP 42367, Butcher to Hill, 26 November 1956; BP 42366, Snow to European marketing associates, 5 December 1956.

72 On restrictions on consumption see BP 59158, Stacey, Note to OELAC Executive Committee, 23 November 1956.

73 PRO POWE 33/2164, Fletcher to oil companies, 7 November 1956; PRO POWE 33/2198, Statement on oil supplies by the Minister of Fuel and Power to the House of Commons, 7 November 1956. Essential services and industries were excluded from the cuts, as were products such as aviation fuels, lubricants and paraffin. These constituted only a small proportion of total British oil consumption.

74 PRO POWE 33/2165, Unsigned note, Restrictions on oil consumption, undated; PRO POWE 33/2198, MOFP to Washington Embassy, 20 November 1956; *ibid.*, MOFP Information Branch, 29 November 1956.

75 BP 59138, Berkin to Astley-Bell, 23 November 1956.

76 BP 62906, Resolution of the Council concerning the establishment of an OEEC Petroleum Industry Emergency Group, 30 November 1956; OEEC, *Europe's Need for Oil*, p. 28.

77 It was agreed that it was not necessary to have a higher-level committee on the lines of the OELAC Main Committee, which was dissolved. See BP 4844, Minutes of OPEG inaugural meeting on 20 December 1956; *ibid.*, OEEC Petroleum Industry Emergency Group, 1 January 1957; BP 59158, Fisher to Coleman, 21 December 1956.

78 Rhodes James, *Anthony Eden*, pp. 582–5.

79 BP 36081, Extract from *The Times*, 1 December 1956; BP 59158, Doremus release, Oil plan in effect, 30 November 1956.

80 Royal Institute of International Affairs (RIIA), *Survey of International Affairs, 1956–1958* (London, 1962), pp. 151–2; Venkataramani, 'Oil and US foreign policy', 140. The troop withdrawals commenced on 5 December 1956 and were completed on 22 December 1956.

81 PRO CAB 128/30, CM (56) 98th conclusions, 11 December 1956.

82 BP 42541, Plan of action (as amended 3 December 1956) under voluntary agreement relating to foreign petroleum supply.

83 BP 20545, Schedule no. 1, pursuant to plan of action dated 10 August 1956 (as amended 3 December 1956) under voluntary agreement relating to foreign petroleum supply, 7 December 1956; *ibid.*, Schedule no. 2, pursuant to plan of action dated 10 August 1956 (as amended 3 December 1956) under voluntary agreement relating to foreign petroleum supply, 7 December 1956; BP 42367, Rathbone statement, 13 February 1957.

84 For examples of inter-company exchanges in November see BP 42366, Jones, Exchange arrangements agreed with Shell since Suez emergency, 21 November 1956.

85 BP 59158, Loudon to Astley-Bell, 13 December 1956.

86 BP 102836 and 102837, Management reports for board meetings on 20 December 1956 and 24 January 1957.

87 BP 42553, Davies to European associates, 3 January 1957.

88 BP 36081, Extract from *Financial Times*, 9 November 1956.

89 On the Texas Railroad Commission, see BP 20549, Background information on control of oil production by the individual states of the United States, 7 February 1957; BP 36081, *Financial Times*, 11 January 1957; R. H. K. Vietor, *Energy Policy in America since 1945: A Study of Business–Government Relations* (Cambridge, 1984), pp. 21–4; G. D. Nash, *United States Oil Policy, 1890–1964* (Pittsburgh, Pa., 1968), pp. 113–20.

90 Vietor, *Energy Policy*, ch. 5, esp. pp. 91–6; BP 59159, Summary of talk by Harvey, 5 February 1957.

91 BP 36081, Extracts from *The Times*, 9 January 1957 and *Financial Times*, 18, 19 and 21 January 1957; *ibid.*, Vignoles talk to Conservative Oil Sub-Committee, House of Commons, 29 January 1957.

92 On concern in Western Europe see, for example, PRO PREM 11/2014, MOFP to Washington Embassy, 31 December 1956; *ibid.*, MOFP to Washington Embassy, 1 January 1957; *ibid.*, CM (57) 1st conclusions, 3 January 1957; *ibid.*, Macmillan to Eden, 8 January 1957; PRO POWE 33/2198, Aubrey Jones to Prime Minister, 4 January 1957; *ibid.*, Brown to Falle, 4 January 1957; BP 42553, OEEC press release, 8 January 1957.

93 BP 102837 and 102838, Management reports for board meetings on 24 January and 28 February 1957.

94 BP 59158, Astley-Bell to Loudon, 3 January 1957.

95 US House of Representatives, *Petroleum Survey*, pp. 34–40; BP 42367, Rathbone statement, 13 February 1957; BP 59224, Planning Group, Review of the year 1957; BP 102837, Management report for board meeting on 24 January 1957.

96 US House of Representatives, *Petroleum Survey*, p. 34; BP 102837, Management report for board meeting on 24 January 1957.

97 BP 102838, Management report for board meeting on 28 February 1957. There was no immediate increase in the published prices which the companies posted for Middle East crudes in the Persian Gulf, though the price of Saudi crude ex-Tapline had already been increased by 23 cents, from $2.46 to $2.69 per barrel in December 1956, reflecting the advantageous freight differential of crude from the Eastern Mediterranean (BP 102836, Management report for board meeting on 20 December 1956).

98 BP 36081, Extracts from *Financial Times*, 18, 19 and 21 January 1957; PRO POWE 33/2198, Washington Embassy to MOFP, 18 and 22 January 1957.

99 PRO POWE 33/2198, Williams to Beckett, 19 January 1957.

100 BP 36081, Extracts from *The Times*, 7 February 1957 and *Daily Telegraph*, 8 February 1957.

101 *Ibid.*, Extract from *Financial Times*, 20 February 1957.

102 BP 41045, Hill, Notes of meeting held in Paris between the Oil Committee of OEEC and representation from British, French and American oil companies, 6 December 1956; BP 4844, Record of 1st joint session of Oil Committee and OPEG on 6 and 7 December 1956; BP 59158, Loudon to Astley-Bell, 7 December 1956.

103 See, generally, BP 4844, 20546, 20549, 41045, 42543, 42553, 59158.

104 OEEC, *Europe's Need for Oil*, pp. 31–2; BP 59158, Loudon to Astley-Bell, 11 December 1956.

105 BP 4844, Record of 1st joint session of Oil Committee and OPEG on 6 and 7 December 1956.

106 See, generally, documents in BP 4844, 41045, 42553.

107 BP 59158, Loudon to Astley-Bell, 18 December 1956.

108 BP 42543, Beckett to Berkin, 7 January 1957; *ibid.*, Minutes of 3rd meeting of

OPEG, 9 January 1957; *ibid.*, Berkin to Beckett, 9 January 1957; BP 59158, Unsigned, Allocation of European oil supplies, Principles of equalization, 22 January 1957; OEEC, *Europe's Need for Oil*, p. 31.

109 OEEC, *Europe's Need for Oil*, pp. 30–1; BP 59158, Unsigned, Allocation of European oil supplies, Principles of equalization, 22 January 1957; BP 42543, Unsigned, Implementation of the memorandum of understanding, 14 February 1957.

110 OEEC, *Europe's Need for Oil*, p. 36; BP 20549, Berkin's introduction of OPEG report, 5 March 1957; BP 42543, OEEC record of 4th and 5th joint sessions of the Oil Committee and OPEG on 7 February and 5 March 1957.

111 OEEC, *Europe's Need for Oil*, pp. 37–8; BP 42543, OEEC record of 4th joint session of the Oil Committee and OPEG on 7 February 1957.

112 BP 58675, Davies to European marketing associates, 14 February 1957; *Parl. Debates*, House of Lords, vol. 201, col. 1082, 20 February 1957, Mills.

113 On the removal of political barriers to the reopening of the Suez Canal, see RIIA, *Survey*, pp. 152–9.

114 BP 20546, Unsigned draft for OEEC/OPEG meeting, 18 March 1957; BP 102840, Management report for board meeting on 25 April 1957.

115 BP 20546, Unsigned draft for OEEC/OPEG meeting, 18 March 1957. Also, BP 102839, Management report for board meeting on 28 March 1957. At the same time, the published prices which the companies posted for Iraq crudes at the Eastern Mediterranean terminals were brought into line with posted prices for Saudi crude at Sidon at $2.69 per barrel, an increase of 23 cents per barrel on the pre-crisis price (BP 41723, BP Trading Limited, Crude oil posted prices effective 14 March 1957; BP 59224, Planning Group, Review of the year 1957).

116 BP 20546, Unsigned draft for OEEC/OPEG meeting, 18 March 1957.

117 BP 20549, Record of meeting between the OEEC Oil Committee and OPEG on 20 March 1957.

118 BP 42543, Record of 90th session of OEEC Oil Committee on 20 March 1957.

119 *Ibid.*

120 BP 20547, OEEC Record of 6th joint session of the OEEC Oil Committee and OPEG on 2 and 3 April 1957.

121 RIIA, *Survey*, pp. 159–161; BP 45680, Bean, Suez Canal, 15 April 1957; *ibid.*, Illegibly signed note on Suez Canal, 13 March 1957; *ibid.*, Reuter message, 26 March 1957; *ibid.*, Bean, Suez Canal, 15 April 1957; *ibid.*, Bean, Meeting with Minister of Transport on 17 April 1957; *ibid.*, Reuter message, 13 May 1957.

122 BP 45680, Suez Canal, unsigned, 13 May 1957; BP 102842, Management report for board meeting on 27 June 1957.

123 BP 59159, US Department of the Interior, Information for release on 18 April 1957.

124 BP 42543, OEEC record of 7th joint session of the OEEC Oil Committee and OPEG on 2 May 1957.

125 BP 102840 and 102842, Management reports for board meetings on 25 April and 27 June 1957.

126 BP 29657, Western Markets Division to European marketing associates, 9 April 1957; *ibid.*, Davies to European marketing associates, 14 May 1957.
127 PRO POWE 33/2165.
128 Venkataramani, 'Oil and US foreign policy', 145; OEEC, *Europe's Need for Oil*, p. 38.

4 'THE ENERGETIC SEARCH FOR NEW SOURCES OF CRUDE OIL'

1 The seven fields were Masjid-i-Suleiman, Haft Kel, Gach Saran and Agha Jari in Iran; Kirkuk in Iraq; Burgan in Kuwait; and Dukhan in Qatar. Percentages calculated from BPX report EXT 57512, Sayer, BP discoveries to the end of 1970 – an old working list, 1971; and BP 102759, Management report for board meeting, January 1951.
2 R. W. Ferrier, *The History of The British Petroleum Company: Volume I, The Developing Years, 1901–1932* (Cambridge, 1982), ch. 1.
3 For an excellent account of exploration in Iran in this period, see BPX report P905, Cox, Exploration for oil in the D'Arcy Concession area of Persia, n.d. For a worldwide view, see Kent papers, PK G 3/1/1–4, Kent, Exploration review 1970, 2 February 1970. On the curtailment of exploration outside Iran to mollify the Shah, see J. H. Bamberg, *The History of The British Petroleum Company: Volume II, The Anglo-Iranian Years, 1928–1954* (Cambridge, 1994), pp. 22 and 144.
4 BPX report P905, Cox, Exploration for oil in the D'Arcy Concession area of Persia, n.d.
5 W. J. Arkell, 'George Martin Lees, 1898–1955', in *Biographical Memoirs of Fellows of the Royal Society* 1 (1955), 163–72; Kent papers, PK G 3/1/1–4, Kent, Exploration review 1970, 2 February 1970.
6 BPX report OC 278, Lees and Richardson, Memorandum on the oil prospects of various countries throughout the world from the point of view of the D'Arcy Exploration Company, 7 February 1930.
7 BPX report P905, Cox, Exploration for oil in the D'Arcy Concession area of Persia, n.d.
8 *Ibid.* On the use of seismic refraction methods for the detection of Asmari limestone oil reservoirs in Iran, see G. M. Lees and F. D. S. Richardson, 'The geology of the oil-field belt of S.W. Iran and Iraq', *Geological Magazine* 77 (1940), 249.
9 For an excellent overall study, see Elizabeth Monroe, *Britain's Moment in the Middle East, 1914–1971*, 2nd edn (London, 1981).
10 Marian Kent, *Oil and Empire: British Policy and Mesopotamian Oil, 1900–1920* (London, 1976); Geoffrey Jones, *The State and the Emergence of the British Oil Industry* (London, 1981), esp. chs. 5–6.
11 Ferrier, *British Petroleum, Volume I*, p. 42 and appendix 1.1; Bamberg, *British Petroleum, Volume II*, p. 48.

12 Bamberg, *British Petroleum, Volume II*, pp. 155–71; Edith and E. F. Penrose, *Iraq: International Relations and National Development* (London, 1978), pp. 56–74 and 138–40; G. W. Stocking, *Middle East Oil: A Study in Political and Economic Controversy* (London, 1971), ch. 2.

13 Kent papers, PK G 3/1/1–4, Kent, Exploration review 1970, 2 February 1970; Bamberg, *British Petroleum, Volume II*, pp. 171–2.

14 A. H. T. Chisholm, *The First Kuwait Oil Concession Agreement: A Record of the Negotiations, 1911–1934* (London, 1975); Bamberg, *British Petroleum, Volume II*, pp. 146–55.

15 Lees and Richardson, 'The geology of the oil-field belt of S.W. Iran and Iraq', 251. See also G. M. Lees, 'The geology of the oilfield belt of Iran and Iraq' in A. E. Dunstan, A. W. Nash, B. T. Brooks and H. Tizard (eds.), *The Science of Petroleum* (Oxford, 1938), vol. I, pp. 140–8.

16 G. M. Lees, 'Foreland folding', *Quarterly Journal of the Geological Society of London* 108 (1952), 2.

17 Bamberg interview with Jenkins, Martin and Thomas, 9 August 1996. Also, BP 114266, Manpower Study Group, Report on Exploration and Production Department, March 1971.

18 For data supporting these points, see A. J. Martin, 'The prediction of strategic reserves', in T. Niblock and R. Lawless (eds.), *Prospects for the World Oil Industry* (Beckenham, Kent, 1985), pp. 16–41.

19 There are many published accounts of these developments. The theme of growing competition is perhaps most central in Neil H. Jacoby, *Multinational Oil: A Study in Industrial Dynamics* (New York, 1974).

20 BP 44391, Coxon to Pattinson and Bridgeman, Exploration, 17 June 1953. Quoted, with underlining, in BP 9214, Central Planning Department to Fraser, The search for new sources of crude oil, December 1953.

21 BPX reports OC 710 parts 1–3, AIOC Geological Division, A review of world oil prospects, October 1953. Summarised in BP 9214, Central Planning Department to Fraser, The search for new sources of crude oil, December 1953.

22 Cox profile, *BP Shield*, February 1963; BP Archive shelf profiles, Unattributed Cox obituary, 1969.

23 A. J. Martin and P. B. Lapworth, 'Norman Leslie Falcon', *Biographical Memoirs of Fellows of the Royal Society* 44 (1998), 163–5.

24 N. L. Falcon and Sir Kingsley Dunham, 'Percy Edward Kent, 1913–1986', *Biographical Memoirs of Fellows of the Royal Society* 33 (1987), 345–73.

25 Kent papers, Memoirs, PK A 10/14–15. For his formal reports on East Africa, see BPX report OC 492, Kent and Kündig, A geological reconnaissance of north eastern Kenya, 2 August 1951; BPX report OC 562, Kent, The oil prospects of British Somaliland, April 1952; BPX report OC 677, Kent and Kündig, Geology and oil prospects of Tanganyika and Zanzibar, 1 July 1952; BPX report OC 12653, Kent and Linton, A reconnaissance flight along the Kenya coast, December 1953; BPX report OC 793, Kent, A visit to western Madagascar,

September 1953; BPX report OC 885, Kent, A brief visit to British Somaliland, October 1954.

26 Bamberg interview with Rickwood, 19 November 1997; Kent papers, Memoirs, PK A 11/11. On the history of exploration in Papua, see Frank Rickwood, *The Kutubu Discovery: Papua New Guinea, its People, the Country, and the Exploration and Discovery of Oil* (Victoria, Australia, 1992).

27 Kent papers, Memoirs, PK A 11/24.

28 Falcon and Dunham, 'Kent', 350; Warman profile, *BP Shield*, April 1974.

29 Kent papers, Memoirs, PK A 12/3.

30 Falcon and Dunham, 'Kent', 352–3; Warman profile, *BP Shield*, April 1974.

31 Falcon retired on 31 December 1965 and Kent took over from the beginning of 1966. See Falcon profile, *BP Shield*, February 1966.

32 Bamberg interview with Rickwood, 19 November 1997.

33 AIOC, *Annual Report and Accounts*, 1950.

34 AIOC/BP, *Annual Reports and Accounts*, 1950–60; BPX report OC 6254, Livingstone, BP's exploration expenditure and reserves 1951–1968, July 1969; BPX report OC 694, Kent, Report on reconnaissance of eastern Tunisia, September 1952; BPX report OC 559, Thomas, Note on the geology and oil prospects of Tunisia, 27 November 1952; BPX report OC 793, Kent, Visit to western Madagascar, September 1953.

35 BPX report OC 6254, Livingstone, BP's exploration expenditure and reserves 1951–1968, July 1969.

36 For a general account, see, for example, J. D. Fage, *A History of Africa* (London, 1978), chs. 13–17.

37 Sarah Ahmad Khan, *Nigeria: The Political Economy of Oil* (Oxford, 1994), p. 16; Augustine A. Ikein, *The Impact of Oil on a Developing Country: The Case of Nigeria* (New York, 1990), pp. 2 and 25; J. K. Onoh, *The Nigerian Oil Economy from Prosperity to Glut* (London, 1983), p. 42.

38 BP 106485, Shell-BP Petroleum Development Company of Nigeria, The Shell-BP story, March 1965; BP 62259, Draft for Drake speech on visit to Nigeria, 7 February 1958; BP 65581, Shell Public Relations Department, 21 years' search for oil, 17 February 1958; BP 49670, Sayer, Nigeria, 9 May 1963.

39 BP 69631, Simpson to Carruthers, n.d.; BP 62259, Draft for Drake speech on visit to Nigeria, 7 February 1958; Skinner's *Oil and Petroleum Yearbook* (1960).

40 BP 65581, Shell press release, First shipment of oil from Nigeria, 17 February 1958; BP 106485, Shell-BP Petroleum Development Company of Nigeria, The Shell-BP story, March 1965; BP 44288, Shell-D'Arcy Petroleum Development Company of Nigeria, Petroleum engineering report, September 1951; BP 44718, Shell-D'Arcy Petroleum Development Company of Nigeria, Geological Department, Monthly progress report for August 1955; *Petroleum Times*, 31 August and 28 September 1956; *Petroleum Press Service*, September 1956; BP 65581, Unattributed note, Oil search in Nigeria, 5 March 1959.

41 BP 106485, Shell-BP Petroleum Development Company of Nigeria, The Shell-BP

story, March 1965; BP 18273, Extract from *The Times*, British Colonies Review, First quarter 1958; BP 49670, Sayer, Nigeria, 9 May 1963; BP 60683, Unattributed notes on BP's operations in Nigeria, October 1963; *Petroleum Press Service*, February 1964; R. C. Graham, 'Nigeria: oil country now', *International Commerce*, 20 July 1964; For full listings of Shell-BP's exploration licences, prospecting licences and mining leases, 1938–60, see BP 34886.

42 Unattributed article, '30 years of Nigerian oil', *West Africa*, 21 February 1970.

43 BPX report EXT 57512, Sayer, BP discoveries to the end of 1970 – an old working list, 1971.

44 BPX report OC 8113, Exploration and Production Department, BP's share of proven and probable reserves of oil as at 1 January 1972, June 1972; DeGolyer and MacNaughton, *Twentieth-Century Petroleum Statistics* (Dallas, Tex., 1994).

45 BPX report EXT 57512, Sayer, BP discoveries to the end of 1970 – an old working list, 1971.

46 BP 69764, Waller, Nigeria – political and economic notes (Spring 1962), 16 March 1962; *Financial Times*, 1 October and 31 December 1964.

47 Judith Gurney, *Libya: The Political Economy of Oil* (Oxford, 1996), ch. 1; Frank C. Waddams, *The Libyan Oil Industry* (London, 1980), ch. 1; J. A. Allan, *Libya: The Experience of Oil* (London, 1981), ch. 1.

48 BP 69650, Matthews, Libya, 28 April 1958; BPX report OC 1485, Shell Overseas Exploration Company and D'Arcy Exploration Company, Geological report on northern Cyrenaica, March 1954; BP 104132, D'Arcy Exploration Company board minutes, 17 January and 17 July 1952.

49 A. N. Thomas profile, *BP Shield*, 1974.

50 BPX report OC 558/1, Thomas, Oil prospects of Tripolitania, 7 April 1952; BPX report OC 556, Thomas, Forecast of the oil prospects of Libya, 19 August 1952.

51 Gurney, *Libya*, pp. 22–3; Waddams, *Libyan Oil*, p. 29.

52 Gurney, *Libya*, ch. 2 and pp. 44–6; Waddams, *Libyan Oil*, ch. 3 and pp. 78–81.

53 Waddams, *Libyan Oil*, pp. 73 and 83–9.

54 BP 70642, Notes for D'Arcy Exploration Company board meeting, 11 May 1954; BP 69650, Matthews, Libya, 28 April 1958.

55 BP 96483, Cox to D'Arcy Exploration Company (Africa) Ltd, 26 July 1957; BP 41080, Amended application submitted 24 November 1955; Waddams, *Libyan Oil*, p. 85; BP 70642, *passim*.

56 Waddams, *Libyan Oil*, p. 29; BP 69650, MacLean, Proposals for exploration programme for 1960 and revised programme for 1959, 26 June 1959; BP 104132, BP Exploration Company board minutes, 10 January 1957; BP 70642, BP Exploration Company notes for the director for board meetings on 14 March, 10 April and 8 May 1957; BP 96483, *passim*.

57 BP 69650, MacLean, Proposals for exploration programme for 1960 and revised programme for 1959, 26 June 1959.

58 *Ibid.*

59 *Ibid.*; *ibid.*, Holroyd, A programme for Libya, 8 June 1959; *ibid.*, Unattributed notes for meeting re Libya, 1 July 1959; *ibid.*, Luard to Macpherson,

13 November 1959; *ibid.*, Dollin to Cox, n.d.; BP 42702, Mann to Crosthwaite, 9 July 1959; BP 69648, Notes for BP Exploration Company board meeting, 9 July and 10 September 1959; Gurney, *Libya*, p. 45; Waddams, *Libyan Oil*, pp. 79, 81 and 89.

60 BP 69650, Macpherson to Cox, 29 August 1959.

61 Daniel Yergin, *The Prize: The Epic Quest for Oil, Money, and Power* (New York, 1991), p. 248.

62 BP 69650, Macpherson to Cox, 29 August 1959.

63 For summaries of the terms, see BP 42701, Libya – concession 65, Summary of agreement with Nelson Bunker Hunt, 12 October 1960; BP 4984, Easton to Steel, 16 October 1965; *ibid.*, Whytock to Luard, 24 November 1965.

64 BP 30917, Pamphlet, 'Oil across the sands', 1969; *Oil and Gas Journal*, 31 August 1964.

65 BP 64272, Libyan Ministry of Petroleum Affairs, 'Petroleum development in Libya 1954–1964' (Tripoli, 1965); Waddams, *Libyan Oil*, pp. 76–7 and 91–7.

66 BP 82487, Unattributed draft history of BP in Libya, n.d.

67 BPX report OC 8113, Exploration and Production Department, BP's share of proven and probable reserves of oil as at 1 January 1972, June 1972; DeGolyer and MacNaughton, *Petroleum Statistics*.

68 BPX report OC 8113, Exploration and Production Department, BP's share of proven and probable reserves of oil as at 1 January 1972, June 1972.

69 Yergin, *The Prize*, pp. 233–7.

70 Bamberg, *British Petroleum, Volume II*, pp. 144–6.

71 BPX report OC 342, Cox, Report on republic of Colombia, August 1938.

72 Bamberg, *British Petroleum, Volume II*, pp. 143–4; BPX report OC 342, Item 8190, Cox, Report on republic of Colombia, August 1938; Kent papers, Memoirs, PK G 3/1/3. On the high rating of oil prospects in Colombia and Venezuela, see BPX report OC 278, Lees and Richardson, Memorandum on oil prospects of various countries throughout the world from the point of view of D'Arcy Exploration Company, 7 February 1930; *ibid.*, Lees, An attempt to assess numerically the relative attractiveness of various countries of the world as fields for oil exploration, 16 October 1934; BP 54220, Lees to Jackson, An enquiry into present opportunities for D'Arcy enterprise in various countries of the world, 17 June 1937.

73 BP 104333, Newman, BP in Trinidad, 19 March 1964.

74 BP 64366, Collection of articles on Trinidad's oil, September 1950 – July 1951.

75 Bamberg interview with Rickwood, 17 November 1997.

76 BPX report EXT 57512, Sayer, BP discoveries to the end of 1970 – an old working list, 1971.

77 BP 104333, Newman, BP in Trinidad, 19 March 1964.

78 BP 104333, 77844, 51817, 47043 and 3117, *passim*; Scott B. MacDonald, *Trinidad and Tobago: Democracy and Development in the Caribbean* (New York, 1986), chs. 6–8.

79 BP 63276, Belgrave, Note on Trinidad, 17 May 1967; *ibid.*, Lindsay letter,

received 28 February 1968; *ibid.*, Fraser, Note on Trinidad, 30 April 1969; *ibid.*, Disposal of BP's producing assets in Trinidad, 8 January 1970.

80 MacDonald, *Trinidad and Tobago*, ch. 8.

81 BPX report OC 1029, Lehner, Review of oil prospects of Colombia, July 1955; BPX report OC 12717, O'Brien, Appraisal of oil prospects of Colombia, 22 December 1955.

82 BPX report OC 12164, Phizackerley, Exploration and production by BP in Colombia 1959–1976, May 1978. The BP/Sinclair alliance came out of long negotiations which are documented in BP 9192, 16585, 16588, 58677, 58946 and 59180.

83 BP 5330, Sinclair and BP Colombian Inc., Company rundown prepared for Thomas, Bush, Elston, Lincoln and Prudhomme, 21 January 1967; BPX report OC 12164, Phizackerley, Exploration and production by BP in Colombia 1959–1976, May 1978.

84 BPX report OC 12164, Phizackerley, Exploration and production by BP in Colombia 1959–1976, May 1978; BPX report OC 12793, Warman, Note on current knowledge of oil prospects in the northern Colombian Llanos, 14 October 1959; *ibid.*, Birks to Milward, 9 November 1959; *ibid.*, Thomas to Milward, 20 November 1959; *ibid.*, Matthews to BP (North America) Ltd, 20 November 1959; Bamberg interview with Rickwood, 17 November 1997.

85 BPX report OC 12164, Phizackerley, Exploration and production by BP in Colombia 1959–1976, May 1978.

86 BP 79236, Bell to Cadman, 18 November 1925; *ibid.*, Cadman to management committee, 24 November 1925; *ibid.*, Bell, Note of interview with Masters, 10 June 1926; BP 79235, Clyne to Watson, 12 November 1926; *ibid.*, Wilson to Watson, 22 December 1926; BP 43198, Wilson and Hearn, Memorandum for management committee, 22 December 1926; *ibid.*, Jackson, Canada, 2 December 1938; BP 79238, Hearn to Cadman, Memoranda on Canada, 18 May 1928 and 12 November 1930; BP 79237, Wilson to Hearn, 10 October 1930; *ibid.*, Lees, Note on Richardson's report on the Nordon Corporation's and British Dominion Land Settlement Corporation's properties in Alberta, 9 October 1930; BP 62759, Jackson, Note on Canada, 9 February 1949; BP 59660, Lees to Pattinson, 19 April 1949.

87 Earle Gray, *The Great Canadian Oil Patch* (Toronto, 1970), chs. 2–6; J. Richards and L. Pratt, *Prairie Capitalism: Power and Influence in the New West* (Toronto, 1979), esp. pp. 46–7; David H. Breen, *Alberta's Petroleum Industry and the Conservation Board* (Alberta, 1993), chs. 1–5; BP 62760, Cox, Report on visit to Canada, August 1952; BP 49853, Bridgeman and Pattinson, Report on visit to western Canada, August 1953.

88 Lees wrote in 1953 that 'Canada is frequently quoted as a second Middle East, but it is far from being so', BP 54220, Lees to Pattinson, 24 June 1953.

89 Kent papers, Memoirs, PK A 12/2.

90 BP 59660, Pattinson to chairman and deputy chairman, 28 August 1953.

91 BP 49853, Bridgeman and Pattinson, Report on visit to western Canada, August 1953.
92 BP 62760, *passim*; BP 44337, Down to deputy chairman, Note on Triad, 12 March 1953; *ibid.*, D'Arcy-Triad heads of agreement, 21 August 1953; BP 59660, Bridgeman, Canada, 18 May 1953; *ibid.*, Pattinson to chairman *et al*, 21 July 1953; BP 49853, Bridgeman and Pattinson, Report on visit to western Canada, August 1953; BP 59442, Pattinson, Memorandum no. 79, 1 September 1953. A copy of the final agreement with Triad is in BP 59463.
93 BP 58686, Down to Bridgeman, 2 April 1954.
94 *Ibid.*, Down to Bridgeman, 30 April 1954.
95 *Ibid.*, Cox, Review of Triad's exploration, 24 October 1955; *ibid.*, Cox to Pattinson and Bridgeman, 26 October 1955; BP 58711, Down to Bridgeman, 2 April and 17 May 1957; *ibid.*, Cox, Canada foothills exploration, 20 May 1957; *ibid.*, Bridgeman to Down, 31 May 1957; *ibid.*, Cox to Bridgeman, 4 October 1957; BP 47351, Information booklet for chairman's visit to Canada, September 1959.
96 BP 60375, Cox to Down, 1 January 1961.
97 BP 58711, Steel, Aide-memoire, Triad, Conclusions from discussions at Britannic House, 8–19 October 1957; BP 58744, Down to Pattinson, 10 June 1958; *ibid.*, Down to Bridgeman, 10 June 1958; *ibid.*, Tanner, Draft interim report to Triad shareholders, August 1958; BP 60375, Cox, Note on Triad, 30 June 1959; *ibid.*, Cox to Down, 11 February 1960; BP 45135, The British Petroleum Company of Canada, 1960 annual review; BP 2422, Extract from Triad board minutes, 7 October 1960; BP 62435, The British Petroleum Company of Canada, 1961 annual review.
98 BP 58686, Bridgeman to Davies, 29 November 1954; *ibid.*, Down to D'Arcy Exploration Company secretary, 10 December 1954; BP 49984, Steel to Bridgeman, 24 August 1961; *ibid.*, Down, Devon-Palmer Oils, 18 October 1961; *ibid.*, Extract from *Daily Telegraph*, 8 January 1962; BP 45161, The British Petroleum Company of Canada, 1964 annual review.
99 BP 27323, BP Canada, Review of operations, 1969.

5 FINANCE AND THE BRITISH GOVERNMENT

1 PRO T 236/4743, Rickett to Bridges, 21 September 1956.
2 PRO T 236/4740, Bridges note, 3 May 1955.
3 *Ibid.*
4 *Ibid.*, Bridges note for record, 3 May 1955.
5 Butler papers, RAB G46/5 (53), Butler, Untitled handwritten note, 3 May [1955].
6 PRO T 236/4740, Butler, Oral message to Lord Strathalmond, 4 May 1955.
7 *Ibid.*, Gilbert note for record, 5 May 1955.
8 J. H. Bamberg, *The History of The British Petroleum Company: Volume II, The Anglo-Iranian Years, 1928–1954* (Cambridge, 1994), pt. 3.

9 *Ibid.*, pp. 317–18.
10 For Butler's budget statement see *Parl. Debates*, House of Commons, vol. 540, cols. 35–63, 19 April 1955, Butler.
11 See generally the debate on the budget in *Parl. Debates*, House of Commons, vol. 540, cols. 63–133, 185–283, 349–479 and 485–580, 19–22 April 1955. Also J. C. R. Dow, *The Management of the British Economy, 1945–1960* (Cambridge, 1964), pp. 78–9; S. Pollard, *The Development of the British Economy, 1914–1990*, 4th edn (London, 1992), p. 356
12 *Parl. Debates*, House of Commons, vol. 545, cols. 202–27, 26 October 1955, Butler.
13 PRO T 236/4740, Bridges to Brittain, 20 October 1955; *ibid.*, Brittain to Bridges, 22 October 1955; *ibid.*, Potter to Armstrong, 1 November 1955; *ibid.*, Bridges to Petch, 7 November 1955.
14 *Parl. Debates*, House of Commons, vol. 545, cols. 1996–9, 10 November 1955, Grimond, Osborne and Collins; *ibid.*, vol. 546, col. 763, 17 November 1955, Gaitskell, and cols. 2109–10, 29 November 1955, Jenkins, Gaitskell and Collins.
15 *Ibid.*, vol. 545, cols. 1996–7, 10 November 1955, Boyle.
16 PRO T 236/4740, Bridges to Petch, 1 May 1956.
17 *Ibid.*, Bridges note for record, 2 May 1956.
18 *Ibid.*, Brittain to Maude, 24 April 1957; *ibid.*, Armstrong note for record, 16 April 1958.
19 *Ibid.*, Armstrong note for record, 9 April 1959.
20 BP, *Annual Report and Accounts*, 1956.
21 BP, *Annual Report and Accounts*, 1957.
22 BP 115920, Middle East crude oil prices, statements 2–6.
23 On the increasingly competitive conditions in the international oil industry at this time, see, for example, Daniel Yergin, *The Prize: The Epic Quest for Oil, Money, and Power* (New York, 1991), chs. 25 and 26 and pp. 437–45, 477 and 546–8; Neil H. Jacoby, *Multinational Oil: A Study in Industrial Dynamics* (New York, 1974), especially chs. 7–9; Christopher Tugendhat and Adrian Hamilton, *Oil: The Biggest Business*, revised edn (London, 1975), chs. 16–17.
24 BP, *Annual Reports and Accounts*, 1956–61.
25 Parra (F. R.) Associates, *The International Oil Industry, 1950–1992: A Statistical History* (Reading, 1993) [privately printed], Table 16.
26 PRO T 236/4741–3, *passim*.
27 PRO T 236/4744–5, *passim*.
28 BP, *Annual Reports and Accounts*, 1957–8.
29 PRO T 236/4745, Makins note, 6 December 1957.
30 BP, *Annual Reports and Accounts*, 1957–60.
31 BP, *Annual Report and Accounts*, 1959.
32 BP 115920, Middle East crude oil prices, statements 7–8.
33 *Ibid.*, Middle East crude oil prices, statement 9. Also Ian Seymour, *OPEC: Instrument of Change* (London, 1981), p. 20.

6 THE ADVENT OF OPEC

1 On these trends in the Middle East see, for example, Elizabeth Monroe, *Britain's Moment in the Middle East, 1914–1971*, 2nd edn (London, 1981); Fawaz A. Gerges, *The Superpowers and the Middle East: Regional and International Politics, 1955–1967* (Boulder, Col., 1994).

2 Edith and E. F. Penrose, *Iraq: International Relations and National Development* (London, 1978), chs. 8–12.

3 A. D. Miller, *Search for Security: Saudi Arabian Oil and American Foreign Policy, 1939–1949* (Chapel Hill, N.C., 1980); Irvine H. Anderson, *Aramco, the United States and Saudi Arabia: A Study of the Dynamics of Foreign Oil Policy, 1933–1950* (Princeton, N.J., 1981); M. B. Stoff, *Oil, War, and American Security: The Search for a National Policy on Foreign Oil, 1941–1947* (New Haven, Conn., 1980).

4 PRO FO 371/133247, Paper on problems of Middle Eastern oil enclosed with Wall to Harvey, 22 April 1958; PRO FO 371/141201, Working party report, The special position of Kuwait as a source of oil, 5 October 1959; BP, *Statistical Review of the World Oil Industry*, 1958; Monroe, *Britain's Moment*, pp. 213–16; J. B. Kelly, *Arabia, the Gulf and the West* (London, 1980), ch. 2; Glen Balfour-Paul, *The End of Empire in the Middle East: Britain's Relinquishment of Power in her Last Three Arab Dependencies* (Cambridge, 1991), ch. 4; David Holden, 'The Persian Gulf: After the British Raj', *Foreign Affairs* 49 (1971), 721–35.

5 James A. Bill, *The Eagle and the Lion: The Tragedy of American-Iranian Relations* (New Haven, Conn., 1988); P. Avery, *Modern Iran* (London, 1965); Richard W. Cottam, *Nationalism in Iran updated through 1978* (Pittsburgh, Penn., 1979); Rose Greaves, 'The reign of Mohammad Riza Shah', in H. Amirsadeghi (ed.), *Twentieth Century Iran* (London, 1977), ch. 3; Mark J. Gasiorowski, *US Foreign Policy and the Shah: Building a Client State* (New York, 1991) and 'US foreign policy toward Iran during the Mussadiq era', in David W. Lesch (ed.), *The Middle East and the United States: A Historical and Political Reassessment* (Boulder, Col., 1996), pp. 51–66.

6 See, for example, PRO FO 371/140858, Pattinson, Record of interview with the Shah on 10 March 1959; PRO FO 371/140859, Harrison, Oil, 18 November 1959; PRO FO 371/140861, Tehran to FO, 14 November 1959; *ibid.*, Record note of audience given by the Shah to Berlin and Snow on 17 November 1959; PRO FO 371/149808, Harrison to Stevens, 6 April 1960.

7 PRO FO 371/141193, Working party report on Middle East oil, January 1959. The working party consisted of officials from the Foreign Office, Treasury, Ministry of Power and Bank of England.

8 See, generally, PRO FO 371/141195.

9 PRO FO 371/150053, State Department comments on the Middle East oil report, 11 May 1960.

10 PRO FO 371/141196, Record of meeting with the British oil companies in Stevens' room on 25 March 1959.

11 PRO FO 371/150059, Fearnley minute on conversations with Miss Jablonski, 12 October 1960. Jablonski was international editor of *Petroleum Week* and had many top-level contacts. She was, for example, able to gain access to the Shah of Iran, the boards of the oil majors and senior Foreign Office officials in Whitehall.

12 For published accounts of the Congress, see Pierre Terzian, *OPEC: The Inside Story* (London, 1985), pp. 26–9; Juan Perez Alfonzo, *Venezuela and OPEC* (Caracas, 1961), p. 103; Ian Seymour, *OPEC: Instrument of Change* (London, 1980), pp. 32–3; Ian Skeet, *OPEC: Twenty Five Years of Prices and Politics* (Cambridge, 1988), pp. 15–16.

13 Seymour, *OPEC*, p. 34; Skeet, *OPEC*, p. 17.

14 BP 62388, Snow to Gass *et al.*, 23 May 1960. Sir David Steel, who was the head of BP's New York office at the time, confirmed to the author on 10 April 1996 that BP was against cutting posted prices because of the effect which this might have on relations with the producing countries.

15 BP 58783, Snow to Belgrave, 26 July 1960.

16 Bennett H. Wall, *Growth in a Changing Environment: A History of Standard Oil Company (New Jersey), Exxon Corporation, 1950–1975* (New York, 1988), pp. 602–3; BP 59226, Comtel, New York to London, 9 August 1960; *ibid.*, Steel to Snow, 9 August 1960.

17 BP 59226, Snow to Steel, 9 August 1960.

18 *Ibid.*, Snow to Bridgeman, 15 August 1960; *ibid.*, Snow to Steel, 15 August 1960.

19 *Ibid.*, Chart of Middle East crude oil price changes – position at 19 September 1960.

20 Seymour, *OPEC*, p. 25.

21 PRO FO 371/150065, Lee, Note on oil prices, 22 August 1960.

22 Seymour, *OPEC*, pp. 34–6; Skeet, *OPEC*, p. 18.

23 PRO FO 371/150065, Stevens minute, 9 September 1960.

24 Seymour, *OPEC*, pp. 36–7; Skeet, *OPEC*, pp. 20–2 and, for the full text of the resolutions passed at the Baghdad conference, Appendix 5, pp. 246–7.

25 PRO FO 371/150061, Nuttall, Note on the second Arab Petroleum Congress, October 1960.

26 BP 58931, Bridgeman to Whiteford, 24 March 1961; BP 59585, Stockwell to Snow, OPEC, 2 January 1961; *ibid.*, Stockwell to the chairman, 15 March 1961; PRO FO 371/150066, Lord Hood to Stevens, 5 November 1960; *ibid.*, Stephens to Stevens, 7 November 1960; *ibid.*, Male, Record of meeting in FO between Stevens, Loudon and Wilkinson on 15 November 1960; PRO FO 371/150057, Powell minute, 18 November 1960; PRO FO 371/150058, Record of meeting with the British oil companies in Stevens' room on 22 November 1960; PRO FO 371/150055, Draft reply to Middleton suggested by Powell, enclosed with Eagers to Male, 28 December 1960; PRO FO 371/158051, Male, OPEC, 3 March 1961; PRO FO 371/158052, FO to various British embassies, 13 January 1961; *ibid.*, Eagers, OPEC and Russian oil, 5 July 1961; *ibid.*, Male to Carter, 30 November 1961; PRO FO 371/164599, 2nd revise, Working party report on

Middle East oil, 1962; PRO T 236/6443, Lucas to Mackay, 3 March 1961; *ibid.*, Extract from minutes of meeting with oil companies on 21 March 1961.

27 Wanda Jablonski, 'Shah of Iran calls proration "impractical"', *Petroleum Week*, 9 December 1960.

28 Wanda Jablonski, 'OPEC won't be a "cartel", Iranians say', *Petroleum Week*, 16 December 1960.

29 Seymour, *OPEC*, p. 41; Skeet, *OPEC*, p. 24.

30 Penrose and Penrose, *Iraq*, ch. 11; Seymour, *OPEC*, p. 8; Skeet, *OPEC*, p. 26; PRO FO 371/158047, Economist Intelligence Unit, Three-monthly review on Middle East oil, September 1961; PRO FO 371/164599, 2nd revise, Working party report on Middle East oil, 1962; PRO FO 371/164606, Intel, FO to certain of Her Majesty's representatives, 13 August 1962.

31 Penrose and Penrose, *Iraq*, p. 287; Seymour, *OPEC*, p. 41; Skeet, *OPEC*, p. 24.

32 PRO FO 371/158052, Tehran to FO, 2 November 1961.

33 G. W. Stocking, *Middle East Oil: A Study in Political and Economic Controversy* (London, 1971), p. 365.

34 Skeet, *OPEC*, p. 27.

35 Stocking, *Middle East Oil*, p. 358.

36 PRO FO 371/164606, Intel, FO to certain of Her Majesty's representatives, 13 August 1962.

37 *Ibid.*

38 PRO FO 371/164220, Harrison to Reilly, 20 August 1962.

39 PRO FO 371/164605, Note of informal talk with Kelly, Assistant Secretary, Department of Interior, on 21 September 1962.

40 BP 100334, Amouzegar to Velders, 11 July 1962; *ibid.*, Draft letter to Finance Minister agreed at Participants' meeting, 26 July 1962.

41 PRO FO 371/164221, Goodchild, Discussions in Paris between the Consortium and the Iranian government on the OPEC resolutions, 10 October 1962. See also, BP 100369, Notes of meeting, 2 October 1962; *ibid.*, Stockwell to Bridgeman, 8 October 1962; *ibid.*, Stockwell to Pattinson, 9 October 1962.

42 PRO FO 371/164221, Stockwell and Warder, Summary of Behnia's informal remarks at 4 October meeting compiled from memory immediately after its end, 8 October 1962.

43 Bill, *The Eagle and the Lion*, p. 104.

44 PRO FO 371/164221, Harrison to Crawford, 3 November 1962.

45 *Ibid.*, Negotiations between the Iran government and the Consortium, unsigned note on Stockwell's report to Crawford on 6 November 1962; *ibid.*, Harrison to Crawford, 9 November 1962; PRO FO 371/164222, Hiller, Iran and OPEC, 12 December 1962. For BP's record of these events, see BP 66814, Stockwell file note on Bridgeman's visit to Tehran, 8 November 1962.

46 PRO FO 371/164221, Negotiations between the Iran government and the Consortium, unsigned note on Stockwell's report to Crawford on 6 November 1962; PRO FO 371/164222, Tehran to FO, 14 November 1962.

47 PRO FO 371/164222, Harrison to Crawford, 3 December 1962.
48 *Ibid.*
49 *Ibid.*; *ibid.*, Chevalier minute, 11 December 1962; *ibid.*, Hiller, Iran and OPEC 12 and 13 December 1962.
50 *Ibid.*, FO to Tehran, 18 December 1962.
51 *Ibid.*, FO to Tehran, 18 December 1962; *ibid.*, Tehran to FO, 23 and 29 December 1962.
52 *Petroleum Press Service*, April, May and August 1963.
53 Skeet, *OPEC*, p. 8.
54 On Yamani, see Jeffrey Robinson, *Yamani: The Inside Story* (London, 1988).
55 PRO FO 371/172515, Record of meeting with Shell and BP in FO on 3 April 1963.
56 PRO FO 371/172516, Lamb minute, 18 April 1963.
57 PRO FO 371/164221, FO to Tehran, 7 December 1962.
58 BP 100372, Audience with the Shah, 5 March 1963.
59 *Ibid.*, Addison to Consortium members, 18 March 1963.
60 *Ibid.*, Sweet-Escott to Müller, 8 May 1963; BP 100330, Sweet-Escott to Stockwell, 11 June 1963.
61 BP 100330, Addison to Consortium members, Khor Musa products terminal with enclosures, 16 August 1963; *ibid.*, Addison to Luard with enclosures, 16 August 1963.
62 BP 66814, Addison to Behnia, 30 August 1963; BP 100383, Pattinson to Page, 6 September 1963; PRO FO 371/172528, Rose, The talks between Rouhani and the Iranian Consortium, 3 October 1963.
63 The '3Ps' also offered to capitalise drilling and exploration expenditure. By this measure, the expenditure would no longer be treated as an operating cost, deducted from income in the year that it was incurred. Instead, it would be treated as capital expenditure, whose deduction from income would be spread (as depreciation) over a number of years. This would have the effect of temporarily raising profits and would therefore increase Iran's revenues. As the benefits of phasing drilling and exploration expenditure in this way would only be temporary, the offer to capitalise the expenditure is not mentioned in the main text.
64 BP 100383, Meeting 16 September 1963; *ibid.*, Page's handwritten notes of 20 September meeting; *ibid.*, Note on discussions of OPEC Resolution 33, 30 September 1963; *ibid.*, Stockwell to Pattinson, OPEC meetings, minutes, 21 October 1963; PRO FO 371/172528, Rose, The talks between Rouhani and the Iranian Consortium, 3 October 1963; PRO FO 371/172529, Mason, OPEC, 15 October 1963; *ibid.*, Lamb, OPEC negotiations, 28 October 1963; PRO FO 371/172528, Lamb minute, 17 October 1963; *ibid.*, Washington to FO, 26 October 1963.
65 PRO FO 371/172528, Lamb minute, 17 October 1963.
66 *Ibid.*, FO to Washington, 1 November 1963; PRO FO 371/172529, Lamb minutes, 5 and 6 November 1963; *ibid.*, FO to Tehran, 5 November 1963; *ibid.*, Lamb, OPEC, 8 November 1963; *ibid.*, Miles, OPEC negotiations, 15

November 1963; BP 66814, Pattinson's notes of meeting between Rouhani and the '3Ps' on 4 November 1963.

67 BP 66814, An account received in Tehran of the OPEC meeting in Beirut (via the FO), 9 December 1963; PRO FO 371/172530, Tehran to FO, 7 December 1963.

68 Penrose and Penrose, *Iraq*, ch. 12; Peter Mansfield, *The Middle East: A Political and Economic Survey*, 4th edn (Oxford, 1973), pp. 324–5.

69 Mansfield, *The Middle East*, pp. 279–83; Bill, *The Eagle and the Lion*, pp. 147–8 and 152.

70 PRO FO 371/172531, Lamb minute, 11 December 1963.

71 *Ibid.*, Draft speaking notes for use by Lord Carrington at Cabinet on 17 December 1963.

72 PRO FO 371/172530, Wright to FO, 9 December 1963.

73 *Ibid.*, Tehran to FO, 10 December 1963.

74 *Ibid.*, Wright to FO, 9 December 1963.

75 PRO FO 371/172529, Lamb, OPEC, 8 November 1963; PRO FO 371/172530, Mason minute, 2 December 1963; *ibid.*, Lamb minute, 10 December 1963; PRO FO 371/172531, Wright to FO, 12 December 1963.

76 PRO FO 371/172530, Wright to FO, 9 December 1963; *ibid.*, Lamb, OPEC, n.d. [probably 11 December 1963]; Parra to Bamberg, 6 March 1997 ('BP delegate').

77 PRO FO 371/172531, Lamb minute, 17 December 1963.

78 *Ibid.*, Lamb minutes, 17 and 18 December 1963; PRO FO 371/172532, Harrison, Middle Eastern oil: OPEC, 18 December 1963; *ibid.*, FO to Tehran, 20 December 1963; IOP 60.04 and BP 100553, Addison's handwritten notes of meeting, 17 December 1963; BP 100553, Notes on '3Ps' meeting with Fallah, 17 December 1963; *ibid.*, Lamb to Stockwell, 24 December 1963 [probably].

79 PRO FO 371/172532, FO to Kuwait, 20 December 1963; *ibid.*, Kuwait to FO, 22 December 1963.

80 *Ibid.*, Tehran to FO, 30 December 1963; PRO FO 371/178153, Miles minute, 14 January 1964; *Petroleum Press Service*, February 1964. The three-man committee consisted of Rouhani, whose term as OPEC Secretary General was due to expire at the end of April 1964; his designated successor, Dr Abdul Rahman Bazzaz, hitherto Iraqi Ambassador in London; and Hisham Nazer of Saudi Arabia.

81 *Petroleum Press Service*, May 1964; PRO FO 371/178154, FO to certain of Her Majesty's representatives, 27 August 1964.

82 PRO FO 371/178154, FO to certain of Her Majesty's representatives, 27 August 1964; Seymour, *OPEC*, p. 49; Stocking, *Middle East Oil*, p. 367; *Petroleum Press Service*, September 1964.

83 Stocking, *Middle East Oil*, pp. 368–9; Seymour, *OPEC*, p. 50.

84 Stocking, *Middle East Oil*, p. 369; IOP 60.04 (13), Addison to Warder, 30 December 1964; *ibid.*, Warder to Addison, 1 January 1965; IOP 10.07 (2), Loan letter and industrial development corporation letter, Consortium to Minister of Finance, 9 January 1965.

85 Stocking, *Middle East Oil*, pp. 369–79.

86 The discounts were to decline to 5½ per cent in 1968, 4½ per cent in 1969, 3½ per cent in 1970, 2 per cent in 1971 and zero in 1972.

87 Stocking, *Middle East Oil*, pp. 379–80; Seymour, *OPEC*, pp. 51–2; Skeet, *OPEC*, p. 67.

7   THE POLITICAL BALANCING ACT

1 John M. Blair, *The Control of Oil* (London, 1976), ch. 5; United States Congress, Senate, Committee on Foreign Relations, *Hearings before the Subcommittee on Multinational Corporations*, 93rd Congress, 2nd session (Washington D.C., 1974–5) (hereinafter abbreviated to *MNC Hearings*), pt. 9, Blair testimony, pp. 192–230; United States Congress, Senate, Committee on Foreign Relations, Subcommittee on Multinational Corporations, *Report on Multinational Oil Corporations and U.S. Foreign Policy*, 93rd Congress, 2nd session, (Washington D.C., 1975) (hereinafter abbreviated to *MNC Report*), ch. 5.

2 See, for example, Peter R. Odell, *Oil and World Power*, 7th edn (Harmondsworth, 1983), p. 23; Louis Turner, *Oil Companies in the International System* (London, 1978), pp. 118–21; BP 38921, Dutton, The political environment, BP management conference paper, May 1968; BP 55782, Butcher, Europe – action for 1972 and 1973, BP management conference paper, 1972; BP 10201, Policy Planning Staff, The role of the international oil companies, 16 May 1974.

3 For an excellent account of Iraqi politics in this period, see Edith and E. F. Penrose, *Iraq: International Relations and National Development* (London, 1978), chs. 8–15.

4 The only part of Iraq not covered by the IPC Group's concessions was an area of 700 square miles which had been transferred in 1913–14 from Iran to Mesopotamia (later Iraq). The Khanaqin Oil Company, a wholly owned subsidiary of BP, held the concession for this area, which included the Naftkhana oil field and a small refinery at Alwand. At the time of the 50:50 agreement between IPC and Iraq in 1952, the Khanaqin Oil Company was given the right to continue operating the Naftkhana field on the basis that the concession would be relinquished if exports from the field had not reached 2 million tons a year (about 40,000 bpd) by February 1959. In the event, the export target was not reached by that time, and the Khanaqin Oil Company gave up its concessionary rights. See BP 36191, Willy to Cox, Former BP concessional interests in the Middle East, 16 February 1961; PRO FO 371/133119, Doyle, Khanaqin Oil Company, 4 September 1958.

5 On Herridge's visit, see papers in BP 100375.

6 Penrose and Penrose, *Iraq*, p. 257.

7 PRO FO 371/133879, Wright minute, 1 August 1958; PRO FO 371/133119, Combs note on Herridge's visit to Iraq, 29 August 1958.

8 For accounts of the negotiations from different viewpoints, see BP 108127, The Iraq oil negotiations, a commentary written at the request of the IPC by an experienced outside observer, February 1962; IPC A-CC/1/1, box 179, Iraqi Oil Ministry pamphlet, The revolution government and oil negotiation, April 1961.

9 PRO FO 371/141064, Sir Roger Stevens, IPC, 27 May 1959.

10 BP 58783, Bridgeman to Belgrave, 16 March 1959.

11 PRO FO 371/133879, Rose minute, 26 June 1958; *ibid.*, Hayter minute, 1 July 1958; PRO FO 371/133120, Gore-Booth, IPC and the Iraq government, 27 October 1958.

12 PRO FO 371/164599, 2nd revise, Working party report on Middle East oil, appendix 8, 1962.

13 PRO FO 371/141066, Profumo to Prime Minister, 15 July 1959.

14 *Ibid.*, Stevens minute, 15 July 1959.

15 PRO FO 371/141070, Fearnley, IPC negotiations, 15 October 1959.

16 Penrose and Penrose, *Iraq*, p. 235.

17 *Ibid.*, pp. 258–60.

18 BP 59364, Stockwell to Snow, 6 September and 16 December 1960; *ibid.*, Herridge to London, 23 and 29 September 1960; *ibid.*, Herridge to Murphy, 23 September 1960; BP 100373, Herridge to London, 3 November 1960; BP 79847, Relinquishment – Item 7.

19 IPC SCC/1/1, box 3, pt. 9, Lawson to Groups, 29 March 1961; PRO FO 371/157720, FO to certain of Her Majesty's representatives, 12 April 1961.

20 PRO FO 371/157720, FO to certain of Her Majesty's representatives, 12 April 1961. See also BP 100722, Willy to Pattinson, 11 April 1961.

21 IPC SCC/1/1, box 3, pt. 9, Herridge to Murphy, 1 April 1961.

22 BP 100722, Stockwell to Bridgeman, 21 April 1961.

23 PRO FO 371/157718, Herridge to Murphy, 3 April 1961; IPC SCC/1/1, box 3, pt. 9, Herridge to Murphy, 4 April 1961.

24 PRO FO 371/157719, Trevelyan to FO, 4 April 1961.

25 PRO FO 371/157720, FO to certain of Her Majesty's representatives, 12 April 1961; IPC SCC/1/1, box 3, pt. 9, Bird to Groups, 7 April 1961.

26 BP 100722, Willy to Pattinson, 11 April 1961.

27 The delegation consisted of H. W. Fisher (a director of Standard Oil (NJ)), J. F. Stephens (chairman of Shell Transport and Trading) and Herridge. See PRO FO 371/164599, 2nd revise, Working party report on Middle East oil, appendix 8, 1962.

28 PRO FO 371/164599, 2nd revise, Working party report on Middle East oil, appendix 8, 1962; PRO FO 371/157728, Inter-departmental working group, Iraq petroleum negotiations, 19 September 1961.

29 PRO FO 371/157727, Trevelyan to FO, 31 August 1961.

30 PRO FO 371/157728, Heath to Macmillan, 30 September 1961.

31 *Ibid.*, Inter-departmental working group, Iraq petroleum negotiations, 19 September 1961.

32 PRO FO 371/157729, Trevelyan to FO, 12 October 1961; PRO FO 371/157730, Male, Record of conversation with Herridge on 18 October 1961. See also BP 8586, Cables from Baghdad, 9–12 October 1961.
33 PRO FO 371/164274, Hiller, The Iraq government's action against the IPC, 19 December 1961.
34 PRO FO 371/157732, Hiller, Action by HMG in support of the IPC, 19 December 1961; PRO FO 371/164274, Hiller, The Iraq government's action against the IPC, 19 December 1961.
35 G. W. Stocking, *Middle East Oil: A Study in Political and Economic Controversy* (London, 1971), pp. 251–2.
36 PRO FO 371/164280, Allen to Crawford, 8 October 1962.
37 BP 4802, Stockwell to Herridge, 23 August 1962.
38 Penrose and Penrose, *Iraq*, p. 382; Stocking, *Middle East Oil*, p. 257.
39 Penrose and Penrose, *Iraq*, p. 325.
40 On the negotiations, see IPC SCC/1/1, box 6, Negotiations 64/65; IPC SCC/1/1, pts. 21–2. For a published account, Penrose and Penrose, *Iraq*, pp. 382–7.
41 IPC SCC/1/1, box 5, pt. 25, Draft summary for briefing operating companies, 13 April 1965; *ibid.*, Boyer's draft summary of the settlement, 25 May 1965; Penrose and Penrose, *Iraq*, pp. 387–8; Stocking, *Middle East Oil*, pp. 262–4.
42 BP 8588, Stockwell to Bridgeman, 6 July 1965.
43 Penrose and Penrose, *Iraq*, p. 389; Stocking, *Middle East Oil*, pp. 265–6.
44 BP 63109, Arber to Steel, 26 January 1966; IPC SCC/1/1, box 6, pt. 27, Jäckli to Dalley, 19 February 1966; Penrose and Penrose, *Iraq*, p. 389; Stocking, *Middle East Oil*, pp. 267–8.
45 For a detailed account of the Syrian dispute, see Stocking, *Middle East Oil*, chs. 12–13.
46 BP 63109, Message from Baghdad, 30 May 1967; *ibid.*, Dalley to Groups, 14 March 1967; BP 8588, Stockwell to Bridgeman, 9 March 1967; IPC SCC/1/1, pts. 31–2, *passim.*
47 IPC RE/3/42, box 245, P. C. K. O'Ferrall, Diary of government negotiations in Iraq, 1961–1967.
48 Ethan B. Kapstein, *The Insecure Alliance: Energy Crises and Western Politics since 1944* (Oxford, 1990), pp. 141–9; Daniel Yergin, *The Prize: The Epic Quest for Oil, Money, and Power* (New York, 1991), pp. 554–8; George Lenczowski, *American Presidents and the Middle East* (Durham and London, 1990), pp. 106–9 and *The Middle East in World Affairs*, 4th edn (Ithaca, 1980), p. 775; Henry Kissinger, *White House Years* (Boston, Mass., 1979), ch. 10.
49 David Birmingham, *The Decolonisation of Africa* (London, 1995), pp. 33–6; Sarah Ahmad Khan, *Nigeria: The Political Economy of Oil* (Oxford, 1994), pp. 9–11; BPX report 2643, A political review of Nigeria, 1960–1970; BP 41996, Dean, Nigeria, Revised information dossier, 26 February 1969; BP 20423, Regional directorate, Middle East and North Africa, BP's activities in Nigeria, January 1973.

50 Kapstein, *The Insecure Alliance*, pp. 141–9; Yergin, *The Prize*, pp. 554–8; BP 38921, McColl, Paper on Supply for BP management conference, May 1968.

51 Fawaz A. Gerges, *The Superpowers and the Middle East: Regional and International Politics, 1955–1967* (Boulder, Col., 1994), ch. 8 and 'The 1967 Arab-Israeli war: US actions and Arab perceptions', in David W. Lesch (ed.), *The Middle East and the United States: A Historical and Political Reassessment* (Boulder, Col., 1996), pp. 189–208; Kissinger, *White House Years*, ch. 10.

52 Penrose and Penrose, *Iraq*, pp. 394–5; Stocking, *Middle East Oil*, p. 305.

53 Stocking, *Middle East Oil*, pp. 306–12.

54 Penrose and Penrose, *Iraq*, p. 396.

55 Arif had by this time given up the post of Prime Minister, to which he had appointed Tahir Yahya again on 10 July 1967. Yahya was also deposed in the Baathist coup.

56 Stocking, *Middle East Oil*, pp. 314–15; Peter Mansfield, *The Middle East: A Political and Economic Survey*, 4th edn (Oxford, 1973) p. 335.

57 BP 62686, Fraser to the chairman, Middle East crude oils – offtake targets, 3 December 1963.

58 *MNC Report*, pp. 102–3; BP 90858, Consortium Agreement, 1954, Article 20.

59 *MNC Report*, p. 96; BP, *Statistical Review of the World Oil Industry*, 1967 and 1970.

60 BP 63021, Brief for Shah's visit to the US on 22/24 August 1967.

61 *MNC Report*, p. 96; BP, *Statistical Review of the World Oil Industry*, 1967 and 1970.

62 BP 18478, Extract from *Middle East Economic Survey*, 18 March 1966.

63 IOP 110.03/2, Audience with the Shah, 12 March 1966; BP 63018, Steel to Bridgeman, 24 March 1966.

64 *MNC Report*, pp. 105–6; *MNC Hearings*, pt. 8, pp. 564–7, Discussions with United Kingdom on Iran, 2–3 November 1966.

65 BP 63018, Iran – meeting between government ministers and Consortium member companies (except CFP), 6 July 1966. Another record of the meeting is in McCloy papers, series 20, box oil 2, folder 31, Bergquist, Report on meeting with Hoveyda, 6 July 1966.

66 McCloy papers, series 20, box oil 2, folder 31, Bergquist, An account of the group audience with the Shah on 6 July 1966; *ibid.*, Bergquist, Additional items of information and general observations, 12 July 1966; *ibid.*, Bergquist, Audience by Dorsey, Law, Bergquist and Moses with the Shah on 7 July 1966.

67 *MNC Report*, p. 107; IOP chron. files, vol. 7, Addison to members, 24 October 1966.

68 IOP chron. files, vol. 7, Extract from Warder to Addison in Addison to members, 24 October 1966.

69 *The Times*, 28 October 1966; *The Economist*, 29 October 1966.

70 *MNC Report*, p. 107; *MNC Hearings*, pt. 8, pp. 571–4, State Department memorandum of conversation, 24 October 1966.

71 *MNC Report*, pp. 108; *MNC Hearings*, pt. 8, pp. 564–7, Discussions with United Kingdom on Iran, 2–3 November 1966.

72 On the negotiations, see IOP 50.15 pt. 2, Addison to members, 17 November 1966; *ibid.*, Addison to Head Office, 27 November 1966.

73 IOP 50.15 pt. 2, PT1140 for Eardley from Addison, 10 December 1966; *ibid.*, IOP to NIOC, 11 December 1966; BP 18478, Reuter, 13 December 1966; Stocking, *Middle East Oil*, p. 193.

74 According to Sir Denis Wright, the British Ambassador in Tehran, the Shah was 'very content and indicated that he had no oil problems at all' in January 1967. See BP 63021, Stockwell to Bridgeman, 13 January 1967.

75 *MNC Report*, p. 103.

76 J. B. Kelly, *Arabia, the Gulf and the West* (London, 1980), chs. 1–2; Glen Balfour-Paul, *The End of Empire in the Middle East: Britain's Relinquishment of Power in her Last Three Arab Dependencies* (Cambridge, 1991), chs. 3–4; Center for Strategic and International Studies, *The Gulf: Implications of British Withdrawal* (Georgetown University, Washington D.C., Special report series, no. 8, 1969).

77 BP 63022, Bridgeman record note, 12 January 1968.

78 *Ibid.*, Audience with the Shah, 31 January 1968.

79 *Ibid.*, Addison to members, 5 February 1968.

80 *Ibid.*, Pennell to Steel, 9 February 1968.

81 IOP 60.04 (20), Meeting with Minister of Finance on 9 March 1968, in Addison to members, 26 March 1968.

82 IOP 110.03 (10), Talks with NIOC, 14 March 1968.

83 *MNC Hearings*, pt. 7, pp. 274–5, Shafer, Meeting with State Department re Iranian problems, 28 March 1968; *MNC Report*, pp. 115–16.

84 On McCloy, see Kai Bird, *The Chairman: John J. McCloy, The Making of the American Establishment* (New York, 1992); James A. Bill, *The Eagle and the Lion: The Tragedy of American-Iranian Relations* (New Haven, Conn., 1988), pp. 334–5; Kissinger, *White House Years*, p. 22.

85 Minutes of the meetings are in McCloy papers, series 20, box oil 2, folders 24–33.

86 State Department Freedom of Information Act papers, 8204102, State Department memorandum, Highlights of McCloy–Hoveyda conversation, 31 March 1968; *ibid.*, State Department telegram, Saudi–Iranian relations, 1 April 1968; *ibid.*, State Department telegram, Persian Gulf, 3 April 1968.

87 BP 63022, Note of audience granted by the Shah to Steel *et al.* on 20 April 1968, attached to Addison's letter dated 22 April 1968.

88 *Ibid.*, Note of meeting with Eghbal and Fallah, 20 April 1968.

89 McCloy papers, series 20, box oil 2, folder 38, Minutes of meeting of counsel with oil company representatives, 5 June 1968.

90 McCloy papers, series 20, box oil 1, folder 40, Unsigned letter [probably by Otto Miller, chairman of Socal] to Ball, 5 June 1968; *ibid.*, Paper on the Middle East enclosed with Moses to McCloy, 26 November 1968.

91 Lenczowski, *American Presidents*, p. 116.
92 Nixon archives, WHCF, CO, boxes 37–8, folders [EX] CO 68 on Iran.
93 Kissinger, *White House Years*, p. 1262.
94 BP 63026, Addison, Note of audience with the Shah on 31 January 1969; personal communication from Steel to Bamberg, April 1996.
95 IOP 110.03 (11), Draft, Level of oil offtake, 19 February 1969.
96 National Security Archive, 00711, State Department telegram, Shah's visit to Washington, 3 April 1969.
97 McCloy papers, series 20, box oil 1, folder 44, Boyer to Hedlund, Meeting with State Department regarding Iran, 16 April 1969.
98 BP 63026, Addison to members, 8 April 1969.
99 *Ibid.*, Eardley to Steel from O'Brien, 7 May 1969.
100 *Ibid.*, Extracts from *The Observer* and *Sunday Telegraph*, 11 May 1969.
101 BP 4970, Account of the statement of Mr Steel enclosed with Addison to members, 20 May 1969.
102 National Security Archive, 00724, State Department, Semi-annual assessment of the political situation in Iran, 4 September 1969.
103 McCloy papers, series 20, box oil 1, folder 42, McCloy, Memorandum of conversation with Nathaniel Samuels, Deputy Under-Secretary of State for Foreign Affairs on 7 August 1969 at the State Department; *ibid.*, Folmar to Hedlund, Moses and Parkhurst, 11 September 1969; *ibid.*, folder 46, Moses to McCloy with enclosure, 27 October 1969.
104 McCloy papers, series 20, box oil 1, folder 44, Moses to McCloy with enclosures, 30 September 1969.
105 Nixon archives, WHCF, subject files, CO, box 6, folder [EX] CO 1–7 Middle/Near East [1969–70], Warner and Jamieson to Nixon, 26 November 1969.
106 *Ibid.*, Kissinger to Jamieson, 2 December 1969; *ibid.*, Jamieson to Nixon, 15 December 1969.
107 Lenczowski, *American Presidents*, pp. 122–3; Kissinger, *White House Years*, pp. 374–7.
108 Nixon archives, WHCF, subject files, CO, box 6, folders [Gen] CO 1–7 Middle/Near East beginning 31 December 1969 and 1 January to 31 December 1970.
109 Nixon archives, WHCF, subject files, CO, box 37, folder [EX] CO 68 1 January 1970–, Flanigan to Kissinger, 23 January 1970; *ibid.*, Flanigan to Ehrlichman, Kissinger and Krogh, 12 March 1970.
110 McCloy papers, series 20, box oil 2, folder 1, Piercy to McCloy with enclosures, 2 April 1970.
111 BP 63028, Addison to Steel, 20 April 1970.
112 *Ibid.*, *Procès verbal*, 6 May 1970.
113 Nixon archives, WHCF, subject files, CO, box 37, folder [EX] CO 68 1 January 1970–, Flanigan to Nixon, 8 May 1970.
114 See, for example, Frank C. Waddams, *The Libyan Oil Industry* (London, 1980),

ch. 12; Ian Seymour, *OPEC: Instrument of Change* (London, 1980), pp. 65–74; Ian Skeet, *OPEC: Twenty Five Years of Prices and Politics* (Cambridge, 1988), pp. 58–61.

8　THE 'HOLY GRAIL' OF EXPLORATION

1　Data derived from BP's 1971 estimates of ultimate recoverable reserves in its oil fields at the end of 1970, backdated to the years when the fields were discovered, and subsequently reduced by production. See BPX report OC 8113, Exploration and Production Department, BP's share of proven and probable reserves of crude oil as at 1 January 1972, June 1972; BPX report EXT 57512, Sayer, BP discoveries to the end of 1970 – an old working list, 1971.

2　Henry Kissinger, *White House Years* (Boston, Mass., 1979), p. 59.

3　Bryan Cooper, *Alaska – The Last Frontier* (London, 1972); John Strohmeyer, *Extreme Conditions: Big Oil and the Transformation of Alaska* (New York, 1993); BP 38078, BP Public Affairs and Information Department, Alaska background notes, April 1975; P. E. Kent, 'Entry into Alaska', *BP Shield,* January 1970.

4　BP 85943, Kent, Comment on proposal to participate with Sinclair in Alaska, 13 February 1959; *ibid.*, Matthews, Alaska, Report on offer by Sinclair, n.d.; *ibid.*, Warman, Geological appraisal of the Sinclair offer to BP to participate in their exploration in Alaska, 6 March 1959.

5　BP 21340, Cox to Bridgeman, Alaska, 23 March 1959; *ibid.*, Cox telegram, 2 April 1959; *ibid.*, Cox to Sinclair and BP Explorations Inc., 3 April 1959.

6　*Ibid.*, Cox to Bridgeman, Alaska, 14 August 1959; *ibid.*, Cox to Pattinson, Alaska, Note after visit in June 1960; BP 90192, Cox to Birks, 7 September 1959; *ibid.*, Cox to Barker, 4 September 1959; *ibid.*, Cox to Kent, 4 September 1959; Kent papers, Memoirs, PK A 12/9.

7　Kent, 'Entry into Alaska', *BP Shield,* January 1970; BP 21340, Cox to Pattinson, Alaska, Note after visit in June 1960.

8　Cooper, *Alaska*, pp. 123–7; BP 117797, Bischoff, Exploration, BP and Alaska, A history, December 1992. Bischoff counted Kuparuk No. 1 and 1A as two wells, arriving at a total of seven, rather than six wells drilled by BP and Sinclair.

9　BP 117797, Bischoff, Exploration, BP and Alaska, A history, December 1992; BP 60376, Thomas to Falcon, 15 April 1964.

10　BP 117797, Bischoff, Exploration, BP and Alaska, A history, December 1992.

11　BP 99300, Spence, Appraisal of state land on northern coast of Alaska, November 1964.

12　*Ibid.*, Thomas to Milward, 20 November 1964; Cooper, *Alaska*, p. 127; BP 117797, Bischoff, Exploration, BP and Alaska, A history, December 1992.

13　Bamberg interview with Jenkins, Martin and Thomas, 9 August 1996; BP 99300, Spence, Notes on the state land on north coast of Alaska, 23 April 1965;

BP 34939, Gough to Haward, 22 July 1965; BP 117797, Bischoff, Exploration, BP and Alaska, A history, December 1992.

14 Cooper, *Alaska*, p. 128.

15 Bamberg interview with Jenkins, Martin and Thomas, 9 August 1996; BP 99300, Spence, Notes on the state land on the North Slope after the drilling of Colville no. 1, 28 March 1966.

16 BP 34939, Gay to Stephens, 15 October 1966; *ibid.*, Stephens to Gay, 18 October 1966; *ibid.*, Rickwood to Stephens, 25 October 1966; *ibid.*, Thomas to Stephens, 19 October 1966; *ibid.*, Gay to Stephens, 4 November 1966; Bamberg interview with Rickwood, 19 November 1997; Cooper, *Alaska*, p. 129.

17 BP 117797, Bischoff, Exploration, BP and Alaska, A history, December 1992; Cooper, *Alaska*, p. 129.

18 BP 117797, Bischoff, Exploration, BP and Alaska, A history, December 1992; Cooper, *Alaska*, pp. 76–83.

19 BP 38078, BP Public Affairs and Information Department, Alaska background notes, April 1975; Bamberg interview with Rickwood, 19 November 1997; Cooper, *Alaska*, pp. 130–1 and ch. 11.

20 BP 57693, Unattributed note for Interdepartmental Committee, Summary of some implications of BP's Alaskan discovery, 5 December 1969; BP 38078, BP Public Affairs and Information Department, Alaska background notes, April 1975.

21 P. E. Kent, 'UK onshore oil exploration', *Marine and Petroleum Geology* 2 (1985), 56–64; BP 45645, Matthews and Parry, Report on future profitability of operations in UK indigenous crude oil, 30 June 1961.

22 Kent papers, PK G 1/1/1, Lees to Taylor, Oil prospects of the continental shelf throughout the world, 1 March 1946.

23 Bryan Cooper and T. F. Gaskell, *North Sea Oil – The Great Gamble* (London, 1966), p. 32 and, by the same authors, *The Adventure of North Sea Oil* (London, 1976), p. 20.

24 BP 60267, Cox to Pattinson, 7 February 1962.

25 Cooper and Gaskell, *Adventure of North Sea Oil*, p. 20; BP 60267 and 46800, *passim.*

26 Cooper and Gaskell, *Great Gamble*, pp. 22, 87–8 and 96. See BP 60267 and 46385 on the consultations between the British government and the oil companies in the drafting of the Continental Shelf Act.

27 Bridgeman hoped that this would be the case. See BP 60267, Lowson to Luard [paraphrasing Bridgeman], 9 October 1963.

28 BP 60267, Malone to Falcon, 30 April 1963.

29 Julia Bruce (BP history researcher) interview with Beckett, 4 February 1987.

30 BP 60267, BP-MOP meeting of 27 August 1964. Memorandum states that 48/6 was the most sought-after block.

31 Cooper and Gaskell, *Great Gamble*, pp. 12–13, 62–70, 76 and 79–80; BPX item 66767, Jackson, A history of offshore drilling, n.d.

32 Cooper and Gaskell, *Great Gamble*, pp. 75–6 and 101–4.
33 Cooper and Gaskell, *Adventure of North Sea Oil*, pp. 24–8.
34 *Ibid.*, pp. 64–5, 69 and 72.
35 BP 60266, Matthews to Lowson, 15 October 1965; *ibid.*, Matthews memorandum, 26 October 1965; *ibid.*, Falcon to Lowson, 29 October 1965; BP 43748, BP application to MOP, 16 September 1965.
36 BP 60266, Matthews memorandum, 3 November 1965; *ibid.*, Luard to MOP, 19 November 1965.
37 BP 46181, Matthews to Kent, 5 May 1967; *ibid.*, Crosthwaite to Matthews, 5 June 1967; *ibid.*, Matthews to Kent, 23 June 1967; *ibid.*, Matthews to Pennell, 4 August 1967.
38 BP 3408, Kirkby, 30/1–1 evaluation, 7 August 1970; *ibid.*, McTear to Skelton, 25 September 1970; BP 59752, Birks to Steel, 19 October 1970; BP press release, 19 October 1970.
39 BPX report OC 8113, Exploration and Production Department, BP's share of proven and probable reserves of crude oil as at 1 January 1972, June 1972; BPX report EXT 57512, Sayer, BP discoveries to the end of 1970 – an old working list, 1971.
40 BP 59756, Exploration and Production Department, BP and competitor companies, January 1972.
41 BPX report OC 8113, Exploration and Production Department, BP's share of proven and probable reserves of crude oil as at 1 January 1972, June 1972; BPX report EXT 57512, Sayer, BP discoveries to the end of 1970 – an old working list, 1971.
42 BPX report OC 8113, Exploration and Production Department, BP's share of proven and probable reserves of crude oil as at 1 January 1972, June 1972; BPX report EXT 57512, Sayer, BP discoveries to the end of 1970 – an old working list, 1971.
43 Kent papers, PK G 3/1/3, Exploration review 1970, 2 February 1970.
44 BP 29994, *passim*; S. H. Longrigg, *Oil in the Middle East: Its Discovery and Development*, 3rd edn (London, 1968), pp. 427–8.
45 BP 118112, Sutcliffe to Steel, 10 April 1969.
46 BP 46100, Notes on chairman's visit to Japan on 28 September to 2 October 1970; BP 62986, Central Planning Department, BP and Japan, April 1970; BP Archive shelf files, BP press release, BP agreement with Japanese companies, 11 September 1970.
47 BPX report OC 8113, Exploration and Production Department, BP's share of proven and probable reserves of crude oil as at 1 January 1972, June 1972; BPX report EXT 57512, Sayer, BP discoveries to the end of 1970 – an old working list, 1971.
48 BPX report EXT 53570, Sayer, A historical look at Geological Division's annual estimates of world oil and gas reserves, production and future discoveries 1971–1988, April 1989.
49 H. R. Warman, 'Future problems in petroleum exploration', *Petroleum Review*

25 (1971), 96–101 and 'The future of oil', *Geographical Journal* 138 (1972), 287–97.

50 Warman, 'Future of oil', 287.

51 BPX report EXT 53570, Sayer, A historical look at Geological Division's annual estimates of world oil and gas reserves, production and future discoveries 1971–1988, April 1989.

52 BP 57693, Kent and Warman, Exploration planning and expenditure levels, 21 June 1971; BP 59756, Exploration and Production Department, Basis for considering future BP expenditure levels on exploration, January 1972.

53 Kent papers, PK G 3/2/3, Exploration review 1970, 6 February 1970.

54 Kent papers, PK G 5/1/1, Mann to Thomas, The urgent need to increase exploration, 23 May 1972.

55 Kent papers, PK G 3/1/1–4, Exploration review 1970, 2 February 1970; *ibid.*, PK G 3/2/1–5, Exploration review 1970, 6 February 1970; *ibid.*, PK G 5/5/1, Requirements of exploration policy 1972–1982, Draft 10 year look ahead, 27 September 1972; *ibid.*, PK G 6/3/1–2, Kent to Birks, The status of BP's exploration, 7 December 1972; BP 57693, Kent and Warman, Exploration planning and expenditure levels, 21 June 1971.

56 BP 43660, Exploration Department, An outline scheme for career development and training in Exploration Department, vols. 2–3, appendix 8, March 1971.

57 Kent papers, PK G 6/3/1–2, Kent to Birks, The status of BP's exploration, 7 December 1972.

58 On the representations see, for example, BP 55775, Bexon, Exploration capital expenditure policy, September 1974.

59 Cooper and Gaskell, *Adventure of North Sea Oil*, p. 76; BPX report OC 6914, Exploration and Production Department annual reports for 1974 and 1975.

60 BP 17389, Unattributed study on exploration, n.d.

9   THE PUSH FOR OUTLETS . . .

1 John G. Clark, *The Political Economy of World Energy: A Twentieth Century Perspective* (Hemel Hempstead, 1990), chs. 4 and 6.

2 BP, *Statistical Review of the World Oil Industry*, 1960 and 1970.

3 BP 40561, Seven majors in the sixties.

4 R. W. Ferrier, *The History of The British Petroleum Company: Volume I, The Developing Years, 1901–1932* (Cambridge, 1982), pp. 217–19 and 470–97.

5 Ferrier, *British Petroleum, Volume I*, p. 511; J. H. Bamberg, *The History of The British Petroleum Company: Volume II, The Anglo-Iranian Years, 1928–1954* (Cambridge, 1994), p. 107.

6 Ferrier, *British Petroleum, Volume I*, p. 512; T. H. Bingham and S. M. Gray, *Report on the Supply of Petroleum Products to Rhodesia* (HMSO, 1975), p. 8.

7 BP 65245, Rees Jenkins to Seddon, Marketing of BP Energol by Atlantic South Africa, 5 June 1956.

8 Bamberg, *British Petroleum, Volume II*, pp. 107–17.

9 Geoffrey Jones, *The State and the Emergence of the British Oil Industry* (London, 1981), pp. 217 and 239; Daniel Yergin, *The Prize: The Epic Quest for Oil, Money, and Power* (New York, 1991), pp. 232 and 274; BP 72502, Agreement between Shell-Mex, British Petroleum and The Scottish Oils Agency, 31 December 1931; *ibid.*, Agreement between Shell Marketing Company, Eagle Oil and Shipping Company, Anglo-Persian Oil Company and Shell-Mex and BP (with attached memorandum, referred to in clause 20 of agreement), 31 December 1931; BP 113643, Shell-Mex and BP, Special Resolutions passed at extraordinary general meetings on 16 March 1934 and 25 October 1945; BP 58500, Rees Jenkins to Butcher, Managers course (Stage 1) no. 24, 12 December 1963.

10 Bamberg, *British Petroleum, Volume II*, pp. 33–50, 57–61, 230–5, 250–7, 278 and 292–307.

11 BP 97334, Minutes of Co-ordination Committee meeting, 2 November 1965; BP 63508, Central Planning Department, Long term policy, 21 December 1966; BP 50302, Bean, General policy, 10 October 1967; *ibid.*, Minutes of Planning Committee, 6 December 1967; *ibid.*, Minutes of Planning Committee, 20 December 1967; *ibid.*, Minutes of Planning Committee, 29 January 1968.

12 BP 10178, Armstrong, Marketing policy, November 1972.

13 Ferrier, *British Petroleum, Volume I*, ch. 5; Jones, *The State and the Emergence of the British Oil Industry*, pp. 144–56 and ch. 6.

14 BP 40561, Seven majors in the sixties.

15 Neil H. Jacoby, *Multinational Oil: A Study in Industrial Dynamics* (New York, 1974), ch. 7; Christopher Tugendhat and Adrian Hamilton, *Oil: The Biggest Business*, revised edn (London, 1975), pp. 147–9 and 157–8; Edith T. Penrose, *The Large International Firm in Developing Countries: The International Petroleum Industry* (London, 1968), pp. 138–40; Frank C. Waddams, *The Libyan Oil Industry* (London, 1980), p. 211.

16 BP 102837 and 103052, Management reports for board meetings in January 1957 and 1971; BP 115941, Monthly report for general management, Supply and Development Division, December 1956.

17 For useful surveys of conditions in different markets, see US Federal Energy Administration, *The Relationship of Oil Companies and Foreign Governments* (Washington D.C., 1975); BP 63695, Policy Planning Staff, Europe, 16 November 1972.

18 BP 38921, Robertson, Marketing, BP management conference, May 1968; BP 65611, Strategic planning: a review, 13 April 1972.

19 For data on BP's market shares by grade of product from 1963 to 1969, see BP 119420, Oil companies sales and market shares, 1969.

20 BP 38921, Robertson, Marketing, BP management conference, May 1968.

21 BP 9352, Halliday, Notes on CEI proposals for the BP 'New Look', 1 November 1957.

22 BP 9284, Hodges to von Pachelbel, 8 October 1965.

23 SMBP archives, Booklet 1713, 'This is Your New Look', 1958.
24 BP 9289, Hodges to Norris, 18 April 1958; BP 9284, Hodges to von Pachelbel, 8 October 1965.
25 BP 56397, Burns, Presentation to BP Trading board, 21 December 1971; BP 100187, An image for the '90s.
26 BP 103052, Management report for board meeting in January 1971.
27 BP 118832, Hooper, History of BP research, vol. 7(a), 1946–1954, Postwar recovery and expansion, section (1), pp. 4–6 and 12–14.
28 *Ibid.*, Hooper, History of BP research, vol. 7(a), 1946–1954, Postwar recovery and expansion, section (1), p. 15; BP 118835, Hooper, History of BP research, vol. 8, 1955–1973, section (2), pp. 108–11; Donald F. Dixon, 'The development of the solus system of petrol distribution in the United Kingdom, 1950–1960', *Economica* 29 (1962), 40 and 45; The Monopolies Commission, *A Report on the Supply of Petrol to Retailers in the United Kingdom* (HMSO, 1965), pp. 13–21.
29 On the development of the solus system, see Dixon, 'The development of the solus system', 40–52, and Monopolies Commission, *Report*, esp. chs. 3–7.
30 BP 10201, Skinner, Continental Oil Company, 15 November 1971.
31 BP 18724, Extract from *Daily Express*, 30 December 1955; BP 49963, Extract from *Financial Times*, 14 December 1960; *ibid.*, Extract from *Daily Telegraph*, 3 January 1961; BP 20898, Market economics and group planning, Castrol note, 4 November 1966. On the genesis and operation of the Monopolies and Restrictive Practices Act, see Helen Mercer, *Constructing a Competitive Order: The Hidden History of British Antitrust Policies* (Cambridge, 1995), chs. 5–6.
32 On the adoption of the Energol brand name, the multiplicity of Energol oils, and the introduction of BP Energol Visco-static, see BP 51711, 61008, 61106, 22485, 22486, 22487, 22488, 55959, 30778, 30779, 30780, 95507. For BP as worst offender, BP 65245, Waller, Record note on C. C. Wakefield, 13 May 1957; *ibid.*, Waller to Barran, Our position vis-à-vis C. C. Wakefield, 28 May 1957.
33 Monopolies Commission, *Report*, ch. 14. Other recommendations provided for increased security of tenure for the tenants of supplier-owned filling stations, and for a loosening of the restrictive conditions on which the suppliers made loans to retailers.
34 BP 56602, Carey to SMBP, 10 December 1966; *ibid.*, Riddell-Webster to Butcher, UK retail market – Monopolies Commission, 13 July 1966; *ibid.*, Riddell-Webster to Dummett, Monopolies Commission, 26 July 1966; *ibid.*, Waller to Dummett, Total Oil Products (GB) Ltd, 9 and 16 November 1966; SMBP archives, 9857 (51/C/0079), General manager (retail) to all retail staff, Undertakings given to the Board of Trade arising out of the Monopolies Commission recommendations, 5 August 1966.
35 See, for example, BP 8618, Berkin to Dummett and Fraser, 29 May 1964.
36 Monopolies Commission, *Report*, p. 109; SMBP archives, bag 6675, 'Retail brand objectives' file, Brand aims and philosophies, appendix 3, 24 July 1967.

37 SMBP archives, bag 6675, 'Retail brand objectives' file, Brand aims and philosophies, appendix 3, 24 July 1967.

38 BP 8618, Davies to Dummett and Berkin, appendix on dealer market brands, 12 June 1964.

39 SMBP archives, bag 6675, 'Retail brand objectives' file, Brand aims and philosophies, appendix 3, 24 July 1967.

40 BP 65856, Findings of key surveys for Policy Committee, [estimated date, 1961/2]; BP 58965, International information panel minutes, 22 April 1963.

41 BP 46148, Mayhew memorandum, Attachment on five-year plan for BP marketing, 22 March 1966.

42 BP 12786, BP's reputation in the UK, The 1964 benchmark study, 27 July 1965.

43 BP 46148, Armstrong to Tritton, 1964 benchmark study on BP's reputation in the UK, 28 May 1965.

44 *Ibid.*, Armstrong to Nasmyth, 27 September 1965.

45 *Ibid.*, Mayhew, A five-year prestige plan for BP advertising and public relations, 9 September 1965.

46 *Ibid.*, Mullaly to Armstrong, Public relations, BP's reputation in the UK, 2 August 1965.

47 *Ibid.*, Cited by Armstrong, 20 April 1966.

48 *Ibid.*, Armstrong to Mullaly, 2 August 1965.

49 BP 58500, Rees Jenkins to Butcher, Managers course (Stage 1) no. 24, 12 December 1963; BP 113643, Agreement between Shell, Eagle and BP, 27 February 1957.

50 BP 97414, Drake to Gass, 18 May 1954.

51 BP 113028, Record note of meeting held at Shell on 16 February 1955.

52 BP 9199, Snow to chairman, Shell-Mex and BP, 8 March 1960.

53 BP 59586, Unsigned note, Shell sale, 6 October 1964; *ibid.*, Bridgeman to Pattinson, Office note, 12 November 1964; BP 9371, Fraser to Bridgeman, 8 October 1964; *ibid.*, Bean, Crude oil for Shell, 8 December 1964.

54 PRO FO 371/187682, IOD 1511/13, Tomkys minute, 28 March 1966.

55 BP 9371, Bean, Record note, Discussion with McFadzean and Bridges of Shell, 18 July 1966.

56 BP 8618, UK: Dealer motor spirit market, Note of points that arose in a Shell meeting re. Davies' report, 29 May 1964; *ibid.*, Fraser to Dummett *et al.*, Shell-Mex and BP, 3 June 1964; BP 22148, Dummett to Bridgeman, Drake and Fraser, Discussion with Barran, 6 June 1968.

57 BP 118850, Waller to Fraser, Shell-Mex and BP, 29 December 1966; BP 14259, Waller to Fraser, 12 January 1967; *ibid.*, Waller and de Quidt, 'Shotgun', 12 January 1967; *ibid.*, Trade Relations Department, Exercise 'Shotgun', July 1967; BP 22148, Dummett to Bridgeman, Drake and Fraser, Discussion with Barran, 6 June 1968; *ibid.*, Barran to Fraser, 'Operation Doon', 9 July 1968; *ibid.*, Milne to Laidlaw, Product quota review, 10 March 1969; BP 113527, Milne, Product quota review, 13 March 1969; *ibid.*, Milne, Record note, Barran/Dummett/Fraser/Eccles/Milne, Shell Centre on 19 March 1969, Product

quota review, 14 April 1969. On BP's dissatisfaction with the situation in South Africa, see next chapter, section on Africa.

58 BP 63531, Tilleard to Laidlaw, Supermix, 14 December 1964; BP 66570, Retail market development (1965–69), 7 August 1969; SMBP archives, bag 4764, BP retail market file, SMBP press release, BP adopts the mixer pump, 17 October 1966.

59 SMBP archives, bag 4764, BP retail market file, SMBP press release, A national network of check-up centres on BP petrol stations – the first BP Autocare centre opened, 21 April 1966; BP 66570, Retail market development (1965–69), 7 August 1969.

60 BP 13954, Withers to Starling, Ford Motor Company UK, Approval Visco-Static Longlife, 12 February 1964; *ibid.*, Extract from *Financial Times*, 7 March 1964; *ibid.*, Hill to Sanderson, Marketing of BP Visco-static Longlife in the UK, 9 November 1964; *ibid.*, BP premier grade crankcase lubricants, Future marketing policy, 14 December 1965.

61 BP 66570, The BP retail automotive lubricant story, 7 August 1969.

62 BP 13954, BP premier grade crankcase lubricants, Future marketing policy, 14 December 1965.

63 *Ibid.*, Press release, Shell Super Motor Oil, 31 March 1965; *ibid.*, BP premier grade crankcase lubricants, Future marketing policy, 14 December 1965; *ibid.*, Hardman, Notes on a meeting held in BP House to discuss multigrade oil requirements on 15 November 1965; *ibid.*, Caldwell record note, BP UK crankcase lubricants policy, 6 December 1965; BP 45732, Lubricants Branch quarterly marketing bulletin, mid-year 1966.

64 BP 66558, Market intelligence summary, Duckhams, 6 December 1967; *ibid.*, Riddell-Webster to Milne with attachment on Duckhams' performance 1960–68, 25 June 1968; *ibid.*, Draft aide-memoire, 24 October 1969; BP 18921, Extracts from *The Observer*, 12 March 1967 and *Financial Times*, 25 October 1969; BP 18957, Extracts from *Financial Times*, 25 November 1969 and *Garage*, 29 November 1969; BP 22441, Milne, Alexander Duckham & Co – general note, 16 January 1970.

65 BP 112294, Shell-Mex and BP group brand separation – the overall picture and implementation report, appendix A, 7 April 1971.

66 BP 104248, Note for Executive Committee, Shell-Mex and BP brand separation operations, 18 December 1973.

67 BP 22148, PQR meeting at Shell Centre, 22 July 1970.

68 *Ibid.*, Dummett file note on meeting, 11 January 1971; *ibid.*, Robertson to Drake, 1 April 1971; BP 113527, Memorandum of agreement on supply support arrangements, 15 September 1971; BP 51275, BP press release, Separate marketing for Shell and BP in Britain, 19 April 1971; BP 112294, Shell-Mex and BP group brand separation – the overall picture and implementation report, November 1971; BP 56187, Supply support arrangements, 14 April 1971; *ibid.*, Ross to Robertson, Brand separation operations and supply support, 19 December 1972; *ibid.*, Baxenden to Greenborough, 21 December

1973, with attachment on brand separation of operations assets; BP 104248, Note for Executive Committee, Shell-Mex and BP brand separation operations, 18 December 1973.

69 When CFP was formed in 1924, it was privately owned. The state acquired its shareholding in 1931.

70 Leslie E. Grayson, *National Oil Companies* (Chichester, 1981), ch. 3.

71 *Ibid.*, ch. 4.

72 BP 18995, Gregory to Dutton, 13 January 1965; BP 38921, Chenevier, The French oil policy and the Common Market, Management conference, May 1968; BP 56404, Le Mesurier, ERAP, 17 July 1969; BP 40055, Unsigned note on French oil regime, 4 December 1969; BP 8864, Unsigned note on French oil regime, 8 November 1972; *ibid.*, Untitled report on the French oil regime, *c.* May 1974; BP 80365, SFBP strategy reappraisal study, Historical report and analysis, 1971–1973, April 1974.

73 BP 80365, SFBP strategy reappraisal study, Historical report and analysis, 1971–1973, April 1974.

74 BP 60902, Sales promotion, BP marketing management conference, May 1956; BP 46148, BP corporate image study in Europe, September 1971.

75 BP 38921, Chenevier, The French oil policy and the Common Market, Management conference, May 1968; BP 56404, Le Mesurier, ERAP, 17 July 1969; BP 8864, Unsigned note on French oil regime, 8 November 1972.

76 A. E. Safarian, *Multinational Enterprise and Public Policy: A Study of the Industrial Countries* (Aldershot, 1993), pp. 322–8.

77 BP 66136, BP Germany, Three-year plan 1964–66; BP 63551, Wood to Dummett, German national oil companies, 1 February 1966.

78 BP, *Statistical Review of the World Oil Industry*, 1970.

79 BP 62829, Dutton to Bean, Developments in EEC oil policy, 26 April 1967; BP 40055, Dutton to Butcher *et al.*, Anglo-French oil talks, 6 June 1967; BP 62829, Dutton, Common Market oil policy, 18 August 1967; BP 8592, Dutton to Dummett *et al.*, EEC, Visit of Spaak and Directorate of Energy staff, 4 January 1971.

80 Waddams, *Libyan Oil*, p. 76.

81 BP 39939, McLintock, Record note, Bonn meeting between Schiller and Dummett on 17 December 1968.

82 BP 62829, Dutton, Common Market oil policy, 18 August 1967; BP 93093, Bean, General policy, 10 October 1967; BP 40766, Dutton to Bean, 17 November 1967; BP 50302, Planning Committee minutes, 6 and 20 December 1967.

83 BP 38921, Sasse, Germany's views on the Common Market, BP management conference, May 1968; BP 40766, Wright to McLintock, Germany, 22 November 1968; BP 39939, Europe, no. 247 (new series), 8 January 1969; *ibid.*, Extract from *Financial Times*, 14 January 1969; BP 42179, Unsigned note on development of German oil policy, 26 February 1969.

84 BP 8831, Statement by BP to the European Parliamentary Committee on Energy, 28 May 1968.

85 BP 65372, Dutton to Dummett and Fraser, Germany, 31 July 1968.
86 BP 39939, McLintock, Record note, Bonn meeting between Schiller and Dummett on 17 December 1968.
87 The shareholders in Deminex were Gelsenberg, Scholven and Wintershall, 18½ per cent each; Wesseling, 13½ per cent; Deutsche Schachtbau, 10 per cent; Saarbergwerke, 9 per cent; Preussag, 7 per cent; and Deilmann, 5 per cent. See *ibid.*, Unsigned note, Item 3, BP and Germany, 17 February 1969; BP 81328, Dutton to Stratton, Germany, 24 February 1969; *ibid.*, Economic Relations Department, Development of Deminex, 14 August 1969; BP 42179, Unsigned note, Second visit to Bonn on 24 April 1969; *ibid.*, Unsigned note, Deminex, 6 June 1969; BP 63007, Stockwell, Record note, Deminex, 3 July 1969; BP 8567, Unsigned note, Deminex, 30 April 1971.
88 BP 63007, Record note, Meeting with Warner on 29 October 1969.
89 *Ibid.*, Memorandum, Possible participation by Deminex in Iranian Oil Consortium, 4 November 1969; *ibid.*, Note on Deminex, n.d.; *ibid.*, Steel, Record note, Deminex – Iran, 21 January 1970; *ibid.*, Steel to Hayman with attachment on Deminex, 26 February 1970; BP 8567, McLintock, Record note, Deminex – Iran, Visit to Bonn, 11 February 1970.
90 The negotiations are documented in great detail in BP 8567, 11344, 21792, 34661, 42179, 63007 and 81328.
91 BP 61428, Brand image study, Germany, November 1970.
92 BP 19897, BP Germany, Summary of 1969/70 historical analysis, 21 July 1972.
93 *Ibid.*, BP Belgium, Summary of 1969/70 historical analysis.
94 BP 46148, BP corporate image study in Europe, September 1971.
95 BP 19897, Downstream profitability study – BP Holland, Summary of 1969/70 historical analysis.
96 BP 46148, BP corporate image study in Europe, September 1971.
97 *Ibid.*
98 BP 6960, ASP regional seminars, May 1968.
99 Bamberg, *British Petroleum, Volume II*, p. 116.
100 *Ibid.*, p. 301; BP 60537, Outline of BP operations in Italy, 6 November 1959; BP 62694, BP Italiana – 3-year integrated plan, June 1962.
101 BP 11362, Downstream profitability study, Italy, April 1973.
102 BP 62694, BP Italiana – 3-year integrated plan, June 1962.
103 BP 2912, Extract from *Financial Times*, 2 April 1973; BP 106285, Extract from *Financial Times*, 26 May 1973.
104 BP 19897, Downstream profitability study – Italy, Summary of 1969/70 historical analysis, 11 July 1972; BP 11362, Downstream profitability study, Italy, April 1973. On market shares, see BP 119420, Oil companies sales and market shares, 1969.
105 BP 8449, Minutes of BP Trading Executive Committee meeting on 7 January 1972.
106 BP 11169, Brief for BP Trading Executive Committee, BP Italiana, pt. 2, 4 May

1973; BP 19897, Downstream profitability study – Italy, Summary of 1969/70 historical analysis, 11 July 1972.

107 BP 11169, Brief for BP Trading Executive Committee, BP Italiana, pt. 2, 4 May 1973.

108 BP 56404, Central Planning Department, ENI, November 1969.

109 BP 21717, Unsigned draft on Europe, 9 November 1972.

110 BP 8449, Minutes of BP Trading Executive Committee meeting on 7 January 1972.

111 BP 21717, Unsigned draft on Europe, 9 November 1972.

112 See, for example, William Carley, 'Oiling the wheels of Rome', *The Guardian*, 21 May 1975; Geoffrey Hodgson, 'The secret power of oil money', *Sunday Times*, 11 April 1976; David Norris, 'BP and Shell in Italian bribes enquiry', *Sunday Telegraph*, 11 April 1976.

113 BP 6076, Adam, Meeting with SEC Washington on 9 April 1976; *ibid.*, Steel, Sunday Times/Granada/Italy, 12 April 1976. See also Special Review Committee of the Board of Directors of Gulf Oil Corporation, *Report* (In the United States District Court of Columbia, December 1975), pp. 122–60.

114 BP 6076, Commission of enquiry for proceeding of indictment. Minutes of interrogation of accused or suspect of a crime, Scarito on 18 April 1974, Boxshall on 16 May 1974, and Holzer on 16 May 1974.

115 *Ibid.*, Adam, Meeting with SEC Washington on 9 April 1976.

116 BP 2912, Extract from *Petroleum Press Service*, March 1973, p. 106; *ibid.*, Extract from *Financial Times*, 2 April 1973; BP 106285, Extract from *Petroleum Times*, 13 July 1973, p. 13; *ibid.*, Extract from *Financial Times*, 26 May 1973; *ibid.*, Extract from *The Guardian*, 7 June 1973; BP 8692, Italian Senate fails to vote on oil product tax changes, 2 February 1973.

117 BP 11169, Boxshall to Executive Committee, BP Italiana, 7 May 1973; BP 61340, Minutes of BP Trading Executive Committee meetings on 8 and 18 May 1973; BP 6076, Laidlaw, Memorandum for board meeting, Proposed sale of Group shareholding in BP Italiana, 17 May 1973; BP 57527, Laidlaw, AGIP, 21 May 1973; *ibid.*, Extract from *Petroleum Intelligence Weekly*, 4 June 1973; BP 106285, BP press release, Sale of BP Italiana, 25 May 1973; *ibid.*, BP press release, Sale of BP Italiana completed, 2 July 1973.

118 BP 57527, Extract from *Petroleum Intelligence Weekly*, 4 June 1973; BP 41306, Extracts from *The Times*, 16 October 1973 and *Financial Times*, 17 October 1973; Grayson, *National Oil Companies*, p. 114; US Federal Energy Administration, *Oil Companies and Foreign Governments*, pp. 86–8.

10   . . . AND MORE OUTLETS

1 BP 102799, Management report for board meeting on 29 April 1954.

2 BP 30580, Monthly bunker statements.

3 T. A. B. Corley, *A History of the Burmah Oil Company: Volume II, 1924–1966* (London, 1988), p. 25; J. H. Bamberg, *The History of The British Petroleum*

Company: Volume II, The Anglo-Iranian Years, 1928–1954 (Cambridge, 1994), p. 107; BP 59545, Burmah-Shell Agreements, 30 March 1955.

4 BP 58499, Agreement between the Consolidated Petroleum Company and the Asiatic Petroleum Company [Royal Dutch-Shell], 22 July 1931; T. H. Bingham and S. M. Gray, *Report on the Supply of Petroleum Products to Rhodesia* (HMSO, 1975), p. 8.

5 For an excellent account of British decolonisation see John Darwin, *Britain and Decolonisation: The Retreat from Empire in the Postwar World* (London, 1988); for a more comparative survey of European decolonisation, R. F. Holland, *European Decolonization, 1918–1981: An Introductory Survey* (London, 1985).

6 Hermann Kulke and Dietmar Rothermund, *A History of India*, revised edn (London, 1990), ch. 7; B. R. Tomlinson, *The Economy of Modern India, 1860–1970*, vol. III of The New Cambridge History of India (Cambridge, 1993), ch. 4; Biplab Dasgupta, *The Oil Industry in India* (London, 1971); Michael Tanzer, *The Political Economy of International Oil and the Underdeveloped Countries* (Boston, Mass., 1969), chs. 13–20; Michael Kidron, *Foreign Investments in India* (London, 1965), pp. 90, 97–103, 166–75 and 191–4; Dennis J. Encarnation, *Dislodging Multinationals: India's Strategy in Comparative Perspective* (Ithaca, N.Y., and London, 1989); Corley, *Burmah Oil, Volume II*, pp. 308–29; BP 36345, Note on Burmah-Shell Oil Storage and Distributing Companies of India and Pakistan in folder on Consolidated accounts, 1957.

7 Corley, *Burmah Oil, Volume II*, pp. 229–40; BP 50252, AP-Dow daily news précis, 2 September 1969; BP 113332, Burmah and Conoco, 5 March 1971.

8 BP 50252, Burmah-Shell new deal, 14 August 1962; *ibid.*, Burmah-Shell arrangement w.e.f. 1 January 1962, 11 June 1964; BP 94804, Burmah Shell arrangements from 1.1.62 (The new deal), 1 July 1964; Corley, *Burmah Oil, Volume II*, pp. 271–2 and 289.

9 See, for example, BP 8523, Fraser to R. P. Smith, 9 February 1965; BP 37644, R. P. Smith to Fraser, 6 April 1965; *ibid.*, Fraser to Bridgeman, Burmah, 27 February 1968; *ibid.*, Burmah agreements, 19 February 1974; BP 43604, Hardy to Bean, Burmah Oil, 18 August 1967; *ibid.*, Forsyth to Sandford, 17 November 1967; *ibid.*, Dufficy, BOC, 16 November 1967; *ibid.*, Forsyth to Dufficy, Burmah – right through sharing, 16 November 1967; *ibid.*, Burmah – right through sharing, 12 January 1968; BP 56403, Sandford to Robertson, 29 June 1971; BP 8280, Sandford to Robertson, Burmah Oil Company – India and Pakistan, 13 July 1971; *ibid.*, Robertson to Laidlaw, Burmah Oil Company – India and Pakistan, 19 July 1971; *ibid.*, Sandford to Dummett, Burmah Oil Company – India and Pakistan, 17 September 1971.

10 BP 104260, Termination of agreements with the Burmah Oil Company, 21 September 1976.

11 BP 36345, Joint marketing areas – Consolidated, 29 August 1957; BP 58499, Rees Jenkins to Dummett and Davies, The Consolidated Petroleum Company, 25 November 1957; *ibid.*, Unsigned draft to the chairman [estimated date

December 1957]; *ibid.*, Rees Jenkins to Dummett, Consolidated Petroleum, 10 January 1958.

12 BP 58499, Agreement between the Consolidated Petroleum Company and the Asiatic Petroleum Company, 22 July 1931; *ibid.*, Sarre to Rees Jenkins, 2 January 1958; *ibid.*, Draft, Consolidated management, 10 March 1960.

13 BP 92779, Gordon to von Stahl and others, Atlantic deal, 17 August 1954; BP 29589, Visit to Atlantic Refining Company of Africa on 11–26 May 1959, 8 June 1959.

14 Philip Curtin, Steven Feierman, Leonard Thompson and Jan Vansina, *African History from Earliest Time to Independence*, 2nd edn (London, 1995), pp. 437–8 and 448–50; Bingham and Gray, *Report*, p. 10; BP 19849, Southern Africa, Area review, 1 November 1976, attached to Savage to Gregory and Belgrave, South Africa – 1977 budget, 2 November 1976.

15 BP 94868, BP in South Africa with particular reference to BP Southern Africa (Pty) Ltd, 24 August 1970; Bingham and Gray, *Report*, p. 8.

16 BP 64042, Dummett, Record note, 13 August 1958; *ibid.*, Dummett to the chairman and others, Consolidated management, 3 October 1958; BP 58499, Draft, Consolidated management, 10 March 1960.

17 Bingham and Gray, *Report*, p. 12; BP 118455, Note to Robertson, Southern Africa, 4 February 1975; BP 8647, Paper on South Africa for Strategic Planning Committee, June 1973.

18 BP 63510, Participation with governments, Interim report, attachment 1, 3 July 1970.

19 For a detailed account, see Bingham and Gray's *Report*.

20 BP 94655, Secretarial minute, The Consolidated Petroleum Company Ltd, January 1981; *ibid.*, Gillam, The Consolidated area, 31 December 1981.

21 Bamberg, *British Petroleum, Volume II*, pp. 6 and 297.

22 AIOC/BP, *Annual Reports and Accounts*, 1952 and 1957.

23 BP, *Annual Report and Accounts*, 1955.

24 Laura E. Hein, *Fueling Growth: The Energy Revolution and Economic Policy in Postwar Japan* (Cambridge, Mass., 1990), chs. 2, 3 and 7.

25 BP 9249, Strictly confidential memorandum, 19 December 1949; BP 106234, Board minutes, 20 December 1949; BP 113811C, Agreement between AIOC and Shell, 31 January 1951.

26 BP 113811, Addison to Phillips, Shell long term agreement, 17 March 1952; *ibid.*, Addison to Snow, 10 September 1952; BP 113028, Linklaters and Paines, Opinion on Shell long-term agreement, 1954; BP 59345, Gass to chairman and deputy chairman with attachments, 25 January 1955; *ibid.*, Shell Far Eastern Agreement, Extract from memorandum dated 17 April 1954 from Drake to Gass; BP 52928, Record note of meeting held at Shell Company's offices on 16 February 1955; BP 113555, Shell long-term agreement – frustration, 14 April 1955; *ibid.*, Prince to Snow, Shell Far Eastern Agreement – date of frustration, 17 February 1956.

27 Daniel Yergin, *The Prize: The Epic Quest for Oil, Money, and Power* (New York, 1991), pp. 545–6; John G. Clark, *The Political Economy of World Energy: A Twentieth Century Perspective* (Hemel Hempstead, 1990), p. 189; BP, *Statistical Review of the World Oil Industry*, 1970; BP 35254, Flower, Japan, 3 October 1975.

28 Hein, *Fueling Growth*, ch. 10; Yergin, *The Prize*, pp. 545–6.

29 BP 43629, Laidlaw, Bygrave and Grey, Report on visit to Japan, 3–21 October 1955; BP 23992, Laidlaw and Bygrave, Report on visit to Japan, 29 April – 14 May 1957; BP 45754, O'Meara, Report on visit to Japan, 13 November – 2 December 1958; BP 62986, Adam and Gresham, Report on visit to Japan, November 1964; *ibid.*, Central Planning Department, BP and Japan, April 1970; BP 101576, Le Mesurier to Greaves, Japan, 6 January 1969.

30 BP 101576, Adam to members of Executive Committee, Japan – future strategy, 11 September 1973.

31 BP 8254, BP press release, BP crude oil and shipping deal in Japan, 22 May 1967; *ibid.*, Stratton, Aide-memoire, Kyushu, 6 November 1967; *ibid.*, Extracts from BP board minutes on 11 January and 22 March 1962; *ibid.*, Stratton to the chairman, Japan, 4 November 1969; *ibid.*, St Clair Erskine, Tanaka, 14 November 1969; *ibid.*, Stratton, Operation Dahlia, 17 November 1969; *ibid.*, St Clair Erskine to Dummett, Operation Dahlia, 30 January 1970; BP 38921, Stratton, Crude oil sales, Management conference, May 1968; BP 46100, Notes on chairman's visit to Japan, 28 September – 2 October 1970.

32 BP 62986, Supply and Development, Japan Arabian Oil, 23 March 1961; *ibid.*, Note for the chairman, Arabian Oil Company, 2 November 1961; BP 101576, Extract from *Japan Petroleum Weekly*, Khafji crude to lose its preference status, 9 February 1970.

33 BP 49504, Extracts from *Japan Petroleum Weekly*, 1 March, 31 May and 13 September 1971.

34 *Ibid.*, St Clair Erskine, Record note, JPDC, 15 December 1971.

35 For summaries of the negotiations, the terms and the completion, see BP 78457, Hart to the chairman and others, Japinex, 15 November 1972; BP 34662, BP press release, BP negotiating sale of share of its interest in Abu Dhabi Marine Areas to Japanese group, 22 December 1971; *ibid.*, BP press release, BP signs $780 million agreement with Japanese group, 27 December 1972; BP 101576, Johnson to Russett, Chairman's visit to Japan, Appendix 1 to briefing paper, 2 February 1973; BP 90232, Draft press release, Sale of ADMA interest completed, 14 February 1973; *ibid.*, Background note, 14 February 1973; *ibid.*, Hart to Laidlaw attaching note on Japinex, 26 January 1973.

36 BP 8499, Fraser, Memorandum no. 34, Maruzen Toyo refinery, Singapore, 16 April 1964; *ibid.*, Press release, BP to buy refinery in Singapore, May 1964; *ibid.*, SE Asia, Establishment of a marketing organisation in Malaysia, 9 September 1964; *ibid.*, Dummett, Memorandum no. 104, BP Malaysia Ltd, 15 October 1964; *ibid.*, Down, The Far East, 20 December 1966; BP 37613,

Malaysia and Singapore three-year plan 1967–1969, 25 April 1967; BP 40457, Malaysia and Singapore, September 1968; BP 23918, Notes on Malaysia and Singapore, 1 January 1973.

37 BP 9247, Unsigned notes on the oil industry in Canada and the USA, March 1957; BP 33500, Brochure on BP in Canada, undated; BP 54014, Introduction to Montreal refinery, undated; BP 72172, BP Refinery Canada Ltd, Prospectus for 5½% first mortgage sinking fund bonds, 1959; BP 45135, The British Petroleum Company of Canada Ltd, Annual review for 1960; BP 45125, The British Petroleum Company of Canada Ltd, Annual review for 1963.

38 BP 45161, The British Petroleum Company of Canada Ltd, Annual review for 1964; Gene T. Kinney, 'BP pegs Canada as market of future', *Oil and Gas Journal*, 27 July 1964; BP 27323, BP Canada, Review of operations, 1969.

39 BP 37354, BP Canada Ltd, Annual report, 1971; BP, *Annual Report and Accounts*, 1971; 'Silver story: BP celebrates 25 years in Canada', *BP Shield*, December 1978.

40 On the story of the pipeline, see James P. Roscow, *800 Miles to Valdez: The Building of the Alaska Pipeline* (Englewood Cliffs, N.J., 1977).

41 BP 10294, Adam, Draft affidavit, In the Superior Court for the State of Alaska Third Judicial District at Anchorage, May 1982.

42 Bamberg interview with Adam, 12 August 1986.

43 Bamberg interview with Belgrave, 1 October 1986.

44 Bamberg interview with Steel, 21 October 1986.

45 BP 10294, Adam, Draft affidavit, In the Superior Court for the State of Alaska Third Judicial District at Anchorage, May 1982.

46 Bamberg interview with Belgrave, 1 October 1986; BP 27809, Frampton to Belgrave, Notes for the Sohio presentation to the main board on 19 February 1976; Matthews, 'Target America', *BP Shield*, May 1978; Harvard Business School, 'British Petroleum Company Limited' (Case Clearing House of Great Britain and Ireland, Cranfield Institute of Technology, Bedford, 1973); *Petroleum Times*, 22 November 1968 and 17 and 31 January, 28 February, 14 March and 10 October 1969; *The Economist*, 8 March and 10 May 1969.

47 BP 10379, The Standard Oil Company (Ohio) approved long-range corporate plan 1974–83; BP 10408, Down to Drake, BP/Sohio, 21 June 1972; *The Economist*, 8 March and 10 May 1969.

48 BP 10408, Strathalmond to Drake, Sohio, 4 May 1972; *ibid.*, Hardcastle, Sohio, 16 June 1972; BP 10406, Walters, Sohio, 2 September 1971.

49 Geoffrey Jones, *The Evolution of International Business: An Introduction* (London, 1996), esp. pp. 46–8; Mira Wilkins, *The Maturing of Multinational Enterprise: American Business Abroad from 1914 to 1970* (Cambridge, Mass., 1974), chs. 12–14.

50 Christopher Tugendhat, 'BP joins the giants', *BP Shield*, December 1968.

51 BP 27809, Frampton to Belgrave, Notes for the Sohio presentation to the main board on 19 February 1976.

52 James P. Roscow, 'Sohio today', *BP Shield,* May 1978; P. D. Phillips, J. C. Groth and R. M. Richards, 'Financing the Alaskan project: The experience at Sohio' (Financial Management Association, 1979); Harvard Business School, 'British Petroleum'.
53 Phillips, Groth and Richards, 'Financing the Alaskan project'.
54 Sohio, *Annual Report and Accounts,* 1968; *The Economist,* 7 June 1969.
55 BP 10513, Adam to Steel, The future of Sohio and its relationship with BP, 18 August 1975.
56 BP 11810, Public Affairs and Information Department, Internal briefing note, BP and Sohio, 27 April 1982; BP 10394, Agreement, The Standard Oil Company (Ohio), The British Petroleum Company Limited and British Petroleum (Overzee) NV, 7 August 1969; BP 10386, Draft notice of annual meeting and proxy statement, Standard Oil Company, 19 February 1973; BP 10384, Draft preliminary prospectus, Sohio Pipeline Company issue of debentures, 20 September 1971.
57 BP 10394, Memorandum of Intent, Standard Oil Company and BP Oil Corporation, 2 June 1969; BP 10394, Agreement, The Standard Oil Company (Ohio), The British Petroleum Company Limited and British Petroleum (Overzee) NV, 7 August 1969.
58 *Ibid.*; BP 81276, Operating agreement for Prudhoe properties between BP Oil Corporation & BP Alaska Inc., 1 August 1969.
59 Bamberg interview with Adam, 12 August 1986. The agreement was unwritten for antitrust reasons.
60 BP 10384, Draft preliminary prospectus, Sohio Pipeline Company issue of debentures, 20 September 1971; Sohio, *Annual Report and Accounts,* 1969.
61 BP 10384, Draft preliminary prospectus, Sohio Pipeline Company issue of debentures, 20 September 1971; BP 27809, Frampton to Belgrave, Notes for the Sohio presentation to the main board on 19 February 1976.
62 Bamberg interview with Adam, 12 August 1986.
63 BP 10394, BP press release, 1 January 1970.

11   REFINING AND SHIPPING

1 Clive Callow, 'Blueprint for the retail network', *BP Shield,* February 1971.
2 Calculated from BP, *Statistical Review of the World Oil Industry,* 1960 and 1971–5.
3 AIOC, *Annual Report and Accounts,* 1950; Walter E. Skinner, *Oil and Petroleum Year Book,* 1950.
4 BP, *Annual Report and Accounts,* 1973; Walter R. Skinner, *Oil and Gas International Year Book,* 1973.
5 *BP Shield,* March 1967 and April 1968; BP, *Annual Report and Accounts,* 1970.
6 *BP Shield,* June, September and October 1967, July 1968, October 1969, February and June 1970, September 1971, October 1972 and November 1973; BP, *Annual Reports and Accounts,* 1970–3.

7 BP 118832, Hooper, History of BP research, vol. 7(a), 1946–1954, section (3), pp. 19–23.

8 Mari E. W. Williams, 'Choices in oil refining: The case of BP, 1900–1960', *Business History* 26 (1984), 307–28; BP 118835, Hooper, History of BP research, vol. 8, 1955–1973, section (1), pp. 23–35.

9 Barker in *Petroleum Times*, 31 July 1970.

10 BP 56396, Minutes of BP Trading executive committee meeting on 22 January 1971.

11 Stephen Howarth, *Sea Shell: The Story of Shell's British Tanker Fleets, 1892–1992* (London, 1992), p. 16.

12 BP 118237, Platt, History of the BP Tanker Company, p. 12.

13 Howarth, *Sea Shell*, pp. 89–91; Norman L. Middlemiss, *The British Tankers*, revised edn (Newcastle, 1995), pp. 8–13; R. F. Macleod, 'The last of the Twelves', *BP Shield*, August 1967.

14 Macleod, 'The last of the Twelves', *BP Shield*, August 1967. See also Middlemiss, *British Tankers*, pp. 36–7.

15 G. A. B. King, *A Love of Ships* (Hampshire, 1991), p. 142. George King joined the British Tanker Company as a junior officer after the war and went on to become managing director of the BP Tanker Company, as the British Tanker Company was renamed in 1956. His autobiography, *A Love of Ships*, gives a personal account of his time with BP in chs. 6–10. On conditions aboard a typical tanker, see also Middlemiss, *British Tankers*, p. 35.

16 J. H. Bamberg, *The History of The British Petroleum Company: Volume II, The Anglo-Iranian Years, 1928–1954* (Cambridge, 1994), pp. 287–90.

17 BP 118237, Platt, History of the BP Tanker Company, p. 245 and table 6.4; Middlemiss, *British Tankers*, p. 38.

18 Howarth, *Sea Shell*, p. 127.

19 BP 118237, Platt, History of the BP Tanker Company, p. 245.

20 *Ibid.*, p. 247 and table 6.4; Middlemiss, *British Tankers*, pp. 39–40 and 49.

21 Committee of Enquiry into Shipping (Rochdale Committee), *Report* (Cmnd 4337, 1970), chs. 2 and 9; Michael Davies, *Belief in the Sea: State Encouragement of British Merchant Shipping and Shipbuilding* (London, 1992), ch. 8 (p. 163, 'distaste').

22 Middlemiss, *British Tankers*, p. 50; King, *Love of Ships*, p. 225 ('handsome').

23 Davies, *Belief in the Sea*, p. 166.

24 See sources for figure 11.2.

25 BP 118237, Platt, History of the BP Tanker Company, p. 264.

26 Howarth, *Sea Shell*, p. 143; BP 118237, Platt, History of the BP Tanker Company, p. 267.

27 BP 118237, Platt, History of the BP Tanker Company, pp. 282 and 292.

28 *Ibid.*, p. 309; Howarth, *Sea Shell*, p. 149.

29 Middlemiss, *British Tankers*, p. 52; R. F. Macleod, 'Giant strides to giant tankers', *BP Shield*, May 1970.

30 Rochdale Committee, *Report*, p. 155.
31 BP, *Annual Report and Accounts*, 1966.
32 BP 8254, BP press release, BP crude oil and shipping deal with Japan, 22 May 1967.
33 BP 118237, Platt, History of the BP Tanker Company, p. 311; Middlemiss, *British Tankers*, p. 54.
34 BP 118237, Platt, History of the BP Tanker Company, pp. 376–7 and 445–6.
35 *Ibid.*, appendices 1 and 3.

12 FINANCIAL STRAINS

1 BP, *Annual Reports and Accounts*, 1961–3.
2 BP, *Annual Reports and Accounts*, 1963–5.
3 BP 37829, BP rights issue, Record of meeting in Figgures' room on 3 December 1965; BP 8402, Searle, Group accounts and finance, BP management conference, May 1966.
4 BP 37659, Down to Bridgeman, 22 February and 24 June 1965.
5 *Ibid.*, Mitchell, Review of capital expenditure, Memorandum to Management Committee, 6 July 1965; *ibid.*, Mitchell to Drake *et al.*, Capital expenditure, 29 July 1965; *ibid.*, Banks to Howes *et al.*, Capital expenditure plans, 27 August 1965; BP 42678, Searle to Banks, Finance, 6 August 1965; *ibid.*, Mitchell to managing directors *et al.*, Group capital expenditure plans, 22 September 1965; BP 37829, Sweet-Escott, Group capital expenditure outside Sterling Area, Note on meeting with Owen of the Treasury on 7 September 1965; BP 23645, *passim*.
6 BP 54428, Unattributed record note of meeting at the Treasury on 16 July 1965; BP 37829, *passim*.
7 R. W. Ferrier, *The History of The British Petroleum Company: Volume I, The Developing Years, 1901–1932* (Cambridge, 1982), pp. 160 and 461.
8 J. H. Bamberg, *The History of The British Petroleum Company: Volume II, The Anglo-Iranian Years, 1928–1954* (Cambridge, 1994), pp. 301–7. For data on the breakdown of sales between crude oil and products from 1946 to 1950, see BP 102729, 102738, 102749, 102760, 102771, 102796, 102810, 102837, 102848, 102860, 102866, 102873, 102875 and 102879, Management reports for board meetings, 1946–60.
9 BP, *Annual Report and Accounts*, 1961.
10 BP 8384, Central Planning Department, Long term policy assessment, 10 November 1966.
11 BP 102882, 102885, 102887, 102890, 102893, 102896, 102898, 102907, 102910, 102913, 102915, 102918, 102921, 102924, 102926, 102940, 102970, and 102982, Management reports for board meetings, 1960–6.
12 BP 37661, Mitchell, Long term supply review, 6 April 1966.
13 BP 8384, Central Planning Department, Long term policy assessment, 10 November 1966.

14 Data in BP 102994, 103016, 103028, 103041 and 103052, Management reports for board meetings, 1966–71.
15 BP 38921, Stratton, Crude oil sales, BP management conference, May 1968,
16 BP 103052, Management report for board meeting, January 1971.
17 BP, *Annual Report and Accounts*, 1961.
18 The Arthur D. Little study was conducted by Francisco R. Parra, who had been employed from 1951 to 1960 as an economic analyst with the Creole Petroleum Corporation, Standard Oil (NJ)'s Venezuelan subsidiary. In 1960 he joined Arthur D. Little as petroleum economist, and in 1962 he joined OPEC as economic adviser. He later became Chief of the Economics Department of OPEC, and, later still, OPEC's Secretary General. He is, at the time of writing, drafting a history of international oil, and has generously showed me copies of successive drafts, including draft chapter 8 on 'Enter OPEC: the early years, 1960–1968' (September 1998). This includes details of Arthur D. Little's study on pp. 10–11 and in appendix 2, pp. 27–8.
19 BP 59340, Sweet-Escott to Bridgeman, 11 December 1961 and 9 January 1962.
20 BP 62346, Unattributed draft discussion paper on Haliq's paper, n.d. [*c.* autumn 1958]; *ibid.*, Belgrave to Drake, Arab League Oil Conference, 16 January 1959; BP 58772, Barker to Bridgeman, 10 March 1959; BP 4811, Duff to Ballou *et al.* with attached memorandum, 28 October 1965; PRO POWE 33/2469, Williams to Stock, 11 and 13 March 1959. The first FNCB study was published as William S. Evans, 'Petroleum in the Eastern Hemisphere' (FNCB, 1959).
21 Edward Symonds, 'Life with a world oil surplus' (FNCB, 1960); Symonds, 'Oil prospects and profits in the Eastern Hemisphere' (FNCB, 1961); Symonds, 'Oil advances in the Eastern Hemisphere' (FNCB, 1962); Symonds, 'Financing oil expansion in the development decade' (FNCB, 1962); Symonds, 'Eastern hemisphere petroleum – another year's progress analysed' (FNCB, 1963).
22 Francisco R. Parra, 'The Eastern Hemisphere – oil profits in focus', *Middle East Economic Survey*, 6 (1963).
23 There is a large collection of documents, too numerous to reference individually, in the BP archive on the downstream profitability studies in the second half of the 1960s. See files BP 10207, 21710, 21711, 21712, 21715, 21723, 37661, 50302, 53554 and 54428.
24 BP, *Annual Reports and Accounts*, 1966–9.
25 BP 40561, Seven majors in the sixties.
26 *Ibid.*
27 BP 44312, Paper on 'Capital requirements in the international petroleum industry' enclosed with Whittlesey to operating companies, 17 March 1971.
28 BP 40561, Seven majors in the sixties.
29 Bamberg, *British Petroleum, Volume II*, p. 509; BP 38921, Searle, Group accounts and finance, BP management conference, May 1968; BP 8831, Searle, Group finance and accounts, BP management conference, June 1970; BP 37828, BP issue of 23,922,954 ordinary shares, Offer document, 13 October 1971.
30 John Chown, *The Corporation Tax – A Closer Look* (London, 1965); BP 18378,

Press extracts; BP 8414, *passim* (for BP's representations to the government on the proposed changes); *Parl. Debates*, House of Commons, vol. 712, col. 210, 10 May 1965 and vol. 716, col. 922, 15 July 1965; BP 8402, Searle, Group accounts and finance, BP management conference, May 1966; BP 8831, Searle, Group finance and accounts, BP management conference, June 1970; BP 37828, BP issue of 23,922,954 ordinary shares, Offer document, 13 October 1971.

31 BP, *Annual Report and Accounts*, 1972; *The Economist*, 23 March 1972.
32 BP, *Annual Report and Accounts*, 1970.
33 BP, *Annual Reports and Accounts*, 1966–9.
34 BP 93538, Draft Treasury press announcement, 2 January 1967.
35 At Callaghan's request, the Treasury was given the right of first refusal to buy any BP shares which Distillers subsequently decided to sell. In the event, the government did not exercise its right to buy when Distillers offered it 10 million BP shares in July 1967. See BP 93538, Adam, Record note of meeting at the Treasury on 14 October 1966; *ibid.*, Down, Record notes (VII) and (VIII), 28 October and 2 November 1966; *ibid.*, Exchange of letters between Armstrong and Cumming, 3 January 1967. The letters were published in *Parl. Debates*, House of Commons, vol. 739, Written answers, cols. 222–3, 24 January 1967. On Distillers' offer of BP shares, see press extracts in BP 113720.
36 BP, *Annual Report and Accounts*, 1969.
37 BP 37828, *passim.*
38 BP, *Annual Report and Accounts*, 1971.
39 BP, *Annual Reports and Accounts*, 1969–71.
40 BP, *Annual Report and Accounts*, 1972.

13   THE MANAGERIAL HIERARCHY

1 BP 106219, Phillips, Discussion with Fraser, 26 March 1969.
2 *Ibid.*, Phillips and Burns, Interview with Down, 16 July 1969.
3 *Ibid.*, Phillips and Burns, Interview with Drake, 18 September 1969.
4 BP staff file on Steel, Billy Fraser's comments in UK staff confidential report, 1956.
5 BP staff file on Drake, Mylles to Lloyd, 9 June 1944.
6 BP staff file on Drake, Various correspondence.
7 Bamberg interview with Drake, 3 July 1990; personal communication Brackley to Bamberg, 25 March 1999.
8 PRO FO 371/187683, Tomkys minute, 1 September 1966; *ibid.*, Pitblado to Figgures, 14 September 1966.
9 BP 106219, Phillips and Burns, Interview with Drake, 18 September 1969.
10 BP staff file on Pennell. See also his obituary in *BP Shield International*, 1982.
11 BP staff file on Laidlaw.
12 Walter Goldsmith and Berry Ritchie, *The New Elite: Britain's Top Executives* (London, 1987), pp. 147–60.
13 BP 106299, Head office staff strengths.

14 See, generally, Staff Department files in BP Archive.

15 BP 106219, Phillips, Discussion with Down, 9 April 1969.

16 *Ibid.*, Phillips, Discussion with Fraser, 26 March 1969.

17 BP 110817, Bridgeman address to staff managers conference, 1963.

18 BP Archive shelf files on organisation.

19 BP 19056, Study of staff preoccupations in BP, January 1966.

20 BP 34714, Design research unit, Britannic House, Moor Lane, 28 June 1967.

21 BP 66156, Luncheon club management committee minutes, 5 July 1967.

22 BP 12986, Managers meeting, Staff administration, 7 November 1967.

23 BP Archive shelf files on organisation; BP 36900, Laidlaw, Terms of reference for organisational study, 7 March 1966; BP 114268, Manpower Study Group, Interim report, 5 November 1969.

24 BP 36900, Chenevier, Staff productivity conditions, Impressions on an experiment, June 1967.

25 BP 56404, Cruikshank (senior vice-president and director of Esso Europe), Organizing for change: Standard Oil Company (New Jersey), 1969 [his emphasis in the quotation]; BP 65372, Waller, Reorganisation of Jersey, 7 January 1966.

26 BP 8270, Laidlaw, Shell reorganisation, 9 June 1966; Graham Turner, *Business in Britain*, revised edn (Harmondsworth, 1971), esp. pp. 106 ('helicopter view'), 113 and 116.

27 BP 106219, Phillips and Burns, Interview with Down, 16 July 1969.

28 BP 93823, Drake, Manpower study project, 19 November 1968. On the genesis of the Manpower Study Group see, generally, documents in BP 93502. On its full terms of reference, BP 114268, Terms of reference and membership of the Manpower Study Group, appendix 1, 7 January 1969.

29 BP 106219, Phillips and Burns, Interview with Drake, 18 September 1969.

30 BP 114268, MSG interim report, 5 November 1969.

31 BP 17512, Guide to reorganisation, October 1970.

32 R. W. Ferrier, *The History of The British Petroleum Company: Volume I, The Developing Years, 1901–1932* (Cambridge, 1982), pp. 129–53, 306–09 and 316–19.

33 BP 106299, Head office staff strengths.

14   ALLIANCES IN PETROCHEMICALS

1 On these developments see, for example, D. W. F. Hardie and J. D. Pratt, *A History of the Modern British Chemical Industry* (Oxford, 1966), esp. pp. 78, 189–90, 195–7, 203–07; W. J. Reader, *Imperial Chemical Industries: A History, Volume II, The First Quarter-Century, 1926–1952* (Oxford, 1975), ch. 19.

2 Peter H. Spitz, *Petrochemicals: The Rise of an Industry* (New York, 1988), ch. 2; Keith Chapman, *The International Petrochemical Industry* (Oxford, 1991), ch. 3.

3 Spitz, *Petrochemicals*, ch. 1. For more detailed studies of IG Farbenindustrie,

see P. Hayes, *Industry and Ideology: IG Farben in the Nazi Era* (Cambridge, 1987) and J. Borkin, *The Crime and Punishment of IG Farben* (London, 1979).

4   Reader, *ICI, Volume II*, especially chs. 17–21.

5   This summary of DCL's entry into chemicals and plastics is drawn from Ronald Weir, *The History of the Distillers Company, 1877–1939* (Oxford, 1995), chs. 16–19; Hardie and Pratt, *Modern Chemical Industry*, pp. 236 and 288; Reader, *ICI, Volume II*, pp. 46–54 and 322–8; BP 111107, DCL, The industrial activities of the Distillers Company Limited, 1877–1957; BP 112269, Unsigned note, Historical record of plastics companies, n.d.; BP 111183, The DCL chemicals and plastics group, Press guide, 2nd edn, June 1965; BP 76464, Review of the chemicals interests of BP and DCL, n.d.

6   J. H. Bamberg, *The History of The British Petroleum Company: Volume II, The Anglo-Iranian Years, 1928–1954* (Cambridge, 1994), p. 193.

7   BP 118829, Hooper, History of BP research, vol. 4, 1930–1934, Gas and gas utilisation, p. 42.

8   The work at Sunbury was recorded in monthly reports entitled 'Special products' or 'Utilisation of cracker gases' in BP Sunbury technical records TB 311. For a brief report, see BP 88334, Thole, Birch and Scott, Special products, 8 August 1931. DCL's interest in this field is mentioned in BP 111183, The DCL chemicals and plastics group, Press guide, 2nd edn, June 1965. A fuller account of the early history of DCL's research department can be found in the series of articles by J. E. Youell in *DCL Gazette*, 1962–3.

9   BP 21977, Unsigned report, Cracker gases – their commercial use, March 1931; Weir, *Distillers*, p. 325.

10 Reader, *ICI, Volume II*, p. 322.

11 *Ibid.* Also, BP 40871, Norris, Notes on meeting held in Britannic House on 10 March 1943.

12 Weir, *Distillers*, p. 370.

13 DCL Epsom Research Centre records 76/18/72, Minutes of special meeting of the chemical and industrial products committee, 8 January 1942. For reference to the talks, see also BP 72266, Spearing to Fraser, 27 November 1946.

14 BP 112109, Coxon, Note on meeting with ICI directorate, 19 February 1943. An account of the meeting and the subsequent negotiations between the two companies can be found in Reader, *ICI, Volume II*, ch. 21.

15 Detailed papers on the meetings of the Collaboration Committee and the draft basis of collaboration are in BP 40871.

16 A series of documents on this scheme can be found in BP 71041.

17 *Ibid.*, Nicholson to Fraser, 11 November 1943; BP 106233, Minutes of AIOC board meeting on 30 November 1943.

18 BP 71042, Jameson to Crichton, 30 August 1944; BP 70641, Gordon, Petroleum Chemical Developments Ltd, Report on preliminary investigations, 25 October 1944; *ibid.*, Gordon, Petroleum Chemical Development, 24 January 1945.

19 BP 70641, Jameson to Pattinson, 13 November 1944. For evidence of the interest taken by Fraser and McGowan, see BP 71041, McGowan to Fraser, 27 October 1944.

20 BP 71042, Jameson to Crichton, 6 November 1944.

21 BP 113572, Idelson, Opinion, 19 February 1945.

22 BP 71041, ICI/AIOC Joint Policy Panel, Minutes of meeting on 30 January 1947.

23 Reader, *ICI, Volume II*, ch. 24.

24 *Ibid.*, p. 398. Also, BP 113572, Jameson and Gass note to Fraser on ICI/AIOC, 4 November 1946; BP 71041, ICI/AIOC Joint Policy Panel, Minutes of meeting on 30 January 1947.

25 BP 71042, Sir Frank Smith, AIOC & ICI re PCD, Notes on meeting held on 17 July 1945. The abandonment of the PCD scheme was confirmed in August 1945, when ICI cancelled the 'Notes of proposed arrangement' which had earlier been initialled by Fraser and McGowan. See BP 113572, Jameson and Gass note to Fraser on ICI/AIOC, 4 November 1946.

26 The proceedings of the Joint Policy Panel are recorded in detail in BP 51068 and 51069.

27 On Smith as chief research chemist at Sunbury, see Bamberg, *British Petroleum, Volume II*, p. 190.

28 BP 51069, D. G. Smith, Note on Project A, 17 September 1946.

29 BP 71041, Fraser to Rogers, 19 November 1946.

30 Bamberg, *British Petroleum, Volume II*, pp. 287–9.

31 BP 51067, Fraser to Rogers, 12 December 1946 enclosing note on manufacture of chemicals from petroleum. It is apparent from the copy of the note in BP 51069 that it was written by D. G. Smith.

32 BP 51067, Fleck and Gordon, The manufacture of chemicals from petroleum, Note on document given to Rogers by Fraser, 30 December 1946.

33 BP 51069, Gass to Jameson *et al.*, 10 January 1947.

34 BP 71041, ICI/AIOC Joint Policy Panel, Minutes of meeting on 30 January 1947.

35 BP 51067, Rogers to Fraser, 26 February 1947; *ibid.*, Fraser to Rogers, 27 February 1947.

36 Reader, *ICI, Volume II*, pp. 407–8.

37 BP 70641, Matthews, Progress report, 5 July 1945.

38 Reader, *ICI, Volume II*, p. 401; Weir, *Distillers*, p. 333.

39 BP 72266, Hayman to Quig, 21 September 1945; *ibid.*, Spearing to Fraser, 27 November 1946.

40 *Ibid.*, Quig to Rogers enclosing draft letter from Quig to Hayman, 3 October 1945; *ibid.*, Fraser to Rogers, 11 October 1945.

41 *Ibid.*, Spearing to Fraser, 27 November 1946.

42 *Ibid.*

43 *Ibid.*

44 *Ibid.*, Eric Stein, managing director of British Industrial Solvents, was also present at the lunch.
45 Again, Stein was also present.
46 BP 72266, Minutes of meeting with Hayman, 13 December 1946.
47 *Ibid.*
48 Documents on these events can be found in BP 51115, 8648, 72324 and 113650.
49 BP 113650, Memorandum covering the broad basis for formation of joint company by AIOC and Distillers to produce chemicals from oil, signed by Gass and Hayman, 28 July 1947.
50 *Ibid.*
51 *Ibid.*, Memorandum and Articles of Association of British Petroleum Chemicals Limited.
52 BP 9210, Coxon to Fraser, 14 January 1952; *ibid.*, Coxon to Fraser, The motor spirit-black oil position, 1 May 1952; *ibid.*, Central Planning Department, The long term growth in demand for products, August 1953; *ibid.*, Hubbard to Fraser, The changing pattern of demand, 3 September 1953; *ibid.*, Hubbard to Gass, Supplementary note to report on the long term growth in demand for products, 30 November 1953; BP 93547, Central Planning Department, The relative growth in demand for motor spirit and black oils in the Eastern Hemisphere, April 1952.
53 On the development of the petrochemical industry and its largest firms in this period, see, for example, Chapman, *International Petrochemical Industry*, chs. 4 and 5; Economic Commission for Europe, *Market Trends and Prospects for Chemical Products* (New York, 1969), 3 vols.; Economic Commission for Europe, *Market Trends and Prospects for Chemical Products* (New York, 1973), 2 vols; W. Molle and E. Wever, *Oil Refineries and Petrochemical Industries in Western Europe* (Aldershot, 1984); V. S. Swaminathan, 'The United Kingdom petroleum chemicals industry', *Industrial Chemist*, March 1961; 'UK heavy organic chemicals industry – plants and producers surveyed', *Chemical Age*, 11 March 1961; C. George, 'Role of the giant firms', *The Statist*, 10 January 1964; 'Jumbo packs for petrochemicals', *The Economist*, 1 February 1964; K. Richardson, 'Why Shell are sure of chemicals', *Sunday Times*, 21 March 1965; W. Rolt, 'PVC: the coming battle', *The Statist*, 4 February 1966; 'Chemicals: America moves into Europe', *The Economist*, 16 April 1966; F. Broadway, 'Boldness needed in chemicals', *The Statist*, 24 June 1966; J. S. Hunter, 'Chemicals and oil', *Petroleum Review*, August 1970; W. C. King, 'Petrochemicals face reality', *World Petroleum*, June 1971; D. P. McDonald, 'The UK petrochemical industry is solving its long-term problems', *Petroleum Times*, 10 September 1971.
54 On the effects on the Company of government controls and the shortages of steel and dollars at this time, see Bamberg, *British Petroleum, Volume II*, ch. 12. Documents on BPC's progress with the construction of its first plants can be found in BP 51034, 51036 and 81312.

55 Documents on the formation of Forth Chemicals can be found in BP 25891, 67176, 71798 and 75214.

56 This account of BHC's expansion from 1951 to 1960 is drawn from BP 54926, BHC pamphlet, September 1958; BP 29176, Howes, Recommendations concerning BP future policy in petroleum chemicals, December 1962; BP 68020, Ball, BPC, A short historical summary, 1947–1955; BP 28707, BHC, A short historical summary, 1947–1962; BP 91173, Hunter, History of BHC, 6 September 1962; *Chemical Age*, 3 May and 25 October 1958, 26 March and 9 July 1960; BP 106309, British Hydrocarbon Chemicals press notice, 29 June 1960.

57 BP 106309, British Hydrocarbon Chemicals press notice, 4 October 1960. On the reasoning behind the decision, see BP 72308, Evans, Memorandum to British Hydrocarbon Chemicals board, 1 July 1960; *ibid.*, Kirkpatrick to Evans, 27 October 1960; *ibid.*, Record of meeting between Evans, Archibald and representatives of Du Pont, 9 November 1960.

58 BP 106309, British Hydrocarbon Chemicals press notice, 17 April 1961.

59 *Financial Times*, 18 October 1963; *Chemical Age*, 19 October 1963.

60 BP 29176, Howes, Recommendations concerning BP future policy in petroleum chemicals, December 1962.

61 BP 69975, Morris to Elgood, 30 September 1948.

62 BP 29176, Howes, Recommendations concerning BP future policy in petroleum chemicals, December 1962.

63 On the break-up of IG Farben, see Raymond G. Stokes, *Divide and Prosper: The Heirs of IG Farben under Allied Authority, 1945–1951* (Berkeley, Calif., 1988).

64 Raymond G. Stokes, *Opting for Oil: The Political Economy of Technological Change in the West German Chemical Industry, 1945–1961* (Cambridge, 1994), esp. chs. 5–7.

65 BP 59642, Dudman to Spearing, Petrochemicals – Germany, 8 May 1957.

66 BP 14769, Howes, Petroleum chemicals – Germany, 24 April 1957.

67 *Ibid.*

68 Rheinische Olefinwerke's first plant started up in 1955 and was the first plant in West Germany devoted exclusively to the manufacture of organic chemicals from petroleum. On its formation, see Stokes, *Opting for Oil*, ch. 5.

69 BP 14767, Distillers, Review of the German aliphatic chemical industry, n.d.

70 *Ibid.* For a published account, see Stokes, *Opting for Oil*, ch. 7.

71 BP 14769, Howes, Petroleum chemicals – Germany, 24 April 1957; BP 59642, Howes, Petroleum chemicals – Germany, 4 October 1956. Scholven was owned by Hibernia and both Hibernia and Gelsenberg had large shareholdings in Hüls.

72 BP 59642, Howes, Petroleum chemicals – Germany, 4 October 1956. See also Stokes, *Opting for Oil*, ch. 6.

73 BP 59642, Howes, Petroleum chemicals – Germany, 28 January 1957; BP 14769, Howes, Petroleum chemicals – Germany, 24 April 1957.

74 BP 14769, Howes, Petroleum chemicals – Germany, 24 April 1957.

75 BP 59642, Pattinson to Haberland, 6 May 1957.
76 *Ibid.*
77 *Ibid.*, Handwritten comment by Bridgeman on note from Pattinson, 10 May 1957.
78 BP 104227, Pattinson, Memorandum no. 75, Petroleum chemicals – Germany, 14 May 1957. See BP 14767 for the board approval.
79 BP 104227, Pattinson, Memorandum no. 121, 4 September 1957; BP 37886, Pattinson, BP Benzin und Petroleum AG, 14 July 1961; Stokes, *Opting for Oil*, p. 170.
80 Stokes, *Opting for Oil*, p. 171; BP 29176, Howes, Recommendations concerning BP future policy in petroleum chemicals, December 1962; BP, *Annual Reports and Accounts*, 1957–60.
81 This was a thermosetting and thermoplastics business which had started in merchanting before entering into plastics manufacture at Feltham, Middlesex, in 1933.
82 BP 112269, Unsigned memorandum, The Distillers Company and its plastics interests, 28 October 1969; *ibid.*, Unsigned memorandum, Historical record of plastics companies, n.d.
83 *Ibid.*, Unsigned memorandum, The Distillers Company and its plastics interests, 28 October 1969; *ibid.*, Unsigned memorandum, Historical record of plastics companies, n.d.
84 *Ibid.*, Unsigned memorandum, The Distillers Company and its plastics interests, 28 October 1969; *ibid.*, Unsigned memorandum, Historical record of plastics companies, n.d.; BP 112265, Woolveridge, Polystyrene, 27 May 1953.
85 BP 112269, Unsigned memorandum, The Distillers Company and its plastics interests, 28 October 1969; *ibid.*, Unsigned memorandum, Historical record of plastics companies, n.d.; BP 112265, Woolveridge, Polystyrene, 27 May 1953.
86 BP 100849, Champion, Topics discussed at a meeting on 21 July 1966.
87 BP 106309, Joint announcements by Distillers and Union Carbide, 4 October and 13 December 1962.
88 For full details of Distillers' chemical and plastics interests, see, for example, BP 100859, Table of DCL interests in chemicals and plastics, March 1959; BP 64844, Table of DCL interests in chemicals and plastics, 1964; BP 36583, The Distillers Company Ltd's chemical interests, 26 November 1964; BP 76464, A review of the chemicals interests of BP and DCL, n.d.
89 BP 111183, The DCL chemicals and plastics group, Press guide, 2nd edn, June 1965.
90 BP 106309, Distillers' press release on reorganisation of chemical and plastics interests, 29 March 1963.
91 BP 100850, Ashford to Management Committee, The Distillers Company Chemicals and Plastics Group, Forward view, 7 February 1966. For an evaluation of the partnership system by another Distillers' manager, see *ibid.*, Champion, The future of the plastics industry, 29 November 1966.
92 BP 112266, LPBM to Hayman, 30 April 1956.

93 BP 112265, Westnedge, Distrene Ltd, 5 June 1961; *ibid*., Westnedge, Record note of discussion between Stein, Stilbert, Gregory, Delafield and Westnedge, 24 July 1962; *ibid*., Stilbert to Stein, 16 August 1962; *ibid*., Westnedge to Ashford, 2 May 1966.

94 BP 76866, Pattinson to Board, 17 December 1962.

95 BP 64844, Down to Bridgeman and Pattinson, 28 July 1964.

96 BP 100850, Champion, The future of the plastics industry, 29 November 1966.

97 *Ibid*. On shortcomings of the partnership system, see also, for example, BP 100850, Ashford to Management Committee, The Distillers Company Chemicals and Plastics Group, Forward view, 7 February 1966; BP 112269, Unsigned memorandum, The Distillers Company and its plastics interests, 28 October 1969; BP 112267, B. F. Goodrich, Notes on visit by Ashford, Westnedge and Mann on 20 and 21 November 1962; BP 64844, Down to Bridgeman and Pattinson, 28 July 1964; BP 100847, Champion, DCL in the chemical industry, 2 December 1964; BP 100849, Champion, Topics discussed at meeting on 21 July 1966.

98 BP 100848, Board, Future of the chemical/plastics interests of DCL, 8 February 1963.

99 See, for example, BP 112269, Unsigned memorandum, The Distillers Company and its plastics interests, 28 October 1969; BP 100858, Ashford, Draft on Z Company, 20 June 1960; BP 100850, Champion, The future of the plastics industry, 29 November 1966.

100 BP 106309, Distillers' press release on reorganisation of chemicals and plastics interests, 29 March 1963.

101 BP 91173, Hunter to Bean, Background to UK operations, 6 January 1971.

102 BP 100858, Ashford, Draft on Z Company, 20 June 1960.

103 BP 100859, Ashford, DCL and BP, Notes on a meeting on 2 April 1959.

104 BP 100858, Ashford, Draft on Z Company, 20 June 1960.

105 *Ibid*.

106 *Ibid*.

107 For a comparison of the production capacities of the main petrochemical firms in Britain in 1961, see *Chemical Age*, 11 March 1961, pp. 407–8.

108 BP 66261, Snow to Anderson, 6 July 1953.

109 *Ibid*.

110 BP 64844, Pattinson to Fraser and Jackson, 19 September 1955. For confirmation that the proposal came from Hayman, see *ibid*., Pattinson to Snow, 12 February 1959.

111 *Ibid*., Ball, Record note of talk with Pattinson on 29 September 1955; *ibid*., Pattinson to Ball, 3 October 1955.

112 This scheme is documented in detail in BP 100858, 100859 and 64844.

113 BP 64844, Pattinson, Note of meeting held with directors of the Distillers Company on 24 April 1959.

114 BP 100858, 100859 and 64844, *passim*.

115 BP 100859, Hayman, Memorandum on discussion with BP on 3 March 1959.

116 *Ibid.*, Hayman, Notes on meeting with BP representatives on 20 July 1960.

117 BP 64844, Banks, Distillers/BP discussions, 23 August 1960. For Hayman's record of the same meeting, see BP 100859, Hayman, Memorandum on meeting with BP on 23 August 1960.

15 INTEGRATION IN PETROCHEMICALS

1 S. Pollard, *The Development of the British Economy, 1914–1990*, 4th edn (London, 1992), pp. 354–75; A. Cairncross, *The British Economy since 1945: Economic Policy and Performance, 1945–1990* (Oxford, 1992), ch. 3.

2 Profits in 1956–7 are shown in table 14.2, chapter 14.

3 'Jumbo packs for petrochemicals' and 'Chemicals: America moves into Europe', *The Economist*, 1 February 1964 and 16 April 1966; *Chemical Age*, 11 April 1964, p. 587; F. Broadway, 'Boldness needed in chemicals' and 'New plant dilemma', *Statist*, 24 June and 30 September 1966; W. C. King, 'Petrochemicals face reality', *World Petroleum*, June 1971.

4 *Chemical Age*, 1 February 1964, pp. 183 and 187–8.

5 *Ibid.*, 10 July 1965, pp. 39–41.

6 *Ibid.*, 2 October 1965, p. 502.

7 *Ibid.*, 4 July 1964, pp. 6 and 9–10; *ibid.*, 27 March 1965, p. 498. The refinery came on stream in 1966. See *ibid.*, 1 October 1966, p. 611 and Christopher Tugendhat, 'Why do oil companies turn to chemicals?', *Financial Times*, 25 January 1967.

8 BP 37916, Down to Bridgeman, 24 November 1964.

9 BP 106309, Distillers' press release, 2 April 1963; BP 76464, A review of the chemicals and plastics interests of BP and DCL, n.d.; *Chemical Age*, 6 April 1963, pp. 487 and 489.

10 BP 76787, Waters to Howes, 21 May 1965. This file also contains full details of the acquisition from Socony. For consultation with the government about the purchase, see PRO BT 64/5303.

11 BP 38832, Shaw, Major companies in the French plastics industry, April 1965.

12 *Chemical Age*, 4 June 1966, p. 984.

13 *Ibid.*, 1 January 1966, p. 7; *ibid.*, 25 March 1967, p. 17; *Financial Times*, 7 October 1968.

14 For the context of Jameson's remark, see chapter 14.

15 On Hayman's career, see *DCL Gazette*, spring 1958, spring 1963 and spring 1966 supplement.

16 *Ibid.*, spring 1963 and autumn 1967.

17 BP 100848, Minute of special meeting of Distillers' Management Committee on 30 May 1963.

18 *Ibid.*

19 BP 100847, Ashford, Notes of conversation with Down on 20 June 1963.

20 *Ibid.*, Ashford, Notes of discussion with Down on 2 March 1964; *ibid.*, Hunter, Notes of discussion with Down and Howes on 11 March 1964.

21 *Ibid.*, Ashford, Notes of meeting on 8 April 1964; *ibid.*, Ashford, Notes of discussion with Down on 20 May 1964; *ibid.*, Ashford, Notes of telephone conversation with Down on 26 May 1964.

22 *Ibid.*, McDonald, Note on discussion with Bridgeman and Pattinson after dinner at Cumming's flat on 30 July 1964.

23 *Ibid.*, Ashford, Notes of a conversation with Down on 18 November 1964; *ibid.*, Ashford, Notes of a discussion on 26 November 1964.

24 BP 75116, Howes, Notes on meeting held on 14 August 1964.

25 Hunter learned about BP's thinking through Charles Evans, the managing director of BHC, who had been told about it by Pattinson. See BP 100847, Hunter to Ashford, BP relationship, 9 September 1964.

26 *Ibid.*, Hunter to Ashford, Partnership problems, 6 November 1964.

27 This last condition reflected Distillers' concern about a project that BP was considering (but later abandoned) for the construction of a major wholly owned petrochemical complex at Rotterdam. Cumming said that he had 'very strong views' about this and that 'DCL could not tolerate the possibility of the enlarged BHC being subjected to competition from a 100% BP owned operation at Rotterdam'. See *ibid.*, Ashford, DCL and BP, Notes of a discussion on 26 November 1964.

28 *Ibid.*, Ashford, DCL and BP, Notes of a discussion on 26 November 1964; *ibid.*, Extract from Distillers' Management Committee minutes, 14/15 December 1964.

29 *Ibid.*, Hunter, Notes on meetings between BP and DCL on 18 and 31 December 1964; BP 93134, Extract from Distillers' Management Committee minutes, 27 January 1965.

30 BP 93134, Down to Ashford, 5 February 1965.

31 BP 100847, Hunter, Notes on meeting between BP and DCL on 31 December 1964; BP 100849, Extract from Distillers' Management Committee minutes, 12/13 January 1965; *ibid.*, Ashford, Notes on discussions with Down on 25 January and 15 February 1965; BP 36563, Howes to Down, 19 January 1965; BP 93134, Extract from Distillers' Management Committee minutes, 27 January 1965; *ibid.*, Down to Ashford, 5 February 1965; *ibid.*, Ashford to Down, 9 February 1965.

32 BP 100849, Ashford, Notes of discussion with Down on 29 March 1965. See also, *ibid.*, Extract from Distillers' Management Committee minutes, 31 March 1965.

33 *Ibid.*, Ashford, Notes of a telephone conversation with Down on 26 April 1965; *ibid.*, Notes of discussion with Down and Howes on 5 May 1965; *ibid.*, Ashford, Notes of discussion with Down on 29 July 1965.

34 *Ibid.*, Extract from Distillers' Management Committee minutes, 14/15 September 1965.

35 BP 100848, Extract from Distillers' Management Committee minutes, 12/13

October 1965. See also BP 100849, Ashford, Notes of meeting with Down on 11 October 1965.

36 BP 59753, BP announcement on rights issue, 31 January 1966.

37 On the working party and the alternatives that were considered, see BP 100849, Champion, Notes of meetings at St James' Square on 17, 22 and 31 December 1965, 17 January and 23 February 1966; BP 100850, Ashford to Distillers' Management Committee, Forward view, 7 February 1966.

38 BP 93538, Down to Bridgeman, 17 February 1966.

39 *Ibid.*, Down to Bridgeman, 8 and 17 February 1966.

40 *Ibid.*, Down, Record note, 20 September 1966.

41 *Ibid.*, Down to McDonald, 30 September 1966.

42 *Ibid.*

43 *Ibid.*, McDonald to Down, 5 October 1966.

44 *Ibid.*, Down to Bridgeman, 6 October 1966.

45 *Ibid.*, Down to McDonald, 11 October 1966.

46 *Ibid.*, Down, Record note (I), 20 October 1966.

47 *Ibid.*

48 *Ibid.*, Down, Record note (II), 20 October 1966.

49 *Ibid.*, Down, Record note (III), 21 October 1966.

50 *Ibid.*

51 *Ibid.*, Down, Record note (IV), 21 October 1966.

52 *Ibid.*, Down to Drake, 24 October 1966.

53 *Ibid.*, Drake, Note of conversation with Cumming on 25 October 1966.

54 On the final negotiations and the agreement, see *ibid.*, Down, Record note (VI), 27 October 1966; *ibid.*, Down, Record note, Hull I, 25 January 1967; *ibid.*, Bridgeman to Cumming, 21 December 1966; *ibid.*, Court, Notes on meeting with Board of Trade on 23 December 1966; *ibid.*, Bridgeman's exchange of letters with Cumming, 3 January 1967.

55 *Ibid.*, BP press release, 3 January 1967; *ibid.*, Cumming circular to Distillers shareholders, 13 February 1967.

56 *Chemical Age*, 7 January 1967.

57 *Ibid.*, 27 July 1968.

58 BP 93538, Down, Note for the board, 12 January 1967.

59 A steering committee under Down was set up to handle the preparations. On its formation, see BP 91173, Down, Acquisition of chemical and plastics interests of DCL, 10 January 1967. Minutes of the steering committee's meetings are in BP 66010.

60 BP 93573, Bridgeman, BP's chemicals and plastics interests, 31 March 1967.

61 BP 100072, Bridgeman, BP Chemicals Ltd, 18 July 1967.

62 BP 112269, Unsigned note on DCL's plastics joint companies, 23 January 1967; BP 66022, Ashford, British Geon, Notes on discussion at Cleveland on 30/31 January 1967; *ibid.*, Ashford, Distrene, Notes on meeting with Dow in Midland, Michigan, on 1 February 1967.

63 BP 76464, Unsigned review of the chemical interests of BP and DCL, n.d.; BP 18331, Telegram, 8 September 1967; *Chemical Age*, 8 July 1967.

64 BP 76464, Unsigned review of the chemical interests of BP and DCL, n.d.; *Chemical Age*, 9 December 1967 and 27 January 1968.

65 BP 76464, Unsigned review of the chemicals interests of BP and DCL, n.d.; *Chemical Age*, 10 June 1967; *Chemical Age International*, 1 June 1973. In 1978 BP purchased Union Carbide's chemicals and plastics interests in Europe, including the Bakelite Xylonite factories in the UK.

66 *Chemical Age*, 17 and 24 September 1966 and, generally, other issues in autumn 1966; F. Broadway, 'New plant dilemma', *Statist*, 30 September 1966.

67 *Chemical Age*, 3 August 1968; *Financial Times*, 27 September 1968.

68 P. Naughton, 'Petrochemicals industry', *The Scotsman*, 21 January 1969.

69 BP 109228, Hunter, Baglan Bay expansion, 5 September 1968; *ibid.*, Ashford, Baglan Bay projects, 23 September 1968; *ibid.*, Down, Memorandum no. 65, Baglan Bay complex, 8 October 1968.

70 BP 91173, Hunter to Bean, Background to UK operations, 6 January 1971; BP 38696, Unsigned background notes for visit of Ashford and Walters to Baglan Bay on 8 August 1972.

71 BP 109228, Hunter, Baglan Bay expansion, 5 September 1968.

72 M. Simmons, 'Plastics makers see the key to stability in co-operation', *Financial Times*, 18 November 1968; *New Scientist*, 13 February 1969.

73 BP 109228, Down, Memorandum no. 65, Baglan Bay complex, 8 October 1968. For the public announcement of the project see, for example, *Chemical Age*, 1 November 1968.

74 BP 109228, Hunter, Baglan Bay expansion, 5 September 1968.

75 BP 57476, BP Chemicals International, Management newsletter, April 1972.

76 *Ibid.*, BP Chemicals International, Management newsletter, August 1973; *Chemical Age International*, 7 and 14 December 1973 and 4/11 January 1974; BP, *Annual Report and Accounts*, 1973.

77 BP 58086, Bean, BPCI strategy, 28 February 1972.

78 See, for example, papers in BP 53501 and 58086.

79 BP 57476, BP Chemicals International, Management newsletter, August 1973.

80 BP, *Annual Reports and Accounts*, 1970–4.

81 BP 36553, BP Chemicals International, Management newsletter, March 1973.

82 BP 75988, Drake, Chemicals reorganisation, 13 November 1970; *ibid.*, Bean, BP Chemicals International reorganisation, 26 July 1971; BP 36553, Down, BP Chemicals International reorganisation, 13 December 1972; *ibid.*, BP Chemicals International, Management newsletter, March 1973; BP 10218, Bean, BP Chemicals International reorganisation, December 1972.

83 The sequence of changes was as follows. By 1964 one of Naphtachimie's founding partners, Kuhlmann, had withdrawn, together with a banking interest of insignificant proportions. This left Péchiney as SFPBP's sole partner in Naphtachimie. In 1969 Péchiney transferred its shareholding to Péchiney-St Gobain, which itself became a 51 per cent subsidiary of France's largest chem-

ical company, Rhône-Poulenc. Another French chemical company, Progil, was also absorbed by Rhône-Poulenc. At the end of 1971 St Gobain withdrew from Péchiney-St Gobain, which was merged with Progil to form Rhône-Progil, which became SFPBP's partner in Naphtachimie. See BP 30138, Bennett, Naphtachimie, 10 July 1973; BP, *Annual Report and Accounts*, 1969.

84 BP 30138, Bennett, Naphtachimie, 10 July 1973.
85 BP, *Annual Reports and Accounts*, 1969–74.
86 BP 11163, Unsigned, BP Chemicals International Ltd, Associated companies coordination – Europe, 18 August 1975.
87 BP 112330, Unsigned, Erdölchemie operations, 28 October 1974.
88 BP 38634, Lachowicz, Meeting with ICI and Shell International Chemicals, 3 May 1968.
89 *Ibid.*, Joint study group, The possible rationalisation of ethylene production between BP Chemicals, Shell Chemicals and ICI, October 1968.
90 *Ibid.*, Carey to Hunter, 29 April 1969.
91 BP 113991, Bean to Ashford, ICI, 10 November 1971.
92 *Ibid.*, Hunter to Bean, ICI, 6 July 1972; *ibid.*, Bean to Ashford, ICI, 10 August 1972.
93 *Ibid.*, Hunter to Bean, ICI, 6 July 1972.
94 *Ibid.*, Lachowicz, Seagull – Present status, 15 September 1972.
95 BP 53501, Champion, Strategic alternatives, 29 August 1972.
96 BP 36553, Ashford, BPCI and ICI, 26 January 1973.
97 *Ibid.*
98 BP 59738, Joint announcement by BP and ICI, 30 April 1974.
99 BP 53501, Champion, Strategic alternatives, 29 August 1972.
100 *Ibid.*

16   COMPUTING

1 BP 8264, Docksey to Banks, 5 July 1961; BP 62436, Bean, Note for Co-ordination Committee, Computer integrated programming of supply operations, 28 March 1962; BP 15608, Deam, The integrated programming of refinery supply and procurement, 18 April 1958; BP 66037, S&D, Use of GRAM for short and medium term refinery/supply estimates, 24 April 1969.
2 BP 66037, S&D, Use of GRAM for short and medium term refinery/supply estimates, 24 April 1969.
3 BP 15608, Deam, The integrated programming of refinery supply and procurement, 18 April 1958; BP 8628, Mitchell, Summary of present planning procedures, 8 October 1965; BP 48851, IMR Research Group, Integrated marketing, refining and supply (IMR), A demonstration problem relating to the BP Group's operations in Australia, June 1968; W. J. Newby, 'An integrated model of an oil company', in *Mathematical Model Building in Economics and Industry*, Papers of a conference organised by C-E-I-R Ltd (London, 1968), p. 64.
4 Bamberg interview with Micklewright, 23 November 1989.

5 *Ibid.*

6 BP 55519, Extract from *Engineering*, 17 November 1961. On operational research generally see C. W. Churchman, R. L. Ackoff and E. L. Arnoff, *Introduction to Operations Research* (New York, 1957).

7 W. J. Newby, 'Planning refinery production', in C. M. Berners-Lee (ed.), *Models for Decision* (London, 1965), p. 43.

8 A linear relationship is one which, when plotted on a graph, will result in a straight line. It is represented by a function such as y = 2x, where for any given value of y, the value of x will be twice as great. Less abstractly, cost could be said to vary linearly with the amount of activity performed where, for example, the cost of shipping two tons of crude was twice that of shipping one ton, etc.

9 Churchman, Ackoff and Arnoff, *Operations Research*, ch. 7; BP 21474, Carruthers, Linear programming in the oil industry, March 1962; BP 54981, Newby, Deam and Sutton, 'Operational research in the petroleum industry', 2nd petroleum symposium, Tehran, September 1962, p. 31; W. J. Newby and R. J. Deam, 'How BP is using linear programming', *World Petroleum* 42 (1971), 230–4.

10 A. Charnes, W. W. Cooper and B. Mellon, 'Blending aviation gasolines', *Econometrica* 20 (1952), 135–59; G. H. Symonds, 'Linear programming for optimum refinery operations', IBM Petroleum Conference, 26 October 1953 cited in BP 54981, Newby, Deam and Sutton, 'Operational research in the petroleum industry', 2nd petroleum symposium, Tehran, September 1962.

11 Churchman, Ackoff and Arnoff, *Operations Research*, pp. 4–8; BP 54981, Newby, Deam and Sutton, 'Operational research in the petroleum industry', 2nd petroleum symposium, Tehran, September 1962, pp. 6–8; W. J. Newby and R. J. Deam, 'Optimisation and operational research', *Transactions of the Institution of Chemical Engineers* 40 (1962), 350.

12 BP 54981, Newby, Deam and Sutton, 'Operational research in the petroleum industry', 2nd petroleum symposium, Tehran, September 1962, p. 61.

13 S. H. Lavington, *Early British Computers: The Story of Vintage Computers and the People who Built them* (Manchester, 1980), ch. 1; Gordon B. Davies, *An Introduction to Electronic Computers* (New York, 1965), p. 2.

14 On the design and construction of the first electronic computers, see, for example, Lavington, *Early British Computers;* M. V. Wilkes, *Automatic Digital Computers* (London, 1956); B. V. Bowden, *Faster than Thought* (London, 1953).

15 Wilkes, *Automatic Digital Computers*, pp. 1–2; Davies, *Introduction to Electronic Computers*, pp. 2–3.

16 Davies, *Introduction to Electronic Computers*, p. 11.

17 Lavington, *Early British Computers*, chs. 5 and 8; Bowden, *Faster than Thought*, p. 135.

18 Bamberg interview with Newby, 29 November 1989.

19 BP staff file on Deam, Deam's personal history sheet; Bamberg interview with Deam, 20 November 1989.

20 BP 52921, Britannica Australia Awards, Newby's citation in support of Deam, July 1965.

21 BP staff file on Deam, Banks to staff manager, 24 September 1956.

22 Bamberg interviews with Deam and Newby, 20 and 29 November 1989. Deam's paper was entitled 'The use of activity analysis in refinery production planning', BP technical conference, 1956, Paper no. 23 in BP 43108.

23 BP staff file on Deam, Godfrey to Deam, 4 October 1956.

24 *Ibid.*, Godfrey to Bevan, 5 October 1956; Bamberg interview with Deam, 20 November 1989.

25 BP staff file on Deam, Allan to telephone manager, 18 June 1957.

26 *Ibid.*, Rawlings, Confidential report on Deam, 12 September 1957; BP 48762, Catchpole, The Deuce electronic computer, Note on a telephone conversation with Adey on 19 March 1957.

27 BP 48762, Deam, The preparation of refinery production programmes on a 'Deuce' electronic computer, 9 April 1957.

28 *Ibid.*, Partridge to Patrickson, 30 April 1957; *ibid.*, Deam, Computer and operations research situation report as at 23 May 1957; Newby, 'Planning refinery production', p. 46; Newby and Deam, 'How BP is using linear programming'.

29 Newby and Deam, 'Optimisation and operational research', 352.

30 BP 48762, Mutch to Rawlings, 4 August 1957; *ibid.*, Button, Notes on discussions held at Llandarcy on 7 and 8 August 1957 with Deam; *ibid.*, Unsigned note, The Computer Group, Situation report as at 30 September 1957.

31 BP staff file on Deam, Rawlings, Confidential report on Deam, 12 September 1957.

32 Newby, 'Planning refinery production', p. 46.

33 BP 48763, Deam to Newby, 20 May 1958.

34 BP 62358, Newby, Memorandum to the Co-ordination Committee, 18 November 1960.

35 Newby, 'Planning refinery production', pp. 46–8; Newby and Deam, 'How BP is using linear programming'; BP 55519, Newby and Deam, Optimisation and operational research, pp. 12–19; BP 62355, Bean *et al.*, Report of Operations Research Group, 5 January 1959; BP 62358, Newby, Memorandum to the Co-ordination Committee, 18 November 1960; BP 8264, Docksey to Banks, 5 July 1961.

36 BP 48763, Deam, Draft summary of the Operations Research and Computer Group activities to date, 24 April 1958; BP 48761, Newby to Banks, 9 January 1959; BP 62355, Bean *et al.* to chairman of Co-ordination Committee, 6 January 1959.

37 BP 48762, Deam, Computer and operations research situation report as at 23 May 1957.

38 BP 55519, Operational research, Information letter, 26 July 1961; BP 37847, Banks, Memorandum no. 100, Revenue expenditure – Head Office, 4 September 1961; Newby, 'BP's new "Mercury" computer', *BP Shield*, August 1961.

39 BP 48761, Perry to Newby, 4 March 1958.

40 *Ibid.*, Adey to Banks, 29 December 1959.

41 BP 55519, Extract from *Engineering*, 17 November 1971.

42 BP 48762, Stephens to Ramsay, 23 December 1957; BP 62349, Boxshall, Newby and Perry, Operations research, 13 March 1958; BP 9173, Davies to Bean, 3 July 1958; BP 62735, Bean to chairman *et al.*, 10 March 1959; BP 55519, Operational research, Information letter, 28 July 1961.

43 BP 62735, Operational research course, April – June 1959, London School of Economics course prospectus.

44 BP 48761, Newby to Banks, 14 and 15 July 1959; *ibid.*, Newby to Cubin, 31 July 1959.

45 The relevant documents are in BP 9173, 48761, 59583, 62349, 62355, 62735.

46 The new department resulted from hiving off the Research and Development Division from the Refineries and Technical Department and giving it separate departmental status.

47 BP 62735, Memorandum to heads of departments no. 60/15, 4 April 1960.

48 BP 61925, Newby to Docksey, 7 December 1960.

49 BP 80804, Operational Research Branch [monthly report], November 1960; BP 52918, Docksey, Notes for the director, 17 November 1960.

50 BP 80804, Operational Research Branch [monthly reports], December 1960, March and April 1961; BP 52918, Notes for the director, Docksey, 9 December 1960; *ibid.*, Hyde, Notes for director, 16 March 1961; BP 8264, Docksey to Banks, 29 June 1961.

51 BP 52934, McColl to Newby, 27 March 1961.

52 BP 80804, Operational Research Branch [monthly report], June 1961.

53 BP 52934, Minutes of meeting held in Britannic House on 19 July 1961.

54 Reported in Newby and Deam, 'Optimisation and operational research', 354; Newby and Deam, 'How BP is using linear programming'.

55 Newby, 'Planning refinery production', p. 48; BP 80804, Newby, Operational Research Division, Report for 1962, 14 May 1963.

56 BP 80804, Operational Research Branch [monthly report], June 1961.

57 BP 65390, S&D, Estimating by computer, May 1964; BP 15609, Cubin and Frampton, Short and medium term estimating for the Five Refinery system, May 1964; BP 62358, Newby, Memorandum to Co-ordination Committee, 18 November 1960; BP 54981, Newby, Deam and Sutton, 'Operational research in the petroleum industry', 2nd petroleum symposium, Tehran, September 1962; BP 55519, Newby and Deam, Optimisation and operational research, n.d.

58 BP 65390, S&D, Estimating by computer, May 1964.

59 BP 15609, Cubin and Frampton, Short and medium term estimating for the Five Refinery system, May 1964.

60 BP 63982, ORDG sub-committee, Development of integrated STE/MTE matrix, 5 June 1968.

61 BP 8264, Docksey to Banks, 5 July 1961; BP 62436, Co-ordination Committee, Minutes of 57th meeting on 3 April 1962; BP 54981, Newby, Deam and Sutton, 'Operational research in the petroleum industry', 2nd petroleum symposium, Tehran, September 1962.

62 Newby and Deam, 'Optimisation and operational research', 354.

63 BP 62436, Bean, Note for Co-ordination Committee, 28 March 1962; *ibid.*, Co-ordination Committee, Minutes of 57th meeting on 3 April 1962; BP 54981, Newby, Deam and Sutton, 'Operational research in the petroleum industry', 2nd petroleum symposium, Tehran, September 1962.

64 BP 63564, Unsigned note, Operational research, January 1964.

65 BP 62436, Bean, Note for Co-ordination Committee, 28 March 1962.

66 BP 37847, Banks, Memorandum no. 100, 4 September 1961; BP 19084, Extract from *The Times*, 3 October 1961; BP 80804, Newby, Operational Research Branch, Report for 1961, 11 May 1962.

67 BP 19084, Extract from *The Times*, 3 October 1961. The first Atlas to be ordered was by the University of Manchester and the second was by the National Institute for Research in Nuclear Science at Harwell.

68 BP 61925. Also *BP Shield*, April 1967.

69 BP 62436, Bean, Note for Co-ordination Committee, 28 March 1962; BP 63982, Marriott to Gillams, 5 June 1968; Newby, 'An integrated model of an oil company', p. 65.

70 BP 63982, Marriott to Gillams, 5 June 1968.

71 BP 15609, Claydon, Micklewright and Cowley, Central European Supply Programme, May 1964.

72 *Ibid.*

73 The inter-departmental team consisted of members of OR Branch and representatives of the Supply, Refineries and Markets Departments. BP 62359, Bean, Note for Co-ordination Committee, n.d.

74 BP 65390, S&D, Estimating by computer, May 1964; BP 15609, Claydon, Micklewright and Cowley, Central European Supply Programme, May 1964.

75 BP 52918, Docksey, Notes for director, 19 October 1962 and 22 February 1963; BP 80804, Operational Research Division [monthly reports], October 1962, July and August 1963; *ibid.*, Newby, Operational Research Division, Report for 1962, 14 May 1963; BP 52916, Operational Research Division, Report for 1963, 9 April 1964. The CESP was reported as coming into operation in 1962 in Newby and Deam, 'How BP is using linear programming', but archive evidence suggests that 1963 was the first year of routine operation.

76 BP 65390, S&D, Estimating by computer, May 1964.

77 BP 52918, Docksey, Notes for director, 23 November 1964; BP 63982, Marriott to Gillams, 5 June 1968.

78 Newby, 'An integrated model of an oil company', p. 65.

79 BP 15609, Claydon, Micklewright and Cowley, Central European Supply Programme, May 1964; BP 61924, Deam to Docksey, 5 May 1965; BP 63972,

Unsigned note for ORDG meeting, 25 May 1966; BP 52918, Docksey, Notes for the director, 17 June 1966.

80 BP 61924, Deam to Docksey, 5 May 1965.

81 Calculated from refining data in BP 95623, Accounts papers, 1967, and from sales data in BP 102994, Management report for board meeting, January 1967.

82 BP 62782, Newby, Support statement for BP/ICT proposal, 3 January 1964.

83 BP 48763, Deam, Discussions held at the London School of Economics on 28 February 1958; *ibid.*, Docksey to Pattinson, 13 May 1958; BP 48761, Bean to Drake, 7 October 1958; BP 62735, Unsigned note, Professor M. G. Kendall – contacts with the Company, 23 March 1959; *ibid.*, Bean to Godfrey, 22 April 1959.

84 BP 19084, Extract from *Data and Control* (December 1966).

85 BP 61925, Newby to Docksey, 7 December 1960; *ibid.*, Newby to Mullaly, 19 May 1961; *ibid.*, Mullaly to Kendall, 30 June 1961.

86 BP 62782, Newby, Support statement for BP/ICT proposal, 3 January 1964.

87 *Ibid.*

88 BP 8264, Memorandum no. 19, 9 May 1962.

89 *Ibid.*, Memorandum no. 18, 9 May 1962.

90 BP 62782, Newby, Support statement for BP/ICT proposal, 3 January 1964.

91 *Ibid.*

92 *Ibid.*

93 *Financial Times, The Guardian, The Times, Evening Standard*, 21 March 1964.

94 BP 54418, Down to Banks, 9 May 1966; *The Times*, 4 December 1964 and 28 June 1966.

95 *Financial Times*, 10 June 1968.

96 BP 19084, Extract from *The Statist*, 19 July 1964. On OR Division's conviction that it was leading the field, see, for example, BP 80804, Newby, Operational Research Division, Report for 1962, 14 May 1963; *ibid.*, Operational Research Division [monthly report], July 1963.

97 BP 19084, Extract from *The Times*, 14 October 1964.

98 BP 59583, Docksey to Banks, 10 October 1963.

99 BP 52921, Gillies to Newby, 24 July 1965.

100 For general accounts, see J. Hendry, 'Prolonged negotiations: the British fast computer project and the early history of the British computer industry', *Business History* 26 (1984), 280–306; Kenneth Flamm, *Creating the Computer* (Washington D.C., 1988), pp. 142–150.

101 BP 80804, Newby, Operational Research Branch, Report for 1961, 11 May 1962.

102 Hendry, 'Prolonged negotiations', 300.

103 BP 52916, Deam, Operational Research Division, Annual report for 1964, 14 June 1965.

104 BP 19084, Extract from *The Observer*, 13 December 1964.

105 *Ibid.*, Extract from University of London, Principal's Report for 1964/5.

106 BP 52917, Operational Research Division, Annual report for 1965, 13 May 1966; BP 52918, Docksey, Notes for the director, 25 January 1965.

107 The problems were partly those of commissioning a new and largely untried machine, but they were compounded by the divergent interests and lack of co-ordination between the University of London, BP and ICT. For the OR Division's account of events, see BP 52917, Operational Research Division, Annual report for 1965, 13 May 1966.

108 BP 52921, Williamson, Review of Operational Research Division activities, January 1966.

109 BP 52918, Deam, Operational Research Division, 16 March 1965.

110 BP 54416, Newby to Banks, 4 March 1966.

111 BP 63972, Newby to chairman of ORDG, 25 May 1966.

112 *Ibid.*, McColl to chairman of ORDG, 3 June 1966.

113 *Ibid.*, Williamson, Atlas computer, 22 July 1966. Also, *BP Shield*, April 1967.

114 BP 63972, Leather, Atlas computer, 11 August 1966; BP 63982, ORDG, Minutes of meetings nos. 33 and 34 held on 26 July and 2 November 1966.

115 BP 42333, C-E-I-R Ltd., Study of data flow in Supply and Development Department, June 1966.

116 The Moving Data File is described in BP 66044, Management Consultancy Group, Interim report to the Corporate Systems Directing Group on management information systems, Attachment F, 7 June 1973.

117 BP 52919, Staff memorandum no. 66/28, 19 July 1966.

118 BP 61924, Newby to Docksey, 10 March 1966.

119 BP 52919, Staff memorandum no. 66/28, 19 July 1966; BP 38963, Staff memorandum no. 66/31, 9 August 1966; *ibid.*, Newby, Note to staff in Computer Department, 31 October 1966; *ibid.*, Staff memorandum no. 67/9, 11 April 1967. In March 1968 the Computer Department was reorganised to include the activities of the former Procedures and Data Processing Division of Finance and Accounts Department. See BP 38963, Searle, Computer Department, Organisation, 25 March 1968.

120 BP 61924, Newby to Docksey, 10 March 1966.

121 BP staff file on Deam, Staff review, 22 September 1967.

122 BP 63972, Extract from minutes of ORDG meeting on 13 February 1967; BP 42691, Marriott to Greaves, 22 March 1967.

123 BP 38921, McColl, Supply, BP management conference, May 1968; BP 63982, ORDG, Minutes of meeting on 20 July 1967.

124 BP 63982, ORDG, Minutes of meetings on 5 June and 20 July 1967.

125 *Ibid.*, Marriott to Gillams, 5 June 1968.

126 *Ibid.* Also, BP 66037, Gillam to Gillams, 9 July 1969.

127 BP 63982, Marriott to Gillams, 5 June 1968.

128 *Ibid.* Also, BP 66037, Gillam to Gillams, 9 July 1969.

129 BP 63982, Newby, Computer Department progress report, 18 July 1967.

130 BP 63972, Marriott, Sub-committee of ORDG for new short and medium term

models, 25 April 1968; BP 63982, ORDG, Minutes of meeting on 22 February 1968.
131 BP 63982, ORDG, Minutes of meeting on 14 June 1968.
132 *Ibid.*, Marriott to Gillams, 5 June 1968.
133 *Ibid.* Also BP 66037, Bowers to Gillams, 24 April 1969.
134 BP 63982, Marriott to Gillams, 5 June 1968; BP 66037, Bowers to Gillams, 24 April 1969.
135 BP 66037, Unsigned note, Computer Department, 10 September 1968; *ibid.*, Marriott, Development of GRAM for STE/MTE purposes, Progress report, 13 September 1968.
136 BP 66037, Marriott, Development of GRAM for STE/MTE purposes, Progress report, 11 November 1968; *ibid.*, Gillam, Development of GRAM for STE/MTE purposes, Progress report, February 1969; *ibid.*, Bowers, Development of GRAM for STE/MTE purposes, Progress report, March 1969; *ibid.*, Gillam to Gillams, 9 July 1969.
137 BP 80804, Newby, Operational Research Branch, Report for 1961, 11 May 1962; *ibid.*, Newby, Operational Research Division, Report for 1962, 14 May 1963; BP 52916, Operational Research Division, Report for 1963, 9 April 1964; *ibid.*, Deam, Operational Research Division, Annual report for 1964, 14 June 1965; BP 52917, Operational Research Division, Annual report for 1965, 13 May 1966.
138 Newby and Deam, 'How BP is using linear programming'.
139 Documents referring to the development of GEM can be found in BP 63972 and 66037. Reference to the introduction of GEM is also made in British Petroleum Company, *Our Industry*, 5th edn (London, 1977), p. 445.
140 BP 52914, Deam to Gillams, 30 November 1966.
141 BP 37661, Deam to Docksey, 14 September 1965.
142 *Ibid.*, Deam, Integrated planning, 14 September 1965; BP 63316, Laidlaw, Marketing plans and objectives, May 1966; BP 52964, Computer Department, IMR, A demonstration problem, November 1967; BP 42691, Stephens, IMR, 29 January 1968; BP 38921, Greaves, A central planning review, BP management conference, May 1968; BP 48851, IMR Research Group, IMR, A demonstration problem relating to the BP Group's operations in Australia, June 1968; BP 19565, Macro-IMR project team, The IMR planning process – a description, October 1969.
143 BP 8291, Long range planning and the IMR method at the British Petroleum Company, Summary of oral presentation to Down on 28 October 1970 prepared by Funkhouser and Laue; BP 10231, Cowley, Paper no. 3 – Group planning, BP Group development course stage III no. 9, 15 May 1972.
144 BP 52916, Operational Research Division, Report for 1963, 9 April 1964.
145 Rosemary Stewart, *How Computers Affect Management* (London, 1971), p. 135.
146 BP 52914, Deam to Gillams, 30 November 1966.

147 For a detailed published account of IMR, see Stewart, *How Computers Affect Management*, ch. 5.

148 BP 38908, Planning Committee, Minutes of 13th meeting on 7 January 1970; BP 93778, Down, Group planning, 4 May 1970.

149 BP 36332, Wyatt, Progress on financial planning model, 5 October 1973; *ibid.*, Clegg, FIRM, 11 December 1973.

150 BP 8291, Long range planning and the IMR method at the British Petroleum Company, Summary of oral presentation to Down on 28 October 1970 prepared by Funkhouser and Laue; BP 36332, Clegg, FIRM, 11 December 1973; Bamberg interview with Deam, 20 November 1989.

151 On Walters, see Walter Goldsmith and Berry Ritchie, *The New Elite: Britain's Top Executives* (London, 1987), pp. 147–60.

152 Bamberg interview with Walters, 17 May 1990.

153 BP 58877, Bell to Fryer, Beneficiation, 14 February 1968.

154 *Ibid.*, Deam and Bell to Barker, 23 February 1968.

155 Bamberg interview with Walters, 17 May 1990.

156 BP 58877, McColl to Fraser, 7 March 1968.

157 BP 17512, Guide to reorganisation, October 1970.

158 BP 54632, Manpower Study Group, Proposals for the organisation of Computer Department, March 1971.

159 BP 52017, Newby to Searle, 27 April 1971.

160 Bamberg interview with Newby, 29 November 1989; BP staff file on Newby, Searle, Finance and planning directorate, 19 May 1971.

161 BP 52017, Walters to Pennell *et al.*, 15 June 1971; BP 66042, Boxshall, Central Developmental Planning Department, 2 June 1971; *BP Shield*, no. 1, 1990.

162 BP staff file on Deam, Down, Finance and planning directorate, 21 May 1971; BP Archive shelf files on organisation, Organigram of the Computer Department, May 1971.

163 BP 52017, MSG, Draft circular on ORDG, 16 February 1971; *ibid.*, Draft minutes of BP Trading Executive Committee meeting on 30 April 1971; *ibid.*, Walters to Pennell *et al.*, 15 June 1971; *ibid.*, ORDG, Minutes of 1st meeting of the newly constituted Group on 24 June 1971.

164 The five BP Trading directors were Christophor Laidlaw, Monty Pennell, Geoffrey Searle, Lahey Bean and Geoffrey Butcher. See BP 52017, Draft minutes of BP Trading Executive Committee meeting on 30 April 1971; BP 8291, unsigned note on ORDG, undated.

165 BP 8291, Phillips to Searle, 19 May 1971.

166 BP 52017, Boxshall to Searle, 1 July 1971.

167 *Ibid.*

168 *Ibid.*, Deam to Searle, 14 July 1971.

169 *Ibid.*, ORDG, Minutes of 1st meeting of the newly constituted Group on 24 June 1971.

170 *Ibid.*, Horton, Report on OR activities, 1 September 1971.
171 *Ibid.*, ORDG, Minutes of 2nd meeting on 6 September 1971; *ibid.*, Walters to Executive Committee, 21 February 1972.
172 *Ibid.*, Walters to Executive Committee, 21 February 1972.
173 *Ibid.*
174 BP 10203, Newby to Searle, 5 November 1971.
175 BP 52017, Deam to Searle, 12 October 1971.
176 The typed minutes can be found in BP 8291.
177 BP 52017, Walters to Executive Committee, 21 February 1972.
178 BP 66044, Management Consultancy Group, Summary of a report (July 1972) on the organisation of the Head Office OR function, February 1973; BP 52017, Walters to Executive Committee, 21 August 1972.
179 BP 66044, Management Consultancy Group, Summary of a report (July 1972) on the organisation of the Head Office OR function, February 1973.
180 BP 52017, Minutes of 6th and 7th meetings of the ORPDG on 16 August and 25 October 1972; *ibid.*, Searle, Organisation of OR in Head Office, 14 September 1972; *ibid.*, Searle, Corporate Systems Directing Group, 15 January 1973.
181 BP staff file on Deam, Searle to Melville, 11 August 1972; *ibid.*, Unsigned note, The world energy model, A joint project between BP and Queen Mary College, 11 August 1972; *ibid.*, Mullaly to Salmon, 1 September 1972; *ibid.*, Salmon to Deam, 22 September 1972; *ibid.*, Pennell to Melville, 1 July 1972.

17 NUTRITION

1 Senez's laboratory was the Centre des Recherches Scientifiques Industrielles et Maritimes (CRSIM).
2 BP 80778, Champagnat interviews, 8 December 1975 and 12 January 1976; BP 91018, Senez interview, January 1976; BP 91019, Rouit interview, 14 January 1976; BP 91022, Vernet interview, 5 May 1976; BP 91014, Vinkhy interview, 9 December 1975; BP 80776, Azoulay interview, 12 January 1976.
3 BP 90936, Note for Co-ordination Committee, Microbiological dewaxing of gas oil and production of a protein-vitamin concentrate, 29 January 1963; British Petroleum Company, *Our Industry*, 5th edn (London, 1977), p. 280.
4 Banks had a BSc. (Tech.) from the College of Technology, Manchester University, and had risen through BP on the technical side of refining.
5 BP 80778, Champagnat interview, 12 January 1976.
6 BP 80871, Banks to Huré, 6 September 1960.
7 BP 116258, Banks, Memorandum no. 69, SFPBP capital expenditure – Lavera, 25 September 1962.
8 BP 80937, Report of discussion on paper 'Microbiological dewaxing with production of a protein-vitamin concentrate', by A. Champagnat, J. Filosa, B. Lainé and C. Vernet, BP technical conference 1962, paper B9.

9   BP 117734, Kermode, Interview with Llewellyn on 20 August 1985.

10 BP 116259, Banks, Memorandum no. 31, BP Refinery (Grangemouth) Limited, Capital expenditure, 24 April 1963, stamped with board approval, 9 May 1963.

11 BP 117736, Elsden to King, 31 July 1963.

12 BP 90936, im Thurn note on meeting on 30 July 1963.

13 BP 46300, Docksey, Biological research and development, 23 October 1963.

14 BP 80798, Docksey interview, October 1975.

15 BP 117736, Gillies to Docksey, 19 April 1963.

16 BP 46301, Gillies to Banks, BP protein, 5 December 1963.

17 BP 80798, Docksey interview, October 1975. Docksey did not specify when Bridgeman gave this ruling.

18 BP 80871, Champagnat memorandum, 'Proteines, Préparation visite M. Banks le 10.3.64', 28 February 1964.

19 BP 91026, Chenevier interview, 6 May 1976.

20 BP 90919, Gillies to Docksey, 27 June 1963.

21 BP 40035, Banks to Chenevier, 10 June 1964; BP 116260, Banks and Docksey, Capital expenditure, Microbiological research, 25 June 1964; BP 116261, Banks, Memorandum no. C2, 30 December 1964, stamped with board approval, 14 January 1965.

22 BP 80878, Record of 'Visite à Sunbury le 5 Juil 1965'.

23 BP 46299, King to Docksey, 2 August 1965.

24 *Ibid.*, King to Docksey, 19 October 1965.

25 BP 90929, Gillies to Banks, Docksey and King, 5 November 1965.

26 *Ibid.*, Gillies to Docksey, 31 January 1966.

27 *Ibid.*, 'Research on protein food source underway', *Journal of Commerce*, 1 July 1966.

28 *Ibid.*, Gillies to Nougaro, 11 July 1966.

29 BP 90961, Mortimer, Notes on meeting with Research and Technical Development Department to discuss protein project, 27 February 1967.

30 BP 90957, Central Planning Department, Study on the commercialisation of BP protein concentrate, October 1967.

31 BP 90961, Knights, Protein project, Assessment of competitors, 8 September 1967.

32 BP 117736, BP press release, BP plans full scale protein production, 20 November 1967.

33 *Chemical Age*, 25 November 1967, p. 12.

34 BP 90928, Gillies to Docksey, 24 November 1967; BP 37910, Gillies to Barker and Down, 3 April 1968; BP 76185, Down to Gillies, 27 May 1968; BP 116264, Down, Memorandum no. 46, 15 July 1968.

35 BP 40082, Watts to Fidler, 19 December 1968; *ibid.*, Dummett, Memorandum no. 4, 1 January 1969.

36 BP 15109, Gillies note on protein, 1 April 1963; *ibid.*, Gillies to Banks, 2 April

1963; *ibid.*, St Clair-Erskine to Belsey, 19 April 1963; BP 117736, Gillies to Docksey, 19 April 1963; *ibid.*, King, Discussion with Professor Elsden at Sheffield University on 20 June 1963; BP 90919, Gillies to Banks, 4 June 1963; BP 90936, Gillies, Protein – Japan, 2 August 1963; BP 8262, Llewellyn, Report on visit to Japan, February 1964; *ibid.*, Banks to Kato, 16 March 1964; BP 90976, Gillies to Docksey, 2 March 1964; BP 22230, Bridgeman to Kato, 18 March 1965; *ibid.*, Kato to Bridgeman, 3 April 1965; *ibid.*, Gillies, Note on discussions with Kyowa Hakko, 10 June 1965; BP 90929, Docksey to Banks and Bridgeman, 11 October 1965; *ibid.*, Telex, Docksey to Belsey, 14 October 1965; *ibid.*, Gillies to Down, 4 May 1966.

37 BP 22230, Gillies to Docksey, 22 August 1967; *ibid.*, Docksey to Gardner, 23 August 1967; *ibid.*, [Bridgeman] to Kato, 14 June 1968; BP 38791, Gillies, King and Nougaro to Bridgeman, 20 November 1967; BP 117736, Nougaro, Report on visit to Japan in November 1967, 22 January 1968.

38 BP 15109, Gillies draft of letter to Kato, 10 June 1968.

39 *Ibid.*, Summary of agreements with Kyowa Hakko, 23 October 1968.

40 BP 37910, Gillies to Barker, 8 July 1968.

41 BP 8570, Gillies to Down, 13 January 1969.

42 BP 38832, Petroleum microbiology: Reflections on future trends and recommendations, May 1969 (translation of French paper).

43 BP 82380, Protein meeting – 7 October 1969.

44 BP 38788, Dummett, Appointments, 1 December 1969.

45 BP 22175, Watts, Review of the Group's protein development, September 1972; BP 107159, Watts to Bean, 22 April 1971.

46 BP 26557, Watts, Paper on BP proteins for BP management conference, May 1974; BP 65636, Lainé, Paper on BP and protein for BP technical conference, 1972; BP 1650, Minutes of BP Proteins board meeting on 10 November 1971.

47 BP 22175, Watts, Review of the Group's protein development, September 1972; BP 26557, Watts, Paper on BP proteins for BP management conference, May 1974.

48 BP 107159, Bean to Watts, 16 June 1971.

49 BP 38786, Memorandum of Agreement, June 1971.

50 *Ibid.*, Watts to Ashford, 7 March 1972; BP 53516, Watts, Italproteine, Notes on meeting with ANIC on 29 February to 1 March 1972.

51 BP 38786, Watts, Note on meeting with ANIC, 10 November 1971; BP 22175, Watts, Review of the Group's protein development, September 1972.

52 BP 8808, Watts to Edwardes, 22 June 1972, with attached text of press release.

53 BP 53517, Ashford, Draft presentation on Italproteine, 7 and 11 November 1972.

54 BP 116268, Ashford, Memorandum no. 15, 31 October 1972.

55 *Gazzetta Ufficiale della Repubblica Italiana*, 27 November 1972.

56 BP 116270, Laidlaw, Memorandum no. 5, 14 January 1974; BP 19597, Italproteine, Project history, 24 November 1977.

57 BP 10931, Note on Italproteine to BP Trading Executive Committee, 30 July 1973; *ibid.*, Extract from Executive Committee minutes, 8 August 1973.

58 BP 116270, Laidlaw, Memorandum no. 5, 14 January 1974.

59 *Ibid.*, Extract from Executive Committee minutes, 14 December 1973.

60 BP 8591, Down to Laidlaw, 20 December 1973.

61 BP 116270, Laidlaw, Memorandum no. 5, 14 January 1974; BP 26557, Watts, Paper on BP proteins for BP management conference, May 1974; BP 19597, Italproteine, Project history, 24 November 1977.

62 BP 8591, Watts, Record note, Objectives: Meeting on 4 January 1974, 15 January 1974; *ibid.*, Belgrave to Watts, 17 January 1974; *ibid.*, Watts, Record note on meeting on BP Proteins – development plan, 4 February 1974; BP 50330, Watts, Policy for proteins, 14 March 1974; BP 26557, Watts, Paper on BP Proteins for BP management conference, May 1974; BP 10476, Boxshall to Executive Committee, BP Proteins, Review of progress and proposals for downstream investment, 23 October 1974; BP 17388, BP Proteins – the way ahead, November 1975; *Chemical Age International*, 26 October 1973.

63 On these and earlier talks with producing countries, see BP 80934, Steel to Gillies, 26 April 1968; BP 38791, Watts to Steel and Ashford, 21 July 1970; *ibid.*, Watts to Steel, 11 February 1972; *ibid.*, Davies to Fallah, 28 May 1972; BP 8591, *passim*; BP 8597, Executive Committee minutes, 7 June, 20 and 27 September, 4 October and 13 December 1974; BP 10931, Watts to Walton, 23 September 1974; *ibid.*, Extract from Executive Committee draft minutes, 25 July 1975; *ibid.*, Extract from Executive Committee minutes, 6 February 1976; BP 115436, *passim*; BP 10476, Boxshall to Executive Committee, BP Proteins – Venezuela, 18 September 1974; *ibid.*, Butcher to Executive Committee, Bioproteinas de Venezuela, 2 February 1976; *ibid.*, Gillams to Executive Committee, 4 February 1976; *ibid.*, Walton to Butcher, 4 February 1976; BP 46299, Docksey to Banks, 6 December 1966; BP 80934, Docksey to Bridgeman, 15 May 1967; BP 38791, Watts to Drake, 27 October 1970; *ibid.*, Watts to Ashford, 9 October 1970; *ibid.*, Catchpole to Ashford, 14 October 1970; *ibid.*, Watts to Ashford and Bean, 3 September 1971; BP 29861, Watts to Boxshall and Laidlaw, 24 May 1973; *ibid.*, Watts to Drake *et al.*, 26 June 1973; BP 4996, Watts, Record note of meeting with Kyrillin and Belayev, 17 September 1976; BP 115432, Watts to Butcher, 20 October 1978.

64 BP 38785, Telex, Gardner to Watts, 10 February 1972; *ibid.*, Watts to Ashford, 10 February 1972; BP 8591, Note on discussions, 29 January 1973.

65 BP 8591, Telex, BP Tokyo to Watts, 16 February 1973; BP 29861, Watts to Laidlaw and Boxshall, 13 March 1973; M. Sherwood, 'Single-cell protein comes of age', *New Scientist*, 28 November 1974.

66 BP 117734, Kermode, Interview with Walker on 1 July 1986.

67 *Ibid.*, Kermode, Interview with Gatumel on 14 April 1986.

68 BP 53516, Righi, Bellani – Watts – ENI, 18 October 1972.

69 BP 4996, Watts to Sutcliffe, Italproteine, 3 March 1976.

70 *Paese Sera*, 8 January 1974.
71 There is a translation of this article in BP 19500.
72 *Avanti*, 11 July 1974. A translation of the article is in BP 19500.
73 BP 19500, Hodges, Note on Italproteine meeting on 30 April 1975.
74 BP 19502, Gambari to Watts, 22 April 1975.
75 *Ibid.*
76 BP 4996, Watts to Sutcliffe, Italproteine, 3 March 1976.
77 BP 19504, Shacklady to Fasella, 27 June 1975.
78 BP 4996, Watts to Sutcliffe, Italproteine, 3 March 1976.
79 BP 19504, Gambari, Record note on Italproteine board meeting on 19 September 1975.
80 BP 19520, Shacklady to Watts, 24 September 1975.
81 BP 19504, d'Amelio to Watts, 28 October 1975. An English translation of the minutes of the board meeting is in BP 19504, 'Information from the Chairman concerning the situation of the statutory authorisation regarding the initiative and the relative discussion'. The Italian minutes of this meeting are in BP 107402, 'Verbale della seduta del consiglio di amminstrazione del 14 Ottobre 1975'. Watts' requested amendments to the Italian version of the minutes are in BP 107402, Telex Watts to Gambari, 19 December 1975.
82 BP 19597, Italproteine, Project history, 24 November 1977.
83 *Ibid.*
84 BP 19501, Gambari to Watts, 22 October 1976; BP 10905, Translation of decree, 21 October 1976.
85 BP 19501, Watts to Sutcliffe with note on Italproteine, 21 February 1977.
86 BP 10932, Telex, Yeo to Snell, 23 March 1977.
87 BP 116273, Finance memorandum, Italproteine, 25 May 1977.
88 BP 19496, Yeo to Watts, 5 September 1977.
89 BP 115432, Butcher, Aide-memoire for board presentation, Italproteine, 1 February 1978.
90 BP 19597, Italproteine, Project history, 24 November 1977; BP 115432, Butcher, Aide-memoire for board presentation, Italproteine, 1 February 1978.
91 BP 107841, BP Proteins, Prospects and strategy review, 1978–1985, November 1977.
92 BP 19495, Green, Note on discussions in Milan on 14 and 15 November 1977.
93 BP 19503, Green, Background notes, 29 November 1977.
94 BP 115432, Extract from minutes of BP board meeting on 2 February 1978.
95 BP 10926, Walton, Note on Italproteine board meeting on 28 February 1978.
96 BP 115432, Watts to Butcher, 20 November 1978. The formal memorandum of agreement, dated 17 November 1978, is also in BP 115432.
97 BP 94681, Butcher, BP Proteins Ltd, 1 December 1978.
98 *Financial Times*, 30 July 1973; M. Sherwood, 'Single-cell protein comes of age', *New Scientist*, 28 November 1974; *Corriere della Sera*, 31 July 1977; J. D. Levi,

Jean L. Shennan and G. P. Ebbon, 'Biomass from liquid n-alkanes', in A. H. Rose (ed.), *Microbial Biomass* (London, 1979), p. 363.

99 Jean L. Shennan, 'Hydrocarbons as substrates in industrial fermentation', in R. M. Atlas (ed.), *Petroleum Microbiology* (New York, 1984), p. 676.

100 BP 50330, Watts, Policy for proteins, 14 March 1974; BP 26557, Watts, Paper on BP Proteins for BP management conference, May 1974.

101 BP 50330, Watts, BP Proteins: Spillers, 26 April 1974; *ibid.*, Watts to Boxshall and Laidlaw, Draft on BP Proteins – Commercial investments, 28 May 1974; *ibid.*, Watts, Note on meeting with Spillers, 6 May 1974; *ibid.*, Watts to Laidlaw and Drake, 6 May 1974; *ibid.*, Philpot to Laidlaw, 10 May 1974; BP 8591, Sutcliffe to Laidlaw, 31 May 1974; BP 4996, Laidlaw to Sutcliffe, 30 May 1974; *ibid.*, Laidlaw to Philpot, 30 May 1974. On BP's earlier contacts with Spillers, see BP 50330, Watts to Boxshall, Attachment, 19 April 1974.

102 *Financial Times*, 18 May 1974; BP 50330, Watts to Boxshall and Laidlaw, Draft on BP Proteins – Commercial investments, 28 May 1974. The purchase of Colborn by Shell went ahead. See BP 10476, Simons to Grassick, 31 October 1974.

103 BP 50330, Watts to Boxshall and Laidlaw, Draft on BP Proteins – Commercial investments, 28 May 1974.

104 *Ibid.*; BP 8591, Executive Committee minute, 1 November 1974; BP 116270, Laidlaw, Memorandum no. 33, 29 October 1974; *ibid.*, Laidlaw to Steel, Down and Drake, 4 November 1974.

105 Full name; Hanseatische Kraftfuttergesellschaft Nielsen and Co KG.

106 BP 10931, Finance memorandum on purchase of HAKRA, 31 January 1975.

107 BP 107166, Steel, Draft on BP Proteins/Nutrition reorganisation, 2 December 1976.

18   'AN AVALANCHE OF ESCALATING DEMANDS'

1 There is a mass of literature on these themes. See, for example, Charles Lipson, *Standing Guard: Protecting Foreign Capital in the Nineteenth and Twentieth Centuries* (Berkeley, Calif., 1985); Kenneth A. Rodman, *Sanctity versus Sovereignty: The United States and the Nationalization of Natural Resource Investments* (New York, 1988); H. Jeffrey Leonard, 'Multinational corporations and politics in developing countries', *World Politics* 32 (1980), 454–82; Charles R. Kennedy, 'Relations between transnational corporations and governments of host countries: a look to the future', *Transnational Corporations* 1 (1992), 67–91; Stephen J. Kobrin, 'Expropriation as an attempt to control foreign firms in LDCs: trends from 1960 to 1979', *International Studies Quarterly* 28 (1984), 329–48 and 'The nationalisation of oil production, 1918–1980', in David W. Pearce, Horst Siebert and Ingo Walter (eds.), *Risk and the Political Economy of Resource Development* (London, 1984), pp. 137–64. On the growth of US multinationals, Mira Wilkins, *The Maturing of*

Multinational Enterprise: American Business Abroad from 1914 to 1970 (Cambridge, Mass., 1974); on the rise of multinationals more generally, Geoffrey Jones, *The Evolution of International Business, An Introduction* (London, 1996).

2 See, for example, Henry Kissinger, *White House Years* (Boston, Mass., 1979), p. 64.

3 W. M. Scammell, *The International Economy since 1945* (London, 1980), ch. 7.

4 George Lenczowski, *American Presidents and the Middle East* (Durham, N.C., and London, 1990), pp. 116–19.

5 Percentages calculated from BP, *Statistical Review of the World Oil Industry*, 1970. For fuller analyses of energy trends, see Joel Darmstadter and Hans H. Landberg, 'The economic background', in Raymond Vernon (ed.), *The Oil Crisis* (New York, 1976), pp. 15–37; John G. Clark, *The Political Economy of World Energy: A Twentieth Century Perspective* (Hemel Hempstead, 1990), chs. 4 and 6.

6 United States Congress, Senate, Committee on Foreign Relations, *Hearings before the Subcommittee on Multinational Corporations*, 93rd Congress, 2nd session (Washington D.C., 1974–5) (hereinafter abbreviated to *MNC Hearings*), pt. 9, pp. 181–2.

7 BP, *Statistical Review of the World Oil Industry*, 1970; Daniel Yergin, *The Prize: The Epic Quest for Oil, Money, and Power* (New York, 1991), pp. 567–8.

8 Henry Kissinger, *Years of Upheaval* (Boston, Mass., 1982), p. 135.

9 By one estimate, more than 300 private companies and more than fifty state-owned firms entered the international oil industry between 1953 and 1972. See Neil H. Jacoby, *Multinational Oil: A Study in Industrial Dynamics* (New York, 1974), p. 120.

10 International crude oil production represents production in the non-communist world, excluding North America. Percentages calculated from Standard Oil (NJ), Socal, Mobil, Texaco and Gulf data from their *Annual Reports and Accounts*; BP data from BP 102770, 102796, 102826, 102837, 102848, 102860, 102879, 102890, Management reports for board meetings, January 1952, 1954 and 1956–61; BP 95610, 95617–21, 95623, 118733, Accounts schedules, 1961–73; Shell data from *Petroleum Press Service* (various issues) and Walter E. Skinner, *Oil and Petroleum Year Book*, 1950–64; continued as Walter R. Skinner, *Oil and Petroleum Year Book*, 1965–1970/71; continued as *Oil and Petroleum International Year Book*, 1971/72; industry data from DeGolyer and Macnaughton, *Twentieth-Century Petroleum Statistics* (Dallas, Tex., 1994).

11 They were Abu Dhabi, Algeria, Indonesia, Iran, Iraq, Kuwait, Libya, Qatar, Saudi Arabia and Venezuela.

12 *BP Statistical Review of the World Oil Industry*, 1970.

13 Antony Jay, *Management and Machiavelli* (London, 1967), p. 159.

14 Malcolm Caldwell, *Oil and Imperialism in East Asia*, Spokesman pamphlet no. 20 (Nottingham, 1971), p. 6.

15 See, for example, Harry Magdoff, *The Age of Imperialism: The Economics of U.S. Foreign Policy* (New York, 1969).

16 Kissinger, *Years of Upheaval*, p. 859.

17 R. F. Holland, *European Decolonization 1918–1981: An Introductory Survey* (London, 1985), pp. 163–75.

18 Bennett H. Wall, *Growth in a Changing Environment: A History of Standard Oil Company (New Jersey), Exxon Corporation, 1950–1975* (New York, 1988), p. 702.

19 *BP Statistical Review of the World Oil Industry*, 1970.

20 On Hammer's remarkable life, see Steven Weinberg, *Armand Hammer: The Untold Story* (Boston, Mass., 1989).

21 The best studies of the development of the Libyan oil industry in this period are Frank C. Waddams, *The Libyan Oil Industry* (London, 1980), chs. 1–10 and Judith Gurney, *Libya: The Political Economy of Oil* (Oxford, 1996), chs. 1–3 and 5–6.

22 BP 96073, BP Libyan crude oil – posted price discussions, section 3; Waddams, *Libyan Oil*, p. 230.

23 For varying translations of this widely cited remark, see, for example, Wall, *Standard Oil Company (New Jersey)*, p. 703; Waddams, *Libyan Oil*, p. 230; Yergin, *The Prize*, p. 578; Ian Seymour, *OPEC: Instrument of Change* (London, 1981), pp. 66–7.

24 Dwight D. Eisenhower, *The White House Years: Mandate for Change, 1953–1956* (London, 1963), p. 159.

25 Dean Acheson, *Present at the Creation: My Years in the State Department* (London, 1970), p. 511.

26 George Philip, *The Political Economy of International Oil* (Edinburgh, 1994), p. 137.

27 BP 118940, Bexon record note, 11 February 1970; *ibid.*, Bexon to Stockwell, 2 March 1970; *ibid.*, Stockwell file note, 4 March 1970; *MNC Hearings*, pt. 6, pp. 2–3.

28 BP 63657, Sutcliffe to chairman, 31 March 1970.

29 BP 118940, Bexon to Stockwell, 12 April 1970; United States Congress, Senate, Committee on Foreign Relations, Subcommittee on Multinational Corporations, *Report on Multinational Oil Corporations and U.S. Foreign Policy*, 93rd Congress, 2nd session, (Washington D.C., 1975) (hereinafter abbreviated to *MNC Report*), p. 122.

30 Seymour, *OPEC*, p. 58; Gurney, *Libya*, p. 117; *MNC Hearings*, pt. 5, p. 214.

31 BP 106619, Bexon to Stockwell, 23 May 1970.

32 Seymour, *OPEC*, p. 67.

33 *Ibid.*, p. 59.

34 *MNC Hearings*, pt. 6, p. 3; Gurney, *Libya*, p. 117.

35 *MNC Report*, p. 123.

36 Seymour, *OPEC*, p. 68; *MNC Hearings*, pt. 6, p. 3.

37 Seymour, *OPEC*, p. 70.

38 BP 106619, Sarre to Steel, 14 September 1970; *MNC Hearings*, pt. 6, p. 4.
39 BP 63029, Account by Van Reeven of his visits to the Shah, the Prime Minister, Alam, Eghbal and Mina, accompanied by Page and Hedlund, attached to Addison letter, 1 October 1970.
40 *MNC Hearings*, pt. 8, pp. 771–3, Barran to Church, 16 August 1974.
41 BP 63657, Pennell to chairman, 21 September 1970; BP 118940, Notes of Shell's meetings at the Ministry of Oil, 19–22 September 1970; *MNC Hearings*, pt. 6, p. 4.
42 McCloy papers, series 20, box oil 3, folder 2, McCloy memorandum, 29 September 1970; BP 63657, Sutcliffe to chairman, 22 September 1970; *ibid.*, Drake to McCloy, 24 September 1970.
43 BP 86197, Report for chairman at board meeting on 24 September 1970; BP 106241, Minutes of BP board meeting on 24 September 1970; BP 106619, Sutcliffe, Libya, Note on meetings held in New York and Washington on 25 September 1970.
44 BP 106619, Sutcliffe, Libya, Note on meetings held in New York and Washington on 25 September 1970.
45 Anthony Sampson, *The Seven Sisters: The Great Oil Companies and the World they Made* (London, 1975), p. 214.
46 Kissinger, *Years of Upheaval*, p. 862.
47 BP 106619, Sutcliffe, Libya, Note on meetings held in New York and Washington on 25 September 1970.
48 *MNC Report*, p. 125; *MNC Hearings*, pt. 8, pp. 771–3, Barran to Church, 16 August 1974.
49 BP 106619, Sutcliffe, Libya, Note on meetings held in New York and Washington on 25 September 1970; McCloy papers, series 20, box oil 3, folder 2, McCloy memorandum, 29 September 1970.
50 BP 106619, Sutcliffe, Libya, Note on meetings held in New York and Washington on 25 September 1970; McCloy papers, series 20, box oil 3, folder 2, McCloy memorandum, 29 September 1970; *MNC Hearings*, pt. 8, pp. 771–3, Barran to Church, 16 August 1974.
51 Seymour, *OPEC*, p. 73; *MNC Hearings*, pt. 6, p. 4; BP 106619, Sutcliffe, Notes on meetings on 3–8 October 1970; *ibid.*, Curtis to Libyan Ministry of Petroleum, 8 October 1970; *ibid.*, Notes on Libya settlement, 4 November 1970.
52 *MNC Hearings*, pt. 8, pp. 771–3, Barran to Church, 16 August 1974.
53 Ian Skeet, *OPEC: Twenty-five Years of Prices and Politics* (Cambridge, 1988), p. 61; BP 63040, Steel to Drake, n.d.
54 *MNC Hearings*, pt. 6, p. 5; Seymour, *OPEC*, pp. 74–5.
55 *MNC Hearings*, pt. 6, p. 6; BP 6087, Resolutions of the 21st OPEC conference, 28 December 1970.
56 BP 63040, Steel to Drake, 11 December 1970.
57 BP 4934, Notes of meeting called by Deputy Prime Minister on 2 January 1971; *ibid.*, Sutcliffe to Drake, 4 January 1971; *ibid.*, Sutcliffe to Boxshall, 4 January

1971; *MNC Hearings*, pt. 6, pp. 7 and 233 [which wrongly gives the date of the meeting as 3 January on p. 7, but gets it right on p. 233].

58 BP 79131, Bexon to Dummett, 11 and 12 January 1971; *MNC Hearings*, pt. 6, p. 7.
59 The leap-frog metaphor was widely used. See, for example, Seymour, *OPEC*, p. 78; Skeet, *OPEC*, p. 62; and various mentions in *MNC Hearings*, pts. 5, 6, 7 and 9.
60 BP 6087, Drake to Heath, 5 January 1971.
61 The agreement can be found in BP 79127 and is reprinted in *MNC Hearings*, pt. 6, pp. 224–8.
62 BP 79127, Message to OPEC, 16 January 1971. Reprinted in *MNC Hearings*, pt. 6, pp. 60–1.
63 *MNC Hearings*, pt. 6, pp. 10 and 234–5; Seymour, *OPEC*, p. 78.
64 BP 8558, The Tehran Oil Agreement, FCO report, June 1971.
65 *MNC Hearings*, pt. 6, pp. 11–12 and 223–4.
66 BP 4934, Piercy and Holmes to Sutcliffe, 19 January 1971; *MNC Hearings*, pt. 6, pp. 60–1.
67 BP 79131, Wright to FCO, 19 January 1971; BP 78672, PT 0095, Van Reeven to Addison, 21 January 1971; reprints in *MNC Hearings*, pt. 6, pp. 63–6; Kissinger, *Years of Upheaval*, p. 864. See also Irwin's testimony in *MNC Hearings*, pt. 5, pp. 145–73.
68 Kissinger, *Years of Upheaval*, p. 865.
69 *MNC Hearings*, pt. 6, pp. 10–11, 13, 61, 73–4, 76–8 and 236–7; BP 67500, TP 142, Addison to Strathalmond and Piercy via Van Reeven, 20 January 1971; *ibid.*, London Group to New York Group, 22 January 1971; BP 78672, VR 0005, Van Reeven to Addison, 21 January 1971; *ibid.*, Schuler and Espey to Hunt, 24 January 1971.
70 *MNC Hearings*, pt. 6, pp. 18 and 89–91; BP 78673, Message from Esso Libya, 27 January 1971; *ibid.*, Esso to Addison, 28 January 1971; *ibid.*, Tripoli to London, 29 January 1971; BP 4519, Recent negotiations, Tab L; BP 67500, JA0072, Addison to Van Reeven, 30 January 1971.
71 Kissinger, *Years of Upheaval*, p. 876; Yergin, *The Prize*, p. 581; Seymour, *OPEC*, p. 80.
72 *MNC Hearings*, pt. 6, pp. 18–19 and 89–93; BP 78673, VR 0030, Strathalmond to Addison, 28 January 1971.
73 BP 78673, VR 0095, Van Reeven to Addison, 2 February 1971.
74 BP 34484, VR 0036, Van Reeven to Addison, 29 January 1971; BP 8558, The Tehran Oil Agreement, FCO report, undated.
75 BP 8558, The Tehran Oil Agreement, FCO report, n.d.
76 Seymour, *OPEC*, p. 85; BP 8558, The Tehran Oil Agreement, FCO report, n.d. Reprints of the cable traffic between Tehran, London and New York in *MNC Hearings*, pt. 6, pp. 92–117 give a blow-by-blow account of the negotiations.

77 BP 8558, The Tehran Oil Agreement, FCO report, n.d. Heath's message reached the Shah via Sir Denis Wright and the Iranian Minister of Court.

78 *MNC Hearings*, pt. 6, p. 228.

79 BP 8558, The Tehran Oil Agreement, FCO report, n.d.

80 BP 34676, Steel to Drake, Aide-memoire, 12 February 1971. Emphasis added by author.

81 They were Abu Dhabi, Iran, Iraq, Kuwait, Qatar and Saudi Arabia.

82 BP 8558, The Tehran Oil Agreement, FCO report, n.d.

83 The full text of the agreement is in BP 8554 and is reprinted in *MNC Hearings*, pt. 6, pp. 169–72.

84 BP CRO 5H 5154, London Policy Group meeting, 13 February 1971. The comment was made by Al Martin of Gulf Oil.

85 BP CRO 5H 5149, TA23, Piercy to Addison, 24 February 1971; *MNC Hearings*, pt. 6, p. 188.

86 BP 106619, Sutcliffe to Steel, 12 October 1970.

87 BP CRO 5H 5147, AG08, Addison to Goerner, 1 March 1971; BP CRO 5H 5149, TA24, Piercy for relay [to] London Group, 24 February 1971; *ibid.*, TA26, Piercy to Addison, 24 February 1971; *MNC Hearings*, pt. 6, pp. 189–91.

88 BP CRO 5H 5149, TA35, Piercy to Addison, 1 March 1971; *ibid.*, TA42, Piercy to Addison, 2 March 1971.

89 *Ibid.*, TA69, Piercy to Addison, 7 March 1971.

90 *Ibid.*, TA67, Piercy to Addison, 7 March 1971; *ibid.*, TA72, Piercy to Addison, 8 March 1971.

91 *Ibid.*, TA 70, Piercy to Addison, 7 March 1971.

92 *Ibid.*, TA73, Piercy to Addison, 8 March 1971.

93 For separate but consistent accounts of this meeting, see *ibid.*, TA82, Piercy to Addison, 10 March 1971 and *MNC Hearings*, pt. 7, p. 378.

94 BP CRO 5H 5149, TA93, Piercy to Addison, 13 March 1971.

95 *Ibid.*, TA96, Piercy to Addison, 14 March 1971.

96 *Ibid.*, TA98, Piercy to Addison, 15 March 1971; *MNC Hearings*, pt. 6, pp. 211–12.

97 BP CRO 5H 5149, TA99, Piercy to Addison, 15 March 1971.

98 BP 79061, Authority to the team, 17 March 1971.

99 BP 34676, Strathalmond to Steel and Steel's pencilled reply, 31 March 1971.

100 The full text of the agreements is reprinted in *MNC Hearings*, pt. 6, pp. 212–15.

101 Seymour, *OPEC*, pp. 93–6.

102 BP 63657, Goerner to Curtis, 3 April 1971.

103 The Libyan price went up from $2.23 to $2.53 on 1 September 1970, to $2.55 by the addition of 2 cents annual escalation on 1 January 1971, to $3.447 on 20 March 1971. The price of 34° Saudi crude went up by the general increase of 35 cents a barrel under the Tehran Agreement, plus 3 cents owing to changes in gravity differentials, also under the Tehran Agreement.

104 Walter J. Levy, 'Oil power', *Foreign Affairs* 49 (1971), 652–68.

19 THE END OF AN ERA

1 United States Congress, Senate, Committee on Foreign Relations, *Hearings before the Subcommittee on Multinational Corporations*, 93rd Congress, 2nd session (Washington D.C., 1974–5) (hereinafter abbreviated to *MNC Hearings*), pt. 6, pp. 169–72 and 212–15.

2 Ian Seymour, *OPEC: Instrument of Change* (London, 1981), pp. 218–19; Ian Skeet, *OPEC: Twenty-five Years of Prices and Politics* (Cambridge, 1988), pp. 49 and 70–1.

3 Mira Wilkins, *The Maturing of Multinational Enterprise: American Business Abroad from 1914 to 1970* (Cambridge, Mass., 1974), pp. 353–71; Stephen J. Kobrin, 'The nationalisation of oil production, 1918–1980', in David W. Pearce, Horst Siebert and Ingo Walter (eds.), *Risk and the Political Economy of Resource Development* (London, 1984), pp. 137–64.

4 McCloy papers, series 20, box oil 3, folder 6, Minutes of meeting of international oil companies with counsel, 19 August 1971; BP 63657, Belgrave, OPEC, 19 August 1971; BP 34720, Belgrave to Down and Steel, 13 December 1971.

5 National Security Archive, 00759, State Department, The international oil industry through 1980, December 1971.

6 Seymour, *OPEC*, pp. 89, 93, 101 and 219.

7 J. B. Kelly, *Arabia, the Gulf and the West* (London, 1980), ch. 2; Glen Balfour-Paul, *The End of Empire in the Middle East: Britain's Relinquishment of Power in her Last Three Arab Dependencies* (Cambridge, 1991), ch. 4; BP 99991, TA357, 8 December 1971.

8 Frank C. Waddams, *The Libyan Oil Industry* (London, 1980), pp. 251–3; Judith Gurney, *Libya: The Political Economy of Oil* (Oxford, 1996), pp. 126–8 and 151; BP 4973, Claim for damages, 7 March 1972; BP 3298, Tottenham-Smith to Drake, 20 November 1974.

9 BP 114000, Enclosure with Gregory to Watts, 26 November 1973. A transcript of the first negotiating session can be found in McCloy papers, series 20, box oil 3, folder 23, Report of meeting held in Geneva at Intercontinental Hotel on 21 January 1972, enclosed with O'Brien to McCloy, 24 January 1972.

10 BP 99997, SNLG-18, 9 March 1972; *ibid.*, NYG 355, 9 March 1972; *ibid.*, MF 404, 21 March 1972; BP 66759, ADMA to Ministry of Petroleum and Industry, 18 March 1972; *ibid.*, Milne to Minister of Oil and Minerals, 18 March 1972; *ibid.*, Pawson to Emir of Qatar, 19 March 1972; Seymour, *OPEC*, pp. 221–2.

11 BP 34721 and 34722 *passim*.

12 Edith and E. F. Penrose, *Iraq: International Relations and National Development* (London, 1978), pp. 405–14; IPC SCC/1/1, pt. 45, box 8; *ibid.*, pt. 46, box 9; *ibid.*, pt. 47, box 9; *ibid.*, pt. 48, box 9; *ibid.*, pt. 50, box 9.

13 Nixon archives, WHCF, subject files, CO, box 37, folder [EX] CO 68 Iran 1 January 1971–, Flanigan to Nixon, 27 April 1972; National Security Archive, 00767, State Department memorandum for the President, Meetings with the

Shah, 12 May 1972; BP 4964, Bexon to Steel via Adam, 20 December 1971; *ibid.*, Attachment to letter no. 174, 24 May 1972; *ibid.*, Attachment to letter no. 175, 22 May 1972; BP 4779, Bexon, Record note, 27 June 1972; BP 4930, Record note and attachments to letter no. 143, 4 April 1972; BP 66818, Clegg to Belgrave, 23 June 1972; BP 4962, Agreement with Iran, 29 June 1972; George Lenczowski, *American Presidents and the Middle East* (Durham, N.C., and London, 1990), p. 118; Henry Kissinger, *Years of Upheaval* (Boston, 1982), p. 869; James A. Bill, *The Eagle and the Lion: The Tragedy of American–Iranian Relations* (New Haven, Conn., 1988), p. 200; Mark J. Gasiorowski, *US Foreign Policy and the Shah: Building a Client State* (New York, 1991), pp. 113–14.

14 BP 34722, Sutcliffe to Drake, 20 June 1972.

15 *Ibid.*, For Gulf producers, 17 July 1972.

16 McCloy papers, series 20, box oil 3, folder 18, Faisal to Nixon, 10 July 1972; Bennett H. Wall, *Growth in a Changing Environment: A History of Standard Oil Company (New Jersey), Exxon Corporation, 1950–1975* (New York, 1988), p. 832.

17 BP 34722, 34731, 34732, 99994, 23721, 63594 *passim.*

18 Seymour, *OPEC*, pp. 224–5; Skeet, *OPEC*, pp. 77–8.

19 Skeet, *OPEC*, p. 77.

20 BP 6060, Steel record note on visit to Kuwait on 19 March 1972; BP 23721, DN-699, 16 September 1972; BP 34732, PC-4777 at Aramco, 27 September 1972; BP 8516, Sutcliffe to Drake, 28 September 1972; BP 34731, PC-9748, 19 December 1972; *ibid.*, PC-9746, 18 December 1972.

21 BP 4930, Record note and attachments to letter no. 143, 4 April 1972; BP 4964, Attachment to letter no. 174, PT 0616, 24 May 1972; *ibid.*, Attachment to letter no. 175, Additional points, 22 May 1972.

22 BP 4935, Meeting with Fallah, 13 October 1972, attachment to letter no. 242.

23 BP 4937, Manson to Taylor, PT 1349, 10 October 1972.

24 National Security Archive, 00807, State Department memorandum of conversation, Iranian oil crisis, 30 January 1973.

25 BP 4935, Audience, 14 January 1973, Attachment 2 to letter no. 298. On the Shah's demands and his talks with the Consortium, BP 4935, 4938, 4958 *passim.*

26 Nixon archives, WHCF, subject files, CO, box 38, folder [EX] CO 68 Iran 1 January–30 June 1973, Memorandum of conversation between Ansari, Agnew and Dunn, 15 January 1973; National Security Archive, 00800, Memorandum of conversation between Ansari, Afshar, Rogers and Rouse, 5 January 1973.

27 McCloy papers, series 20, box oil 2, folder 11, Memorandum on Iranian oil problem, 17 January 1973; National Security Archive, 00801, Armstrong to the Secretary, Iranian negotiations with Oil Consortium, 18 January 1973; *ibid.*, 00806, Katz to Armstrong, Iran oil negotiations, 29 January 1973; Nixon archives, WHSF, SMOF, Flanigan, box 4, folder Iran file [1954, 1957 (December 1971) – February 1973], Memorandum on Iranian oil problem, 29 January 1973.

28 National Security Archive, 00804, State Department intelligence note, Iranian

oil negotiations update, 24 January 1973; *ibid.*, 00805, State Department memorandum for the President, Iran's negotiations with the oil companies, 26 January 1973.

29 National Security Archive, 00815, Rogers memorandum for the President, Iranian Oil Consortium negotiations, 17 February 1973.

30 National Security Archive, 00815, Rogers memorandum for the President, Iranian Oil Consortium negotiations, 17 February 1973; *ibid.*, 00807, State Department memorandum of conversation, Iranian oil crisis, 30 January 1973; McCloy papers, series 20, box oil 2, folder 11, Attachment 1 to letter 320, Manson to members, 8 February 1973; *ibid.*, series 20, box oil 1, folder 44A, Proposed letter from Nixon to the Shah, n.d.

31 BP 2720, Iran – new agreement, 31 May 1973; BP 26072, Agreement, 19 July 1973.

32 Nixon archives, WHCF, subject files, CO, box 38, folder [EX] CO 68 Iran 1 January-30 June 1973, State Department memorandum for the President, Shah of Iran and Oil Consortium reach agreement, 1 March 1973.

33 Sarah Ahmad Khan, *Nigeria: The Political Economy of Oil* (Oxford, 1994), pp. 13, 18 and 22.

34 BP 48112, Llewellyn, Report on visit to Nigeria in March 1972, 27 April 1972.

35 Seymour, *OPEC*, p. 96; Skeet, *OPEC*, p. 82.

36 *MNC Hearings*, pt. 6, p. 56.

37 *Ibid.*, pt. 7, pp. 507–8.

38 Reuven Hollo, 'Oil and American foreign policy in the Persian Gulf 1947–1991', PhD thesis, University of Texas at Austin (1995), p. 230.

39 *MNC Hearings*, pt. 7, pp. 504–5.

40 *Ibid.*, pp. 412–13 and 509.

41 *Ibid.*, pp. 415 and 508–13; American University of Beirut, Speech delivered by Page to the AUB Alumni Association on 19 June 1973.

42 Daniel Yergin, *The Prize: The Epic Quest for Oil, Money, and Power* (New York, 1991), p. 597.

43 *MNC Hearings*, pt. 7, p. 542; *Newsweek*, 10 September 1973.

44 Donella H. Meadows, Dennis L. Meadows, Jorgen Randers and William Behrens III, *The Limits to Growth: A Report for the Club of Rome's Project on the Predicament of Mankind* (London, 1972); Louis Turner, *Oil Companies in the International System* (London, 1978), pp. 171–2; Seymour, *OPEC*, pp. 99–101.

45 James E. Akins, 'The oil crisis: this time the wolf is here', *Foreign Affairs* 51 (1973), 462–90.

46 *New York Times*, 24 and 31 May and 28 June 1973.

47 The best-known dissenter was M. A. Adelman. See his article 'Is the oil shortage real? Oil companies as OPEC tax collectors', *Foreign Policy* (Winter 1972–3), 69–107.

48 Seymour, *OPEC*, pp. 103–9; BP 6087, Drake to Heath with enclosure, 4 July 1973.

49 BP 78896, VL-04, 8 October 1973; BP 34728, VL-06, VL-09 and VL-10, 9 October 1973; BP 99995, Handwritten note of Piercy/Benard meeting with Yamani, 10 October 1973; *ibid.*, Piercy/Yamani meeting, 11 October 1973; BP 34825, Steel, OPEC – Tehran Agreement, 1 November 1973; BP 78896, VL-19, VL-20 and VL-21, 12 October 1973; BP 22098, Posted prices, 16 October 1973; Skeet, *OPEC*, p. 89.

50 McCloy papers, series 20, box oil 2, folder 10, Memorandum to the President, enclosed with McCloy to Haig, 12 October 1973; *ibid.*, Haig to McCloy, 15 October 1973. Reprinted in *MNC Hearings*, pt. 7, pp. 546–7.

51 On the absence of a reply from Nixon, see *MNC Hearings*, pt. 7, p. 450.

52 Kissinger, *Years of Upheaval*, pp. 520, 709 and 713–14; Joan Garratt, 'Euro-American energy diplomacy in the Middle East, 1970–80: the pervasive crisis', in Steven L. Spiegel (ed.), *The Middle East and the Western Alliance* (London, 1982), pp. 83–4; Lenczowski, *American Presidents*, p. 130.

53 George Lenczowski, 'The oil-producing countries', in Raymond Vernon (ed.), *The Oil Crisis* (New York, 1976), pp. 60–5; Kissinger, *Years of Upheaval*, pp. 536–8, 545 and 873; Penrose and Penrose, *Iraq*, pp. 414–15.

54 Kissinger, *Years of Upheaval*, pp. 538–91.

55 BP 106242, BP board minutes, 8 November 1973.

56 BP 34741, Sutcliffe, Meeting at FCO, 7 November 1973; Kissinger, *Years of Upheaval*, pp. 718, 745, 879 and 881; Masahiro Sasagawa, 'Japan and the Middle East', in Steven L. Spiegel (ed.), *The Middle East and the Western Alliance* (London, 1982), p. 36; Janice Gross Stein, 'Alice in Wonderland: The North Atlantic Alliance and the Arab-Israeli dispute', in *ibid.*, p. 55; Romano Prodi and Alberto Clo, 'Europe', in Raymond Vernon (ed.), *The Oil Crisis* (New York, 1976), p. 106; Yoshi Tsurumi 'Japan', in *ibid.*, p. 124; Kazushige Hirasawa, 'Japan's tilting neutrality', in J. C. Hurewitz (ed.), *Oil, the Arab-Israel Dispute, and the Industrial World: Horizons of Crisis* (Boulder, Col., 1976), p. 140.

57 Robert J. Lieber, *Oil and the Middle East War: Europe in the Energy Crisis*, Harvard Studies in International Affairs 35 (1976), pp. 14–15; Garratt, 'Euro-American energy diplomacy', p. 85.

58 Yergin, *The Prize*, p. 617; Prodi and Clo, 'Europe', p. 100; BP 11211, Boxshall, Supply emergency – government measures, 22 November 1973.

59 Turner, *Oil Companies*, p. 153.

60 BP 6087, Drake, Dinner at Chequers, 21 October 1973; Edward Heath, *The Course of My Life* (London, 1998), pp. 502–3.

61 BP 19049, OECD Oil Committee, Summary record of 26th session on 25 and 26 October 1973; BP 36471, Gregory, Note on OECD meetings on 25/26 October; Ulf Lantzke, 'The OECD and its International Energy Agency', in Vernon, *Oil Crisis*, pp. 217–27.

62 BP 114000, Walters, Meeting with Carrington, 29 October 1973; *ibid.*, Drake to Walker, 29 October 1973; BP 6087, Walters, Meeting with Carrington with attached note on BP Group, UK and European oil supplies, 30 October 1973;

BP 34741, Sutcliffe, Meeting at the FCO, 7 November 1973; BP 106242, BP board minutes, 8 November 1973.

63 BP 114000, Butcher to Walters, 31 October 1973; *ibid.*, Robertson, Record notes, 31 October 1973; BP 6087, Steel to Drake, 9 November 1973.

64 Robert B. Stobaugh, 'The oil companies in the crisis', in Vernon, *Oil Crisis*, pp. 190–1; BP 36471, Burchell, Reduction in supply commitment to SFBP, 30 October 1973; BP 34741, Burchell, Meeting of French Prime Minister and heads of oil companies in France, 16 November 1973; BP 114000, Chenevier, Record note, 16 November 1973; BP 106242, BP board minutes, 22 November 1973.

65 BP 114000, Walters, Meeting with minister [name withheld by author], 13 November 1973.

66 BP 6087, Laidlaw to Drake, 6 December 1973; *ibid.*, Drake to minister [name withheld by author], 6 December 1973.

67 Heath, *Course of My Life*, p. 503.

68 Sasagawa, 'Japan and the Middle East', p. 36; Stein, 'Alice in Wonderland: The North Atlantic Alliance and the Arab-Israeli dispute', p. 55; Lieber, *Oil and the Middle East War*, pp. 30–1; Tsurumi 'Japan', p. 124; Turner, *Oil Companies*, pp. 179–80; Hirasawa, 'Japan's tilting neutrality', p. 141.

69 BP 34741, Walters, Record note, 26 November 1973; Seymour, *OPEC*, p. 122.

70 There are many fuller accounts of this decision. See, for example, Yergin, *The Prize*, p. 625.

71 *MNC Hearings*, pt. 7, p. 186.

72 Kissinger, *Years of Upheaval*, p. 885.

73 *Ibid.*, p. 888; Yergin, *The Prize*, p. 626.

74 Kissinger, *Years of Upheaval*, ch. 20; Lantzke, 'The OECD and its International Energy Agency', pp. 217–27.

75 Jill Crystal, *Kuwait: The Transformation of an Oil State* (Boulder, Col., 1992), p. 43; Seymour, *OPEC*, p. 228; Yergin, *The Prize*, pp. 647–52; Khan, *Nigeria*, pp. 18–19; Waddams, *Libyan Oil*, pp. 258–9.

76 Kobrin, 'The nationalisation of oil production, 1918–1980', p. 137.

77 Kenneth A. Rodman, *Sanctity versus Sovereignty: The United States and the Nationalization of Natural Resource Investments* (New York, 1988), chs. 1–7; Charles Lipson, *Standing Guard: Protecting Foreign Capital in the Nineteenth and Twentieth Centuries* (Berkeley, Calif., 1985); Charles R. Kennedy, 'Relations between transnational corporations and governments of host countries: a look to the future', *Transnational Corporations*, 1 (1992), 67–91; Kobrin, 'The nationalisation of oil production, 1918–1980'; Kobrin, 'Expropriation as an attempt to control foreign firms in LDCs: trends from 1960 to 1979', *International Studies Quarterly* 28 (1984), 329–48; and Kobrin, 'Diffusion as an explanation of oil nationalization: or the domino effect rides again', *Journal of Conflict Resolution* 29 (1985), 3–32.

78 A. E. Safarian, *Multinational Enterprise and Public Policy: A Study of the Industrial Countries* (Aldershot, 1993), pp. 379–82.

79 *Ibid.*, pp. 222–8.
80 *Ibid.*, pp. 333–9.
81 BP 9527, BP New York to BP London, 5 April 1977.
82 *MNC Hearings*, pt. 7, esp. pp. 1–12 and 470–8.
83 Hollo, 'Oil and American foreign policy', p. 240.
84 *MNC Hearings, passim.*; Turner, *Oil Companies*, p. 190.
85 Turner, *Oil Companies*, pp. 191–4.
86 Tony Benn, *Against the Tide: Diaries 1973–1976* (London, 1989), p. 419.
87 *Ibid.*, p. 449.

RETROSPECT AND CONCLUSION

1 BP 102837 and 103052, Management reports for board meetings, January 1957 and January 1971; BP 115941, Monthly report for general management, Supply and Development division, December 1956; BP 118733, Accounts papers.
2 BP 102837, Management report for board meeting, January 1957; BP 118733, Accounts papers. Iran is counted as one of the countries because the Consortium (in which BP held a 40 per cent share) operated the oil fields in its area, although they were formally nationalised.
3 Walter E. Skinner, *Oil and Petroleum Year Book*, 1955 and 1970–1. Iran is counted as one of the countries because the Consortium (in which BP held a 40 per cent share) operated the Abadan refinery, although it was formally nationalised. Rhodesia (later Zimbabwe) is not counted because the Umtali refinery (in which BP held a 21 per cent share) had been shut down since early 1966 owing to the imposition of sanctions against the white settler regime.
4 For general studies covering these trends in the international economy, see A. G. Kenwood and A. L. Lougheed, *The Growth of the International Economy, 1820–1990*, 3rd edn (London, 1992); Angus Maddison, *The World Economy in the Twentieth Century* (Paris, 1989); James Foreman-Peck, *A History of the World Economy: International Economic Relations since 1850*, 2nd edn (Hemel Hempstead, 1995); W. M. Scammell, *The International Economy since 1945* (London, 1980). On the spread of US multinationals, see Mira Wilkins, *The Maturing of Multinational Enterprise: American Business Abroad from 1914 to 1970* (Cambridge, Mass., 1974). On the environment for international business, Geoffrey Jones, *The Evolution of International Business, An Introduction* (London, 1996), esp. chs. 2 and 8. On decolonisation, John Darwin, *Britain and Decolonisation: The Retreat from Empire in the Postwar World* (London, 1988); R. F. Holland, *European Decolonization 1918–1981: An Introductory Survey* (London, 1985).
5 BP 56397, Policy Planning Staff, OPEC, 7 October 1971.
6 BP 103052, Management report for board meeting, January 1971.
7 BP 63529, Belgrave to Down, Long term impacts of OPEC settlement, 5 February 1971.

8 For a clear statement of BP's volume-driven downstream strategy, see BP 63695, Policy Planning Staff, Europe, 16 November 1972.

9 BP 10231, Grassick, Draft Euroday talk, 4 December 1973.

10 BP 118237, Platt, History of the BP Tanker Company, p. 348.

11 BP 38866, Drake, Strategic Planning Committee, 10 February 1971. On the consideration of strategy, see, for example, documents in BP 38866, 56385 and 65611.

Notes to the tables, graphs and diagrams

TABLES

0.1 Christopher Schmitz, 'Introduction', in Schmitz (ed.), *Big Business in Mining and Petroleum* (Aldershot, 1995), p. xiii.

1.1 BP 96455, Interim report of the Future Programme Committee, 25 July 1951.

1.2 The income statements in the published AIOC and BP *Annual Reports and Accounts* before 1956 exclude a large number of subsidiaries and do not, therefore, represent the whole of the BP group. To remedy that problem, a retro-consolidation of the published accounts and excluded subsidiaries was carried out for this book by J. K. Taggart, a former BP employee. His data form the basis of this table, in conjunction with BP 80540, Kuwait income tax schedule, 1951–70. The price index used is that for plant and machinery in C. H. Feinstein, *National Income, Expenditure and Output in the United Kingdom, 1855–1965* (Cambridge, 1972), table 63.

3.1 BP 102846–102854, Management reports for board meetings, November 1957–July 1958.

5.1 AIOC and BP, *Annual Reports and Accounts*, 1950–60; BP 95578–95582, Accounts schedules, 1950–4. The price index used is that for retail prices in B. R. Mitchell, *British Historical Statistics* (Cambridge, 1988), pp. 740–1.

5.2 On 1955 profits, see note to table 1.2. Post-1955 profits are from BP, *Annual Reports and Accounts*, 1956–60. The price index used is that for plant and machinery in Feinstein, *National Income, Expenditure and Output*, table 63.

6.1 PRO FO 371/141201, Working party report, The special position of Kuwait as a source of oil, 5 October 1959.

8.1 BP 59756, Exploration and Production Department, BP and competitor companies, January 1972.

11.1 AIOC and BP *Annual Reports and Accounts*, 1950–75; Walter E. Skinner, *Oil and Petroleum Year Book*, 1950–64; continued as Walter R. Skinner, *Oil and Petroleum Year Book*, 1965–1970/71; continued as *Oil and Petroleum International Year Book*, 1971/72; continued as *Oil and Gas International Year Book*, 1973–1976/77.

11.2 BP 4944, Strength and weakness profile of oil majors, March 1974.

13.1 BP staff files on directors; BP Archive shelf profiles; *Who's Who* and *Who was Who* (various volumes); R. W. Ferrier, 'Sir Maurice Richard Bridgeman', in D. J. Jeremy (ed.), *Dictionary of Business Biography*, vol. I (London, 1984), pp. 440–3; Ferrier, 'Sir Arthur Eric Courtney Drake and William Milligan Fraser, 1st Lord Strathalmond of Pumpherston', in *ibid.*, vol. II (London, 1984), pp. 169–72 and 423–7.

13.2 BP Archive shelf profiles; *Who's Who* and *Who was Who* (various volumes); R. P. T. Davenport-Hines, 'Cameron Fromanteel Cobbold, 1st Lord Cobbold, in D. J. Jeremy (ed.), *Dictionary of Business Biography*, vol. I (London, 1984), pp. 707–10; Peter W. Brooks, 'Ronald Morce Weeks, Lord Weeks of Ryton, in *ibid.*, vol. V (London, 1986), pp. 715–18.

14.1 BP 91173, Hunter, History of British Hydrocarbon Chemicals, Table A, 6 September 1962.

14.2 BP 50862, 51031, 91173. British Hydrocarbon Chemicals' subsidiaries, Forth Chemicals and Grange Chemicals, are included.

15.1 BP 51031. The data are for British Hydrocarbon Chemicals' financial years, whose end-date was changed from 30 April to 31 March in 1965. The source data for 1964 were therefore for only 11 months. They have been grossed up to 12 months in the table to give a closer comparison with other years. British Hydrocarbon Chemicals' subsidiaries, Forth Chemicals and Grange Chemicals, are included.

15.2 BP 2721, 38551, 38553, 57679, 58446, 58448 and 65848.

16.1 W. J. Newby and R. J. Deam, 'How BP is using linear programming', *World Petroleum* 42 (June 1971), 234; BP 102860, Management report for board meeting, January 1959.

20.1 *Fortune*, August 1971; Christopher Schmitz, 'Introduction', in Schmitz (ed.), *Big Business in Mining and Petroleum* (Aldershot, 1995), p. xiii.

GRAPHS AND DIAGRAMS

0.1 Adapted from United States Congress, Senate Committee on Foreign Relations, *Hearings before the Subcommittee on Multinational Corporations*, 93rd Congress, 1st and 2nd sessions (Washington D.C., 1974–5), pt. 5, p. 290.

1.1 BP 102770, 102796 and 102826, Management reports for board meetings, January 1952, 1954 and 1956.

1.2 BP 102759, 102770, 102781, 102796 and 102826, Management reports for board meetings, January 1951–4 and 1956; BP 95578–95582, Accounts schedules, 1950–4; AIOC, *Annual Reports and Accounts*, 1950–4.

1.3 BP 102759 and 102837, Management reports for board meetings, January 1951 and 1957; BP 95578, Accounts schedules, 1950; AIOC, *Annual Report and Accounts*, 1950; BP 115941, Monthly report for general management, Supply and Development Division, December 1956.

2.1 BP 59585, Pattinson to Brandon Grove, 28 November 1960.

3.1   BP 30580, Monthly bunker statements.

4.1   BP 17470, Statistical review of the world oil industry, 1952.

4.2   BP 9228.

6.1   BP 40561, Seven majors in the sixties.

7.1   BP, *Statistical Review of the World Oil Industry*, 1967 and 1970.

8.1   BP 59756, Exploration and Production Department, BP and competitor companies, January 1972.

8.2   BP 59756, Exploration and Production Department, BP and competitor companies, January 1972.

8.3   World data are based on BP's 1971 and 1976 estimates of the ultimate recoverable reserves of giant oil fields (fields of more than 500 million barrels) backdated to the year of discovery, with discoveries in non-giant fields estimated at $1\frac{1}{4}$ billion barrels a year for 1900–35 and $3\frac{3}{4}$ billion barrels a year for 1935–75. This was the approximation used by BP geologist H. R. Warman in, for example, his article, 'Future problems in petroleum exploration', *Petroleum Review* 25 (1971), 96–101. BP data for 1908–70 are BP estimates of ultimate recoverable reserves at the end of 1970, backdated to the year of discovery; and for 1971–5, the data are BP Exploration and Production Department's annual estimates of recoverable reserves after taking account of new discoveries, revisions and other changes. See BPX report EXT 63781, Sayer, Giant oil fields of the world discovered by 1970. Ultimate recoverable reserves as estimated by Geological Division in 1971 and 1981; BPX report EXT 63782, Sayer, Giant oil fields of the world, April 1976; BPX report EXT 57512, Sayer, BP discoveries to the end of 1970 – an old working list, 1971; BPX report OC 6914, Exploration and Production Department annual reports, 1971–5.

8.4   As for figure 8.3, with some data also drawn from BPX report OC 6274, BP share of worldwide proven and probable recoverable oil reserves at 1 January 1970; BPX report OC 8113, BP share of worldwide proven and probable recoverable oil reserves at 1 January 1972; DeGolyer and MacNaughton, *Twentieth-Century Petroleum Statistics* (Dallas, Tex., 1994).

8.5   BP 57693, Kent and Warman, Exploration planning and expenditure levels, 21 June 1971; BPX report OC 6254, Livingstone, BP's exploration expenditure and reserves 1951–68, July 1969; BPX report OC 6914, Exploration and Production Department annual reports, 1970 and 1976. Inflation index from D. Butler and G. Butler, *British Political Facts 1900–1985*, 6th edn (London, 1986), pp. 381–2. Expenditure on gas exploration is not included.

8.6   BP 55775, Bexon, Exploration capital expenditure policy, September 1974.

8.7   The data for 1909–70 are BP's 1971 estimates of ultimate recoverable reserves at the end of 1970, backdated to the year of discovery. For 1971–8 the data are BP Exploration and Production Department's annual estimates of recoverable reserves after taking account of new discoveries, revisions and other changes. See BPX report EXT 57512, Sayer, BP discoveries to the end of 1970 – an old working list, 1971; BPX report OC 6274, BP share of worldwide proven and probable recoverable oil reserves at 1 January 1970; BPX report OC 8113, BP share of worldwide proven and probable recoverable oil reserves at 1 January 1972; BPX item 2340599, BP share of worldwide proven and probable recoverable oil reserves at 1 January 1974; BPX report OC 6914, Exploration and Production Department annual reports, 1971–8.

9.1 Standard Oil (NJ), Socal, Mobil, Texaco and Gulf data are from their *Annual Reports and Accounts*. BP data are from BP 102770 and 102837, Management reports for board meetings, January 1952 and 1957; BP 95617, 95620 and 118733, Accounts schedules, 1961, 1966 and 1968–73. Shell data are from *Petroleum Press Service* (various issues) and Walter E. Skinner, *Oil and Petroleum Year Book*, 1950–64; continued as Walter R. Skinner, *Oil and Petroleum Year Book*, 1965–1970/71; continued as *Oil and Petroleum International Year Book*, 1971/72 (various issues).

9.2 Standard Oil (NJ), Socal, Mobil, Texaco and Gulf data are from their *Annual Reports and Accounts*. BP data are from BP 102759, 102837 and 102907, Management reports for board meetings, January 1951, 1957 and 1962; BP 95578, 95617, 95620 and 118733, Accounts schedules, 1950, 1961, 1966 and 1968–73; BP 115941, Monthly report for general management, Supply and Development Division, December 1956; AIOC, *Annual Report and Accounts*, 1950. Shell data are from Walter E. Skinner, *Oil and Petroleum Year Book*, 1950–1964; continued as Walter R. Skinner, *Oil and Petroleum Year Book*, 1965–1970/71; continued as *Oil and Petroleum International Year Book*, 1971/72 (various issues).

9.3 Standard Oil (NJ), Socal, Mobil, Texaco and Gulf data are from their *Annual Reports and Accounts*. BP data are from BP 102771, 102837, 102882, 102885, 102887, 102890, 102982 and 103052, Management reports for board meetings, (selected months) 1952–71; BP 30576, Marine bunker data. Shell data are from *Petroleum Press Service* (various issues) and Edith T. Penrose, *The Large International Firm in Developing Countries: The International Petroleum Industry* (London, 1968), p. 107.

9.4 BP 40561, Seven majors in the sixties.

9.5 BP 12918, 119420 and 119421. The data are for inland trade in motor spirits, kerosenes, automotive and heating gas oils, diesel oils and fuel oils.

9.6 Monopolies Commission, *A Report on the Supply of Petrol to Retailers in the United Kingdom* (HMSO, 1965), p. 23; SMBP archives, Statistics files.

9.7 Monopolies Commission, *A Report on the Supply of Petrol to Retailers in the United Kingdom* (HMSO, 1965), p. 23; SMBP archives, Statistics files.

10.1 BP 54603.

10.2 T. H. Bingham and S. M. Gray, *Report on the Supply of Petroleum Products to Rhodesia* (HMSO, 1975), p. 293.

11.1 AIOC and BP, *Annual Reports and Accounts*, 1950–1975; E. Skinner, *Oil and Petroleum Year Book*, 1950–64; continued as Walter R. Skinner, *Oil and Petroleum Year Book*, 1965–1970/71; continued as *Oil and Petroleum*

International Year Book, 1971/72; continued as *Oil and Gas International Year Book*, 1973–1976/77.

11.2 Calculated from data in BP 118237, Platt, History of the BP Tanker Company, appendices 1, 3 and 4.

12.1 Sales data from BP 102771, 102796, 102810, 102837, 102848, 102860, 102866, 102873, 102875, 102879, 102882, 102885, 102887, 102890, 102893, 102896, 102898, 102907, 102910, 102913, 102915, 102918, 102921, 102924, 102926, 102940, 102970, 102982, 102994, 103016, 103028, 103041, 103052, Management reports for board meetings, 1952–71 (selected months); BP 30576, Marine bunker data; BP 115941, Monthly report for general management, Supply and Development Division, December 1956. On profits for 1950–5, see note to table 1.2. Post-1955 profits are from BP, *Annual Reports and Accounts*, 1956–71. The price index used is that for plant and machinery in Feinstein, *National Income, Expenditure and Output*, table 63; and calculated from Central Statistical Office, *National Income and Expenditure 1963–1973* (HMSO, 1974), tables 54 and 55.

12.2 *Financial Times*, 1950–1971; AIOC and BP, *Annual Reports and Accounts*, 1950–60; BP 95578–95582, Accounts schedules, 1950–4. The price index used is that for retail prices in B. R. Mitchell, *British Historical Statistics* (Cambridge, 1988), pp.740–1.

12.3 Parra (F. R.) Associates, *The International Oil Industry, 1950–1992; A Statistical History* (Reading, 1993) [privately printed], Table 16.

12.4 *Ibid*.

12.5 BP 10231, Board, The Group's trading structure (attachment), 11 May 1972; BP 38921, Searle, Group accounts and finance (appendix 11), BP management conference, May 1968; BP, *Annual Reports and Accounts*, 1961–72.

13.1 BP 106299.

14.1 Adapted from chart in BP 54926, BHC pamphlet, September 1958, with added information from BP 91173, Hunter, History of British Hydrocarbon Chemicals, Table A, 6 September 1962.

14.2 BP 91173 and 51031. Other chemical intermediates include butadiene, cumene, detergent alkylate, phenol, propylene tetramer and styrene.

14.3 Adapted from BP 29176, Howes, Recommendations concerning BP future policy in petroleum chemicals, section 2, fig. 1, December 1962, with information added from BP 100859, Table of DCL interests in chemicals and plastics, March 1959 and BP 76464, HRBW, The Distillers Company Limited, 8 November 1962.

15.1 BP 51031. The data are for British Hydrocarbon Chemicals' financial years, whose end-date was changed from 30 April to 31 March in 1965. The source data for 1964/5 were therefore for only 11 months. They have been grossed up to 12 months in the graph to give a closer comparison with other years. Other chemical intermediates include butadiene, cumene, detergent alkylate, ethylene dichloride, isobutylene, methanol, phenol, polybutene, propylene tetramer and styrene.

15.2 *Chemical Age*, 27 July 1968.
16.1 W. J. Newby, 'Planning refinery production', in C. M. Berners-Lee (ed.), *Models for Decision* (London, 1965), p.47.
16.2 W. J. Newby, 'Running a large company by computer', *New Scientist*, 4 June 1964.
16.3 BP 42333, Study of data flow in Supply and Development department, C-E-I-R Ltd, June 1966.
19.1 Charles R. Kennedy, 'Relations between transnational corporations and governments of host countries: a look to the future', *Transnational Corporations* 1 (1992), 69.
20.1 BPX 13376, 12123, 12124, 12127, 2340599, 11423, 11424, 11420.
20.2 BP, *Statistical Review of the World Oil Industry*, 1960 and 1971–5; AIOC and BP, *Annual Reports and Accounts*, 1950–75; Walter E. Skinner, *Oil and Petroleum Year Book*, 1950–64; continued as Walter R. Skinner, *Oil and Petroleum Year Book*, 1965–1970/71; continued as *Oil and Petroleum International Year Book*, 1971/72; continued as *Oil and Gas International Year Book*, 1973–1976/77.
20.3 BP, *Annual Reports and Accounts*, 1965 and 1970–5; BP, *Statistical Review of the World Oil Industry*, 1965 and 1975; Walter R. Skinner, *Oil and Petroleum Year Book*, 1965 and 1970/71; continued as *Oil and Petroleum International Year Book*, 1971/72; continued as *Oil and Gas International Year Book*, 1973–1976/77.

Select bibliography

ARCHIVES

BP archives (BP), BP Archive, University of Warwick.
BP Exploration records (BPX), BP Research Centre, Sunbury-on-Thames.
BP Sunbury technical records, BP Research Centre, Sunbury-on-Thames.
R. A. Butler papers, Trinity College, Cambridge.
Cambridge University Appointments Board archives, Cambridge University Library.
Distillers Company Epsom Research Centre records, BP Archive, University of Warwick.
Federation of British Industries archives, Modern Records Centre, University of Warwick.
Iranian Oil Participants (Consortium) archives (IOP), c/o BP Archive, University of Warwick.
Iraq Petroleum Company archives (IPC), c/o BP Archive, University of Warwick.
Sir Peter Kent papers, University of Nottingham.
John J. McCloy papers, Amherst College, Massachusetts.
National Security Archive, George Washington University, Washington D.C.
Richard M. Nixon archives, National Archive, Washington D.C.
Public Record Office (PRO), Kew, London.
Shell-Mex and BP archives (SMBP), c/o BP Archive, University of Warwick.
Sir Roger Stevens papers, Churchill Archive Centre, Churchill College, Cambridge.
State Department Freedom of Information Act papers, Washington D.C.

INTERVIEWS

Robin Adam, Deputy Chairman, British Petroleum.
Sir Lindsay Alexander, Non-executive Director, British Petroleum.
George Ashford, Director, Distillers; Managing Director, British Petroleum.
J. Angus Beckett, Chairman, OEEC Oil Committee; Permanent Under-Secretary, Ministry of Power.
Robert Belgrave, Oil and Middle East Desk, Foreign Office; Policy Planning Adviser, British Petroleum.

602

Roger Bexon, Deputy Chairman, British Petroleum.
Sir John Browne, Group Chief Executive, BP.
Basil Butler, Managing Director, British Petroleum.
Sir Peter Cazalet, Deputy Chairman, British Petroleum.
Rodney Chase, Deputy Group Chief Executive, BP.
Professor R. J. (Bob) Deam, Manager, Operational Research Division, British Petroleum.
Sir Alastair Down, Deputy Chairman, British Petroleum.
Sir Eric Drake, Chairman, British Petroleum.
Lord Greenhill of Harrow, Permanent Under-Secretary of State, Foreign and Commonwealth Office, and Head of the Diplomatic Service; HM Government Director, British Petroleum.
Frank Gulliver, Manager, Group Internal Audit, British Petroleum.
Sir Frederic Harmer, HM Government Director, British Petroleum.
Sir Robert Horton, Chairman, British Petroleum.
John Hunter, Director, Distillers Company; Director, BP Trading.
David Jenkins, Chief Executive, Technology, BP Exploration.
Sir Christophor Laidlaw, Deputy Chairman, British Petroleum.
Alistair Macleod Matthews, Assistant General Manager, Exploration and Production, British Petroleum.
Dr A. John Martin, General Manager, Exploration, British Petroleum.
Stuart McColl, General Manager, Supply, British Petroleum.
Michael Micklewright, Supply and Development, British Petroleum.
H. Steven Mullaly, Director, BP Trading.
W. J. (Bill) Newby, General Manager, Computer Department, British Petroleum.
Francisco Parra, Secretary General, OPEC.
Commander Edward Platt, Director, BP Tanker Company.
Sir Peter Ramsbotham, British Ambassador to Iran.
Frank Rickwood, Managing Director, BP Exploration and Production.
Fuad Rouhani, Secretary General, OPEC.
Lord Simon of Highbury, Chairman, British Petroleum.
Charles Spahr, Chairman, Standard Oil Company (Sohio).
Sir David Steel, Chairman, British Petroleum.
Glyn Thomas, Regional Geophysicist, British Petroleum.
Ken Walder, Group General Counsel, British Petroleum.
Sir Peter Walters, Chairman, British Petroleum.
Harry Warman, Chief Geologist, British Petroleum.
Sir Denis Wright, British Ambassador to Iran.

OFFICIAL PUBLICATIONS

Bingham, T. H. and S. M. Gray, *Report on the Supply of Petroleum Products to Rhodesia* (HMSO, 1975).
Committee of Enquiry into Shipping (Rochdale Committee), *Report* (Cmnd 4337, 1970).

Monopolies Commission, *A Report on the Supply of Petrol to Retailers in the United Kingdom* (HMSO, 1965).

Parliamentary Debates (Hansard), House of Commons, 1950–75.

Special Review Committee of the Board of Directors of Gulf Oil Corporation, *Report* (In the United States District Court of Columbia, December 1975).

United States Congress, House of Representatives, *Petroleum Survey: 1957 Outlook, Oil Lift to Europe, Price Increases*, 85th Congress, 1st session (Washington D.C., 1957).

United States Congress, Senate, Select Committee on Small Business, Subcommittee on Monopoly, *The International Petroleum Cartel: Staff Report to the Federal Trade Commission*, 82nd Congress, 1st and 2nd sessions (Washington D.C., 1952).

United States Congress, Senate, Committee on Foreign Relations, *Hearings before the Subcommittee on Multinational Corporations*, 93rd Congress, 1st and 2nd sessions (Washington D.C., 1974–5).

United States Congress, Senate, Committee on Foreign Relations, Subcommittee on Multinational Corporations, *Report on Multinational Oil Corporations and U.S. Foreign Policy*, 93rd Congress, 2nd session (Washington D.C., 1975).

United States, Federal Energy Administration, *The Relationship of Oil Companies and Foreign Governments* (Washington D.C., 1975).

NEWSPAPERS AND JOURNALS

Avanti
BP Shield
BP Shield International
Chemical Age
Chemical Age International
Corriere della Sera
Daily Express
Daily Telegraph
DCL Gazette
The Economist
Engineering
Evening News
Evening Standard
Financial Times
Fortune
Garage
Gazzetta Ufficiale della Repubblica Italiana
The Guardian
Industrial Chemist
Japan Petroleum Weekly

Middle East Economic Survey
New Scientist
Newsweek
New York Times
The Observer
Oil and Gas Journal
Paese Sera
Petroleum Intelligence Weekly
Petroleum Press Service
Petroleum Review
Petroleum Times
Petroleum Week
The Scotsman
The Statist
Sunday Telegraph
Sunday Times
The Times
West Africa
World Petroleum

PUBLISHED SOURCES OF DATA

Anglo-Iranian Oil Company/British Petroleum Company, *Annual Reports and Accounts*, 1950–75.

British Petroleum, *Statistical Review of the World Oil Industry*, 1955–75.

Butler, D. and G. Butler, *British Political Facts 1900–1985*, 6th edn (London, 1986).

Central Statistical Office, *National Income and Expenditure 1963–1973* (London, 1974).

DeGolyer and MacNaughton, *Twentieth-Century Petroleum Statistics* (Dallas, Tex., 1994).

Feinstein, C. H., *National Income, Expenditure and Output in the United Kingdom, 1855–1965* (Cambridge, 1972).

Gulf Oil Corporation, *Annual Reports and Accounts*, 1950–75.

Mitchell, B. R., *British Historical Statistics* (Cambridge, 1988).

Parra (F. R.) Associates, *The International Oil Industry, 1950–1992: A Statistical History* (Reading, 1993) [privately printed].

Skinner, Walter E., *Oil and Petroleum Year Book*, 1950–64.

Skinner, Walter R., *Oil and Petroleum Year Book*, 1965–70/1.

Skinner, Walter R., *Oil and Petroleum International Year Book*, 1971/2.

Skinner, Walter R., *Oil and Gas International Year Book*, 1973–6/7.

Socony-Vacuum Oil Company/Socony Mobil Company/Mobil Oil Corporation, *Annual Reports and Accounts*, 1950–75.

Standard Oil Company of California, *Annual Reports and Accounts*, 1950–75.

Standard Oil Company (New Jersey)/Exxon Corporation, *Annual Reports and Accounts*, 1950–75.

Texas Company/Texaco, *Annual Reports and Accounts*, 1950–75.

BOOKS, ARTICLES AND DISSERTATIONS

Acheson, Dean, *Present at the Creation: My Years in the State Department* (London, 1970).

Adelman, M. A., *The World Petroleum Market* (Baltimore, Md., 1972).

Adelman, M. A., 'Is the oil shortage real? Oil companies as OPEC tax collectors', *Foreign Policy* (Winter 1972–3), 69–107.

Akins, James E., 'The oil crisis: this time the wolf is here', *Foreign Affairs* 51 (1973), 462–90.

Alford, B. W. E., *British Economic Performance, 1945–1975* (London, 1988).

Allan, J. A., *Libya: The Experience of Oil* (London, 1981).

Amuzegar, Jahangir, 'The oil story: facts, fiction and fair play', *Foreign Affairs* 51 (1973), 676–89.

Anderson, Irvine H., *Aramco, the United States and Saudi Arabia: A Study of the Dynamics of Foreign Oil Policy, 1933–1950* (Princeton, N.J., 1981).

Arkell, W. J., 'George Martin Lees, 1898–1955', *Biographical Memoirs of Fellows of the Royal Society* 1 (1955), 163–72.

Avery, P., *Modern Iran* (London, 1965).

Balfour-Paul, Glen, *The End of Empire in the Middle East: Britain's Relinquishment of Power in her Last Three Arab Dependencies* (Cambridge, 1991).

Bamberg, J. H., *The History of The British Petroleum Company: Volume II, The Anglo-Iranian Years, 1928–1954* (Cambridge, 1994).

Benn, Tony, *Against the Tide: Diaries 1973–1976* (London, 1989).

Bharier, Julian, *Economic Development in Iran, 1900–1970* (London, 1971).

Bill, James A., *The Eagle and the Lion: The Tragedy of American-Iranian Relations* (New Haven, Conn., 1988).

Bird, Kai, *The Chairman: John J. McCloy, The Making of the American Establishment* (New York, 1992).

Birmingham, David, *The Decolonisation of Africa* (London, 1995).

Blair, John M., *The Control of Oil* (London, 1976).

Borkin, J., *The Crime and Punishment of IG Farben* (London, 1979).

Bowden, B. V., *Faster than Thought* (London, 1953).

Breen, David H., *Alberta's Petroleum Industry and the Conservation Board* (Alberta, 1993).

British Petroleum Company, *Our Industry*, 5th edn (London, 1977).

Brooks, P. W., 'Ronald Morce Weeks', in D. J. Jeremy (ed.), *Dictionary of Business Biography*, vol. V (London, 1986).

Cairncross, A., *The British Economy since 1945: Economic Policy and Performance, 1945–1990* (Oxford, 1992).

Caldwell, Malcolm, *Oil and Imperialism in East Asia*, Spokesman pamphlet no. 20 (Nottingham, 1971).

Center for Strategic and International Studies, *The Gulf: Implications of British Withdrawal* Georgetown University, Special report series, no. 8 (Washington D.C., 1969).

Chandler, Alfred D., *Strategy and Structure: Chapters in the History of the American Industrial Enterprise* (Cambridge, Mass., 1962).

Chandler, Alfred D., *Scale and Scope: The Dynamics of Industrial Capitalism* (Cambridge, Mass., 1990).

Channon, Derek F., *The Strategy and Structure of British Enterprise* (London, 1973).

Chapman, Keith, *The International Petrochemical Industry* (Oxford, 1991).

Charnes, A., W. W. Cooper and B. Mellon, 'Blending aviation gasolines', *Econometrica* 20 (1952), 135–59.

Chernow, Ron, *Titan: The Life of John D. Rockefeller* (New York, 1998).

Chisholm, A. H. T., *The First Kuwait Oil Concession Agreement: A Record of the Negotiations, 1911–1934* (London, 1975).

Chown, John, *The Corporation Tax – A Closer Look*, Institute of Economic Affairs (London, 1965).

Churchman, C. W., R. L. Ackoff and E. L. Arnoff, *Introduction to Operations Research* (New York, 1957).

Clark, John G., *The Political Economy of World Energy: A Twentieth Century Perspective* (Hemel Hempstead, 1990).

Cooper, Bryan, *Alaska – The Last Frontier* (London, 1972).

Cooper, Bryan, and T. F. Gaskell, *North Sea Oil – The Great Gamble* (London, 1966).

Cooper, Bryan, and T. F. Gaskell, *The Adventure of North Sea Oil* (London, 1976).

Corley, T. A. B., *A History of the Burmah Oil Company: Volume II, 1924–1966* (London, 1988).

Cottam, Richard W., *Nationalism in Iran updated through 1978* (Pittsburgh, Pa., 1979).

Crystal, Jill, *Kuwait: The Transformation of an Oil State* (Boulder, Col., 1992).

Curtin, Philip, Steven Feierman, Leonard Thompson and Jan Vansina, *African History from Earliest Times to Independence*, 2nd edn (London, 1995).

Darmstadter, Joel, *Energy in the World Economy: A Statistical Review of Trends in Output, Trade, and Consumption since 1925* (Baltimore, Md., 1971).

Darmstadter, Joel, and Hans H. Landberg, 'The economic background', in Raymond Vernon (ed.), *The Oil Crisis* (New York, 1976), pp. 15–37.

Darwin, John, *Britain and Decolonisation: The Retreat from Empire in the Postwar World* (London, 1988).

Dasgupta, Biplab, *The Oil Industry in India* (London, 1971).

Davies, Gordon B., *An Introduction to Electronic Computers* (New York, 1965).

Davies, Michael, *Belief in the Sea: State Encouragement of British Merchant Shipping and Shipbuilding* (London, 1992).

Dixon, Donald F., 'The development of the solus system of petrol distribution in the United Kingdom, 1950–1960', *Economica* 29 (1962), 40–52.

Dow, J. C. R., *The Management of the British Economy, 1945–1960* (Cambridge, 1964).

Economic Commission for Europe, *Market Trends and Prospects for Chemical Products*, 3 vols. (New York, 1969).

Economic Commission for Europe, *Market Trends and Prospects for Chemical Products*, 2 vols. (New York, 1973).

Eden, A., *Full Circle* (London, 1960).

Eisenhower, Dwight D., *The White House Years: Mandate for Change, 1953–1956* (London, 1963).

Encarnation, Dennis J., *Dislodging Multinationals: India's Strategy in Comparative Perspective* (Ithaca, N.Y., and London, 1989).

Evans, William S., 'Petroleum in the Eastern Hemisphere' (First National City Bank of New York, 1959).

Fage, J. D., *A History of Africa* (London, 1978).

Falcon, N. L., and Sir Kingsley Dunham, 'Percy Edward Kent, 1913–1986', *Biographical Memoirs of Fellows of the Royal Society* 33 (1987), 345–73.

Farmanfarmaian, Khodadad, Armin Gutowski, Saburo Okita, Robert V. Roosa and Carroll L. Wilson, 'How can the world afford OPEC oil?', *Foreign Affairs* 53 (1975), 201–22.

Ferrier, R. W., *The History of The British Petroleum Company: Volume I, The Developing Years, 1901–1932* (Cambridge, 1982).

Fesharaki, Fereidun, *Development of the Iranian Oil Industry: International and Domestic Aspects* (New York, 1976).

Flamm, Kenneth, *Creating the Computer* (Washington D.C., 1988).

Foreman-Peck, James, *A History of the World Economy: International Economic Relations since 1850*, 2nd edn (Hemel Hempstead, 1995).

Frankel, P. H., *Essentials of Petroleum* (London, 1940).

Galambos, Louis, *The Public Image of Big Business in America, 1880–1940* (Baltimore, Md., 1975).

Garratt, Joan, 'Euro-American energy diplomacy in the Middle East, 1970–80: the pervasive crisis', in Steven L. Spiegel (ed.), *The Middle East and the Western Alliance* (London, 1982), pp. 82–103.

Gasiorowski, Mark J., *US Foreign Policy and the Shah: Building a Client State* (New York, 1991).

Gasiorowski, Mark J., 'US foreign policy toward Iran during the Mussadiq era', in David W. Lesch (ed.), *The Middle East and the United States: A Historical and Political Reassessment* (Boulder, Col., 1996), pp. 51–66.

Gerges, Fawaz A., *The Superpowers and the Middle East: Regional and International Politics, 1955–1967* (Boulder, Col., 1994).

Gerges, Fawaz A., 'The 1967 Arab-Israeli war: US actions and Arab perceptions', in David W. Lesch (ed.), *The Middle East and the United States: A Historical and Political Reassessment* (Boulder, Col., 1996), pp. 189–208.

Girvan, Norman, 'Economic nationalism', in Raymond Vernon (ed.), *The Oil Crisis* (New York, 1976), pp. 145–58.

Goldsmith, Walter, and Berry Ritchie, *The New Elite: Britain's Top Executives* (London, 1987).

Gray, Earle, *The Great Canadian Oil Patch* (Toronto, 1970).

Grayson, Leslie E., *National Oil Companies* (Chichester, 1981).

Greaves, Rose, 'The reign of Mohammad Riza Shah', in H. Amirsadeghi (ed.), *Twentieth Century Iran* (London, 1977).

Greenhill, Denis, *More by Accident* (York, 1992).

Gurney, Judith, *Libya: The Political Economy of Oil* (Oxford, 1996).

Hardie, D. W. F., and J. D. Pratt, *A History of the Modern British Chemical Industry* (London, 1966).

Hayes, P., *Industry and Ideology: IG Farben in the Nazi Era* (Cambridge, 1987).

Heath, Edward, *The Course of My Life* (London, 1998).

Hein, Laura E., *Fueling Growth: The Energy Revolution and Economic Policy in Postwar Japan* (Cambridge, Mass., 1990).

Hendry, J., 'Prolonged negotiations: the British fast computer project and the early history of the British computer industry', *Business History* 26 (1984), 280–306.

Hidy, Ralph W. and Muriel E., *Pioneering in Big Business: A History of Standard Oil Company (New Jersey), 1882–1911* (New York, 1955).

Hirasawa, Kazushige, 'Japan's tilting neutrality', in J. C. Hurewitz (ed.), *Oil, the Arab-Israel Dispute, and the Industrial World: Horizons of Crisis* (Boulder, Col., 1976), pp. 138–46.

Hofstede, Geert, *Cultures and Organizations: Software of the Mind* (London, 1991).

Holden, David, 'The Persian Gulf: after the British Raj', *Foreign Affairs* 49 (1971), 721–35.

Holland, R. F., *European Decolonization 1918–1981: An Introductory Survey* (London, 1985).

Hollo, Reuven, 'Oil and American foreign policy in the Persian Gulf, 1947–1991', PhD thesis, University of Texas at Austin (1995).

Howarth, Stephen, *Sea Shell: The Story of Shell's British Tanker Fleets, 1892–1992* (London, 1992).

Howarth, Stephen, *A Century in Oil: The 'Shell' Transport and Trading Company, 1897–1997* (London, 1997).

Ikein, Augustine A., *The Impact of Oil on a Developing Country: The Case of Nigeria* (New York, 1990).

Jacoby, Neil H., *Multinational Oil: A Study in Industrial Dynamics* (New York, 1974).

Jay, Antony, *Management and Machiavelli* (London, 1967).

Jodice, David A., 'Sources of change in Third World regimes for foreign direct investment, 1968–1976', *International Organization* 34 (1980), 177–206.

Jones, Geoffrey, *The State and the Emergence of the British Oil Industry* (London, 1981).

Jones, Geoffrey, *The Evolution of International Business: An Introduction* (London, 1996).

Kahler, Miles, 'Political regime and economic actors: the response of firms to the end of colonial rule', *World Politics* 33 (1981), 383–412.

Kapstein, Ethan B., *The Insecure Alliance: Energy Crises and Western Politics since 1944* (Oxford, 1990).

Karshenas, Massoud, *Oil, State and Industrialisation in Iran* (Cambridge, 1990).

Kelly, J. B., *Arabia, the Gulf and the West* (London, 1980).

Kennedy, Charles R., 'Relations between transnational corporations and governments of host countries: a look to the future', *Transnational Corporations* 1 (1992), 67–91.

Kent, Marian, *Oil and Empire: British Policy and Mesopotamian Oil, 1900–1920* (London, 1976).

Kent, P. E., 'UK onshore oil exploration', *Marine and Petroleum Geology* 2 (1985), 56–64.

Kenwood, A. G., and A. L. Lougheed, *The Growth of the International Economy, 1820–1990*, 3rd edn (London, 1992).

Khan, Sarah Ahmad, *Nigeria: The Political Economy of Oil* (Oxford, 1994).

Kidron, Michael, *Foreign Investments in India* (London, 1965).

King, G. A. B., *A Love of Ships* (Hampshire, 1991).

Kissinger, Henry, *White House Years* (Boston, Mass., 1979).

Kissinger, Henry, *Years of Upheaval* (Boston, Mass., 1982).

Kobrin, Stephen J., 'Expropriation as an attempt to control foreign firms in LDCs: trends from 1960 to 1979', *International Studies Quarterly* 28 (1984), 329–48.

Kobrin, Stephen J., 'The nationalisation of oil production, 1918–1980', in David W. Pearce, Horst Siebert and Ingo Walter (eds.), *Risk and the Political Economy of Resource Development* (London, 1984), pp. 137–64.

Kobrin, Stephen J., 'Diffusion as an explanation of oil nationalization: or the domino effect rides again', *Journal of Conflict Resolution* 29 (1985), 3–32.

Kulke, Hermann, and Dietmar Rothermund, *A History of India*, revised edn (London, 1990).

Kunz, Diane B., *The Economic Diplomacy of the Suez Crisis* (Chapel Hill, N.C., 1991).

Kyle, Keith, *Suez* (London, 1991).

Lantzke, Ulf, 'The OECD and its International Energy Agency', in Raymond Vernon (ed.), *The Oil Crisis* (New York, 1976), pp. 217–27.

Lavington, S. H., *Early British Computers: The Story of Vintage Computers and the People Who Built Them* (Manchester, 1980).

Lees, G. M., 'The geology of the oilfield belt of Iran and Iraq', in A. E. Dunstan, A. W. Nash, B. T. Brooks and H. Tizard (eds.), *The Science of Petroleum* (Oxford, 1938), vol. I, pp. 140–8.

Lees, G. M., 'Foreland folding', *Quarterly Journal of the Geological Society of London* 108 (1952), 1–34.

Lees, G. M., and F. D. S. Richardson, 'The geology of the oil-field belt of S. W. Iran and Iraq', *Geological Magazine* 77 (1940), 227–52.

Lenczowski, George, 'The oil-producing countries', in Raymond Vernon (ed.), *The Oil Crisis* (New York, 1976), pp. 59–72.

Lenczowski, George, *The Middle East in World Affairs*, 4th edn (Ithaca, 1980).

Lenczowski, George, *American Presidents and the Middle East* (Durham, N.C., and London, 1990).

Leonard, H. Jeffrey, 'Multinational corporations and politics in developing countries', *World Politics* 32 (1980), 454–82.

Levi, J. D., Jean L. Shennan and G. P. Ebbon, 'Biomass from liquid n-alkanes', in A. H. Rose (ed.), *Microbial Biomass* (London, 1979), pp. 360–419.

Levy, Walter J., 'Oil power', *Foreign Affairs* 49 (1971), 652–68.

Levy, Walter J., 'World oil cooperation or international chaos', *Foreign Affairs* 52 (1974), 690–713.

Lieber, Robert J., *Oil and the Middle East War: Europe in the Energy Crisis*, Harvard Studies in International Affairs 35 (Harvard, 1976).

Lipson, Charles, *Standing Guard: Protecting Foreign Capital in the Nineteenth and Twentieth Centuries* (Berkeley, Calif., 1985).

Longrigg, S. H., *Oil in the Middle East: Its Discovery and Development*, 3rd edn (London, 1968).

Lubbell, H., 'Middle East crises and world petroleum movements', *Middle Eastern Affairs* 9 (1958), 338–48.

Lubbell, H., *Middle East Oil Crises and Western Europe's Energy Supplies* (Baltimore, Md., 1963).

MacDonald, Scott B., *Trinidad and Tobago: Democracy and Development in the Caribbean* (New York, 1986).

Macmillan, H., *Riding the Storm, 1956–1959* (London, 1971).

Maddison, Angus, *The World Economy in the Twentieth Century* (Paris, 1989).

Magdoff, Harry, *The Age of Imperialism: The Economics of U.S. Foreign Policy* (New York, 1969).

Mansfield, Peter, *The Middle East: A Political and Economic Survey*, 4th edn (Oxford, 1973).

Martin, A. J., 'The prediction of strategic reserves', in T. Niblock and R. Lawless (eds.), *Prospects for the World Oil Industry* (Beckenham, Kent, 1985), pp. 16–41.

Martin, A. J., and P. B. Lapworth, 'Norman Leslie Falcon', *Biographical Memoirs of Fellows of the Royal Society* 44 (1998), 159–74.

McKie, James W., 'The United States', in Raymond Vernon (ed.), *The Oil Crisis* (New York, 1976), pp. 73–90.

Meadows, Donella H., Dennis L. Meadows, Jorgen Randers and William W. Behrens III, *The Limits to Growth: A Report for the Club of Rome's Project on the Predicament of Mankind* (London, 1972).

Mejcher, Helmut, *Imperial Quest for Oil: Iraq, 1910–1928* (London, 1976).

Mercer, Helen, *Constructing a Competitive Order: The Hidden History of British Antitrust Policies* (Cambridge, 1995).

Middlemiss, Norman L., *The British Tankers*, revised edn (Newcastle, 1995).

Mikdashi, Zuhayr, 'The OPEC process', in Raymond Vernon (ed.), *The Oil Crisis* (New York, 1976), pp. 203–15.

Miller, A. D., *Search for Security: Saudi Arabian Oil and American Foreign Policy, 1939–1949* (Chapel Hill, N.C., 1980).

Molle, W., and E. Wever, *Oil Refineries and Petrochemical Industries in Western Europe* (Aldershot, 1984).

Monroe, Elizabeth, *Britain's Moment in the Middle East, 1914–1971*, 2nd edn (London, 1981).

Musgrave, Richard A. and Peggy B., 'Fiscal policy' in R. E. Caves (ed.), *Britain's Economic Prospects* (London, 1968), pp. 21–67.

Nash, G. D., *United States Oil Policy, 1890–1964* (Pittsburgh, Pa., 1968).

Newby, W. J., 'An integrated model of an oil company', in *Mathematical Model Building in Economics and Industry*, Papers of a conference organised by C-E-I-R Ltd (London, 1968), pp. 61–9.

Newby, W. J., 'Planning refinery production', in C. M. Berners-Lee (ed.), *Models for Decision* (London, 1965), pp. 43–52.

Newby, W. J., and R. J. Deam, 'How BP is using linear programming', *World Petroleum* 42 (1971), 230–4.

Newby, W. J., and R. J. Deam, 'Optimisation and operational research', *Transactions of the Institution of Chemical Engineers* 40 (1962), 350–5.

Nowell, Gregory P., *Mercantile States and the World Oil Cartel, 1900–1939* (New York, 1994).

Nutting, Anthony, *No End of a Lesson: The Story of Suez* (London, 1967).

Odell, Peter R., *Oil and World Power*, 7th edn (Harmondsworth, 1983).

Onoh, J. K., *The Nigerian Oil Economy from Prosperity to Glut* (London, 1983).

Organisation for European Economic Co-operation (OEEC), *Europe's Growing Needs of Energy: How Can They Be Met?* (Paris, 1956).

Organisation for European Economic Co-operation (OEEC), *Oil: The Outlook for Europe* (Paris, 1956).

Organisation for European Economic Co-operation (OEEC), *Europe's Need for Oil: Implications and Lessons of the Suez Crisis* (Paris, 1958).

Painter, David S., *Oil and the American Century: The Political Economy of US Foreign Oil Policy, 1941–1954* (Baltimore, Md., 1986).

Parra, Francisco R., 'The Eastern Hemisphere – oil profits in focus', *Middle East Economic Survey* 6 (1963).

Parsons, Anthony, *The Pride and the Fall, Iran 1974–1979* (London, 1984).

Penrose, Edith T., *The Large International Firm in Developing Countries: The International Petroleum Industry* (London, 1968).

Penrose, Edith T., 'The development of crisis', in Raymond Vernon (ed.), *The Oil Crisis* (New York, 1976), pp. 39–57.

Penrose, Edith T. and E. F., *Iraq: International Relations and National Development* (London, 1978).

Perez Alfonzo, Juan Pablo, *Venezuela and OPEC* (Caracas, 1961).

Philip, George, *Oil and Politics in Latin America* (Cambridge, 1982).

Philip, George, *The Political Economy of International Oil* (Edinburgh, 1994).

Political and Economic Planning, *Graduate Employment: A Sample Survey* (London, 1956).

Pollack, Gerald A., 'The economic consequences of the energy crisis', *Foreign Affairs* 52 (1974), 452–71.

Pollard, S., *The Development of the British Economy, 1914–1990*, 4th edn (London, 1992).

Prodi, Romano, and Alberto Clo, 'Europe', in Raymond Vernon (ed.), *The Oil Crisis* (New York, 1976), pp. 91–112.

Quandt, William B., 'The Western Alliance in the Middle East: problems for US foreign policy', in Steven L. Spiegel (ed.), *The Middle East and the Western Alliance* (London, 1982), pp. 9–17.

Reader, W. J., *Imperial Chemical Industries: A History, Volume II, The First Quarter-Century, 1926–1952* (Oxford, 1975).

Rhodes James, R., *Anthony Eden* (London, 1987).

Richards, J., and L. Pratt, *Prairie Capitalism: Power and Influence in the New West* (Toronto, 1979).

Rickwood, Frank, *The Kutubu Discovery: Papua New Guinea, its People, the Country, and the Exploration and Discovery of Oil* (Victoria, Australia, 1992).

Robinson, Jeffrey, *Yamani: The Inside Story* (London, 1988).

Rodman, Kenneth A., *Sanctity versus Sovereignty: The United States and the Nationalization of Natural Resource Investments* (New York, 1988).

Roscow, James P., *800 Miles to Valdez: The Building of the Alaska Pipeline* (Englewood Cliffs, N.J., 1977).

Rouhani, F., *A History of OPEC* (New York, 1971).

Safarian, A. E., *Multinational Enterprise and Public Policy: A Study of the Industrial Countries* (Aldershot, 1993).

Sampson, Anthony, *The Seven Sisters: The Great Oil Companies and the World they Made* (London, 1975).

Sasagawa, Masahiro, 'Japan and the Middle East', in Steven L. Spiegel (ed.), *The Middle East and the Western Alliance* (London, 1982), pp. 33–46.

Scammell, W. M., *The International Economy since 1945* (London, 1980).

Schmitz, Christopher (ed.), *Big Business in Mining and Petroleum* (Aldershot, 1995).

Selwyn Lloyd, J., *Suez 1956: A Personal Account* (London, 1978).

Seymour, Ian, *OPEC: Instrument of Change* (London, 1980).

Shennan, Jean L., 'Hydrocarbons as substrates in industrial fermentation', in R. M. Atlas (ed.), *Petroleum Microbiology* (New York, 1984), pp. 643–83.

Skeet, Ian, *OPEC: Twenty-five Years of Prices and Politics* (Cambridge, 1988).

Smith, David N., and Louis T. Wells, 'Mineral agreements in developing countries: structure and substance', *American Journal of International Law* 69 (1975), 560–90.

Spitz, Peter H., *Petrochemicals: The Rise of an Industry* (New York, 1988).

Stein, Janice Gross, 'Alice in Wonderland: The North Atlantic Alliance and the Arab-Israeli dispute', in Steven L. Spiegel (ed.), *The Middle East and the Western Alliance* (London, 1982), pp. 49–81.

Stewart, Rosemary, *How Computers Affect Management* (London, 1971).

Stobaugh, Robert B., 'The oil companies in the crisis', in Raymond Vernon (ed.), *The Oil Crisis* (New York, 1976), pp. 179–202.

Stocking, G. W., *Middle East Oil: A Study in Political and Economic Controversy* (London, 1971).

Stoff, M. B., *Oil, War, and American Security: The Search for a National Policy on Foreign Oil, 1941–1947* (New Haven, Conn., 1980).

Stokes, Raymond G., *Divide and Prosper: The Heirs of IG Farben under Allied Authority, 1945–1951* (Berkeley, Calif., 1988).

Stokes, Raymond G., *Opting for Oil: The Political Economy of Technological Change in the West German Chemical Industry, 1945–1961* (Cambridge, 1994).

Strohmeyer, John, *Extreme Conditions: Big Oil and the Transformation of Alaska* (New York, 1993).

Symonds, Edward, 'Life with a world oil surplus' (First National City Bank of New York, 1960).

Symonds, Edward, 'Oil prospects and profits in the Eastern Hemisphere' (First National City Bank of New York, 1961).

Symonds, Edward, 'Financing oil expansion in the development decade' (First National City Bank of New York, 1962).

Symonds, Edward, 'Oil advances in the Eastern Hemisphere' (First National City Bank of New York, 1962).

Symonds, Edward, 'Eastern hemisphere petroleum – another year's progress analysed' (First National City Bank of New York, 1963).

Tanzer, Michael, *The Political Economy of International Oil and the Underdeveloped Countries* (Boston, Mass., 1969).

Terzian, Pierre, *OPEC: The Inside Story* (London, 1985).

Tomlinson, B. R., *The Economy of Modern India, 1860–1970*, vol. III of The New Cambridge History of India (Cambridge, 1993).

Tsurumi, Yoshi, 'Japan', in Raymond Vernon (ed.), *The Oil Crisis* (New York, 1976), pp. 113–27.

Tugendhat, Christopher, and Adrian Hamilton, *Oil: The Biggest Business*, revised edn (London, 1975).

Turner, Graham, *Business in Britain*, revised edn (Harmondsworth, 1971).

Turner, Louis, *Oil Companies in the International System* (London, 1978).

Turner, Louis, 'The European Community: factors of disintegration; politics of the energy crisis', *International Affairs* 50 (1974), 404–15.

Venkataramani, M. S., 'Oil and US foreign policy during the Suez crisis, 1956–1957', *International Studies* 2 (1960–1), 105–52.

Vernon, Raymond, 'An interpretation' and 'The distribution of power', in Raymond Vernon (ed.), *The Oil Crisis* (New York, 1976), pp. 1–14 and 245–57.

Vernon, Raymond, *Storm over the Multinationals: The Real Issues* (London, 1977).

Vietor, R. H. K., *Energy Policy in America since 1945: A Study of Business–Government Relations* (Cambridge, 1984).

Waddams, Frank C., *The Libyan Oil Industry* (London, 1980).

Wall, Bennett H., *Growth in a Changing Environment: A History of Standard Oil Company (New Jersey), Exxon Corporation, 1950–1975* (New York, 1988).

Warman, H. R., 'Future problems in petroleum exploration', *Petroleum Review* 25 (1971), 96–101.

Warman, H. R., 'The future of oil', *Geographical Journal* 138 (1972), 287–97.

Watt, D. C., 'Britain and the Suez Canal', Report for Royal Institute of International Affairs (London, 1956).

Weinberg, Steven, *Armand Hammer: The Untold Story* (Boston, Mass., 1989).

Weir, Ronald, *The History of the Distillers Company, 1877–1939* (Oxford, 1995).

Wilkes, M. V., *Automatic Digital Computers* (London, 1956).

Wilkins, Mira, 'The oil companies in perspective', in Raymond Vernon (ed.), *The Oil Crisis* (New York, 1976), pp. 159–78.

Wilkins, Mira, *The Maturing of Multinational Enterprise: American Business Abroad from 1914 to 1970* (Cambridge, Mass., 1974).

Williams, Mari E. W., 'Choices in oil refining: The case of BP, 1900–1960', *Business History* 26 (1984), 307–28.

Williams, M. L., 'The extent and significance of the nationalization of foreign-owned assets in developing countries, 1956–1972', *Oxford Economic Papers* 27 (1975), 260–73.

Yergin, Daniel, *The Prize: The Epic Quest for Oil, Money, and Power* (New York, 1991).

Index